Law, Governance and Technology Series

Issues in Privacy and Data Protection

Volume 37

Issues in Privacy and Data Protection aims at publishing peer reviewed scientific manuscripts that focus upon issues that engage into an analysis or reflexion related to the consequences of scientific and technological developments upon the private sphere, the personal autonomy and the self-construction of humans with data protection and privacy as anchor points. The objective is to publish both disciplinary, multidisciplinary and interdisciplinary works on questions that relate to experiences and phenomena that can or could be covered by legal concepts stemming from the law regarding the protection of privacy and/or the processing of personal data. Since both the development of science and technology, and in particular information technology (ambient intelligence, robotics, artificial intelligence, knowledge discovery, data mining, surveillance, etc.), and the law on privacy and data protection are in constant frenetic mood of change (as is clear from the many legal conflicts and reforms at hand), we have the ambition to reassemble a series of highly contemporary and forward-looking books, wherein cutting edge issues are analytically, conceptually and prospectively presented

More information about this series at http://www.springer.com/series/8808

Olga Mironenko Enerstvedt

Aviation Security, Privacy, Data Protection and Other Human Rights: Technologies and Legal Principles

Springer

Olga Mironenko Enerstvedt
Aviation and ICT Law Consulting
Langhus, Norway

ISSN 2352-1902 ISSN 2352-1910 (electronic)
Law, Governance and Technology Series
ISSN 2352-1929 ISSN 2352-1937 (electronic)
Issues in Privacy and Data Protection
ISBN 978-3-319-58138-5 ISBN 978-3-319-58139-2 (eBook)
DOI 10.1007/978-3-319-58139-2

Library of Congress Control Number: 2017941081

Printed on acid-free paper

This Springer imprint is published by Springer Nature
The registered company is Springer International Publishing AG
The registered company address is: Gewerbestrasse 11, 6330 Cham, Switzerland

Preface

When I was finalizing this work, the worst incident in Russian/Soviet history of civil aviation occurred. On 31 October 2015, terrorists blew up Airbus A321 during its flight from Sharm el-Sheikh to St. Petersburg, killing 224 persons on board, mostly Russians. Tragically, I found I needed to update my research with new sad facts and figures. When people are killed, it hurts. When fellow citizens are killed, the pain is doubled.

As a participant of a student exchange programme, I was in the USA at the moment of the 9/11 attack. I flew from New York to Moscow in October 2001 and experienced the more stringent security measures. A pair of small scissors was confiscated from my hand baggage. I was living in Moscow when the worst terrorist events occurred there, such as the Dubrovka Theatre siege in 2002 and bombing of two flights by suicide terrorists in 2004. On 22 July 2011, I was living in Norway and personally heard the explosion of bombs in government buildings in Oslo. One of the guests invited to my wedding party a week later was unable to attend; her close relative had been killed in the bombings of the government quarter in Oslo. There have been other terrorist-related events in my life.

Turning from the personal to a more general perspective, I must say that although not all these catastrophic events were aviation-related, almost all of them had some influence on security regimes, including aviation. Flying from Oslo to Moscow and back quite frequently, as well as other routes, I could not help but notice direct or indirect consequences such as restrictions on liquids, having to remove shoes, metal detectors at the entrance of the airport or body scanners. As a researcher in the field of aviation security and data privacy, however, I know that these measures are only the tip of the iceberg. The emerging mass of security-related data and the corresponding impact on individual privacy are only in the beginning stage of development.

My interest in aviation grew as a result of my work as a lawyer in a business aviation company, Moscow Sky, operating internationally. I worked there during 2005–2008. I have fond memories of my colleagues and the time I spent there. When I then took the LLM degree in ICT law at the University of Oslo, I was very fortunate to be able to combine aviation and data protection issues in my master's thesis. The topic concerned passenger name record agreements between the EU and the USA. My master's thesis was published by the Norwegian Research Center for Computers and Law (NRCCL). I was privileged to have Professor Jon Bing as my supervisor. Very sadly, he passed away in 2014.

Inspired by conversations with Jon Bing and the director of NRCCL, Professor Dag Wiese Schartum, I decided to further develop the topic. The aim was to research broader issues in civil aviation security versus privacy, data protection and other human rights and – on a global level – to write a book. I completed the work in 2016.

First of all, I would like to thank Professor Dag Wiese Schartum. He has always been very supportive, available for help and advice at any time and willing to provide valuable comments and ideas. I am also very thankful to Professor Lee Andrew Bygrave, for his invaluable advice and comments, in particular on data privacy issues.

I would like to thank Peter Burgess from PRIO for his encouragement of my work at the initial stage, inviting me to relevant seminars and sharing contacts in the civil aviation field. I would like to thank the Norwegian Civil Aviation Authority for hosting a meeting at which they presented their organization and work. Many thanks go to the personnel of the Norwegian airport operator Avinor, in particular to senior security advisor Ole Folkestad, who was especially helpful in providing me with an of understanding the use of technologies in modern airports. My gratitude also goes to other relevant entities and persons, in particular from Norway and Russia, who took the time to answer my questions.

I would also like to thank all colleagues at the NRCCL who have always been very kind and friendly towards me. Special thanks go to the head of administration of the Institute of Private Law, Eva Modvar, and then to her successor, Eli Knotten, for their kindness and help with every practical issue I faced and for always being available to provide support and advice. I would like to specially thank Tim Challman for proofreading the final text and for offering valuable comments and advice regarding the English language.

I am indebted to my family, especially to my mother, my husband Håken and our children, Maria, Anna and Aleksander. Håken has been very kind and extremely patient. He has been a willing babysitter and took good care of our small children, while I was busy with the book and was unable to spend as much time with them as I wished. This work is dedicated to my family. I love you.

Langhus, Norway Olga Mironenko Enerstvedt

Contents

Table of Cases

European Court of Human Rights

A. and Others v. the United Kingdom, No. 3455/05, 19 Feb 2009.
Ahmet Arslan and Others, No. 41135/98, 23 Feb 2010.
Aksoy v. Turkey, No. 21987/93, 18 Dec 1996.
Aktas v. France, No. 43563/08, declared inadmissible 30 Jun 2009.
Amann v. Switzerland, No. 27798/95, 16 Feb 2000.
Bayatyan v. Armenia, No. 23459/03, 7 Jul 2011.
Brannigan and McBride v. the United Kingdom, No. 14553/89, 25 May 1993.
Campbell v. the United Kingdom, No. 13590/88, 25 Mar 1992.
Case *"Relating to certain aspects of the laws on the use of languages in education in Belgium" v. Belgium*, No. 1474/62; 1677/62; 1691/62; 1769/63; 1994/63; 2126/64, 23 July 1968
Česnulevičius v. Lithuania, No. 13462/06, 10 Jan 2012.
Cyprus v. Turkey, No. 25781/94, 10 May 2001.
Dahlab v. Switzerland, No. 42393/98, declared inadmissible 15 Feb 2001.
El Morsli v. France, No. 15585/06, declared inadmissible 4 Mar 2008.
Erçep v. Turkey, No. 43965/04, 22 Nov 2011.
Fedorov and Fedorova v. Russia, No. 31008/02, 13 Oct 2005.
Gillan and Quinton v. the United Kingdom, No. 4158/05, 12 Jan 2010.
Huvig v. France, No. 11105/84, 24 April 1990.
Ireland v. the United Kingdom, No. 5310/71, 18 Jan 1978.
Kervanci v. France, No. 31645/04, 4 Dec 2008.
Khan v. the United Kingdom, No. 35394/97, 12 May 2000.
Klass and others v. Germany, No. 5029/71, 6 Sep 1978.
Kopp v. Switzerland, 13/1997/797/1000, 25 Mar 1998.
Kruslin v. France, No. 11801/85, 24 Apr 1990.
Lawless v. Ireland, No. 332/57, 1 July 1961.
Leander v. Sweden, No. 9248/81, 26 Mar 1987.
Leyla Sahin v. Turkey, No. 44774/98, 10 Nov 2005.

Liberty and Others v. the United Kingdom, No. 58243/00, 1 July 2008.
Lingens v. Austria, No. 9815/82, 8 July 1986.
Luordo v. Italy, No. 32190/96, 17 Mar 2013.
Malone v. the United Kingdom, No. 8691/79, 2 Aug 1984.
Marper v. the United Kingdom, No. 30562/04 and 30566/04, 4 Dec 2008.
McFeeley v. the United Kingdom, No. 8317/78, 15 May 1980.
Miazdzyk v. Poland, No. 23592/07, 24 Jan 2012.
Ollson v. Sweden, No. 10465/83, 24 May 1988.
Peck v. the United Kingdom, No. 44647/98, 28 Jan 2003.
Prescher v. Bulgaria, No. 6767/04, 7 Jun 2011.
Rangelov v. Germany, No. 5123/07, 22 Mar 2012.
Riener v. Bulgaria, No. 46343/99, 23 May 2006.
S and Marper v. the United Kingdom, No. 30562/04 and 30566/04, 4 Dec 2008.
Sejdić and Finci v. Bosnia and Herzegovina, No. 27996/06 and 34836/06, 22 Dec 2009.
Shimovolos v. Russia, No. 30194/09, 21 Jun 2011.
Soering v. the United Kingdom, No. 14038/88, 7 Jul 1989.
Stamose v. Bulgaria, No. 29713/05, 27 Nov 2012.
Uzun v. Germany, No. 35623/05, 2 Sep 2010.
Paul and Audrey Edwards v. the United Kingdom, No. 46477/99, 14 March 2002.
Bromiley v. the United Kingdom (dec.), No. 33747/96, 23 Nov 1999.
Finogenov and others v. Russia, No. 18299/03 and 27311/03, 20 Dec 2011.

European Court of Justice

Case C–70/10, *Scarlet Extended SA v. Société belge des auteurs, compositeurs et éditeurs SCRL (SABAM)*, 24 Nov 2011.
Case C-73/07, *Satamedia*, 16 Dec 2008.
Case C 362/14, *Maximillian Schrems v. Data Protection Commissioner*, 6 Oct 2015.
Case C-101/01, *Lindqvist*, 6 Nov 2003.
Case C-131/12, *Google Spain SL, Google Inc. v. Agencia Española de Protección de Datos, Mario Costeja González*, 13 May 2014.
Case C-465/00, *Rechnungshof decision*, 20 May 2003.
Case C-524/06, *Heinz Huber v. Bundesrepublik Deutschland*, 16 Dec 2008.
Case T-13/99, *Pfizer Animal Health SA v. Council of the European Union*, 11 Sep 2002.
Joined Cases C 293/12 and C 594/12, *Digital Rights Ireland Ltd v. Minister for Communications, etc.*, 8 Apr 2014.
Joined Cases C-317/04 and C-318/04, *European Parliament v. Council of the European Union, European Parliament v. Commission*, 30 May 2006.
Joined cases C-402/05 P and C-415/05 *P, Yassin Abdullah Kadi and Al Barakaat International Foundation v Council and Commission*, 3 Sept 2008.

Joined Cases C-92/09 and C-93/09 *Volker und Markus Schecke GbR and Hartmut Eifert v. Land Hessen*, 9 November 2010.
Joined Cases C-141/12 and C-372/12, *YS v Minister voor Immigratie, Integratie en Asiel and Minister voor Immigratie, Integratie en Asiel v M and S.*, 17 July 2014.

USA

Arizona v. Evans, 514 U.S. 1 (1995), 1 March 1995.
EPIC v. DHS, 653 F.3d 1 (D.C. Cir. 2011), 15 July2011.
Gilmore v. Gonzales, 435 F. 3d 1125 - Court of Appeals, 9th Circuit 2006, 26 January 2006.
Hasbrouck v. U.S. Customs and Border Protection, Case number C 10-03793 RS, U.S. District Court, Northern District of California, filed 25 August 2010.
Pavesich v. New England Life Insurance Co. Supreme Court of Georgia, 122 Ga. 190; 50 S.E. 68; 1905 Ga., 3 March 1905.
Schmerber v. California, 384 U.S. 757 (1966), 20 June 1996.
Sima Prods. Corp. v. McLucas, 612 F.2d 309, 312–13 (7th Cir. 1980), 3 January1980.
United States v. Aukai, 497 F.3d 955-963 (9th Cir. 2007), 10 August 2007.
United States v. Dionisio, 410 U.S. 1, 14 (1973), 22 January 1973.
United States v. Epperson, 454 F.2d 769, 771 (4th Cir. 1971), 7 February 1972.

UK

Durant v. FSA [2003] EWCA Civ 1746, Court of Appeal (Civil Division) decision of Lord Justices Auld, Mummery and Buxton (8 December 2003).
R v. Rooney [2006] EWCA Crim 1841 (12 July 2006).

List of Figures

List of Tables

Chapter 1
Introduction

1.1 Background and Subject Matter

As the title of this work explicitly reflects, the discussion will mainly deal with two groups of issues: aviation security and privacy, data protection and other human rights. While aviation security is mainly aimed at preventing acts of unlawful interference with civil aviation[1] and relates to the state's undertakings to protect persons, aircraft, airport facilities and other material assets from harm, there are also various fundamental human rights instruments aimed at protecting not only the right to life, but also other rights and freedoms of individuals.

"Individuals" or "persons"[2] in both cases, within the civil aviation field, mean people: air passengers, crew, and people on the ground.[3] This research concentrates on air passengers, since this category is the principal concern when it comes to security screenings, personal data collection and transfer, as well as other methods that impact on human rights. Although statistically only a small part of the worlds'

[1] It is common to differentiate between two types of aviation: civil aviation and state aviation. Civil aviation mainly refers to commercial and other needs of the citizens and economy. In this research, civil aviation will mainly imply the commercial aspect and will refer to transportation of passengers.

[2] Used as synonymous.

[3] People on the ground (non-passengers) may include different persons: those meeting their relatives or friends at the airport arrival halls (they may be subject to a part of security control such as metal detectors at the airport entrance, e.g. in Moscow), airport staff (are usually subject to special control) or just people who can suffer being in the neighbourhood of a terrorist attack (e.g. if the bomb is put in a car near the airport, or if the exploded airplane falls onto civil buildings, etc.). Aviation security certainly endeavours to protect all these people as well. As I will explain in Sect. 4.5.3, aviation security covers the airport infrastructure too, thus, it extends to such elements as airport train stations, taxi areas, etc. However, people on the ground are usually subject to a much lesser degree of control/surveillance than passengers.

O.M. Enerstvedt, *Aviation Security, Privacy, Data Protection and Other Human Rights: Technologies and Legal Principles*, Law, Governance and Technology Series 37, DOI 10.1007/978-3-319-58139-2_1

population travels by air, this number is constantly growing,[4] thus, aviation security measures may potentially concern a substantial part of people.

Screenings and other aviation security measures at airports have become a part of day-to-day life and culture for millions of people all around the world, so that people seem to accept that there is no other way to keep them safe in the air.[5] The scope of these procedures, their obligatory nature, as well as possible negative consequences in the event of failure to be subject to the security measures are outlined for instance, in the airlines' carriage conditions available on the websites:

> Passengers and/or their baggage are subject to security screening, including but not limited to, security profiling, physical pat-downs and inspections, x-ray screening, manual bag searches, questioning of Passengers, and use of electronic or other detectors or screening or security devices, in the sole discretion of the government, airport or UA (United Airlines), and with or without the Passenger's presence, consent or knowledge. Neither UA nor its employees or agents is liable for any damage, loss, delay (including refusal to transport), confiscation of property, injury or other harm relating to or arising out of security screening conducted by an agent of the airport or any local, state, or federal agency or a Passenger's failure to submit to or comply with such security screening.[6]

On the one hand, the general public often considers aviation security measures that are designed to enhance security as positive developments. On the other hand, it is clear that they have a heavy impact on the processing of passengers at the airport. For most travellers screening and other procedures constitute an additional inconvenience and the most uncomfortable part of their entire journey.[7] They are time-consuming: before 9/11, the average checkpoint processed 350 passengers per hour; today fewer than 150 per hour are processed,[8] in some cases intrusively, with increasing costs that are ultimately passed on to passengers.

The inconveniences to passengers definitely may include problems connected with their rights as human beings. This is particularly important for this work: some aviation security measures, technologies, or methods may have impact on the individuals' rights, so that these rights may be interfered with, limited or violated. Civil rights organizations and advocates cry out against violations of individuals' liberties and freedoms in aviation security.[9]

It should be noted that despite the fact that human rights are under threat in general, i.e. this is a concern of all persons, only some individuals will ultimately bear the full (or any) cost of the violations.[10] Clearly, only a small selection of the air passengers will be subject to additional screening due to profiling techniques or will

---

[4] E.g. 3.3 billion passengers were transported in 2014 worldwide. See Chap. 4.

[5] Simmons. *Searching for Terrorists: Why "Public Safety" is not a Special Need* (2009) p. 912.

[6] Rule 20 Screening of passengers and baggage. United Air Lines, Inc. Contract of Carriage (revised 31 December 2015), https://www.united.com/web/format/pdf/Contract_of_Carriage.pdf.

[7] Ghee. *Recommendations for airports and airlines to improve baggage, security, immigration, arrivals and more in 2015* (2015).

[8] IATA. Remarks of Tony Tyler at the IATA Ops Conference in Vienna, 15 April 2013. http://www.iata.org/pressroom/speeches/Pages/2013-04-15-01.aspx.

[9] See, e.g. websites of ACLU: http://www.aclu.org and EPIC: http://epic.org.

[10] McDonagh. *Risk, human rights and the bureacratisation of counter-terrorism* (2011) p. 10.

be prohibited from flying due to their names appearing on a black list. However, hypothetically, no one can be guaranteed that their name might appear on such a list by mistake, or they might otherwise be targeted due to technologies' false positives. Thus, the fact that the potential rights violation pertains to only a few does not make the problem of possible violations less important.

The key word for security measures is surveillance, which can be defined as "the focused, systematic and routine attention to personal details for purposes of influence, management, protection, or direction".[11] There are multiple forms of surveillance; for aviation security, most common forms include camera surveillance (CCTV – Closed Circuit Television),[12] the use of biometrics, the use of data mining and profiling.[13]

At the same time, a dual nature of surveillance can be noted: on the one hand, surveillance, if properly used, can contribute to security, protection from crimes, but on the other hand, surveillance itself may present a threat to individuals, communities and societies because of its ubiquity, intensity and use of personal data; moreover, it is argued that "broad surveillance is a mark of bad security".[14] Government surveillance is directed at all of persons, including innocent people, and this is normally done in secret, with little or no oversight.[15] Although there are democratic procedural constraints, effective oversight and control may be very difficult to ensure as long as the power is exercised by large companies, state organizations and intelligence agencies.[16]

As a result, citizens are often unaware of the most questionable surveillance methods, such as camera surveillance, creation and storage of different dossiers and black lists that may affect the ability to fly, require additional screening, etc. In a state of mass surveillance, personal rights and freedoms are severely limited, as clearly predicted in Orwell's book "1984".

In the public debates, the right to privacy is frequently the first to be mentioned in this regard. Many security measures and technologies provoke "anxiety, protest, and resistance"[17] specifically in relation to privacy. "Security versus privacy" has become a common expression. In 2013, the enhanced discussion of privacy, mainly in connection with PRISM and the Edward Snowden revelations as well as the removal of "naked" body scanners from American airports and other events, made Dictionary.com declare "privacy" word of the year.[18] However, is it merely a *word*

---

[11] Lyon. *Surveillance studies: An overview* (2007a) p. 14.

[12] In the literature, "CCTV" is used as generic term camera surveillance. Although CCTV does not amount to camera surveillance, as explained in Chap. 6, these terms will be used as synonymous in this research.

[13] All these terms will be elaborated in detail in Chap. 6.

[14] Wright [et al.] *Questioning surveillance*. In: Computer Law & Security Review. Vol. 31 (2015) pp. 281–282, Schneier *Schneier on security* (2008) pp. 8, 69.

[15] ACLU. Surveillance and Privacy. http://www.aclu.org/national-security/surveillance-privacy

[16] Wright et al. (2015) p. 282.

[17] Nissenbaum. *Privacy in context* (2010) p. 3.

[18] Dictionary.com: http://blog.dictionary.com/author/dictionary-com-blog/. 17 December 2013.

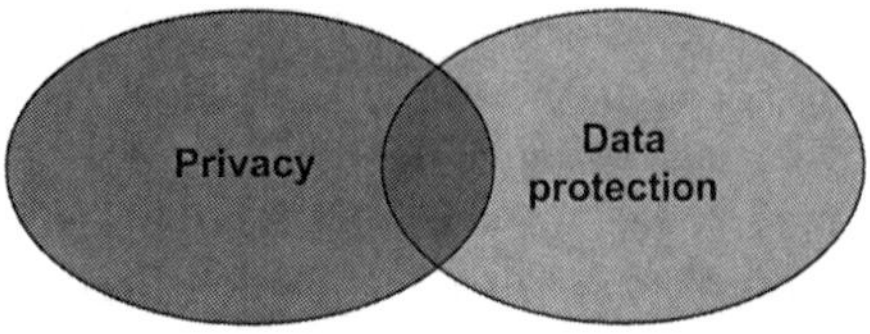

Fig. 1.1 Privacy and data protection: overlap

of the year? Schneier calls security versus privacy "the battle of the century:" which of these values is more important, how much privacy are you willing to give up for security?[19]

In this respect, it should be stressed that the rights to privacy and data protection are interrelated and partly overlap, as shown in Fig. 1.1.

The key words for this research from Lyon's definition of surveillance are "personal details". The point is that personal data have direct relation to the rights of privacy, which protects, among other things, private information[20] pertaining to individuals, and data protection, which can be seen as one of the dimensions of privacy or as a separate right.

It is not surprising, then, that the term "data privacy" is commonly used, or the term "dataveillance" used in reference to the phenomenon of data being used to monitor and surveil citizens.[21] Through extensive data protection regulation, at least in the EU and some other states, it is quite clear that data – if they satisfy the requirements of being personal and being processed[22] – should be protected. The term "privacy", however, is less clear.

Consequently, a question arises as to what privacy is. Numerous theoretical conceptions have been advanced as definitions for the term, but none of them has become common or universally accepted. Moreover, persons' views on this value do not remain static and can change according to the context, other needs and other factors. In many books and articles, the death or the end of privacy is predicted,[23] mainly due to total surveillance and pervasive databases. Everything depends, of course, on what one means by the term privacy. Clearly, privacy as total secrecy is not the most suitable today,[24] nor applicable, especially in public places such as airports. However, privacy nevertheless can be protected, but this should be done taking the new contexts into account. In the airports, the context – providing aviation

[19] Schneier (2008) p. 69.

[20] Data and information are used as synonyms in this research.

[21] Clarke. *Information technology and dataveillance*. In: Communications of the ACM. Vol. 31 (1988). p. 498–512.

[22] See Chap. 2.

[23] E.g. Garfinkel. *Database nation: the death of privacy in the 21st century* (2000), Rubenfeld. *The End of Privacy*. In: Stanford law review (2008). In January 2015, Science dedicated its special issue to "The end of privacy".

[24] Solove. *The end of privacy?* In: Scientific American. Vol. 299 (2008b) p. 104.

security to the passengers – and distinguishing between different types of privacy are key factors for defining privacy.[25]

From a broader perspective, privacy can be considered a component of other human rights, a necessary condition for them to function: without privacy, "it is much harder for dissent to flourish or for democracy to remain healthy and robust".[26] The status of the right to privacy in a particular society can demonstrate, first, what is being sacrificed to security, and secondly, the status of all human rights as the whole, since this right is "the fundamental precondition so that other rights and liberties can be exercised".[27] The right to privacy is not only an individual value, but a common, political value as well, since the abuse of privacy by surveillance technologies may lead to authoritarian regimes having an impact on the population as a whole.[28] In this situation, on the one hand, privacy can be seen as an antidote to surveillance[29]: raising its status can contribute to upholding all human rights. On the other hand, the weakness of privacy in a particular society is as a rule accompanied by other human rights concerns.

The point is that all human rights are interdependent, indivisible and interrelated.[30] This means that violating one right may often impair the enjoyment of other human rights, and vice versa. This idea can be well illustrated by the aviation security measures: in addition to privacy/data protection concerns, they often have impact on other fundamental human rights, in particular, the right to equal treatment and non-discrimination, freedom of movement, etc. Accordingly, the right to privacy alone cannot accommodate all of the negative effects of aviation security.

Hence the scope of the issue is argued to be broader – human rights versus security or liberty versus control[31] – since humans are interested in both security and other human rights, and it is the state and other agencies' control and surveillance functions that interfere.

It should be noted that the term "aviation security" can be seen in two dimensions. First, aviation security is ultimately aimed at protecting, among other things, the right to life of passengers and other persons. In other words, all human beings have the right to be safeguarded from harms – it is also possible to speak about the individuals' right to security.[32] Therefore, security involves elements of human rights. Effective security in this sense contributes to the right to life. The latter belongs to peremptory norms or *jus cogens*, that is, not being subject to any

---

[25] See Chap. 2.

[26] Vermeulen and Bellanova. *European 'smart'surveillance: What's at stake for data protection, privacy and non-discrimination?* In: Security and Human Rights. Vol. 23 (2013) p. 305.

[27] Poullet. *Data protection legislation: What is at stake for our society and democracy?* In: Computer Law & Security Review. Vol. 25 (2009) p. 226.

[28] Vermeulen and Bellanova (2013) p. 305.

[29] Amicelli. *Report on Theoretical Frameworks and Previous Empirical Research* (2012a) p. 14.

[30] See Vienna Declaration and Programme of Action (A/CONF.157/23), adopted by the World Conference on Human Rights, held in Vienna, 14–25 June 1993.

[31] Schneier (2008) p. 70.

[32] See Lyon. *Surveillance after september 11* (2003a).

derogation. Taking into account the threats of terrorism and the amount of harm that may cause a successful attack on an aircraft, the right to life may be used as a strong argument for the enhanced security.

Secondly, aviation security, as defined in legal instruments, involves in general different measures and human and material resources intended to safeguard civil aviation against acts of unlawful interference that may jeopardize the security of civil aviation.[33] In the civil aviation sector, the necessity of safeguarding people from terrorism and crime is obvious. Moreover, this sector has immense importance to the world economy. Its assets, aircraft, airports and infrastructure, constitute considerable economic value and should be protected as well. At the same time, the threat of terrorist attacks against civil aviation is deemed to be heightened. Hence, the connection between the threats and advanced security measures seems to be clear. The latter is a logical consequence of the former.

The aviation security-privacy dilemma is quite visible here: implementation of different measures, procedures and methods is designed to protect individuals, but may have tensions with the individuals' rights. A clear result is what Lyon calls a paradox – the side effect that many aviation security measures justified by security risks and fight against terrorism actually produce risks themselves – in relation to privacy, data protection and other liberties.[34] Compromises to privacy/data protection/other human rights can be considered as the costs to be paid for a security decision.

Actual surveillance may start long before you enter the airport, for instance, at the moment you reserve a plane ticket and fill out various personal data fields. Different systems collect personal data of passengers, use and share them with other systems for further checks, data mining, profiling and so on. The opportunities for surveillance are constantly growing along with technological developments: high-tech "smart" camera surveillance using biometrics, body scanners, security tunnels and so on.

What is next? How much security is enough? How far can the states and security organs go in their attempts to find the most effective measures? To illustrate an absurd situation, security professionals ironically demonstrate a picture of "security of the future" showing naked passengers on board the aircraft, i.e. deprived of any privacy at all. In general, "If passengers are divesting items and holding their arms up like criminals, then the terrorists have won by fundamentally changing our core values".[35]

A common suggestion is the idea of finding a balance between the privacy and security values. In other words, we should balance privacy or, broadly stated, liberty against security using the so-called trade-off model. More details will be elaborated in Chap. 5, but in general, it is not possible to increase one of the values without decreasing the other.

---

[33] Article 3 of Regulation No 300/2008.

[34] Lyon. *Globalizing Surveillance Comparative and Sociological Perspectives*. In: International Sociology. Vol. 19 (2004) p. 137.

[35] Sagi-Dolev. *Manufacturer Perspectives*. In: Aviation Security International. Vol. 19 (2013) p. 26.

This is easy to understand. It is obvious that security, which is vital to survival, is more important than privacy, which is a social and political need but is not an immediate matter of survival.[36] To the contrary: in the worst case, privacy can be seen as "a mere abstraction, a luxury with little concrete value".[37] Only when the security needs are satisfied is it possible to consider the privacy needs at all.[38] Thus, frightened by terrorist attacks, people tend to choose security over privacy. In these circumstances, the process of balancing became a strategy of the security measures proponents,[39] where the risks of terrorism and crime are supposed to rationalize the need for enhanced security measures, typically exemplified by the enhanced measures introduced after the 9/11 attacks.

The balancing concept is subject to criticism, first because it justifies the compromises to privacy and other human rights.[40] It is proposed that instead, the test of proportionality should be used. By using the latter, it can be evaluated whether a concrete security measure is suitable, necessary, without any other less intrusive alternatives, non-excessive, and so on. Applying these criteria allows us to assess whether the privacy risks caused by this measure are proportionate to its security benefits and other factors.[41] In addition, mechanisms of protection, including redress for abuses, are actually provided by the concept of human rights in general and privacy/data protection regimes in particular.[42]

To summarize: aviation security as an element of human rights contributes to the right to life (or freedom from harm), constituting a good argument for enhanced security. At the same time, aviation security as such, i.e. as broader security demands including measures, resources, security decisions, evaluation of threats, risk assessment, etc. – overall leads to enhanced surveillance and control and, consequently, to human rights concerns.

The dilemma to be examined in this work conveyed through the widespread expression "security versus privacy" is the aviation-security-versus-privacy-data--protection-and-other-human-rignts dilemma. The rights to privacy and data protection are the focus of this research and are used to evaluate the selected security measures. But, since other human rights may be relevant, and, as discussed, aviation security is a matter of human rights, they will also be discussed on an additional basis to better reflect the situation: simultaneous violation of a number of human rights makes the whole aggregated impact of aviation security measures on an individual more serious. For the convenience of the reader, further, the term "security

---

[36] Schneier (2008) p. 70.

[37] Spencer. *Security Versus Privacy: Reframing the Debate*. In: Denver University Law Review. Vol. 79 (2002) p. 519.

[38] Aquilina. *Public security versus privacy in technology law: A balancing act?* In: Computer Law & Security Review. Vol. 26 (2010) p. 135.

[39] Spencer (2002) p. 519.

[40] See Chap. 5.

[41] See Chap. 5.

[42] See Chaps. 3 and 2 respectively.

versus privacy" will be used as synonymous to "aviation security versus privacy, data protection and other human rights".

Since this research will use the proportionality test rather than balancing as an approach to the dilemma, the latter cannot be considered exclusively as a tension between opposites. Both values are important for society; what is needed is to find a way to ensure all of them, without losing one or the other. The "ideal" model of interactions between aviation security and privacy, therefore, will be aviation security *plus* privacy, aviation security *and* privacy.

1.2 Aims of the Research

As a result of existing research, it is quite well known that security measures, including those in aviation security, have tensions with human rights in general and with privacy and data protection specifically. However, the pro-privacy side focuses on privacy interests rather than aviation security interests. The latter tends to remain in the "shadows". In addition, substantial research covers mainly the EU, the USA and the Western states in general. There is some lack in the research at the international level and at the national level of non-Western states.

Using existing research extensively, this work analyses different security measures, methods and technologies in civil aviation from the perspective of their impact on the rights to privacy and data protection as well as related individuals' rights, including the right to health, freedom of movement, the right to equal treatment and non-discrimination, freedom of thought, conscience and religion, and rights of the child. In short, the main objective is to discuss what can be considered proportionate security, taking into account privacy/data protection other human rights concerns.

The following sub-aims can be indicated.

1. To discuss whether the use of the selected aviation security measures: body scanners, camera surveillance and Passenger Name Record (PNR) systems[43] ("selected measures"), in particular their impact on privacy and data protection, corresponds to legal principles of privacy and data protection. Thus, to some extent, the research will analyse the situation *de lege lata* – but subject to limitations and on an illustrative basis.[44]
2. To describe the selected measures, their background, emergence and contemporary usage. In other words, the situation will be analysed *de facto*. This is particularly important, since for privacy/liberty-security debate, not only privacy/liberty interests, but also security interests should be better understood and more

[43] Transfer of air passenger data from airlines to state authorities, with further analysis of these data for security and other purposes. See more details in Chap. 6.

[44] See more details below.

meaningfully evaluated.[45] This is the reason for the fairly detailed analysis of aviation security background, importance of civil aviation, aviation security concepts, effectiveness matters etc. and the specific aviation security measures in this research.
3. To analyse options for rectification and recommendation – i.e., to discuss legal policies, how the situation should be – *de lege ferenda*. Legal principles of privacy and data protection will be used as the basis. I stress that this aim refers mainly to general insights rather than to proposing a comprehensive framework for the reform.

I will also draw attention to specific issues relating to legal, technological and other developments in aviation security and privacy/data protection which, in my opinion, present the most controversial and problematic issues relating to aviation security-privacy dilemma.

1.3 Approach of the Research

This research differs from traditional legal research. While it involves three levels of regulation (international, regional and national), deep and comprehensive analysis of legal regulation at all these levels would be unrealistic. Thus, I have chosen to use legal principles as a basis for analysis, and to use legal regulation to first explain principles and secondly for their illustrations. In addition, the use of legal principles has other advantages.

As will be explained in Chap. 5, legal principles can be applied to any jurisdictions considered in this work. Since technological development is rapid and legislation is not, to evaluate selected aviation security measures, we can apply legal principles to aviation security measures on a case-to-case basis in an effort to identify problems and deficiencies, to identify and analyse the options for rectification and recommendation. Moreover, principles enable an easier exchange of views across legal regimes and between various professions involved in the aviation security-privacy debate. Thus, for research like this, which implies various jurisdictions and interests, the analysis of legal principles rather than legal regulation has a pragmatic goal, and is aimed to suggest an effective model for approaches to the aviation security-privacy dilemma.

Respective legal regulation, in particular, applicable international and EU instruments, along with relevant case law, will be used as the sources of the legal principles. For illustration of principles, "regulatory cases" from the EU and some national regimes: the United Kingdom of Great Britain and Northern Ireland (UK), Norway, the United States of America (USA), and Russian Federation (Russia, RF) ("selected states") will be used.

---

[45] Solove. *Nothing to hide: The false tradeoff between privacy and security* (2011) p. 3.

Civil aviation can be characterized by strong international character and increased travel. Globalization[46] and technological developments have led to increased availability and circulation of personal data worldwide. Due to these and other global trends, and keeping in mind that fundamental human rights belong to every person independently of any citizenship, the analysed aviation security-privacy dilemma is (or should be) a relevant topic anywhere in the world where civil aviation and aviation security exists. The analysis of international dimensions is therefore of a great importance. In order to improve privacy and data protection under the new security regimes, it is necessary to have the knowledge of the international context and international efforts in both privacy and security.

The choice of the EU's framework is not accidental. First, as mentioned above, legal principles of privacy and data protection which are central tools for this research derive from international and/or EU law. Secondly, in the field of data protection, Europe is known as the world leader. The EU Data Protection Directive has been implemented across EU Member States and served as model law for many other states.

National perspectives are important too. National illustrations will ensure a broader and a more comprehensive picture. Clearly, the aviation security measures are concentrated at the airports related to definite states; passenger personal data ends up at national law enforcement, security and intelligence organs, etc. Thus, for concrete illustrations of aviation security regimes, definite national regimes will be used as examples. This concerns both regulatory and technological "cases".

Four states were chosen for this purpose: the UK, Norway, the USA, and Russia. The choice of the states was greatly influenced by my language capabilities (English, Russian and Norwegian),[47] which played a key role in providing access to the materials available.

I acknowledge that illustrations from national regimes will be given without carrying out a legal comparative analysis that includes a systematic and extensive comparison of the jurisdictions. I also acknowledge that a problem arose pertaining to the availability of materials at different levels, especially national ones that differ greatly both quantitatively and qualitatively, leading to asymmetry of information at different levels and consequently asymmetry in the analysis. The difference also may occur due to the need to cover specific issues in more detail.

The selected aviation security measures were chosen due to a number of factors: they relate to air passengers; they all are argued to enhance security; their use is growing, and they all may have significant impact on privacy/data protection/other human rights. Moreover, they all are emerging technologies, that is, technologies

---

[46] Globalization is a multidimensional phenomenon, which involves a deepening and broadening of rapid transboundary exchanges due to developments in technology, communications, and media. Such exchanges create a more interdependent world. See Shelton. *Protecting Human Rights in a Globalized World*. In: BC Int'l Comp. L. Rev. Vol. 25 (2002) p. 275.

[47] Whenever possible, the research uses English versions of Russian and Norwegian sources or their official translations into English, with indication of sources. In all other cases, unless specified otherwise, the translations are the author's.

under development; they may present central future security measures, not only within civil aviation. These measures are hence central aviation security measures for this research.

At the same time, there are other security techniques and methods that may serve a part of the selected measures, or be connected to them directly or indirectly. What is important is that they may have influence on the aviation security and privacy dilemma, for instance, by increasing effectiveness or by raising additional concerns, contributing to the measure's aggregated impact on passengers' rights. Hence, they will be discussed as well, in the amount necessary to understand the technique in general, focusing on some specific features in the context of the selected measures. This includes biometrics, profiling, and behaviour analysis. Moreover, more general trends and approaches in modern aviation security may have additional impact on the dilemma and will therefore be discussed too.

It should be noted that in contrast to other legal works whereby the legal requirements influence and determine the relevance of technological features,[48] in this research, technological descriptions are determined by the technology's functions with potential impact on individuals' rights. In other words, the technology description is first functional, with the primary aim to show the "meeting point" of the technologies and human beings.

Obviously, for these purposes, not *all* technical and operative capabilities are essential, but only those which may have impact on human rights, including technical safeguards to protect these rights. Hence, only selected relevant features are chosen and analysed, with subsequent evaluation of the applicability of human rights and operation of legal principles: how effective is the technology for protection of the right to life?; which other concerns for human rights arise?; how serious are they?; are there technical safeguards that are able to protect the rights? etc. The technology may entail different applications, different safeguards and thereby produce different impacts. These different variants should also be described.

As a result, the technological and operational description in this research is broader than is usually found in legal works. This can be considered an advantage, since details about aviation security technologies are not thoroughly addressed by legal research. In addition, people in general are unaware of many aviation security technologies' capabilities and usage, and not all passengers regularly read Aviation Security International and similar journals. Thus, this work can contribute to better awareness, at least for its readers.

At the same time, there is a problem of missing information in respect of aviation security. An important aspect here is the degree of openness of information in aviation security. This issue will be discussed in more detail in Chap. 5 on the subject of the principle of legality, but in short, aviation security, which falls within national security, is characterized by a high amount of restricted information. This can be understandable, since security must be safeguarded. It is essential for the security system to keep many of its elements confidential: the more information about the

---

[48] See e.g. *Ibid.*

security system is made available, the more vulnerable it is. Release may constitute a threat of security.

As the result of limited information, a legal researcher with specific requirements to the study faces a problem with description of some facts.[49] In the case of this research, information about many processes and issues in aviation security is missing, in particular, where exactly some of the discussed technologies are used, the details about technical and operative characteristics of the measures, criteria of risk assessment and effectiveness, etc. Indeed, it can be difficult to rely on facts lacking some essential details or that are unclear or controversial. This leads to a degree of uncertainty when analysing the proportionality of the regimes in this work.

Another challenge is that aviation security measures as well as technologies and operational procedures designed to ensure the individual's rights are subject to quick and constant changes and further development. Thus, while finalizing this research, I tried to update references to all legal sources, facts, figures, technological features, etc. on which the research is built. The presentation of law in this book is current law as of 5 October 2016. Unless otherwise specified, all cited websites were last accessed on that date.

However, not all sources could be sufficiently updated. This creates a potential risk that some facts/numbers used in this research might be outdated. Nevertheless, this problem can be seen mitigated by a global and general character of this work, where details are used mainly as illustrations, and the fact that changes in the analysed spheres are inevitable anyway.

In summary, due to the presented challenges and reasons, the research cannot be considered an all-encompassing and comprehensive evaluation of the current regimes capturing all the relevant details and presenting "the ultimate truth", but should rather be seen as a contributing perspective, suggesting that further research is needed.

At the same time, the research is aimed to concentrate on principal issues on aviation security versus privacy rather than details, thus, can be applicable to future regimes as well as regimes beyond aviation security field, wherever/whenever the individual's rights may be compromised under the motto of security needs/precautions.

As a matter of contribution to the existing research, a number of features of this research can be highlighted.

1. This research endeavours to present the aviation security interests more thoroughly than is usually done in privacy-security debates. Thus, it is an attempt to go deeper into aviation security perspectives than is usually done in security-privacy debates. Civil aviation is a sector in which the risks of terrorism are highlighted, and with the most advanced and sophisticated security measures if compared with other transport modes or other public places. For better understanding of how the situation became what it is, why this sector requires enhanced protection, as well as the logic of aviation security decisions (which exactly

[49] Andersen. *Edb og ansvar.* Jurist-og Økonomforbundets Forlag (1988) p. 29.

aviation security measures should be used and why), aviation security will be discussed more thoroughly, taking into account the importance of civil aviation for the world economy, historical background, threats and risks concepts, etc.
2. This research addresses a number of the most controversial aviation security measures used worldwide; it discusses their additional features, connections between them as well as global trends such as the risk-based approach, the combination of various technologies, global information sharing, etc. Technological and operational descriptions of the measures are broader than what is usually done in legal research.
3. I also include, along with examples from the UK, Norway and the USA, examples from Russia. In contrast to others, belonging to the West and "old democracies", Russia is a non-Western and post-authoritarian state, a "new democracy". Since there are many more countries outside the West that have similar features and traditions and may face similar problems, Russia can be considered as an example from a substantial part of the world.[50] The conclusions on Russia, therefore, may have global dimensions, e.g. as is the case that not only Russia, but also more and more non-Western countries demand PNR for security purposes.

I believe that this research may contribute to further research in security versus privacy/liberty, not only in respect of civil aviation, but security in a broader sense, since many issues are common or similar. This work may also assist in providing a more detailed analysis of some particular issues and problems discussed in this work.

1.4 Overview of the Structure

This book is divided into three substantial parts. Part I – General Part – includes Chaps. 2, 3, 4 and 5 and is dedicated to presenting general issues for this research, going from more broad and general to more specific issues.

Chapter 2 is dedicated to privacy and data protection in aviation security. First, it discusses the essence of the rights to privacy and data protection, relation and connection between each other. Since there are many different approaches and concepts of what privacy is, this work will present those which can be used as tools to analyse applicability of privacy right to specific aviation measures in Part II. Similarly, the concept of data protection will be presented.

Secondly, Chap. 2 provides an overview of general legal regulation of privacy and data protection as well as technology-specific privacy/data protection regulation and soft law which may be applicable in aviation security. Three levels of the selected states will be discussed: international, EU and national. Although the research will use legal principles of privacy and data protection (Chap. 5) as a basis for analysis of privacy/data protection concerns in civil aviation, discussion of legal

[50] Including post-Soviet states, many "developing" states in Asia, Latin America, and Africa.

regulation serves as a necessary basis for the understanding of principles, the basis for illustrations used in Part II.

Chapter 3 will discuss the human rights concept with general mechanisms of human rights protection as well as grounds for their restrictions and limitations. Furthermore, it is a discussion of a number of particular rights which are relevant in the aviation security area.

Indicating the status of the whole picture of security versus human rights in addition to privacy/data protection is particularly important, since, as mentioned above, (i) security serves to protect the passengers' right to life, i.e. contributes to human rights positively, and (ii) privacy alone cannot reflect all of the negative effects of the security measures on human rights other than the right to life – other human rights concerns are applicable too. Accordingly, the other human rights perspectives enables the visualization of a broader picture, to better understand the extent and actuality of the security-privacy problem, since impact of aviation security measures on other human rights ultimately have an effect on the proportionality of aviation security measures and privacy interests.

Chapter 3 will provide an indication of the applicability of the respective human rights with respect to the selected measures, and in so doing, determining particular concerns. It will also contribute towards finding and developing approaches for substantiating human rights concerns. These particular concerns the principle of proportionality and other principles that will be discussed further in Chap. 5, and then applied in Part II.

Chapter 4 will analyse what aviation security is, the reason why civil aviation gets so much security, and why and how the security measures which have an impact on privacy and data protection were developed and implemented. This can be considered from different perspectives. This chapter will start with discussing the term aviation security along with the related and sometimes confusing term safety. Then, it will analyse the importance of civil aviation to the world economy, growth in aviation and in aviation security in numbers. It will discuss historical development of aviation security measures, the rationale of aviation security decisions, including threats and risks, principles of aviation security and finally, general regulation – as discussed above, on a limited basis.

First, this chapter constitutes an important background for understanding the role and essence of aviation security. Secondly, it provides necessary tools for analysis of proportionality of aviation security regimes. For instance, threats, especially terrorism, are usually used to substantiate the interference to human rights, while effectiveness is a prerequisite for evaluating if benefits of a measure are proportionate to the harms. At the same time, privacy can be considered as one of the aviation security costs: if this cost is excessive, a measure may be cancelled or amended, which may be seen as disadvantage for security.

Chapter 5 is dedicated to legal principles of privacy and data protection. For research like this, which implies various disciplines, jurisdictions and languages, the analysis of applicable legal principles has a pragmatic goal; it is aimed to suggest an effective model for approaches to the privacy-aviation security dilemma.

This objective makes this chapter a key source for the Special Part, where the principles will be used with reference to particular aviation security measures.

Accordingly, the General Part will provide an introduction and general knowledge, core dimensions of the relationship between aviation security and privacy/data protection. Information, terms, principles and findings from the General Part will serve as a necessary and important basis, vocabulary and tools for the Special Part, for a further and more detailed analysis of the rights to privacy and data protection as well as other individual rights applicable to particular aviation security measures.

Part II – Special Part – includes two chapters – 6 and 7. It will analyse threats to privacy and data protection from a narrower, but deeper perspective, indicating concrete problems.

Chapter 6 will describe the selected aviation security technologies. It will focus on three of them – body scanners, camera surveillance, and PNR systems. Nevertheless, some related technologies will be discussed as well, including biometrics, in particular, facial recognition, profiling and behaviour analysis, as well as common modern trends: pro-active, risk-based and intelligence-led aviation security, including trusted traveller programmes, global information sharing, randomness in screening, and a combination of aviation security measures. The point is that any of these additional items may have additional effects on the dilemma.

The discussion will focus on technological and operative features having impact on human rights, provision of safeguards, and issues of effectiveness, since the latter may have direct impact on proportionality of the regimes. Simultaneously, the applicability of privacy/data protection/other human rights will be assessed using tools from Chaps. 2 and 3.

All measures are different, with different levels of intrusiveness from the perspective of privacy and data protection, but all these will be "screened" by the privacy and data protection principles in Chap. 7 in order to indicate which specific problems arise and if there are adequate safeguards, etc.

Importantly, the general principles discussed in Chap. 7 are common for evaluation of interference regarding all the applicable human rights. Thus they can be used not only in respect of aviation security versus privacy/data protection, but also in respect of aviation security versus human rights. Accordingly, the ultimate question is whether the selected measures can be considered proportionate, i.e. whether intrusiveness is justified by the level of threat in civil aviation and the corresponding benefits of the aviation security measure.

Part III is Chap. 8 – the conclusion. It will return to the "general" level with additional knowledge from Part II. It will discuss selected findings relating to regulative, technological and other development in aviation security and privacy/data protection which, in my opinion, present the most controversial and problematic issues relating to aviation security-privacy/liberty dilemma. They will reflect current and future prospects.

Part I
General Part

Chapter 2
Protection of Privacy and Data Protection in Aviation Security

2.1 Introduction

As discussed, the notions of privacy and data protection are central for this research. In order to analyse the impact of a particular aviation security measure on privacy and data protection and particular consequences in the Special Part, it should be explained how the terms "privacy" and "data protection" will be used in this research, how they are connected and how the interests that they denote are regulated.

Since Chap. 6 will determine applicability of privacy and data protection rights with respect to particular aviation security technologies, it is important to establish approaches and necessary tools. That is why the description of these approaches/tools in this chapter is quite detailed.

The first problem is that there may be some confusion because of the common use of the term "data protection" in the EU through which the law normally distinguishes these two terms, whereas in the USA, the term "privacy" tends to be used instead. In addition, on both sides of the Atlantic, the term "data privacy" is increasingly used.[1] Data privacy sounds in some ways more logical than "data protection", since it refers to the right itself rather than the "protection" process. In comparison with the other two terms, it reflects more accurately "the focus, thrust, and rationale of the relevant norms".[2]

Secondly, there are difficulties with the term "privacy". The problem concerns not only vagueness of this term and absence of its universally agreed and/or widely acceptable definition, i.e. the formal indication of a specific human right or freedom, but also the broad use of this term outside the legal or human rights sphere. It is often used to denote political, social, psychological privacy, among other aspects.

---

[1] See Bygrave. *Data privacy law: an international perspective* (2014) pp. 23–26.

[2] *Ibid.* p. 26.

O.M. Enerstvedt, *Aviation Security, Privacy, Data Protection and Other Human Rights: Technologies and Legal Principles*, Law, Governance and Technology Series 37, DOI 10.1007/978-3-319-58139-2_2

Thus, apart from human rights, "privacy" may imply a concept, idea, value, interest, or something else. As the result, there are many different interpretations of what privacy is. This work will present those that can be used as tools to discuss the problems in this research.

Thirdly, data protection may involve disputable issues as well, including what exactly makes data "personal" and what amounts to "processing". Additionally, a part of aviation security is information security, which is aimed at protecting corresponding systems against cyber-attacks and cyber-terrorism, creating a certain overlap.

The next part of this chapter is dedicated to an overview of privacy and data protection regulation, in particular in the aviation security context. In general, aviation security is dedicated to protection of passengers' lives and other assets from acts of unlawful interference.[3] However, a basic rule is that if the states want to use an aviation security technology, they must comply with other relevant legal regulation, including that which applies to privacy/data protection. The question that arises is how privacy and data protection requirements should be applied to security technology practices, especially new ones.

As will be discussed in Chap. 4, civil aviation has a strong global character, and there are endeavours to achieve more harmonized international regulation, including of aviation security. At the same time, as discussed above, globalization and technological developments caused personal data issues to become international concerns. Thus, international instruments on privacy/data protection in aviation appear to be necessary. Applicable instruments will be discussed first.

Then, the level of the EU law will be discussed, as a region with the most "fundamentalist" approach to data protection in the world, where data protection law is driven from the human rights perspective, and personal data protection is a fundamental right.[4] As Newman argues, although Europe does not always prevail in international regulatory debates, it has acquired "regulatory capacity" in the field of data privacy, creating and expanding rules in Europe and around the world.[5]

It is a fact that during the past decades many countries have established regimes based on the EU model. The point is that processes of integration and globalization dictate their need to bring national legislation and practice into line with international standards: otherwise, they can be isolated in the data protection field, in particular in personal data flows. In addition to these national endeavours, the EU influences data protection rules of other countries by exporting its data protection standards via Europol cooperation agreements, PNR agreements with the USA,

[3] See Chap. 4.

[4] See, e.g. Privacy Law and Business. *EU Commissioner Věra Jourová says protection of personal data more than a "European" fundamental right.* http://www.privacylaws.com/Int_enews_30_10_15 (30 Oct 2015) and Cline. *Global CRM Requires Different Privacy Approaches* (2005).

[5] Newman. *Protectors of privacy: regulating personal data in the global economy* (2008) pp. 8–9.

Canada, and Australia,[6] ongoing negotiations for an EU-US Umbrella Data Protection Agreement,[7] etc.

The next section is dedicated to national regulation in the selected states. Issues relating to privacy/data protection are solved in many countries at the level of law, soft law, including administrative measures, and mechanisms of self-regulation. In general, the following models of legal regulation can be distinguished[8]:

- The Continental (European) model, designed to ensure a balance between state control and private initiative. The latter is strictly regulated: any code of practice developed by industry group cannot derogate from standards of data protection law. The use of industry-developed codes of practice is limited.[9]
- The American model is characterized by liberalization of the market of information technologies, minimum control of the state, encouraged private initiative and very flexible legislation. The key idea is self-regulation implying that it is up to the market to come with regulation initiative, not through legislation.
- The Asian model: the state is involved in the provision of large-scale investment in the development of information technologies; creation of both material and social structures; regulation is based on a hierarchical society.[10]

At the national level, the perceived level of surveillance, which may have effect on both privacy and aviation security regimes, will be mentioned too. Additionally, some specific national features – in particular, those that concern privacy regulation in the USA specifically, and in Russia, a state outside the Western world – will be mentioned. These additional needs justify to some extent a degree of asymmetry in national-level discussions.

Another point – common to all levels – is that globalization, technological developments, widespread use of enhanced security and surveillance measures, as well as other factors, substantiated the fact that application of general privacy and data protection norms in aviation security can be a problem without more specific and detailed rules on how these general provisions should be applied in this particular area. Thus, in addition to providing an overview of general privacy and data protection regulations, this chapter will also analyse relevant technology-specific laws or other instruments.

---

[6] See Special Part.

[7] Agreement between the USA and the EU on the protection of personal information relating to the prevention, investigation, detection, and prosecution of criminal offences. As of September 2015, negotiations on this agreement were finalized. http://europa.eu/rapid/press-release_MEMO-15-5612_en.

[8] Kovaleva. *Informational Law of Russia* (2007) p. 14.

[9] Bygrave. *Privacy and Data Protection in an International Perspective*. In: Scandinavian Studies in Law (2010) p. 189.

[10] For more detail on data privacy laws across Asia see Greenleaf. *Asian Data Privacy Laws: Trade & Human Rights Perspectives* (2014).

2.2 What Is Privacy and Data Protection

2.2.1 The Concept of Privacy

In the aviation security sector, with reference to a number of measures, one may often hear: where does privacy come into the picture here and what is the problem with interfering with privacy? We are looking for terrorists and protecting your lives; what are you talking about?

For instance, the main purpose of the body scanners is to see objects carried under clothes, an invasion of a very intimate and personal sphere; a tension between protection and privacy seems to be clearly present. Nevertheless, with regard to other security measures, e.g. collecting and analysing personal data, the invasion of privacy may not be so clear: for ordinary air passengers, physical privacy is definitely more visible than informational privacy. This uncertainty leads to the question of what constitutes air passengers' privacy in each particular situation, in particular, with reference to technologies.

The concept of privacy has broad historical roots in sociological, philosophical and anthropological discussions; first of all, in the context of contrasting private and public spheres. Probably the first philosopher who tried to separate the private and the public was Aristotle (384–322 B.C.).[11] While the public realm is normally associated with political activity of citizens, the community and the state, the private realm is associated with family and personal life.

There were many attempts to define privacy, to indicate whether privacy as such is best characterized as a state/condition, a claim, or a right,[12] an interest, a value, a preference,[13] with different conceptions developed.

This research is legal, thus, it is logical that privacy as a right is the most important dimension. The point is that if it is established as a human right or constitutional right privacy would require protection by the state, as the guarantor of human rights, and serve as an instrument of protection against abuses by the state,[14] even if a person is doing nothing wrong at the time of surveillance.[15]

In aviation security, it is the state's security organs that intrude into passengers' private sphere during screening, surveillance and other procedures, and privacy protection can be seen as a way of drawing the line for how far they can go. Thus, privacy is considered in this research as primarily a human right.

In contrast to other "traditional" fundamental rights that were formulated in the XVIII century, privacy was moved from the sphere of philosophy into the legislative sphere only in the late XIX century. It was first defined as "a right" in 1890, in the

[11] Aristotle. *Politics* ([ca. 330 BC], 1983).

[12] Bygrave. *Privacy Protection in a Global Context–A Comparative Overview*. In: Scandinavian studies in law. Vol. 47 (2004) p. 323.

[13] Nissenbaum (2010) p. 2.

[14] See Chap. 3 – the concept of human rights.

[15] Schneier (2008) p. 63.

USA, in an article by Warren and Brandeis.[16] This article also addressed theoretical and legal issues about the relationship between technology and privacy in relation to portable photography equipment.[17] Fifteen years later, the right to privacy was expressly recognized by American courts.[18] In the middle of the twentieth century, the right to privacy began to be recognized as one of the fundamental human rights in official legal instruments, international, regional and national. Today, it is recognized around the world in diverse nations, regions and cultures.

However, having deep philosophical roots and due to its vagueness and complicated scope, the right to privacy is very difficult to define and understand. There is no clear legal or commonly accepted legal definition of privacy; the process of legal formation of this right is not finalized, and privacy is "a concept in disarray".[19] Thus, for a better understanding of privacy, in addition to analysis of legal regulation and case law addressing the right to privacy, the theoretical concepts of privacy will be considered too.

According to Bygrave, there are four principal ways of defining privacy: non-interference, limited accessibility, information control, and incorporation of various elements of the other three sets but linking privacy exclusively to intimate or sensitive aspects of persons' lives.[20] These four conceptions can be considered as basic, having many variants and interpretations.

Moor and Tavani, for instance, distinguish nonintrusion, seclusion, limitation, and control theories of privacy,[21] developing an incorporative theory, so-called Restricted Access/Limited Control (RALC). The "incorporative" theories include the RALC theory, which, according to the authors, has three components: an account of the concept of privacy, an account of the justification of privacy, and an account of the management of privacy.[22] Solove mentions, in addition, the concepts of privacy-as-secrecy and privacy-as-intimacy,[23] also falling into the basic four approaches. The key theories will be briefly discussed below in relation to aviation security.

The first one – non-interference or nonintrusion – is presented by what Warren and Brandeis term the "right to be let alone".[24] According to the seclusion theory of privacy, privacy is also identified with "being alone"[25]; e.g. Gavison describes a

---

[16] Warren and Brandeis. *The Right to Privacy*. In: Harvard Law Review. Vol. 4 (1890) p. 193.

[17] Finn [*et al.*] *Seven types of privacy*. In: European data protection: coming of age (2013) p. 3.

[18] *Pavesich v. New England Life Insurance Co.* Supreme Court of Georgia, 122 Ga. 190; 50 S.E. 68; 1905 Ga., 3 March 1905.

[19] Solove. *Understanding privacy* (2008c) p. 1.

[20] Bygrave. *Data protection law: approaching its rationale, logic and limits* (2002) pp. 128–129.

[21] See, e.g.; Moor. *The ethics of privacy protection*. In: Library Trends. Vol. 39 (1990) p. 69; Tavani. *Philosophical theories of privacy: Implications for an adequate online privacy policy*. In: Metaphilosophy. Vol. 38 (2007) p. 2; Solove (2008c).

[22] Tavani (2007) pp. 9–10.

[23] Solove (2008c).

[24] Warren and Brandeis (1890) pp.193, 205.

[25] Tavani (2007) p. 5.

person as enjoying "perfect privacy" when that person is "completely inaccessible to others" that is, when no one has "physical access to [the individual]".[26]

However, if applied to technologies in general and those used in aviation security in particular, obviously, this "ideal" condition can never be fulfilled. Airports and aviation security as a whole belong to areas where anonymity may constitute threat or result in undesirable behaviour (see more detail below). Thus, persons who wish to have 100 per cent privacy and be let alone should sit at home rather than travel by means of public transportation.

This means inadequacy of the non-interference concept for this sphere. Apparently, instead of a radical "all or nothing" approach, other approaches should be used. Moreover, all the cited scholars recognise that privacy is relative rather than absolute, and you cannot treat privacy according to an all-or-nothing approach. Accordingly, it is more appropriate to speak about *some* control, *some* privacy, etc.

Within the limited-access conception, the core idea is that privacy is a matter of restricted access to persons or information about them – "limited access" to the self.[27] According to Gavison, privacy is a condition of "limited accessibility" consisting of three elements: "secrecy" – "the extent to which we are known to others", "solitude" – "the extent to which others have physical access to us", and "anonymity" – "the extent to which we are the subject of others' attention".[28]

The privacy-as-secrecy conceptions, e.g. "concealment of personal information"[29] can be understood as a subset of limited access to the self, but this conception is narrower because secrecy involves only one dimension of access to the self – the concealment of personal facts.[30]

The control-over-information conception is presented, for example, by Westin's definition of privacy as "the claim of individuals, groups, or institutions to determine for themselves when, how, and to what extent information about them is communicated to others".[31] According to Fried, "Privacy is not simply an absence of information about us in the minds of others, rather it is the control we have over information about ourselves".[32] Accordingly, privacy is a matter of one's control over oneself without external interference. As Solove notes, this concept can be viewed as a subject of the limited-access conception.[33]

The theory of intimacy also states that privacy consists of some form of limited access or control, but it locates the value of privacy in the development of personal relationships; all relationships are formed with differing degrees of intimacy and self-revelation, and people value privacy so that they can maintain the desired levels

[26] Gavison. *Privacy and the Limits of Law*. In: Yale LJ. Vol. 89 (1979) p. 428.

[27] Solove (2008c) p. 18.

[28] Gavison (1979) pp. 421, 428–436.

[29] Posner. *The economics of justice* (1983) p. 268.

[30] Solove (2008c) p. 22.

[31] Westin. *Privacy and freedom* (1970) p. 7.

[32] Fried. *Privacy* In: Philosophical dimensions of privacy (1984) p. 209.

[33] Solove (2008c) p. 25.

of intimacy for each of relationships.[34] Iness defines privacy as "the state of possessing control over a realm of intimate decisions, which includes decisions about intimate access, intimate information, and intimate actions".[35]

As noted above, privacy is relative rather than absolute. A good illustration of such relativeness is the situation with aviation security measures: the passenger's choices are limited: to fly or not to fly, to book a ticket and release personal details or not to book, to be scanned or manually searched, to go through the airport and be watched by CCTV cameras or not go, etc. If he or she opts to fly, he/she must undergo exposure to security measures, typically without knowing the details of the used technologies and without access to collected data. At that point, the passenger loses control over herself.[36] Hence, body scanners, for instance, harm privacy because they require people to give up control over their own body.[37] However, within the control theory, the options for rectifying the situation are quite radical and not very practical: in order to return 100% control, the recommendation is not to fly at all.

Although describing the situation of aviation security versus privacy, these theories focus on harms that already take place and must be stopped.[38] Thus, they fail to suggest more realistic solutions on how to prevent harms, and there may appear problems when applying them specifically to aviation security. The point is that even in the latter sphere, where security needs may limit some human rights deemed to be less important than the right to life/security, privacy should still be provided, and therefore defined.

This leads to conclusion that specifically the aviation security sector, with specific technologies, probably requires some specific privacy considerations, capturing the complexity of privacy issues within the challenges that technologies present.[39]

Two helpful tools can be noted: first, types of privacy, which categorize privacy in logical, structured and coherent ways,[40] secondly, consideration of privacy in the context.

Four basic types of privacy are privacy of person (physical privacy), of personal communication, of personal behaviour (or psychological) and of personal data.[41] Physical privacy is the degree to which a person is physically accessible to others; it is based on the human biological need for personal space.[42] Communicational

---

[34] *Ibid.* p. 34.

[35] Inness. *Privacy, intimacy, and isolation* (1992) p. 140.

[36] Tirosh and Birnhack. *Naked in Front of the Machine: Does Airport Scanning Violate Privacy?* In: Ohio State Law Juornal. Vol. 74 (2013) p. 1301.

[37] Tirosh and Birnhack (2013) p. 1268.

[38] Finn et al. (2013) p. 6.

[39] *Ibid.* p. 3.

[40] *Ibid.* p. 4.

[41] It is believed that four types were first categorized by Clarke. *Introduction to Dataveillance and Information Privacy, and Definitions of Terms* (1997).

[42] Joinson and Paine. *Self-disclosure, privacy and the Internet.* In: The Oxford handbook of Internet psychology (2007) p. 241.

privacy is an individual's ability and effort to control social contacts: control of the participants of, the frequency of, the length of and the content of an interaction.[43] Psychological privacy is the ability to control cognitive and affective inputs and outputs, to form values, and the right to determine with whom and under what circumstances thoughts will be shared or intimate information revealed.[44] Informational privacy overlaps with Westin's control-over-information conception of privacy cited above, i.e. the ability to determine how, when, and to what extent information about the self will be released to others.

Finn et al. argue that these four types are no longer sufficient to address concerns the emerging technologies introduce seven types of privacy as well as dimensions that combine with them in different ways. Seven types of privacy are[45]:

1. Privacy of the person (physical body) – the right to keep body functions and body characteristics such as biometrics private.
2. Privacy of behaviour and action – "the ability to behave in public, semi-public or one's private space without having actions monitored or controlled by others". This includes sexual preferences/habits, political activities and religious practices.
3. Privacy of communication amounting to communicational privacy above.
4. Privacy of data and image amounting to informational privacy above, but stressing the inclusion of images (a type of personal data).
5. Privacy of thoughts and feelings – "a right not to share thoughts and feelings or to have those thoughts or feelings revealed".
6. Privacy of location and space – "the right to move about in public or semi-public space without being identified, tracked or monitored" and the right to solitude and a right to privacy in privacy places.
7. Privacy of association (including group privacy) – the right to associate with whomever they wish without being monitored.

It can be seen that categories overlap or fall within the types above. Nonetheless, it is argued that when using more specific types, it is possible to be more precise about the impact of particular technologies on different types of privacy.[46] According to the developers, the core issue is pro-active, protective nature, since distinguishing privacy types allows one to "maintain an awareness that there are different types of privacy in order to ensure adequate protection of individuals (and society) in relation to existing and emerging technologies, applications and practices".[47] Accordingly, with this approach, consideration of the selected aviation security

[43] Burgoon [*et al.*] *Maintaining and restoring privacy through communication in different types of relationships*. In: Journal of Social and Personal Relationships. Vol. 6 (1989) p. 132.

[44] Joinson and Paine (2007) p. 241.

[45] Finn et al. (2013) pp.8–10.

[46] Raab. *Effects of surveillance on privacy, autonomy and dignity* In: Deliverable D1.1: Surveillance, fighting crime and violence (2012a) p. 261.

[47] Finn et al. (2013) p. 26.

technologies along with specific privacy types may allow one to identify both privacy risks and possible protection opportunities.

The second helpful theory is the theory of contextual integrity, which suggests that the context is crucial for the discussion of where privacy is.[48]

Accordingly, privacy is a context-specific concept, and its notion cannot be divorced from the context in which it is valued and is thus worth protecting.[49] Using this concept, it is possible to identify the impact of a new technology on existing norms within a specific context.[50] Apparently, within the theory of contextual integrity, the questions for the determination of each type of privacy would differ.

For instance, with regard to informational privacy, according to the European Court of Human Rights (ECtHR), the mere storing of data relating to the private life of an individual amounts to privacy interference. The subsequent use of the stored information has no bearing on that finding.[51] However, in determining whether the personal information retained by the authorities involves any of the private-life aspects mentioned above, the Court will have due regard to the specific context in which the information at issue has been recorded and retained, the nature of the records, the way in which these records are used and processed and the results that may be obtained.

Similarly, Nissenbaum suggests considering the following items:

- Who is gathering the information?
- Who is analysing it?
- Who is disseminating it and to whom?
- What is the nature of the information?
- The relationship among the various parties?
- Other institutional and social circumstances may also be important.[52]

For the aviation security sector, some special features can be identified that can play a role in determining the privacy. They include the public character of airports, application of measures to passengers who are a priori considered innocent people, the level of terrorism and crime threats and thereby the security needs, the technological design of the measure, etc.

With reference to the location's public character, it is suggested that crowd density estimators can be suitable indicators regarding privacy expectations and risks of infringement: high crowd density can imply that people do not expect a high level of privacy; low crowd densities mean that individuals would not expect to be seen by many others and may therefore act slightly differently.[53]

---

[48] Nissenbaum. *Privacy as contextual integrity*. In: Wash. L. Rev. Vol. 79 (2004) pp. 154–155.

[49] Badii [et al.] *Visual context identification for privacy-respecting video analytics* (2012) p. 366.

[50] Tirosh and Birnhack (2013) p. 1267.

[51] See *Leander v. Sweden*, No. 9248/81, 26 March 1987; *Kopp v. Switzerland*, 13/1997/797/1000, 25 March 1998; *Amann v. Switzerland*, No. 27798/95, 16 February 2000.

[52] Nissenbaum (2004) p. 154–155.

[53] Zoufal. "*Someone to watch over me?*" *Privacy and governance strategies for CCTV and emerging surveillance technologies* (2008) p. 46.

A place like an airport implies a crowd, where one person is mainly a stranger to anybody else and people are constantly on the move. But, since airports are subject to heavy protection from terrorism and crimes (see Chap. 4), in order to assess who poses a threat and who does not, knowledge about the travellers has to be generated; as a result, passengers are subject to a regime that encodes individuals into algorithmic risk calculations.[54] This entails the use of personal data.

In addition, in civil aviation, the core idea is that airlines carry *identified* passengers: in contrast to other transport modes such as metro, buses, etc. A passenger, in order to purchase a ticket, has to provide personal details. We do not have the absolute right to conceal our identity when boarding an airplane, entering a country, etc.[55] This means that at least some of the people at airports have already given away their personal details and it is possible to re-identify them. Thus, in any case, complete anonymity can hardly be ensured, and airports are usually considered a public places where people do not expect much privacy.

For instance, CCTV is used to collect images and monitor identified and/or identifiable persons as they are moving in public places, hence, nullifying the public privacy previously enjoyed by anonymous citizens in a public space. According to US Department of Homeland Security (DHS) Privacy Impact Assessment (PIA), this concerns persons who enter into or are near federal property.[56] In addition, according to the US case law, although a government search requires conformity with the Fourth Amendment when it invades a person's reasonable expectation of privacy, there is no infringement of a person's reasonable expectation of privacy with regard to physical characteristics that are constantly exposed to the public, including facial features and voice.[57] Thus, the government's taking an image of a person's face that the person shows in a public place does not constitute a search.[58]

However, this does not mean that privacy is not desired and need not be protected. On the contrary, people can expect privacy in public places including airports: there is a right to privacy in public places, which is "the protection of being identified or anticipating the freedom of not being identified or falling under scrutiny".[59] This idea is supported by the case law: in *Peck v. the United Kingdom*,[60] the European Convention on Human Rights (ECHR) accepted that certain incidents that take place in public may fall within someone's private life, and people can reasonably expect that they will have privacy concerning these aspects. The Working

[54] Leese. *Blurring the dimensions of privacy? Law enforcement and trusted traveler programs*. In: Computer Law & Security Review. Vol. 29 (2013) p. 485.

[55] Scarfo. *Biometrics: achieved assured authentication in the digital age*. In: Aviation Security International. Vol. 20 (2014) p. 35.

[56] Department of Homeland Security. *Privacy Impact Assessment for the DHS CCTV Systems* (2012b).

[57] *United States v. Dionisio*, 410 U.S. 1, 14 (1973), 22 January 1973.

[58] Woodward. *Privacy vs. Security: Electronic Surveillance in the Nation's Capital* (2002) p. 3.

[59] EPIC. *EPIC Framework for Protecting Privacy & Civil Liberties If CCTV Systems Are Contemplated.* https://epic.org/privacy/surveillance/epic_cctv_011508.pdf, 2008.

[60] Application No. 44647/98, 28 Jan 2003.

Party notes that individuals in transit may expect a lesser degree of privacy, but not expect to be deprived of it altogether.[61] The UK Surveillance Code also admits that in a public place there is a zone that may fall within the scope of private life.[62] Thus, people can expect privacy in the airports, and the fact that the security routines take place in public does not eliminate the privacy interest.

What is problematic is the low possibility for passengers to negotiate privacy in aviation security. It is argued that airports are example of spaces where individuals possess little or no leverage.[63] Indeed, any refusal to be screened or to provide personal data will lead to denial of the right to fly. Thus, despite the existence of the right to privacy, it can become blurred in the aviation security context.

Further, aviation security measures are primarily taken on a preventive basis. They are not directed solely at suspected terrorists and criminals just because their identities are unknown; they are directed at everybody. Apart from some types of CCTV used for monitoring and reacting to real time crimes, their usage does not constitute a law enforcement activity reacting to a specific danger or threat posed by the individual. Thus, the measures are applied to people who are generally considered innocent.

Privacy advocates argue that the idea that innocent individuals must give up some liberties for the "greater good", in fact means a repression of rights and freedoms in the name of security.[64] From a security point of view, experts agree that aviation security is more effective when it is targeted, risk-based and pro-active, rather than if all measures are applied universally on all passengers.[65]

For instance, one of the purposes of CCTV is detection of dangerous items and persons, criminals or potential offenders. It is argued that measures pertaining to images of people in airports are limited to comparing them with databases of terrorists and known criminals, then there are probably no information privacy concerns.[66] A problem is that these technologies may lack accuracy and reliability and lead to failures, i.e. become highly intrusive for innocent people.

In addition, even if targeted application is sometimes valid, and there are some methods within risk-based approach (e.g. trusted traveller programmes when passengers receive different treatment, see Chap. 6), the majority of all aviation security measures are still used universally for all passengers.

Moreover, in addition to the CCTV's purpose of detecting dangerous persons – i.e. specific individuals – CCTV is also aimed at area observation and monitoring and tracking persons. For observation and monitoring, different visual and other

---

[61] Article 29 Data Protection Working Party *Opinion 4/2004 on the processing of personal data by means of video surveillance* (2004).

[62] Home Office. *Surveillance Camera Code of Practice* (2013).

[63] Leese (2013) p. 483.

[64] Carli. *Assessing CCTV as an Effective Safety and Mangement Tool for Crime-solving, Prevention and Reduction* (2008) p. 11.

[65] COPRA. *Aviation Security Research Roadmap* (2013) http://www.copra-project.eu/Results_files/D5.1%20Copra%20Aviation%20Security%20Roadmap%20V1.0_download.pdf.

[66] Woodward (2002) p. 3.

sensor technologies are used to monitor human conduct. Instead of trying to identify individuals, they focus on conduct and acts within definite areas in order to prevent crime in real time. Here, there is little need to store data for lengthy periods. No storage would mean little privacy impact.[67]

Moreover, in order to detect suspicious behaviour, recognize specific persons, etc., the systems have to analyse behaviour, personal data, including biometrics, of many people.[68] Since cameras recording people cannot discriminate between criminals and non-criminals, they have to collect data on all passengers. Thus, harm to the privacy of the public may be the result.

One more feature is the need for security, i.e. the very purpose that is itself human-rights related: in order to avoid bombs being carried on board, the passengers and their baggage are screened, including via body scanners; CCTVs are also installed for public safety or security. The states are interested in ensuring that no terrorist or criminal enters the country, and they thus require personal data on arriving passengers in advance. All this is understandable, but certain limits to searches, screening, surveillance, data processing and so on still exist – if not, it would be possible to make all passengers take off all clothing before they arrive at the boarding gates. Ideally, such limits should be determined by the desired level of privacy, otherwise, the security measures, which are increasing every day, would constitute a threat to privacy.

Further, an understanding of the concrete aviation security measures in terms of technology and operations is important too. The used technologies may demonstrate both negatives and positives for privacy, e.g. transparency of the systems can serve to address concerns over privacy abuses[69] (will be discussed in Special Part).

It is also argued that the technology has changed the nature of what can reasonably be expected to be private,[70] usually implying less privacy expectations. However, similar to the assumption below that any cultural or personal difference in attitude to privacy does not make the right to privacy as such less important – the technologies are unable to diminish the expectation of privacy. Although the technology may have great influence on the context, it cannot decrease what can be expected as private. However, due to the described aviation security contextual features, it can be problematic for passengers to try to negotiate a higher degree of privacy.

It should be also noted that attitudes to privacy vary greatly on many levels: individual, cultural, national, in both time and space, making privacy being "in constant flux".[71] First, individually: privacy means different things to different people, and what is acceptable to one person, may not be acceptable to another. Indeed, every person has his or her own thoughts and associations pertaining to "privacy".

---

[67] Zoufal (2008) p. 10.

[68] Badii et al. (2012) p. 367.

[69] Zoufal (2008) p. 46.

[70] Gallagher. *CCTV and human rights: The fish and the bicycle? An examination of Peck v. United Kingdom (2003) 36 EHRR 41*. In: Surveillance & Society. Vol. 2 (2002) p. 271.

[71] Bygrave (2004) p. 327.

Secondly, they may vary due to differences in historical and cultural conditions that are quite difficult to change. In many cultures, the right to privacy is not a "mainstream" right. For instance, the researchers note a gap between the common law and the various (Western and Post-Communist) civil law privacy regimes.[72] In post-authoritarian societies, people may be less sensitive to privacy issues and surveillance practices.[73] In Russia, for instance, the right to privacy is not among the traditionally recognized and protected intangible moral rights (see Sect. 2.3.3.4).

Thirdly, attitudes towards privacy may vary with regard to definite historical periods. The modern notion of privacy, for instance, can be considered paradoxical: people feel more and more threatened by surveillance systems and large databases and at the same time voluntarily disclose detailed personal information in social networks.[74]

Nevertheless, a lower privacy expectation for a particular individual or society "does not decrease the responsibility of those who introduce or operate surveillance systems, with special regard to globalization trends that decrease differences among surveillance techniques, practices and ideologies worldwide".[75] Thus, for the aviation-security-versus-privacy dilemma in this study, any cultural or personal difference in attitude towards privacy does not make the right to privacy any less important.

At the same time, it is important to remember that the possible negative effects on privacy and the other human rights are not inevitable or likely; they may be of different severity and may be addressed by regulators.[76] Accordingly, the knowledge of technological capabilities does not determine the risk of harm to privacy. The latter must be empirically investigated bearing in mind many variables, including the context.[77]

A special issue in this regard is the relations between the law and privacy. Clearly, privacy is not "a resource existing in the state of nature that the law must act to conserve"; it is about what people desire, not just about what people expect. The law should thus be a tool to create the amount of desired privacy,[78] in our case – in aviation security. Privacy protection in aviation security should contribute to determine the desired – i.e. acceptable – limits for aviation security measures.

Security undertakings in civil aviation, in contrast to privacy, never existed naturally, and were introduced by human beings. Most critically, they are subject to constant change and enhancement due to technological developments and new threats. In this situation, it is extremely difficult to determine the amount of *expected* privacy that was, is, or should be expected, if any, and with reference to every

---

[72] Increasing Resilience in Surveillance Societies (IRISS). *Deliverable D2.3: The Legal Perspective* (2013). p. 47.

[73] Wright et al. (2015) p. 290.

[74] Badii et al. (2012) p. 366.

[75] Wright et al. (2015) p. 290.

[76] Raab (2012a) p. 261.

[77] *Ibid.*

[78] Solove (2008c) p. 74.

particular security practice. It is possible, however, to try to use the law to define *desired* privacy, to determine the limits how far aviation security measures may go.

The purpose of the law in this situation is to protect privacy when passengers feel a lack of privacy and a desire to rectify the situation. Every particular aviation security measure, therefore, should be analysed for its privacy impact taking into account all the relevant features of the measure – in the preferred case, the aviation security technologies' features should protect or even enhance privacy, and the possibilities of rectification should be considered and implemented. The law is thus able to define both the amount of desired privacy and the amount/limits of the security measure involved.

2.2.2 The Concept of Data Protection

It is often said that information[79] is power. "Someone who knows things about us has some measure of control over us, and someone who knows everything about us has a lot of control over us".[80] Data, especially personal, is of a special significance; access to it indeed determines opportunities of influence and power and those who control it control people's lives.

"Data protection" commonly means a set of norms that serve not only privacy protection, but a broader range of interests.[81] Data protection is closely connected to information technology (IT) developments that allow storing and using huge amounts of personal data. In today's world, characterized by globalization and digitalization, there is increased global availability of personal data. There are huge volumes of personal data being collected, stored, analysed, shared and transferred, with enhanced number of actors. The biggest challenge, for example, is "Big Data", which is the massive merging of data to generate more data, more assumptions and more knowledge about persons; examples of the production of Big Data include digital CCTV, mobile phones, scanning of machine-readable travel passes, social media postings, etc.[82]

After 9/11, processing of personal data, including commercial, by the state for security purposes increased greatly. In civil aviation, this includes the data originating from various intelligence databases (black lists, watch lists, no fly lists etc.), airlines' systems, data produced by camera surveillance systems as well as other aviation security technologies. In some cases, the passenger's body can present data as well (biometrics). The point is that using passenger's personal data collected by

[79] Data and information are used as synonyms in this research.

[80] Schneier. *Data and Goliath: The Hidden Battles to Collect Your Data and Control Your World* (2015).

[81] Bygrave (2014) p. 26.

[82] Kitchin. *Big Data, new epistemologies and paradigm shifts*. In: Big Data & Society 1.1 (2014) p 2.

all these means may determine how thoroughly he/she will be searched or whether he or she will fly at all.[83]

At the same time, the global information society is based on a fundamental compromise: on the one hand, its existence requires the transfer of huge amounts of information, on the other hand, if adequate controls are not provided, individual rights relating to personal data can be significantly violated.[84] Accordingly, the right to data protection is about finding such global compromise, allowing circulation of data, but on certain conditions.

Along with the right to privacy, the right to data protection is relatively new if compared with other human rights. As underlined by the EU Court of Justice, this right is not an absolute right, but must be considered in relation to its function in society[85] and be balanced with other fundamental rights, in accordance with the principle of proportionality.[86] Similarly to privacy, the right to data protection can be considered an instrument to defend individuals against surveillance by the state.[87]

The key issues are the concepts of "personal data" and "processing", since processing of personal data are two the main criteria for the applicability of data protection rules.

2.2.2.1 The Concept of Personal Data

The term "personal data" is defined in Article 2(a) of the DPD: it means "any information relating to an identified or identifiable natural person ("data subject"); an identifiable person is one who can be identified, directly or indirectly, in particular by reference to an identification number or to one or more factors specific to his physical, physiological, mental, economic, cultural or social identity." The General Data Protection Regulation (GDPR) stipulates the factors specific "to the physical, physiological, genetic, mental, economic, cultural or social identity of that natural person".[88]

The DPD contains some specific rules about images and sound. Its recitals and the relevant articles clarify that the DPD applies, in principle, to sound and image data relating to natural persons by also having regard to the importance of the

---

[83] Schneier (2008) p. 61.

[84] Kovaleva (2007) p. 11.

[85] Joined Cases C-92/09 and C-93/09 *Volker und Markus Schecke GbR and Hartmut Eifert v. Land Hessen*, 9 Nov 2010, § 48.

[86] C-101/2001, *Lindqvist*, 6 Nov 2003, §§ 82–90; C-73/07, *Satamedia*, 16 December 2008, § 50–62.

[87] De Hert. *The data protection framework decision of 27 November 2008 regarding police and judicial cooperation in criminal matters–A modest achievement however not the improvement some have hoped for*. In: Computer Law & Security Review. Vol. 25 (2009) pp. 403–404.

[88] Article 4(1) of Regulation (EU) 2016/679 of the European Parliament and of the Council of 27 April 2016 on the protection of natural persons with regard to the processing of personal data and on the free movement of such data, and repealing Directive 95/46/EC (General Data Protection Regulation). In force 24 May 2016, shall apply from 25 May 2018.

developments of the techniques used to capture, manipulate and otherwise use the specific category of personal data collected in this way (Recital 14). Specificity and sensitivity of the processing of sound and image data are acknowledged in Article 33: "The Commission shall examine, in particular, the application of this Directive to the data processing of sound and image data relating to natural persons and shall submit any appropriate proposals which prove to be necessary, taking account of developments in information technology and in the light of the state of progress in the information society".

Further, there is a category of data that deserves special treatment – sensitive data. The list of such data, according to DPD Article 8, includes personal data revealing racial or ethnic origin, political opinions, religious or philosophical beliefs, trade-union membership, data concerning health or sex life. The GDPR also includes genetic data and biometric data used in order to uniquely identify a person (Article 9(1)).

Convention 108 (Article 6) also contains a list of sensitive data, with proposed amendments including genetic data, "biometric data uniquely identifying a person" (Article 6(1)).

Although the list of what can be sensitive data is fixed in DPD, the Council of Europe stresses that the list in Convention 108 is not exhaustive, and states may include in domestic law other categories of sensitive data; the degree of sensitivity depends on the legal and sociological context of the country concerned; "domestic law" includes not only legislation but also appropriate or specific regulations or administrative directives.[89]

Sensitive data are subject to stricter rules than other personal data, a higher level of protection.[90] The rule, pursuant to DPD Article 8, is that the processing of sensitive data is prohibited, however, with a number of exceptions, e.g. if the data subject has given his explicit consent; Member States may, for reasons of substantial public interest, lay down additional exemptions, etc. Council of Europe's Recommendations on a particular sector or technology may also provide exceptions. All this widens the exemptions field greatly.

Article 6 of the Convention for the Protection of Individuals with regard to Automatic Processing of Personal Data requires that sensitive personal data may not be processed automatically unless domestic law provides appropriate safeguards. However, this guarantee may be restricted under certain conditions, as defined in Article 9 of the Convention: "any derogation must be provided for under national legislation and must constitute a necessary measure in a democratic society in the interests of protecting State security, public safety, the monetary interests of the State or the suppression of criminal offences; or of protecting the data subject or the rights and freedoms of others".

Another special category is biometric data. It is not contained in the DPD, but can be found in Article 4(14) of the GDPR: it means "personal data resulting from specific technical processing relating to the physical, physiological or behavioural

[89] See Council of Europe. *Data Protection. Compilation of Council of Europe texts* (2010) p. 23.

[90] Bygrave (2014) p. 165.

characteristics of a natural person, which allow or confirm the unique identification of that natural person, such as facial images or dactyloscopic data". A special feature is that biometrics may include personal data in material form – human body. Biometrics is subject to enhanced protection, since processing operations with such data present specific risks (see GDPR Recitals 53, 91, Article 9).

It should be noted that in aviation security and in a broader security context, there exists another special category of information – intelligence. Intelligence is information that is so sensitive or gained through covert methods or from secret sources that it becomes exempted from public dissemination. It is more closely held and controlled and subject to rigorous requirements.[91] It can be seen that "sensitive" here has a different meaning in comparison with sensitive personal data. However, an overlap is possible, since intelligence may constitute sensitive personal data.

When considering the aviation security area, the key question will be whether a definite measure or procedure deals with personal data. It is quite obvious that a check of identification documents or boarding passes, use of biometrics-based identification systems, use of air carriers' customer data contained in Computer Reservation Systems (CRS), may all deal with personal data. With regard to some other measures, it may be very difficult to establish whether personal data are concerned. For instance, body scanners either present images as mannequin with blurred intimate zones, or are claimed to be "image-free"[92] – is there personal data?

Different tests can be applied. An important approach is proposed by Article 29 Working Party in its Opinion 4/2007 on the concept of personal data.[93] It should be noted that according to some views, this text is a compromise document and reflects a divergent set of views of what is "personal data", although clearly written and takes a logical and analytical approach to the definition of personal data in the DPD.[94]

According to Article 29 Working Party, there are "four main building blocks" which the definition of the DPD contains: "any information", "relating to", "an identified or identifiable", and "natural person".[95] All these items "are closely intertwined and feed on each other, but together determine whether a piece of information should be considered "personal data". Each element is discussed in the Opinion, with appropriate criteria established.

The element "any information" can be interpreted very wide, regardless of nature, content or technical format of data. In conjunction with the security field, the Opinion, for instance, provides that images of individuals captured by a video surveillance system can be personal data to the extent that the individuals are recognizable.

---

[91] Raffel. *Intelligence and Airports*. In: FBI Law Enforcement Bulletin. Vol. 76 (2007) p. 2.

[92] See Chap. 6.

[93] Article 29 Data Protection Working Party. *Opinion 4/2007 on the concept of personal data* (2007).

[94] Grant. *Data protection 1998–2008*. In: Computer Law & Security Report. Vol. 25 (2009). p. 45.

[95] Article 29 Data Protection Working Party (2007) section III.

The element "relating to" plays a crucial role in determining the substantive scope of the concept, especially in relation to objects and new technologies. The Opinion provides three alternative elements (content, purpose or result) to determine whether information "relates to" an individual. This also covers information that may have a clear impact on the way an individual is treated or evaluated.

An important UK decision on "relating to" is *Durant v. FSA,*[96] which differs from this concept. In its paragraph 28, it was stated that "mere mention of the data subject in a document held by a data controller does not necessarily amount to his personal data". This depends upon whether the context and information is connected to an individual and affects their privacy. Opinions about an individual or indications of intent, however, would constitute personal data.

Nevertheless, this approach is an exception rather than a rule. In another UK case, the defendant was unable to avoid conviction for breach of data protection law by running an argument based on an interpretation of what personal data is similar to the interpretation given in *Durant*. In particular, information about the place of residence contained on the database was considered personal data.[97] According to case law of European Court of Justice (ECJ), with reference to minutes of applications for residence permits, the names of applicants contained in the minute are information relating to a natural person identified by name, thereby constituting "personal data".[98] This can be considered a more common point of view.

Nevertheless, there are concerns that there might be a general tendency towards considering an increasingly broader scope of data as personal. For instance, on 28 October 2014 the German Federal Court referred to the European Court of Justice (ECJ) a question as to whether or not dynamic IP addresses[99] are personal data.[100] The decision of the Court will be important because on the one hand, it is difficult to link IP addresses to a person without additional information, and on the other hand, general categorization may result in broadening the scope of personal data.

The element "an identified or identifiable" focuses on the means to identify a person. "Identified" relates to a situation where the identity of the person is already determined or is clear, "identifiable" – to a situation when identification has not happened yet but is possible, e.g. by combining the information being processed with other information; accordingly, the possibility of associating certain information with a particular individual is sufficient to consider criteria "identifiable" fulfilled.[101]

---

[96] *Durant v. FSA* [2003] EWCA Civ 1746, Court of Appeal (Civil Division) decision of Lord Justices Auld, Mummery and Buxton (8 December 2003).

[97] *R v. Rooney* [2006] EWCA Crim 1841 (12 July 2006), § 13.

[98] Joined Cases C-141/12 and C-372/12, *YS v Minister voor Immigratie, Integratie en Asiel and Minister voor Immigratie, Integratie en Asiel v M and S*. 17 July 2014, §§ 38 and 48.

[99] Dynamic IP addresses refer to situations in which a new address is assigned to each computer every time it connects to the Internet.

[100] Clifford and Schroers. *Personal data and dynamic IPs – time for clarity?* 23 Jan 2015. http://blogs.lse.ac.uk/mediapolicyproject/2015/01/23/personal-data-and-dynamic-ips-time-for-clarity/.

[101] Esayas. *The role of anonymisation and pseudonymisation under the EU data privacy rules: beyond the 'all or nothing' approach*. In: European Journal of Information Technology and Law. Vol. 6 (2015) p. 2.

DPD Recital 26 stipulates that identifiability of a person should be assessed taking into account "all the means likely reasonably to be used either by the controller or by any other person to identify the said person" and "the principles of protection shall not apply to data rendered anonymous in such a way that the data subject is no longer identifiable".

If the purpose of the processing implies the identification of individuals, it can be assumed that the controller or any other person involved has or will have the means reasonably likely to identify the data subject. Therefore, the information should be considered as relating to identifiable individuals. With reference to camera surveillance, it may be that identification occurs only in a small percentage of the material collected. However, since the purpose is to identify the persons in the video images in all cases, the whole application as such has to be considered as processing data about identifiable persons, even if some persons recorded are not identifiable in practice.

The fourth element ("natural person") deals with the requirement that "personal data" are about "living individuals", which is self-explainable.

After applying the "four blocks" test, it may be the case that the data do not fall within "personal data". In general, the DPD is not applicable in these cases (Article 3 DPD). However, national data protection law may still apply. According to the ECJ, nothing prevents Member States from extending the scope of the national legislation implementing the provisions of the DPD to areas not included within the scope of the DPD, provided that no other provision of community law precludes it.[102] Thus, certain situations not involving processing of personal data may nevertheless be subject to protective measures under national law. In addition, certain activities may still constitute an interference with the right to privacy. Moreover, the rules of torts law, criminal law, antidiscrimination law, etc. may also provide protection.[103]

2.2.2.2 The Concept of Processing

If there is personal data, then the next step will be to analyse whether the collection, storage, sharing, or any other activity under these data takes place. Article 2 of DPD defines "processing of personal data" ("processing") as "any operation or set of operations which is performed upon personal data, whether or not by automatic means, such as collection, recording, organization, storage, adaptation or alteration, retrieval, consultation, use, disclosure by transmission, dissemination or otherwise making available, alignment or combination, blocking, erasure or destruction". This definition is therefore quite broad and may include different operations. Moreover, analysis of the DPD reveals that it applies to both automated and non-automated processing (Article 3(1)).

---

[102] C-101/2001, *Lindqvist*, 6 Nov 2003, § 98.

[103] Article 29 Data Protection Working Party (2007).

If it is determined that the processing of personal data takes place, the next step will be to determine whether these actions correspond to the data protection rules. At the same time, with regard to data protection requirements, laws make many exemptions available to the state authorities based on a wide range of grounds. In the context of processing personal data for the purposes of aviation security, the applicable grounds may be safety, national security, transport security and security needs in general. For instance, pursuant to these needs, the right of the data subject to access his/her personal data may be restricted; the controller can be released from the obligations to notify data protection authority about the processing[104] or to obtain the data subject's consent even when processing sensitive data. This will be further analysed with respect to selected aviation security measures in Special Part.

2.2.2.3 Information Security in Aviation

As mentioned above, an additional aspect connected to "information" should be discussed. A type of aviation security which is close to airport security (see Chap. 4) is communications and information security which is aimed at protecting against attacks on the corresponding infrastructures: attacks on data integrity of communications networks, attacks on the information and personnel management systems (cyber-attacks and cyber-terrorism), etc. Hackers may exploit vulnerabilities and get access to the computer systems of control towers and aircraft, shut down airport administrative systems, suspend security measures, disengage communication lines between the aircraft and the ground or tamper and manipulate information between them; 'spoof' the Global Positioning System, as well as create a "ghost plane" or negate the existence of an aircraft, etc.,[105] all of which may cause serious harms. Thus, there is an increase of dependency upon information technology by the aviation industry.

Accordingly, this creates an overlap with data protection. As will be shown in this research, aviation security uses personal data extensively. With more of the world becoming interconnected, with the phenomenon of Big Data and the challenges connected with it, information systems and personal data are becoming increasingly vulnerable to cyber criminals. Thus, it is a crucial question for aviation security to protect its systems and all data used. Accordingly, common security undertakings, i.e. measures to protect the systems and keep the data secured can be used (see further principle of security in Chap. 5).

---

[104] E.g. in Russia.

[105] Tanti-Dougall. *Cyber security in aviation: legal aspects*. In: Aviation Security International. Vol. 21 (2015) p. 14. See also Asllani and Ali. *Securing information systems in airports: A practical approach*. In: ICITST (2011).

2.2.3 Relation Between Privacy and Data Protection

Although this work follows the European tradition and distinguishes these rights separately, their connection is very close – from a theoretical, regulatory, and judicial point of view. All these views, differences and similarities, will be discussed below.

First, it should be noted that from the chronological perspective, the views on this connection are not static. They tend to change with time due to different processes and developments. For instance, if considering the scope of these rights, first, the right to data protection was mainly "included" in the right to privacy: as discussed above, some theoretical conceptions of privacy, such as the control-over-information conception, cover "information" directly. It is obvious that data protection often refers to the informational dimension of privacy.

Later, due to increased use of personal data by different actors, development of databases and technologies, it became tendency to separate the two rights, to adopt specific regulation on data protection.[106] As a result, it became common to treat the rights as separate.

However, at the same time, an overlap exists: the scope of personal data may include, among other things, very material items, physical features: biometrics, genetic material, such as DNA and fingerprints (and the use of such data is growing), which would refer to both physical and informational privacy. Hence, today, data protection and privacy are still very close.

As a rule, the violation of the right to data protection leads to a violation of the right to privacy, at least its informational dimension, but a privacy violation does not necessarily result in a violation of the right to data protection.[107] They overlap in such a manner that data protection can be both broader and narrower than privacy: narrower since the scope of privacy is wider than informational privacy, and broader, since it applies to the personal data processing even if this does not impact privacy – e.g. if the data are not "essentially private".[108]

This is especially relevant for the aviation security area. The technologies based on biometrics may include violation to both physical privacy and data protection. Violations to informational privacy, and therefore data protection, may apply as separate violations or/and as in addition to violation to other aspects of privacy. For instance, collecting, storage, sharing and transfer of passenger personal information may present a case of intrusion into informational privacy, where data protection norms would apply as well.

The connection between privacy and data protection can be analysed also from the legislative/regulative perspectives and case law. Some human rights conventions treat the protection of personal data as an extension of the right to privacy. Although

[106] See below and Sect. 2.3.

[107] Increasing Resilience in Surveillance Societies (IRISS) (2013) p. 32.

[108] Gellert. *The legal construction of privacy and data protection*. In: Computer Law & Security Review. Vol. 29 (2013). p. 526.

such human rights instruments as the Universal Declaration of Human Rights (UDHR), the International Covenant on Civil and Political Rights (ICCPR), or the European Convention on Human Rights (ECHR) do not contain an explicit and autonomous right to data protection (see Sect. 2.3), they provide much of the formal normative basis for law and policy on data protection; the right to privacy is commonly regarded as forming the central basis for these principles.[109] However, a number of the EU instruments stipulate the right to data protection independently from the right to privacy (see Sect. 2.3.2).

Many of the data protection laws also expressly recognize the connection between rights to privacy and data protection in their texts. For example, Article 1 of Convention 108 states that its main object is "to secure… for every individual… respect for his fundamental rights and freedoms, and in particular his right to privacy…"; Article 1(1) of the DPD stipulates that that Member States "shall protect the fundamental rights and freedoms of natural persons, and in particular their right to privacy with respect to the processing of personal data".

The principal feature of data protection, in comparison with privacy, is that data protection, especially in the EU, is extensively regulated. Privacy, as discussed above, is not only a right, but also a concept, interest, value. In contrast to that, data protection is procedural and legalistic; it is about policy, legal and administrative aspects of personal data processing – i.e. it is the process by which the right is upheld, including legislation, technologies, organizations and individuals. The right to data protection can thus be seen as a set of "fair information practices".[110]

Since overlap between the two rights exists, norms on privacy can be applied by the courts with respect to data protection, and data protection mechanisms may in some cases be relevant for privacy too. In each respective case, therefore, in order to achieve a maximum protection, it should be analysed on a case-to-case basis which protection is afforded by each of the regimes, which one offers the best protection, and if possible, apply both.

The case law connects these rights as well. Pursuant to the jurisprudence of the ECtHR and ECJ, data protection serves the purpose of the enforcement of the right to privacy,[111] and at the same time, one cannot interpret data protection law without looking at the Article 8 ECHR case law.[112]

On the one hand, the courts' broad interpretations of the term "privacy" covers the right to data protection; on the other hand, data protection emerges from the case law as an aspect of privacy protection, establishing (or confirming/interpreting) important principles and approaches of data protection law. For instance, case law developed around Art 17 of the ICCPR indicates that the right to privacy in

---

[109] Bygrave. *Data protection pursuant to the right to privacy in human rights treaties*. In: International Journal of Law and Information Technology. Vol. 6 (1998). p. 1.

[110] Gellert (2013) p. 525.

[111] Increasing Resilience in Surveillance Societies (IRISS) (2013) p. 32.

[112] *Rechnungshof* decision, C-465/00 of 2003, §91.

international law includes key data protection principles.[113] The ECJ has also interpreted DPD in numerous rulings.

Keeping in mind all the above, for the purposes of this research, the discussion will be based on the assumption that the rights to privacy and data protection are distinct, not identical, but are closely related and overlap (see Fig. 1.1 in Introduction).

Despite the overlap, in a situation when different security measures may have impact on both privacy and data protection, it is important to keep these rights separate and consider them as autonomous human rights. It is crucial to consider both the rights on a case-by-case basis in order to see which one offers the best protection.[114] Therefore, it is necessary to determine the protection that is offered by both of these rights individually and under what.

Specifically, in order to establish violation to privacy, the "interference test" (see below) should be applied, since the law normally does not allow interference with privacy; in order to establish violations of data protection law, data protection rules and requirements should be analysed. Comparing these two processes, it can be noted that since data protection applies to any processing of personal data, it is more specific than privacy,[115] thus, a data protection regime may be easier to apply in comparison with privacy, which, as discussed, is much less defined.

Similar to the general overlap between privacy and data protection, the impact of aviation security measures may concern both privacy and data protection rights, creating overlap here as well. In order to establish maximum protection from any negative impact, the preferable option is to consider application and protection regimes of both rights. This will be done in the Special Part.

2.3 Privacy and Data Protection Regulation

2.3.1 International

The right to privacy is protected in a number of international human rights instruments. The basic principles can be found in Article 12 of the Universal Declaration of Human Rights (UDHR)[116]:

[113] See, e.g. General Comment 16, issued 23.03.1988 (UN Doc A/43/40, 181–183; UN Doc CCPR/C/21/Add.6; UN Doc HRI/GEN/1/Rev. 1, 21–23). For more details, see Bygrave (1998) p. 1.

[114] Gellert (2013) p. 530.

[115] Increasing Resilience in Surveillance Societies (IRISS) (2013) p. 32.

[116] Adopted by General Assembly resolution 217 A (III) of 10 December 1948. UDHR is not a treaty itself, it was explicitly adopted for the purpose of defining the words "fundamental freedoms" and "human rights" in the United Nations Charter, which is binding on all 153 member states.

> No one shall be subjected to arbitrary interference with his privacy, family, home or correspondence, nor to attacks upon his honour and reputation. Everyone has the right to the protection of the law against such interference or attacks.

Similar provisions are contained in Article 17 of International Covenant on Civil and Political Rights of 1966 (ICCPR).[117] These instruments contain both positive and negative obligations of the state: positive – to provide and ensure the right to privacy, negative – to refrain from interfering.

As for the right to data protection, both UDHR and ICCPR treat this right as extension to the right to privacy. As an individual right, it is specifically protected by a number of other international instruments.

The key instrument is the Organization for Economic Cooperation and Development (OECD)[118] Guidelines on the Protection of Privacy and Transborder Flows of Personal Data of 23 September 1980 (OECD Guidelines). They served as the first privacy principles agreed by the international community.[119] This document is in the form of recommendations and is not legally binding on OECD members, but in reality plays an important role in developing data protection laws in many countries of the world, especially non-EU states.

The Guidelines is an example of general regulation that has been reviewed and updated according to new realities. Clearly, after its adoption in 1980, there were substantial changes. As a report indicated, this specifically concerned the volume of personal data being collected, used and stored; the range of analysis of personal data; the value of the benefits enabled by new technologies; the extent of threats to privacy; the number and variety of actors who put privacy at risk or protect it; the frequency and complexity of interactions with personal data that individuals are expected to understand and negotiate; and the global availability of personal data.[120]

As the result, OECD Guidelines was updated in 2013.[121] The new OECD Privacy Framework addressed two themes: a focus on the practical implementation of privacy protection through an approach grounded in risk management, and the need to address the global dimension of privacy through improved interoperability.[122]

---

[117] Adopted by General Assembly resolution 2200A (XXI) of 16 December 1966, was opened for signature on 19 December 1966, entry into force 23 March 1976.

[118] The OECD member countries are: Australia, Austria, Belgium, Canada, Chile, the Czech Republic, Denmark, Estonia, Finland, France, Germany, Greece, Hungary, Iceland, Ireland, Israel, Italy, Japan, Korea, Luxembourg, Mexico, the Netherlands, New Zealand, Norway, Poland, Portugal, the Slovak Republic, Slovenia, Spain, Sweden, Switzerland, Turkey, the UK and the USA. The EU takes part in the work of the OECD.

[119] OECD. 2013 OECD Privacy Guidelines. http://www.oecd.org/internet/ieconomy/privacy-guidelines.htm.

[120] OECD. Privacy Expert Group Report on the Review of the 1980 OECD Privacy Guidelines, 2013.

[121] OECD Recommendation Concerning Guidelines Governing the Protection of Privacy and Transborder Flows of Personal Data (2013) (OECD Privacy Framework 2013).

[122] OECD. 2013 OECD Privacy Guidelines. http://www.oecd.org/internet/ieconomy/privacy-guidelines.htm.

However, it failed to fully address all the issues indicated during review, leaving them for possible future study.[123]

Another instrument is the United Nations Guidelines Concerning Computerized Personal Data Files of 14 December 1990. This is also a recommendation document that played, however, a minor role in developing data protection laws in comparison with the OECD Guidelines.[124]

Regional instruments which have international influence, specifically, the EU ones, will be discussed in Sect. 2.3.2 below. Another important regional international instrument is the Asia-Pacific Economic Cooperation (APEC)[125] Privacy Framework of December 2005. It is a set of "Information Privacy Principles" based on the main principles of the OECD Guidelines rather than the EU instruments. Its data protection standards are generally lower than those of the EU.[126] APEC is sometimes considered a competitor to the EU as the centre of data privacy initiatives; however, there is yet little evidence that the APEC is able to compete with the EU in this regard.[127]

With reference to technology-specific rules that aim to protect the rights to privacy and data protection with respect to particular technologies, they derive mainly from representatives of industry organizations.

The key player is the International Civil Aviation Organization (ICAO) – an intergovernmental organization charged with coordinating and regulating international air travel.[128] An example of a document that deals with both security and privacy issues is the Guidelines on PNR,[129] which establish uniform measures for PNR data transfer and the subsequent handling by the states. Although this document is influential, it is only of an advisory nature (see more detail in Special Part).

Other international aviation organizations representing the aviation and/or security industry such as International Air Transportation Association (IATA) and Airports Council International (ACI), as well as other organizations on international and regional levels develop specific soft law too. For instance, in order to harmonize PNR usage worldwide, IATA issued Recommended Practice PNRGOV,[130] containing data protection standards.

The European Organization for Security (EOS) published the EOS Position Paper on Privacy and Data Protection, initiating a dialogue among key stakeholders on policies on privacy and data protection and their impact on the implementation

[123] *Ibid.*

[124] See more detail in Bygrave (2014) pp. 51–53.

[125] Members include Australia, Brunei, Canada, Chile, China, Hong Kong, China, Indonesia, Japan, Korea, Malaysia, Mexico, New Zealand, Papua New Guinea, Peru, Philippines, the Russian Federation, Singapore, South Korea, Taiwan, Thailand, the USA and Vietnam.

[126] Bygrave (2014) p. 76.

[127] Bygrave (2010) p. 200.

[128] See Chap. 4.

[129] ICAO Document 9944 – Guidelines on Passenger Name Record (PNR) data of 2010 (ICAO PNR Guidelines).

[130] IATA Recommended Practice 1701a, 2011 (PNRGOV).

of security solutions. According to EOS, "The objective, for the industrial sector, is to remain competitive in the security market while providing solutions that do not infringe EU citizens' rights".[131]

The Security Industry Association (SIA), an American-based organization, adopted its own privacy framework with three intended purposes: (i) to identify privacy principles to serve as a guideline for manufacturers, integrators, and distributors of security technologies, including biometrics, CCTV, and video analysis; (ii) to inform policymakers about how the security industry protects privacy when collecting, securing, and storing personally identifiable information; and (iii) to help educate end-users and customers about implementation of privacy protections.[132] These industry-developed forms of self-regulation can contribute too.

Overall, the attention of influential international organizations is increasingly being turned towards privacy/data protection issues in aviation security, including more comprehensive recommendations that mainly correspond to international and EU data protection requirements. They are especially important in aviation security, where it is a challenge for the regulators to catch up with constantly developing technologies and methods.

2.3.2 EU

In Europe, the right to privacy is protected by Article 8 of the European Convention on Human Rights (ECHR)[133] and Articles 7 and 8 of Charter of Fundamental Rights of the European Union (CFREU).[134]

According to Article 8 of ECHR,

> 1. Everyone has the right to respect for his private and family life, his home and his correspondence. 2. There shall be no interference by a public authority with the exercise of this right except such as is in accordance with the law and is necessary in a democratic society in the interests of national security, public safety or the economic well-being of the country, for the prevention of disorder or crime, for the protection of health or morals, or for the protection of the rights and freedoms of others.

[131] EOS. Position Paper on EU policies on privacy and data protection and their impact on the implementation of security solutions. September 2010.

[132] Security Industry Association. Privacy Framework https://www.securityindustry.org/SiteAssets/GovernmentRelations/gr-privacy-framework.pdf (Last modified 28 March 2014).

[133] Convention for the Protection of Human Rights and Fundamental Freedoms signed at Rome on 4 November 1950, effective 3 September 1953.

[134] Charter of Fundamental Rights of the European Union, proclaimed on 7 December 2000, entry into force from the date of entry into force of the Treaty of Lisbon on 1 December 2009, Official Journal of the European Union (2010/C 83/02), 30 March 2010. The Treaty on European Union (TEU) Article 6(1) (Official Journal of the European Union (C 83/13), 30 March 2010) provides it "same legal value" as EU's Treaties thus it is legally binding. However, the CFREU applies only if and to the extent that EU member states implement or enforce EU law over their respective national laws.

It can be seen that in addition to provisions of UDHR and ICCPR mentioned above, Article 8 of ECHR also stipulates exceptions when interference is allowed. Substantial case law has been generated around Article 8 of ECHR, in which the interpretations of the right to privacy are very important for both privacy and data protection rights.

While the ECHR treats the protection of personal data as an extension of the right to privacy, the novelty of CFREU is that Article 8 specifically protects the right to data protection as a fundamental right that is distinct from respect for private and family life pursuant to Article 7.

However, as many authors note, all these documents evidence a considerable degree of inconsistency; they generally do not define the term "privacy", and its scope interleaves with a range of other freedoms and rights.[135] The case law thus became an essential instrument in interpreting and determining what the term "privacy" includes and how it can be applied.

The European Court of Human Rights (ECtHR) has given ECHR Article 8 a very broad interpretation. For instance, with reference to respect of private and family life (i.e. key concepts for privacy), the ECtHR stated that private life is a broad term not susceptible to exhaustive definition; it covers the physical and psychological integrity of a person; it can embrace multiple aspects of the person's physical and social identity; elements such as gender identification, name and sexual orientation and sexual life fall within the personal sphere protected by Article 8; beyond a person's name, his or her private and family life may include other means of personal identification and of linking to a family; information about the person's health is an important element of private life; an individual's ethnic identity must be regarded as another such element; the concept of private life moreover includes elements relating to a person's right to their image.[136]

Case law in general established three definite "steps" in determining the scope and application of the right to privacy. The first stage is to establish whether the concrete case or situation falls within the scope of one of the rights protected by ECHR Article 8 (1), i.e. it must concern interest(s) protected by that provision, namely private life, family life, home or correspondence. It is also established that the state may be under a positive obligation to protect persons from infringements on their physical and moral integrity.[137]

Once it is established that the dispute concerns an Article 8 right, the next stage is to determine whether there has been interference. The applicant must then establish the fact of interference, but if the applicant cannot establish the certainty of the material damage which would constitute the interference, it will be sufficient to demonstrate the likelihood that the interference has occurred.[138]

---

[135] Clarke. *What's Privacy?* Prepared for a Workshop at the Australian Law Reform Commission on 28 July 2006. Version of 7 August 2006. http://www.rogerclarke.com/DV/Privacy.html.

[136] *S and Marper v. the United Kingdom*, No. 30562/04 and 30566/04, 4 Dec 2008, §66.

[137] Kilkelly. *The right to respect for private and family life*. In: Human Rights Handbook, Council of Europe (2001) p. 42.

[138] *Campbell v. the United Kingdom*, No. 13590/88, 25 March 1992, §33.

Moreover, only the existence of legislation permitting interference, such as secret surveillance legislation, may be sufficient to interfere with private life. For example, the ECtHR concluded that the existence of laws and practices which permit and establish a system for effecting secret surveillance of communications alone amounted to an interference with the exercise of the applicant's rights under Article 8 "apart from any measures actually taken against him".[139]

If it is determined that interference took place, in the final step it should be considered whether the interference fulfils all the criteria listed in Article 8(2),[140] stating that limitations to the right to privacy are admitted if the restrictive measure in question: (i) is in accordance with law; (ii) pursues a legitimate aim; and (iii) is necessary in a democratic society in order to pursue that legitimate aim.[141]

In summary, if these steps are applied to aviation security measures, it is possible to consider: (i) whether ECHR Article 8 can be applied to a concrete measure; (ii) whether there is interference; and (iii) whether the interference is justified or there is a violation of ECHR Article 8. Accordingly, more specific analysis will be done in the Special Part of this study.

As mentioned above, in Europe, the right to data protection is protected as part of the right to privacy in Article 8 of the ECHR. At the same time, it is a separate right pursuant to Article 8 of the CFREU and Article 16 of the Treaty on Functioning of the European Union (TFEU).[142] The latter is designed to be the central source of data protection in the EU.[143]

CFREU Article 8 and TFEU Article 16 contain identical provisions: "1. Everyone has the right to the protection of personal data concerning them", thus gives additional value to this right. Moreover, TFEU adds additional norms:

> 2. The European Parliament and the Council, acting in accordance with the ordinary legislative procedure, shall lay down the rules relating to the protection of individuals with regard to the processing of personal data by Union institutions, bodies, offices and agencies, and by the Member States when carrying out activities which fall within the scope of Union law, and the rules relating to the free movement of such data. Compliance with these rules shall be subject to the control of independent authorities.

This article has a direct effect on all the above-mentioned actors in the data protection field universally, including security. It is thus very important for "a new era for data protection",[144] that does not depend on old "pillar" structure (will be discussed below). The recognition of data protection as an autonomous fundamental right with full legal validity as part of primary EU law means that data protection

[139] *Malone v. The United Kingdom*, No. 8691/79, 2 Aug 1984, §64.

[140] Kilkelly (2001) p. 25.

[141] The details about applying these criteria in general will be discussed in Chap. 5.

[142] Consolidated Version of the Treaty on the Functioning of the European Union, Official Journal of the European Union (C 83/47), 30 March 2010.

[143] Hijmans. *Recent developments in data protection at European Union level* (2010) p. 220.

[144] *Ibid.*

will play a more important role when balanced with other values and interests, such as security, and when being defined by the EU law and by the ECJ.[145]

An important regional instrument is the Council of Europe Convention for the Protection of Individuals with regard to Automatic Processing of Personal Data of 1981[146] (known as Convention 108) as well as Additional Protocol to this Convention regarding supervisory authorities and transborder data flows.[147] It is applicable to aviation security, since it applies to automated processing of personal data in both the public and private sectors.[148] This Convention is legally binding, but not self-executing: it obliges members of Council of Europe to incorporate its principles into their national legislation; therefore it may not have direct effect for individuals before courts. It is ratified by all EU Member States and, in addition, is open for ratification by other states. Thus, it plays an important role and contributes to development of national data protection regimes outside Europe.

However, the Convention provides wide-ranging exceptions for Member States, including in the interests of security; some formulations are vague and broad; it relates only to automatically processed personal data, thus its scope excludes non-automated data; and it can be considered outdated with the advent of new technologies. Thus, in 2011, the Council of Europe initiated the revision process, which pursues two main objectives: to deal with challenges for privacy resulting from the use of new technologies and to strengthen the Convention's follow-up mechanism.[149] On 3 December 2014, the Ad hoc Committee on data protection of the Council of Europe approved the modernization proposals of Convention 108.[150] However, it is not finally adopted yet.

The Directive 95/46/EC of the European Parliament and of the Council of 24 October 1995 on the protection of individuals with regard to the processing of personal data and on the free movement of data (DPD) is considered to be the most comprehensive of the European instruments. It is based on the principles established by the other mentioned documents (Convention 108 being a point for departure) and actually constitutes the most important basis for data protection initiatives in and outside the EU.[151] DPD is also applicable in the European Economic Area (EEA).[152]

---

[145] FRA. *Data Protection in the European Union: the role of National Data Protection Authorities* (2010a) p. 18.

[146] ETS No.108, 28 Jan 1981 (Entry into force: 1 Oct 1985).

[147] ETS No. 181 (Opened for signature on 8 November 2001).

[148] Hijmans. *Shortcomings in EU data protection in the Third and the Second Pillars, Can the Lisbon Treaty be expected to help*. In: Common Market Law Review. Vol. 46 (2009) p. 1489.

[149] The Consultative Committee of the Convention for the protection of individuals with regard to automatic processing of personal data [Ets No. 108], Modernisation of Convention 108: new proposals, T-PD-BUR(2012)01Rev2_en, Strasbourg, 27 April 2012.

[150] Ad hoc Committee on data protection of the Council of Europe (CAHDATA), Abridged Report, Strasbourg, 3 Dec 2014.

[151] Bygrave (2014) p. 53.

[152] It applies to the EEA member states which are not members of the EU (Norway, Lichtenstein and Iceland). The DPD was incorporated into the 1992 Agreement on the EEA (Decision of the

Despite common rules, DPD provides a certain freedom to implement provisions of national rules, which may lead to divergences, weakening data protection in security. Another problem is that the scope of the DPD is still based on an internal market legal basis, specifically excluding public security, defence, national security and criminal matters – so-called second and third "pillars".[153] However, with the adoption of the above-mentioned Article 16 TFEU, a new legal basis was established, abolishing the three-pillar structure and applying general data protection principles to all processing in the private and in the public sector, including law enforcement and security areas.[154]

An important step towards a general framework for all the areas is presented by a new regulation intended to substantiate the DPD – the EU General Data Protection Regulation (GDPR).[155] Moreover, since it is adopted in the form of Regulation rather than Directive, meaning more common mandatory rules with direct effect – it will provide more harmonization within the EU.

Another step towards a general framework is Framework Decision 2008/977 covering personal data concerning police and judicial cooperation in criminal matters.[156] However, it establishes a lower level of protection in comparison with the DPD. Thus, due to the need to modernize this instrument too, the Directive on the protection of individuals with regard to the processing of personal data by competent authorities[157] – the so-called "Police" Directive was adopted.

The GDPR and "Police" Directive are considered as packages of data protection reforms. These instruments were supposed to reflect the changes connected to the Lisbon Treaty, to better address the modern trends such as the growth of digital

---

EEA Joint Committee No 83/1999 of 25 June 1999 amending Protocol 37 and Annex XI (Telecommunication services) to the EEA Agreement; OJ L 296/41, of 23 Nov 2000).

[153] See DPD Article 3(2), Recital 16. The exclusions and limitations of DPD are based on the pre-Lisbon so-called three "pillar" structure of the EU. The First Pillar governed the regulation of the common market, where the EU acquired a lot of power, and the Member States had lost a lot of power. The Second (common foreign and security policy) and the Third Pillars (the area of police and judicial cooperation) applied to defence and other types of foreign policy and fighting crime and protecting against internal security threats like terrorism. The EU had powers in these areas, but it was limited by Member States preserving national sovereignty (see the Maastricht Treaty: Treaty on European Union, 7 February 1992, OJ C 191, 29 July 1992.)

[154] See more details in Hijmans (2009).

[155] Regulation (EU) 2016/679 of the European Parliament and of the Council of 27 April 2016 on the protection of natural persons with regard to the processing of personal data and on the free movement of such data, and repealing Directive 95/46/EC (General Data Protection Regulation). In force 24 May 2016, shall apply from 25 May 2018.

[156] Council Framework Decision 2008/977/JHA of 27 November 2008 on the protection of personal data processed in the framework of police and judicial cooperation in criminal matters.

[157] Directive (EU) 2016/680 of the European Parliament and of the Council of 27 April 2016 on the protection of natural persons with regard to the processing of personal data by competent authorities for the purposes of the prevention, investigation, detection or prosecution of criminal offences or the execution of criminal penalties, and on the free movement of such data, and repealing Council Framework Decision 2008/977/JHA. In force 5 May 2016; EU Member States have to transpose it into their national law by 6 May 2018.

information, the phenomenon of Big Data, internationalization of personal data exchange, the use of commercial data for law enforcement purposes, etc.[158] Although some progress was made, privacy advocates note that the package, especially GDPR, "was subject to a huge amount of lobbying, much of which was based on misunderstandings and misrepresentations. The result is that the overall package is less clear and less protective of personal data than it could – and should – have been".[159]

Although the GDPR is not yet applicable, Article 16 TFEU is valid. Thus, this work will follow the approach supported in, for example, different opinions of the Article 29 Working Party[160] that although DPD is a pre-Lisbon instrument, its principles should to be generally applicable in the area of data protection.

Technology-specific regulation which mentions privacy/data protection and is initiated by aviation security stakeholders is presented, for instance, by the Commission Regulation 1141/2011 of 10 November 2011, which added "security scanners which do not use ionising radiation" to the list of allowed methods for passenger screening for aviation security purposes. It states in Recital 7 that it "respects fundamental rights and observes the principles recognized in particular by the Charter of Fundamental Rights of the European Union, including respect for human dignity and for private and family life, the right to the protection of personal data, the rights of the child, the right to freedom of religion and the prohibition of discrimination". The aim of the Regulation is thus to allow the use of those scans which do not use ionising radiation and at the same time are equipped with different functions and operation procedures designed to satisfy the legislation requirements on privacy and data protection. Specifically, if EU states decided to use them, minimum operational conditions and performance requirements aimed at safeguarding privacy and other passenger rights will apply.[161]

There are also instruments on some other measures, such as the PNR Directive[162] and the EU Proposal on Regulation establishing Registered Traveller Programme,[163] all containing some privacy/data protection norms. However, as will be further discussed in this work, this legislation is quite fragmented, and privacy/data protection safeguards are not always satisfactory.

Specific instruments on privacy/data protection with reference to a particular technology or practice include mainly soft law can also be initiated by data protection

---

[158] Hijmans (2010) p. 223.

[159] EDRi-gram 13.24, 16 December 2015, https://edri.org/edri-gram/13-24/.

[160] Article 29 Data Protection Working Party. *Opinion 01/2014 on the application of necessity and proportionality concepts and data protection within the law enforcement sector* (2014a).

[161] See more details: European Commission. Security Scanners. http://ec.europa.eu/transport/modes/air/security/aviation-security-policy/scanners_en.htm.

[162] Directive (EU) 2016/681 of the European Parliament and of the Council of 27 April 2016 on the use of passenger name record (PNR) data for the prevention, detection, investigation and prosecution of terrorist offences and serious crime.

[163] Proposal for a Regulation of the European Parliament and of the Council establishing a Registered Traveller Programme, COM (2013) 97 final – 2013/0059 (COD), Brussels, 28 Feb 2013.

stakeholders. On the basis of Convention 108, Council of Europe developed opinions, guiding principles and recommendations adopted by Committee of Legal Co-operation and the Committee of Ministers[164]; based on DPD, various Article 29 Working Party Opinions laying down additional rights and safeguards which can be relevant for different aviation security measures were issued.

In summary, these initiatives are not mandatory, fragmented and quite outdated. Nevertheless, they provide the conformity and special sets of data protection principles for a specific sector, a specific technique or technology. In the Special Part, I will explore whether these recommendations are important in practice: whether they are used by the selected states as models for adopting national rules, or otherwise taken into account. In any event, they will be helpful in evaluating the existing regimes since they specify/clarify principles and other privacy/data protection issues in respect of technologies.

With reference to mechanisms of voluntary self-regulation, the core idea is provided in DPD Article 27. Codes of conduct are intended to contribute to the proper implementation of the national provisions pursuant to the DPD, taking account of the specific features of the various sectors. It is noteworthy that such Codes of practice developed by industry groups cannot derogate from standards of DPD. The Council of Europe also encourages sector-specific professionals, firms, bodies, public authorities, etc. to adopt and publish codes of ethics that meet the principles set out in Council of Europe's recommendations.[165] In the civil aviation sector, full of professional unions at different levels, establishing codes of conduct may certainly contribute to data protection. Examples of realization of this in practice will be discussed in national regulation sections below.

Overall, as mentioned above, the EU is a global leader in data protection, and many states outside the EU use the DPD as a model. At the same time, it is argued that surveillance practices in the EU are increasing, and the security agenda overrides some of the basic principles of the EU.[166] It is reported that according to analysis of confidential European Council documents, EU governments are making "secret" efforts to weaken EU privacy reforms.[167] It is argued that the EU may go even further than the US in establishing a global system of surveillance.[168] For instance, despite the fact that the EU is blaming the USA for inadequate data protection standards, the EU is planning to introduce its own PNR system; some of

---

[164] E.g., European Committee on Legal Co-operation *Report containing guiding principles for the protection of individuals with regard to the collection and processing of data by means of video surveillance* (2003); Recommendation CM/Rec(2010)13 of the Committee of Ministers to member states on the protection of individuals with regard to automatic processing of personal data in the context of profiling (23 November 2010); Recommendation No.R(91) 10 on the communication to third parties of personal data held by public bodies (9 September 1991), Recommendation No.R(87) 15 regulating the use of personal data in the police sector (17 September 1987), etc.

[165] Council of Europe (2010) p. 50.

[166] Privacy International. *European Privacy and Human Rights* (2011).

[167] Cunningham. *Big Brother is watching EU: As the US moves towards privacy reform, Europe enacts sweeping new spying powers*, 20 May 2015, http://www.politico.eu/article/eu-privacy-surveillance-us-leads-act/.

[168] *Ibid.*

the most extreme control schemes have been implemented in European countries (similar to the USA and Russia), and the DPD accommodates them via exemptions for security reasons, "because any privacy-abusive arrangement is feasible, provided that it has legal authority".[169] In the Special Part, the relation between these tendencies in the aviation security field will be explored.

2.3.3 National

2.3.3.1 UK

The UK has no a codified constitution. However, it has an unwritten one consisting of Acts of Parliament, court judgments and conventions. The constitution dates from the Magna Carta of 1215, the "Great Charter of the Liberties of England".[170] The Human Rights Act of 1998 provides for a limited incorporation of the ECHR into domestic law, including Article 8 – the right of privacy. There is no comprehensive privacy law: privacy is protected through a number of statutes, legal regulations, common law rules, and soft law, guidance by regulators and codes of practice as well as other instruments.[171]

It is often argued that the UK privacy law is underdeveloped and that national authorities fail to show a willingness to provide adequate safeguards for individual privacy.[172] Weaknesses of the UK national legislation give rise to a great number of court decisions on the right of privacy.[173] A major part of important ECtHR decisions on privacy originates in the UK.

The UK data protection law is represented by the Data Protection Act 1998 (DPA 1998). However, the text of the DPA 1998 is quite complex, which results in it being often incorrectly cited as a justification for the mishandling of data by public and private organs.[174] In June 2010, the research funded by the European Commission stated that the DPA 1998 should be amended to better implement DPD, including as regards the monitoring and enforcement powers of the national Data Protection Authority (DPA) – Information Commissioner's Office (ICO) (e.g. the ICO could not perform random checks of data controllers and processors, or to enforce penalties following such checks.)[175] As a result, there were some changes to DPA 1998

[169] Clarke (2006).

[170] Blackburn. *Britain'sunwrittenconstitution*. 2015. http://www.bl.uk/magna-carta/articles/britains-unwritten-constitution.

[171] Increasing Resilience in Surveillance Societies (IRISS) (2013) p. 57.

[172] *Ibid.* p. 47.

[173] Grant (2009) p. 49.

[174] See Bichard Inquiry, *Final Report regarding problems of police understanding the DPA rules* (2004), http://dera.ioe.ac.uk/6394/1/report.pdf.

[175] Study European Privacy and Human Rights by Privacy International (2010), the Electronic Privacy Information Center (EPIC) and the Center for Media and Communications Studies (CMCS), funded by the European Commission's Special Programme "Fundamental Rights and Citizenship", 2007–2013. Page 810.

that may "prove that it is possible to rise up from a surveillance state".[176] The ICO is independent and has the ability to impose fines.

Technology-specific instruments are developed. Notable codes of practices include the Surveillance Camera Code of Practice 2013 and a CCTV Code of Practice 2014[177]; Code of Practice for the Acceptable Use of Security Scanners in an Aviation Security Environment,[178] codes for data-sharing, policing, data-matching practices, etc.

The history of self-regulation in the UK dates back to the early nineteenth century. Despite a shift towards various forms of statutory regulation in the twentieth century, mechanisms of self-regulation are encouraged.[179] In aviation, for instance, some individual airports have CCTV Codes of practices. However, it will be analysed in the Special Part whether this presents advantages or weaknesses regarding the level of data protection.

It is believed that the level of surveillance in the UK is the highest in Europe and close to the countries like China and Russia.[180] For instance, the UK is a leader in camera surveillance numbers (see Chap. 6). The UK Government uses the TEMPORA program for tapping, storing and analysing all electronic data passing into or out of the UK.[181] As a result, there is an increased risk of privacy and data protection violations in the UK, and it is indicated as one of EU's "top three offenders in undermining citizens' rights".[182] It is a pioneer in intrusive aviation security measures too (see Special Part).

2.3.3.2 Norway

The Norwegian Constitution of 1814 (Grunnlov)[183] contains general provisions on privacy in Article 102. Moreover, the ECHR and ICCPR are incorporated into Norwegian law.[184]

---

[176] Privacy International (2011).

[177] Information Commissioner's Office. *In the picture: A data protection code of practice for surveillance cameras and personal information* (2014).

[178] UK Department for Transport. *Code of Practice for the Acceptable Use of Security Scanners in an Aviation Security Environment* (2015a).

[179] Bartle and Vass. *Self-Regulation and the Regulatory State: A Survey of Policy and Practice*. The University of Bath (2005) p. 1.

[180] Study European Privacy and Human Rights by Privacy International (2010), the Electronic Privacy Information Center (EPIC) and the Center for Media and Communications Studies (CMCS), funded by the European Commission's Special Programme "Fundamental Rights and Citizenship", 2007–2013. Page 806.

[181] Killock. *Legal challenge to UK Internet surveillance*. 10 March 2013. https://www.privacynotprism.org.uk/news/2013/10/03/legal-challenge-to-uk-internet-surveillance.

[182] Cunningham (2015).

[183] Kingdom of Norway's Constitution of 17 May 1814.

[184] Act of 21 May 1999 No 30 relating to the strengthening of the status of human rights in Norwegian law (the Human Rights Act).

In 1952, the Norwegian Supreme Court held that in Norwegian law there is general legal protection of "personality", which incorporates a right to privacy. This protection of personality exists independently of statutory authority but helps form the basis of the latter (including data protection legislation), and can be applied by the courts on a case-by-case basis.[185] Bygrave notes that Norwegian data protection laws are concerned with more than safeguarding privacy, extending into regulatory bodies, and promoting various interests such as ensuring adequate quality of personal information, "citizen-friendly" administration, proportionality of control, and rule of law.[186]

The key data protection law is the Personal Data Act of 2000.[187] Together with regulations issued pursuant to it,[188] it constitutes the central legislation on protection of personal data. The legislation protects the right to privacy by setting out safeguards to ensure that personal data are processed in accordance with fundamental respect for the right to privacy, including the need to protect personal integrity and private life and to ensure adequate quality of personal data (Personal Data Act Section 1).

Sector-based law includes some provisions on data protection and privacy, for example the Police Registry Act 2010[189] and Health Register Act 2014.[190] Norwegian DPA is the Norwegian Data Protection Authority (Datatilsynet) accordingly, an independent body having the authority to impose fines.

In Norway, there is a tendency to adopt technology-neutral laws.[191] Although there are no Codes of Practices dedicated to technologies similar to those in the UK, the DPA's website contains many practical recommendations on the use of specific technologies such as CCTV, biometrics, drones, etc. within the law.[192]

Norwegian data protection law is generally considered strong; there is increasing supervision of security agencies and surveillance measures entailing data retention were not realized.[193] Nevertheless, there are more calls for increased surveillance, especially after the attack of 22 July 2011. For instance, in 2013, the police tested surveillance drones planning to use them routinely as a part of daily police work.[194] As a result, the opportunities for surveillance in Norway are growing. Norway uses

---

[185] Bygrave and Aarø. *Privacy, Personality and Publicity – An Overview of Norwegian Law*. In: International privacy, publicity and personality laws (2001) p. 340.

[186] Bygrave (2010) pp. 172–173.

[187] Act of 14 April 2000 No. 31 relating to the processing of personal data.

[188] Regulation of 15 December 2000 No. 1265 on the processing of personal data.

[189] Act of 29 May 2010 No. 16 regarding the processing of personal data by the police and the prosecutors.

[190] Act of 20 June 2014 No. 43.

[191] Teknologirådet and Datatilsynet. *Persovern – tilstand og trender* (2014) p. 42.

[192] Datatilsynet, Technologies. https://www.datatilsynet.no/Teknologi/.

[193] Privacy International (2011).

[194] Statewatch. Police testing surveillance drones in Oslo. 26 October 2013. http://www.statewatch.org/news/2013/oct/drones-police-oslo1.htm.

EU-approved aviation security measures and body scanners will appear soon, and probably also the PNR system (see Special Part).

2.3.3.3 USA

In the USA, the term "privacy protection" or "information privacy law" is used instead of "data protection". While the EU has historically enacted broad legislative protection of personal data, with strict legal requirements, protection afforded by the US law is different. Despite the longest history of privacy regulation, there is no comprehensive privacy or data protection law covering all the sectors: specifically, they are absent in the private sector. In the latter, legal protection "takes the form of ad hoc, narrowly circumscribed, sector-specific legislation, combined with recourse to litigation based on the tort of invasion of privacy and/or breach of trade practices legislation".[195]

The US Constitution and interpretations of case law provide general protection. An important role here is played by the Fourth Amendment to the Bill of Rights in the US Constitution protecting the citizens' constitutional rights against warrantless searches and seizures: "The right of the people to be secure in their persons, houses, papers, and effects, against unreasonable searches and seizures, shall not be violated, and no Warrants shall issue, but upon probable cause, supported by Oath or affirmation, and particularly describing the place to be searched, and the persons or things to be seized".[196] As the Supreme Court has recognized in *Schmerber v. California*, "the overriding function of the Fourth Amendment is to protect personal privacy and dignity against unwarranted intrusion by the State".[197]

A substantial case law has developed around the Fourth Amendment in this regard, specifically on what can be deemed "reasonable"/"unreasonable". In general, for a search to be "reasonable", there should be some degree of suspicion with regard to the person to be searched.[198] But with regard to aviation security, the "suspicion" requirement became a minor criterion in relation to the possible degree of threat, thus, the courts tend to view airport searches as reasonable per se.[199] As a result, the courts approved, finding reasonable, a number of aviation security procedures, for instance the use of metal detectors,[200] the identification requirement,[201]

---

[195] Bygrave (2010) p. 191.

[196] The Fourth Amendment to the Bill of Rights in the U.S. Constitution (1791).

[197] *Schmerber v. California*, 384 U.S. 757 (1966), 20 June 1996, page 384 U.S. 767.

[198] Woodward (2002) p. 3.

[199] Kornblatt. *Are emerging technologies in airport passenger screening reasonable under the fourth amendment*. In: Loy. LAL Rev. Vol. 41 (2007) pp. 393, 396.

[200] *United States v. Epperson*, 454 F.2d 769, 771 (4th Cir. 1971), 7 February 1972.

[201] *Gilmore v. Gonzales*, 435 F. 3d 1125 – Court of Appeals, 9th Circuit 2006, 26 January 2006, 1137–39.

x-raying carry-on baggage,[202] airport searches of passengers,[203] use of body scanners.[204] The question that arises is whether the Fourth Amendment's privacy guarantees can be adapted to deal with increasingly invasive technologies.[205] Some researchers argue that today's approach is inadequate and no longer sufficient.[206]

There are a number of legal acts that protect the privacy and integrity of persons and limit the use of data, for example the Privacy Act of 1974.[207] The Privacy Act governs the collection, maintenance, use, and dissemination of personally identifiable information about individuals that is maintained in systems of records by federal agencies. However, it applies only where the government creates a system of records or if private databases are created under governmental contract.

There are some other sector-specific laws,[208] but they provide only fragmentary protection ("frequent, knee-jerk legislation on specific issues"[209]) and are often not applicable in a security context. For instance, Video Voyeurism Prevention Act of 2004 prohibits knowingly videotaping, photographing, filming, recording by any means, or broadcasting an image of a private area of an individual, without that individual's consent, under circumstances in which that individual has a reasonable expectation of privacy.[210] However, the Act makes such prohibition inapplicable to lawful law enforcement, correctional, or intelligence activity.

With regard to specific security measures, the DHS conducted a Privacy Impact Assessment (PIA) on body scanners[211] and CCTV.[212] Some privacy provisions on CCTV are also contained in the guidelines of the Transportation Security Administration (TSA) called "Recommended Security Guidelines for Airport Planning, Design and Construction" issued in 2011, which covers airport CCTV. However, weaknesses regarding the level of privacy/data protection can be noted, as will be analysed in the Special Part.

---

202 *Sima Prods. Corp. v. McLucas*, 612 F.2d 309, 312–13 (7th Cir. 1980), 3 January 1980.

203 *United States v. Aukai*, 497 F.3d 955–963 (9th Cir. 2007), 10 August 2007.

204 *EPIC v. DHS*, 653 F.3d 1 (D.C. Cir. 2011), 15 July 2011.

205 Brenner. *Fourth Amendment in an Era of Ubiquitous Technology, The*. In: Miss. LJ. Vol. 75 (2005) p. 3.

206 *Ibid.* pp. 82–83.

207 Privacy Act of 1974, 5 U.S.C. § 552a (2000).

208 E.g. Fair Credit Reporting Act 1970, Right to Financial Privacy Act 1978, Health Insurance Portability and Accountability Act 1996.

209 Clarke (2006).

210 According to the Act, "private area" is the naked or undergarment clad genitals, pubic area, buttocks, or female breast of an individual (§1801 (3)).

211 Department of Homeland Security. *Privacy Impact Assessment Update for TSA Advanced Imaging Technology* (2011).

212 Department of Homeland Security (2012b). However, this PIA concerns CCTV at DHS' facilities rather than CCTV in airports.

Another point is that while many Americans agree that privacy is a fundamental right, it is frequently considered to be so only when threatened by governmental abuse of personal information.[213] The "commercial" context is viewed differently; personal information is treated as a commodity.[214] An important issue for this work is air passenger data. Accordingly, in the USA, passenger personal data deriving from airlines are considered the "property" of airlines, which can do anything they please with the data, e.g. share with the US government without the knowledge or consent of the data subjects. Passengers have no control over such data.

In addition, there is no independent agency at the federal level functioning as Data Protection Authority (DPA). Although many agencies have a Chief Privacy Officer (CPO) tasked with performing tasks similar to DPA, they are appointed by and report to the head of their agency, thus cannot be considered independent. For instance, CPO of DHS is appointed by and reports directly to the Secretary of the DHS.[215] Thus, the Conference of Data Protection Commissioners refused to admit the CPO as DPA.[216]

Consequently, "in the USA – often portrayed as the citadel of liberal ideals – legal protection of personal data falls short in significant respects of the protection levels in many other countries...".[217]

On the wake of the *Schrems* case, where the ECJ found that the "Safe Harbour" agreement on transferring personal data to the USA[218] is invalid,[219] there is a call for re-evaluation of the whole privacy regime in the USA. Electronic Privacy Information Centre (EPIC) for instance recommended that the USA should take four steps to update domestic privacy law: "(1) enact the Consumer Privacy Bill of Rights,[220] (2) Modernize the Privacy Act, (3) establish an independent data protection agency, and (4) ratify the International Privacy Convention".[221]

The level of surveillance in the USA is high, as reflected in numerous scandals such as the NSA's secret global collection and use of personal data pursuant to the

---

[213] Manny. *Personal privacy—transatlantic perspectives: European and American privacy: commerce, rights and justice—part II*. In: Computer Law & Security Review. Vol. 19 (2003) p. 4.

[214] Pawlak. *Made in the USA? The Influence of the US on the EU's Data Protection Regime*. In: Brussels, CEPS (2009) p. 3.

[215] Section 222 (a) of the Homeland Security Act of 2002. 6 U.S.C. 142.

[216] Steinhardt. *The EU-US PNR Agreement – A Legal Analysis of Its Failures*. 26 December 2011. http://www.papersplease.org/wp/2012/01/03/the-eu-us-pnr-agreement-a-legal-analysis-of-its-failures/.

[217] Bygrave (2004) p. 328.

[218] Commission Decision 2000/520/EC of 26 July 2000 pursuant to Directive 95/46/EC of the European Parliament and of the Council on the adequacy of the protection provided by the safe harbour privacy principles and related frequently asked questions issued by the US Department of Commerce.

[219] *Maximillian Schrems v. Data Protection Commissioner*, Case C-362/14, 6 Oct 2015, § 106.

[220] This incorporates Fair Information Practice Principles to govern public-sector processing of personal data in commercial contexts. See more details in Bygrave (2014) p. 115.

[221] EPIC Alert, Volume 22.21, 11 Nov 2015, http://www.epic.org/alert/epic_alert_22.21.html

surveillance program PRISM. Similar to the UK, the USA is a pioneer in intrusive aviation security measures. Moreover, as will be discussed in Chap. 4, American aviation security policies have great influence in world regimes.

2.3.3.4 Russia

It is noteworthy that the term "privacy" as one word has never existed in the Russian language, "privacy" is normally (literally) translated as "неприкосновенность частной жизни" ("inviolability of private life").

The Constitution of Russian Constitution[222] recognizes the rights of privacy, data protection and secrecy of communications. The right to privacy here is provided in a special way. It is stipulated as such,[223] but also is revealed through a content of other constitutional rights.[224] The right to privacy is also protected by sector-specific federal laws. As for case law, practice of national courts on this category of cases is very small (there were some case concerning publication of private materials, photographs in mass media), but there is a tendency for new kinds of lawsuits due to emergence of new threats to privacy.[225]

Similar to many other states, data protection law is based on international and EU data protection law.[226] Russia is a member of the Council of Europe and signed Convention 108 in 2001. However, the process of ratification and implementation took years, and the Convention was ratified with several reservations, among other things, that it will not be applied to personal data constituting state secrets. Russia

---

[222] Constitution of the Russian Federation of 12 December 1993.

[223] Article 23 (1) incorporates the right to the inviolability of private life, personal and family secrets, and the right to protection of honour and good name. Here and further, the English translation of the Constitution is done by Garant-Service. http://www.constitution.ru/en/10003000-01.htm.

[224] Article 23 (2) provides the right to privacy of correspondence, of telephone conversations, postal, telegraph and other messages; only in relation to this paragraph, the Constitution determines possible limitations ("Limitations of this right shall be allowed only by court decision."). Article 22 provides the right to freedom and personal immunity. The right to inviolability of home is enshrined as a separate right (Article 25); no one shall have the right to get into a house against the will of those living there, except for the cases established by a federal law or by court decision. The right to access the documents and materials directly affecting rights and freedoms is provided by Article 24 (2). Article 24 (1) stipulates that "the collection, keeping, use and dissemination of information about the private life of a person shall not be allowed without his or her consent". According to Article 21, "human dignity shall be protected by the State. Nothing may serve as a basis for its derogation. No one shall be subject to torture, violence or other severe or humiliating treatment or punishment. No one may be subject to medical, scientific and other experiments without voluntary consent".

[225] Tsadykova. *The constitutional right to privacy* (2007) Abstract.

[226] For a more detailed overview of Russian data protection law, see Enerstvedt. *Russian PNR system: Data protection issues and global prospects*. In: Computer Law & Security Review. Vol. 30 (2014).

reserved the right to impose restrictions on the right of data subject to have access to his/her personal data in order to protect national security and public order.[227]

The Law on Personal Data of 2006[228] was designed to fulfil Russia's obligation to implement Convention 108 into national law. In addition, national data protection rules are also contained in other acts[229] and sector-specific federal laws.[230] Formally, the final stage of the Convention ratification was completed in 2013, with necessary amendments to federal laws.[231] However, not all areas processing huge amounts of personal data are covered, and there is still a lack of amendments to some sector-specific laws.[232]

Consequently, according to some researchers, individuals' rights to personal data were more strongly protected before adopting any data protection laws – when the confidentiality of personal data was protected by general provisions on privacy in the national Constitution, normally prohibiting data processing. However, by adopting specific data protection laws, the state created a legal basis for data processing, "making life of a person fully transparent and his personal data unsecured".[233] However, this research focuses on legal rather than political considerations.

The Russian DPA is a Federal Service for Supervision in the Sphere of Telecom, Information Technologies and Mass Communications (Roskomnadzor). Its status, role and powers are close to those of the European DPA. However, it cannot be considered as a fully independent body, since it is established under the Ministry of Communications,[234] it is a body subordinated to the Ministry (structurally), thus it is under its control. In addition, the Government, the Federal Security Service (FSB), and other executive agencies acquired substantial powers in the personal data field. For instance, as will be noted later in this work, FSB establishes extensive requirements on data security and controls their fulfilment.

---

[227] Federal law of 19 December 2005 N 160-FZ On ratification of the Council of Europe Convention for the protection of individuals with regard to automatic processing of personal data.

[228] Federal law of 27 July 2006 N 152-FZ On personal data.

[229] Federal law of 27 July 2006 N 149-FZ On information, information technologies, and the protection of information, Presidential decree of 6 March 1997 N 188 On approval of the List of confidential information (stipulates that the latter covers personal data, with a few exceptions), Resolutions of Government, etc.

[230] E.g. Labour Code of 30 December 2001 N 197-FZ (Chapter 17), Tax Code of 31 July 1998 N 146-FZ (Art. 84), Federal law of 27 December 1991 N 2124-1 On mass media, Federal law of 12 August 1995 N 144-FZ On operational-search activities, etc.

[231] Federal Law of 7.05.2013 N 99-FZ On Amendments to certain legislative acts of the Russian Federation in connection with the adoption of the Federal Law On ratification of the Council of Europe Convention for the Protection of Individuals with regard to Automatic Processing of Personal Data, and Federal Law On Personal Data.

[232] Chernova. *We protect personal data through multi-stakeholder approach*. In: Personal data (2013).

[233] Yakovleva. *New technologies and human rights* (2012) p. 12.

[234] §2 of Resolution of Government of 16 March 2009 N 228 About Federal service for supervision in the sphere of telecom, information technologies and mass communications.

At the same time, Roskomnadzor faces a number of other difficulties: according to experts; it possesses insufficient resources and personnel; it cannot initiate administrative proceedings, and does not receive help from other organs such as the Ministry of Internal Affairs which considers data protection offences as not serious.[235] The relatively low amount of fines is also a challenge.

Russian regulation is technology-neutral and regulation of specific technologies is not developed. With regard to specific aviation security measures, the regulators tend to repeat or refer to general data protection law (see Special Part). Mechanisms of self-regulation are not developed either.

Researchers note that privacy and data protection rights in Russia are subject to numerous violations and restrictions, and practical protection of this right is not effective enough due to a number of factors, such as the historical background (the public interests prevailed over private for many centuries), vulnerability of the private sphere, "democratic underdevelopment" of the state and social institutions, a relatively short period of legal regulation, the introduction of security restrictions, weaknesses of legislation, lack of an appropriate theoretical framework, lack of enforcement mechanisms and lack of judicial practice.[236]

Similar to the UK and USA, the Russian level of surveillance is high. For instance, a Russian parallel to the US PRISM is the System of Operative-Investigative Measures (SORM), a national system of lawful interception of all electronic utterances, the use of which by FSB has skyrocketed over recent years.[237] In aviation security, many intrusive measures are used.

All these factors lead to the conclusion that from the Western perspective, the situation with privacy and data protection rights in Russia is critical, creating a potential for privacy and data abuse. At the same time, from an internal perspective, the amount of public debates and discussions on privacy in general and security versus privacy in particular is insignificant in comparison with those in the West, indicating that privacy problems are considered less important than other more actual problems such as economic/political ones.

Although the security undertakings, including in aviation security, are growing, the security-versus-privacy dilemma has not yet been addressed in Russia. People in general are not so concerned with these surveillance practices and intrusive techniques. However, at least a part of the Russian population attributes importance to privacy; it is becoming a subject of public discussion, education and propaganda, a civil issue.

A more general point is that in contrast to the UK, USA and Norway, Russia lies outside the Western world, with its own mentality and own rationales. While the

---

[235] Kovrigin. *Total non-compliance with data protection law in Russia* (2012) http://can-work.ru/index.php/neews/press-tsentr-kompanii/145-law-on-personal-data-if-it-works.

[236] Izmailova. *Privacy in civil law: the law of the UK, the USA and Russia* (2009) Abstract, and Omelchenko and Kosterina. *Private life: a sociological perspective* (2005) http://www.pgpalata.ru/reshr/privacy/deb_02.shtml.

[237] Soldatov and Borogan. *Russia's Surveillance State* (2013) http://www.worldpolicy.org/journal/fall2013/Russia-surveillance.

former states can be considered "old" democracies, the Russian political regime, although democratic according to the Constitution, has roots in the Soviet authoritarian regime. As noted above, citizens of former dictatorships or authoritarian regimes may be less sensitive to surveillance practices, and the human rights perspective, as well as the understanding of both privacy and surveillance in Russia differs greatly from the Western approaches.

As an American security expert notes, "[Russia] is a unique place with special problems related to a decade-old internal Islamist insurgency. They have a different mindset there as well as a different human rights and political framework and I have always felt that we in the West had best leave their internal policies alone".[238] In general, I support this view. In Russia, there is an ongoing several-centuries-old dispute between so-called Westerners and Slavophils debating whether Russia should follow the Western way or its own, unique way of development respectively. Nevertheless, history shows that although some Western experience can be useful in Russia, not all Western concepts, ideas or values can be successfully imported and implemented there. In some cases, such import may have a destructive effect on the existing systems. As a Russian proverb says, "what is good for the Russian, is death for the German" (here, the latter means any non-Russian). And vice versa.

This does not mean that the situation with privacy should be left *status quo*, but it should be taken into account that privacy/data protection issues, in particular in the security context, are treated differently in Russia and require a specific, careful approach. Among proposed improvements, commentators suggest establishing a new independent data protection authority, substantially increasing penalties for data protection offenses and imposing more stringent criminal sanctions, turning to the theory and practice of other states in which privacy rights have longer traditions, and whose experience could be useful, including the UK and the USA.[239] However, as discussed above, privacy protection in these countries is not perfect either, and simple copying can be dangerous.

Some problems cannot be solved by improving data protection law only. For instance, the problem of the illegal database market is mainly caused by, first, economic reasons (low salaries of state officials) and, secondly, by lack of legal methods to obtain information, for example via special private firms such as in the USA.[240] Therefore, a broader, complex approach to the solutions is needed: from education and propaganda to repairing civil society systems.

---

[238] Huey. *When This Stuff Gets Personal! A Response to the Moscow Airport Bombing* 27 January 2011 (2011b), http://www.noquarterusa.net/blog/55860/when-this-stuff-gets-personal-a-response-to-the-moscow-airport-bombing/.

[239] Izmailova (2009) Abstract.

[240] See Beroeva. *Who and how do they steal databases?* In: Komsomolskaya Pravda 2006.

2.4 Concluding Remarks

The lack of common legal definition of "privacy" as a term presents difficulties for establishing this right in the security environment. However, this fact provides opportunities for a flexible approach: "privacy" can be construed by law according to a particular context.

For the privacy concept in this research, the most important is the requirement to determine the context and the distinction of four or – if more detailed – seven privacy dimensions, which are designed to specify not only the possible harms, but also opportunities for protection. It is the role of law to determine the amount of desired privacy for every particular measure.

For the application of the right to data protection, each aviation security measure should be analysed whether it deals with personal data or with processing. Accordingly, in Chap. 6, while describing aviation security technologies, concepts of privacy and data protection will be used to determine if these rights are applicable and which concerns may arise.

Protection mechanisms of both rights are essential. The point is that many of the aviation security measures cause tensions with both the right to privacy and data protection. In order to protect both rights at the highest possible level, it is necessary to consider every aviation security privacy/data protection concern on a case-by-case basis, in order to determine which protection is offered by both of the rights, and, if possible, apply both.

The law normally does not allow interference with privacy, but restrictions and limitations *are* possible if they correspond to definite requirements. As for data protection, the regulation is extensive; in order to establish concerns and violations, data protection rules and requirements should be analysed. As noted in Chap. 1, the approach of this work is using legal principles of privacy and data protection rather than all regulation at all levels for analysis of concrete problems. Chap. 5 will discuss the applicable principles, while Chap. 7 will apply them to the selected technologies.

Nevertheless, examples from applicable regulation will be used to illustrate the principles. With this regard, as shown in this Chapter, general legislation on privacy and data protection exists on international, regional, national levels and there are endeavours of harmonization. In particular, many states use the EU DPD as a model. However, in general, not all the states are equally successful in privacy and data protection regulation and their enforcement. There are local divergences and the lack of regulation of privacy and data protection.

With reference to challenges deriving from new technologies, there are two regulative tendencies: (i) the need to update existing general law and to adopt more appropriate general, technology neutral laws, and (ii) since specific technology poses specific threats to privacy and data protection, there is the need to develop specific rules to protect these rights with respect to particular technologies, or technologies in specific areas or situations, as well as regulate the use and output of personal data processed by the technologies,[241] in other words, adopt technology-specific laws. The latter may originate from aviation security regulators, privacy/data protection regulators or other relevant entities. They can also derive from industry-developed norms via mechanisms of self-regulation. Thus, it can be a matter of initiative: of which regulators or other stakeholders endeavour to adopt relevant rules.

On the national levels, different approaches can be noted. In the USA and the UK, technology-specific laws are more developed, in particular in a form of soft law. In Norway and Russia, there is a tendency to adopt technology-neutral laws. The same concerns mechanisms of self-regulation: developed in the USA and the UK, they are not developed in Norway and in Russia. A summary of these and other important features of national regimes can be seen in Table 2.1.

With reference to the general models of legal regulation mentioned above, two of them are presented by the selected states: Continental and American. The first one is presented by the UK, Norway and Russia. The DPD obliges EU/EEA Member States, including the UK and Norway, to incorporate the DPD rules into national law and enforce them. Despite common law traditions, the UK can be considered as being closer to the continental model than to the American. However, some features are similar to the American model, e.g. self-regulation mechanisms. Russia used the EU data protection law as model for its own data-protection law and, in general, can

Table 2.1 Summary of national regulation and other features

	UK	Norway	USA	Russia
Legal system	Common law	Civil law	Common law	Civil law
Political system	"Old democracies"			"New democracy"/ post-authoritarian
Surveillance level	High	Less, but increasing	High	High
General privacy law	Human Rights Act 1998, common law rules	Constitution, Human Rights Act 1999	Constitution, case law, common law mechanisms, Privacy Act of 1974 (covers not all sectors)	Constitution
General data protection law	Data Protection Act 1998	Personal Data Act 2000	No general data protection law	Personal Data Law 2006
Technology-specific or neutral law	Specific	Neutral	Specific	Neutral
Mechanisms of self-regulation	Yes	No	Yes	No

241 Carli (2008) p. 11.

be considered being within the continental model too. The regime in the USA is specific and falls within the American model – there is no general data protection law, mechanisms of self-regulation are developed, etc.

It can be seen that there are national differences within basic models. They derive from many factors. For instance, similar to differences in attitudes to privacy, the respective legislations on national levels vary due to differences in historical and cultural conditions, and there can be noted gaps between various privacy regimes: both between civil and common law and within civil law regimes (Western and Post-Communist).

The difference in approach to data protection may cause a violation of cross-border exchange of information.[242] For instance, both the Russian and the US data protection regimes cannot be considered as adequate from the EU perspective, thus, transfer of personal data from the EU to USA or Russia, without additional guarantees, will be illegal. Differences may also lead to country-specific problems relating to air passengers' rights protection.

At the same time, the level of surveillance in the USA, the UK and Russia is high. This can be partly explained by the perceived level of terrorist threat (see Chap. 4), but as a result, the most advanced aviation security measures are used. Although Norway lies behind in this regard, the surveillance level is increasing, and similar aviation security technologies are used or are planned to be used.

Taking into account all these general tendencies and specific features, the Special Part, on the basis of illustrations, will further explore how successful the regulations at all the levels are in addressing challenges for privacy/data protection presented by surveillance in general and the selected aviation security technologies in particular. It will discuss which particular problems arise, which approaches are the most promising, etc.

---

[242] Kovaleva (2007) p. 11.

Chapter 3
Other Human Rights in Aviation Security

3.1 Introduction

As discussed, the rights to privacy and data protection are inadequate to capture all the aspects of aviation security, and all human rights are interdependent, indivisible and interrelated. Thus, aviation security versus privacy dilemma is actually broader than privacy and data protection rights only.

The first rather common association that comes to mind with the words "air passenger rights" is primarily the rights of a passenger as a customer. Air transportation in a legal sense involves the relationship between air transportation service providers (airlines) and consumers; traditionally, these relations are governed by civil law.[1] According to the contract of carriage, the airline undertakes to deliver passengers and cargo to an agreed destination for a set fee, under certain conditions.

The rights of a passenger as a consumer may vary from airline to airline,[2] but in general, they relate to such issues as flight cancellation, delays, baggage liability, etc. They will not be analysed thoroughly since they do not fall within basic human rights category and, although the security undertakings may have impact on consumers' interests and may result in negative economic consequences, they do not *violate* consumer rights. Typical examples here are denial of boarding, long security lines, the restrictions on liquids allowed on board aircraft, property risks. For instance, when being subjected to hand-held metal detection or body scanning, one can no longer watch his/her hand baggage and items such as a wallet or phone can disappear.

The second association is that air passengers use the airlines' services exercising their right to travel, in other words, freedom of movement, one of the basic fundamental human rights. In addition, however, as a human being and individual, a

[1] Bordunov. *International Air Law* (2007) p. 29.

[2] See "Customer Bill of Rights", "Conditions of Carriage", or "Contract of Carriage" published by some of the carriers.

O.M. Enerstvedt, *Aviation Security, Privacy, Data Protection and Other Human Rights: Technologies and Legal Principles*, Law, Governance and Technology Series 37, DOI 10.1007/978-3-319-58139-2_3

passenger possesses all the fundamental human rights enshrined in different human rights conventions. The first of these is the right to life.

Taking into account the terrorist threat and other risks in civil aviation, a person's right to life is directly related to aviation security undertakings. Everybody wants to travel safely. It is believed that different security measures and technologies are intended to provide the highest level of security for passengers, to protect their lives. Passengers, legitimately expecting states to ensure that security is maintained by such measures, become subject to security rules and regulations governing different screening procedures at the airport and on board the aircraft, as well as to customs and immigration requirements.

As discussed previously, some of these procedures entail risks to privacy/data protection as well as other relevant rights: the right to health, freedom of movement, the right to equal treatment and non-discrimination, freedoms of thought, conscience and religion, and rights of the child.

This chapter will include several parts. First, it is a general discussion on the whole human rights concept. Such a broad analysis is necessary in order to indicate general principles and mechanisms of human rights protection as well as grounds for their restrictions and limitations, which are mainly common for all human rights and are applicable in the aviation security sector. The key issue is to ascertain the circumstances under which the law may allow interference and subsequent limitations of human rights in the course of security undertakings.

Secondly, the chapter will analyse the relevant rights more concretely, in particular, general regulation (the discussion is mainly limited to international and the EU instruments), important case law, general applicability of case law in aviation security, possible interference and basic rules for substantiation.

More importantly, a broader overview of human rights applicable in aviation security may allow the construction of a common general model to evaluate whether the measure's interference with human rights is justified, and consideration of whether the scheme which will be further developed in terms of the right to privacy can be basically applied to other human rights.

Moreover, the correlative impact of aviation security measures with regard to all applicable human rights may have an impact on evaluation of the proportionality of the regime and, accordingly, may contribute in the analysis of the dilemma "aviation security versus privacy". The rule is that the more negative impacts on different human rights, the more "weighty" the cost of aviation security measures in the form of human rights concerns and accordingly, the more security benefits and other factors are needed to justify the impact. Therefore, impact on other rights in addition to privacy/data protection may be additional grounds for the justification or non-justification of the measure and will be taken into account in the Special Part.

I acknowledge that discussion in this chapter is more cross-investigative rather than comprehensive, in-depth analysis of all applicable human rights; thus, the legal regulation and literature used will be quite limited. In addition, respective norms from national Constitutions may be used as examples; thus, there is a mix of human rights law and constitutional law.

3.2 The Concept of Human Rights

3.2.1 *Protection of Human Rights*

The importance of human rights cannot be underestimated. Human rights constitute instruments of protection against humiliation and governmental abuses; they perform important functions to restrict the government in order to protect human dignity and to encourage the government to protect it; the essence of human rights is that they support human dignity and ensure external recognition and respect of it.[3] On the formal side, fundamental human rights are enshrined in different international and regional human rights conventions, treaties and other instruments, as well as in the national constitutions and legislation of many countries.

In the aftermath of the First World War, under the established League of Nations, attempts were made by the international community to develop a system of protection of human rights and international monitoring mechanisms.[4] After the Second World War, such a system was established, when the United Nations General Assembly adopted of the Universal Declaration of Human Rights (UDHR) on 10 December 1948.

Formally the Declaration is not a treaty, but it determined basic civil, political, economic, social and cultural rights, and it served as the foundation for two binding UN human rights covenants: International Covenant on Civil and Political Rights of 1966 (ICCPR)[5] with its two Optional Protocols and the International Covenant on Economic, Social and Cultural Rights of 1966 (ICESCR).[6] The UDHR, together with the ICCPR and the ICESCR, form the so-called International Bill of Human Rights. In addition, on the basis of the UDHR, other international human rights treaties and instruments were adopted.

At the regional level, such instruments were adopted as well, reflecting the particular human rights concerns of the region and providing for specific mechanisms of protection. In Europe, the most important instruments include the European Convention on Human Rights of 1950 (ECHR),[7] the Charter of Fundamental Rights of the European Union of 2000 (CFREU), and the Treaty on Functioning of the European Union (TFEU).

At the national level, the states which become parties to international human rights treaties and ratify them, usually adopt respective norms in national

[3] Averkiev. *Why does a man need human rights?* In: Civitas. Vol. 1 (2003) p. 19.

[4] United Nations. The United Nations Human Rights Treaty System: An introduction to the core human rights treaties and the treaty bodies. Fact sheet No.30. United Nations Office in Geneva, 2005.

[5] Adopted by General Assembly resolution 2200A (XXI) of 16 December 1966, was opened for signature on 19 December 1966, entry into force 23 March 1976.

[6] Adopted and opened for signature, ratification and accession by General Assembly resolution 2200A (XXI) of 16 December 1966, entry into force 3 January 1976.

[7] Convention for the Protection of Human Rights and Fundamental Freedoms signed at Rome on 4 November 1950, effective 3 September 1953.

Constitutions and/or acts on human rights. But only adopting norms on paper is not enough. The states are obliged to promote, respect, protect and fulfil human rights of the people living in their territory.[8] These elements require closer scrutiny.

The term "people" here is used as a synonym for "individuals" or "persons". However, when applied to the state, they include own and foreign citizens as well as persons without citizenship. In general, all of them, being human beings, enjoy all the human rights within the territories of all states which are parties to respective international treaties (with a few exceptions, such as the rights to vote or to stand for public election, which are usually reserved for citizens). Aviation security deals with air passengers, crew and people on the ground; all of them are human beings thus enjoy the rights as well.

As for the "territory", in general, state territory is a part of the globe that is under the sovereignty of a particular state. It includes the land (all the dry land within the state boundaries), the waters (both inland and territorial), and the air space over both the land and the waters (the troposphere, stratosphere, ionosphere, and a considerable part of adjacent outer space). Within its territory, a state has territorial supremacy.[9] In other words, state territory is an area under the jurisdiction of a particular state. There are also territories which do not belong to a particular state, with special, mainly international status (Antarctica etc.).

The aviation security can be provided on the ground, in the air and beyond.[10] It is clear that the ground falls within the term "land" thus constituting a state's territory. Difficulties with jurisdiction may arise in conjunction with international aviation, where aircraft cross state boarders and fly in the air space above different territories. However, jurisdiction can be determined according to international and national air law.[11]

A broad interpretation of these principles allows the conclusion that protection and enforcement of the human rights of air passengers and other people on board as well as on the ground must be provided no matter where a person is, subject to ratification of the human right treaties by applicable states.

The state's obligation to promote human rights is self-explanatory. The obligation to respect means that states must refrain from interfering with or curtailing the enjoyment of human rights. The obligation to protect requires states to protect individuals and groups against human rights abuses, including ensuring that non-state parties do not infringe upon human rights. The obligation to fulfil means that states must take positive action to facilitate the enjoyment of basic human rights.[12]

---

[8] See, for example, Preambles of UDHR, ICCPR and ICESCR, Art. 1 (3), 2 of ICCPR, Art. 1 (3), 2 of ICESCR.

[9] The Great Soviet Encyclopedia, 3rd Edition (1970–1979). http://encyclopedia2.thefreedictionary.com/State+Territory

[10] See Chap. 4.

[11] See Chap. 4.

[12] United Nations. International Human Rights Law. http://www.ohchr.org/EN/ProfessionalInterest/Pages/InternationalLaw.aspx

The states must take steps, individually and in cooperation, in order to progressively achieve the full realization of the rights by all appropriate means.[13] The states must therefore put the rights in effect in national law, that is implement domestic laws and other measures as may be necessary.[14] The human rights are often protected by the main national legal instrument – the Constitution, or by other laws.

Furthermore, the states must ensure that any person whose rights or freedoms are violated shall have an effective remedy, including judicial remedy; the rights to a remedy must be determined by competent judicial, administrative or legislative authorities, or by any other competent authority.[15]

Taking into account different factors and processes affecting the protection and realization of human rights, an effective scheme of protection of any human right or freedom should include the following[16]: (i) existence of laws protecting this right from violations; (ii) assurance of effective protection mechanisms, including judicial remedy. Such implementation and enforcement cannot be effective without transparency of government activities, availability of information about the rights and possibilities of their protection, and existence of independent protection institutions; (iii) existence of self-regulatory mechanisms, such as corporate principles and policies, binding corporate rules, etc.. In other words, this right should be "rooted" in the culture, traditions and customs; (iv) existence and availability of means of self-defence for individuals.

Such an "ideal" protection scheme can be fully realized only in the conditions of the rule of law, with an effective judicial system, responsible and competent officials, law enforcement agencies following the law, and higher standards of legal consciousness of the citizens who respond actively to violations and use all the means of protection.[17]

The human rights concept looks good in the theory: there are certain rights, and the states must promote, respect, protect and fulfil these rights. But in reality there are different factors, circumstances, interests, values and so on which may have impact on human rights; some of them can be described as threats.

Namely, the existing protection mechanisms co-exist with current public conditions (social and political conditions, including the priorities of domestic and foreign policies and features of international relations, the level of technological progress, development of civil society, social communication, etc.) and subjective factors (relevance of the law, level of social activity, etc.). Despite the formal existence of protection possibilities, the right may not actually be protected due to prevailing attitudes in the particular society, the features of national traditions and cultural context, the presence or absence of social experience in the right protection, the degree of public confidence in social institutions, etc.

---

[13] Art. 2(1) of ICESCR.

[14] See, for example, ICCPR Art. 2(2).

[15] ICCPR Art. 2 (3).

[16] Kochenova. *Respect to privacy in Russia: diagnosis and prognosis* (2007) http://www.pgpalata.ru/reshr/privacy/art10.shtml#n77

[17] *Ibid.*

Fortunately, national law and national features are not the only factors that play a role. Many national constitutions and laws include provisions stipulating that in the case of conflict between the provisions of national law and those of international treaties, the provisions of international treaties shall prevail.[18] The incorporation into domestic laws of international instruments recognizing the human rights can significantly strengthen the scope and effectiveness of remedial measures. It enables courts to adjudicate violations of the rights by direct reference to the international treaties.[19]

But is it possible to control how the states fulfil the international obligations concerning human rights, to hold governments accountable if they violate the rights? Monitoring and holding states accountable are procedures taking place at national, regional and international levels and involving a variety of actors, such as the state itself, non-governmental organizations, national human rights institutions or international treaty bodies.

The concept of accountability of the state and redress implies that the rights should be realized and monitored through various mechanisms, such as reviews of policy, budgets or public expenditure, governmental monitoring mechanisms, impact assessments, democratic processes, monitoring and advocacy by civil society organizations. Moreover, at the international level, different committees exist, such as treaty bodies established by the UN, which also monitor the states' compliance with the international instruments.

Finally, when the states fail to ensure an effective remedy for human rights abuses at the national level, it is possible for an individual to apply for redress at the regional and international level, for instance to the ECtHR.

3.2.2 Limitations of Rights

Despite the mechanisms discussed above, a negative impact of different factors on human rights still exists, taking various forms: from "little concern" to limitations, restrictions and violations. In addition, there always appear "internal" conflicts of values between human rights themselves, entailing difficulty in finding an appropriate balance: how to enforce one right or freedom without limiting or even violating another one? What can be done where two or more rights compete and result in none of them being fully realized? A classic example is the relation between the freedom of speech and the right to private life.

A general rule is that rights may not be exercised contrary to the purposes and principles of the UN, or if they are aimed at destroying any of the human rights (UDHR Art. 30). It can hardly be stated that the security process, protecting the

[18] E.g. Article 15(4) of Constitution of the Russian Federation of 12 December 1993; §3 of Norwegian Human Rights Act of 21 May 1999.

[19] United Nations. *Office of the United Nations, High Commissioner for Human Rights, World Health Organization. The right to health. Fact sheet No.31* (2008).

right to life, involves the aim of destroying any other human right. The impact occurs latently, in the course of definite security measures, mainly due to special features, technical capabilities, applied methods, etc. Can the security's "ultimate aim" as well as terrorism threat, the risks justify all the concerns? In the human rights context, the main question is whether the protection of every relevant human right involved allows, in definite circumstances, a lawful exception or limitation of this right.

Scheinin and Vermeulen note that there is nothing new in governments engaging in legal argumentation to justify a measure that at first sight amounts to a human rights violation; the point is that post 9/11, international terrorism and counter-terrorism measures have been transformed from legal arguments into a combination of arguments, doctrines, concepts and interpretations.[20] They suggested several types of such unilateral exceptions[21]; for the purposes of this research, the most relevant are derogation during times of emergency and overly broad use of permissible limitations or restrictions.

In order to evaluate them, one should again refer to the international instruments on human rights, providing that in definite circumstances the human rights can be derogated or limited. Here, the problem of finding appropriate balance is seen from the priority perspective. Accordingly, the derogation can be implemented in time of emergency; limitation is allowed in order to protect national security, public order, public health or morals or the rights and freedoms of others.

According to ECHR Article 15, during times of war or other public emergency "threatening the life of the nation", the states may take measures derogating from their obligations under this Convention "to the extent strictly required by the exigencies of the situation, provided that such measures are not inconsistent with its other obligations under international law".

However, paragraph 2 provides that a number of rights are not subject to such derogations: no derogation from Article 2 (right to life), except in respect of deaths resulting from lawful acts of war, or from Articles 3, 4 § 1 and 7 – to be free from torture, to be free from slavery and servitude, to be free from punishment without law respectively. These rights are usually referred to peremptory norms or *jus cogens*.

Article 15 was used by some states in the context of terrorism: by Ireland and the UK due to terrorist violence connected to Northern Ireland; by Turkey in respect of south-east Turkey due to activities of a terrorist organization Workers' Party of Kurdistan; by the UK in 2001 after the September 11 terrorist attacks in the USA.[22]

---

[20] Scheinin and Vermeulen. *Unilateral Exceptions to International Law: Systematic legal analysis and critique of doctrines to deny or reduce the applicability of Human Rights norms in the fight against terrorism*. In: Essex Human Rights Review. Vol. 8 (2011) p. 28.

[21] *Ibid.* pp. 28–29.

[22] The ECtHR addressed these derogations in respective cases *Lawless v. Ireland* (1 July 1961), *Ireland v. the UK* (18 Jan 1978), *Brannigan and McBride v. the UK* (25 May 1993), *Aksoy v. Turkey* (18 Dec 1996), *A. and Others v. the UK* (19 Feb 2009, Grand Chamber).

These types of exceptions, therefore, may refer to terrorism. Since civil aviation is particularly attractive as a terrorist target (see 4.5.4), aviation security is designed specifically to protect against terrorist attacks. But can this fall under "public emergency" derogation?

Emergency always implies something extraordinary. The above-mentioned cases dealt with threats from particular terrorist groups in particular territories, requiring individual, respective measures. But routine screening of passengers in airports, repeatedly, day after day, can hardly be seen as a measure of national emergency. It is normally carried out in the interest of *routine* national security and public safety. Thus, this derogation is probably not applicable.

As for the second type of limitations, the ECHR contains no general provision applicable to all the rights authorizing restrictions on their exercise.[23] However, restrictions are provided with regard to concrete rights, e.g. Art.8 (2) (right to privacy) states: "There shall be no interference by a public authority with the exercise of this right except such as is in accordance with the law and is necessary in a democratic society in the interests of national security, public safety or the economic well-being of the country, for the prevention of disorder or crime, for the protection of health or morals, or for the protection of the rights and freedoms of others". This contains exhaustive lists of aims which are regarded as legitimate for the purpose of restricting this particular right.

CFREU Article 52 states more generally: "Any limitation on the exercise of the rights and freedoms recognized by this Charter must be provided for by law and respect the essence of those rights and freedoms. Subject to the principle of proportionality, limitations may be made only if they are necessary and genuinely meet objectives of general interest recognized by the Union or the need to protect the rights and freedoms of others". The formulation of aims here is quite vague, e.g. "objectives of general interest recognized by the Union".

The limitation and conditions which are applicable to all rights and freedoms are often contained in national law, e.g. Article 55 (3) of the Russian Constitution[24]: "The rights and freedoms of man and citizen may be limited by the federal law only to such an extent to which it is necessary for the protection of the fundamental principles of the constitutional system, morality, health, the rights and lawful interests of other people, for ensuring defence of the country and security of the State".

Limitations on human rights are not something extraordinary. Most human rights are not absolute, i.e. they permit restrictions or limitations if the following conditions are satisfied: this serves a legitimate aim, is prescribed by the law in a precise and foreseeable manner, and is both necessary and proportionate.[25]

According to the Human Rights Committee's General Comment 27, "States should always be guided by the principle that the restrictions must not impair the essence of the right…; the relation between right and restriction, between norm and

---

[23] Fact Sheet No.2 (Rev.1), The International Bill of Human Rights. United Nations, Geneva, June 1996.

[24] Constitution of the Russian Federation of 12 December 1993.

[25] Scheinin and Vermeulen (2011) p. 50.

exception, must not be reversed".[26] Any limitation hence must remain the exception and must respect the protection of the right in question.[27] It is proposed that with regard to the legitimate interest of public security in the fight against terrorism, any measure restricting the right to privacy should be assessed for its permissibility, through a step-by-step process, including the following conditions:

(a) The essence of a human right is not subject to restrictions.
(b) Any restrictions must be provided by the law.
(c) Restrictions must be necessary in a democratic society.
(d) Any discretion exercised when implementing the restrictions must not be unfettered. The deeper the intrusion, the stronger is the need for judicial review.
(e) For a restriction to be permissible, it is not enough that it serves one of the enumerated legitimate aims; it must be necessary for reaching the legitimate aim.
(f) Restrictive measures must conform to the principle of proportionality; they must be appropriate to achieve their protective function; they must be the least intrusive instrument amongst those which might achieve the desired result; and they must be proportionate to the interest to be protected.
(g) Any restrictions must be consistent with other human rights.[28]

The Council of Europe also stresses that "All measures taken by States to fight terrorism must respect human rights and the principle of the rule of law, while excluding any form of arbitrariness, as well as any discriminatory or racist treatment, and must be subject to appropriate supervision".[29] If applied to each human right which is or may potentially be violated by aviation security measure, these rules provide a strong proportionality test, which takes into account many factors (will be discussed in Chap. 5).

The difficulties arise with the categories "proportionality" and "necessity", which are broad and quite unclear. It is obvious that almost any measure can be argued by security professionals and other interested persons to be "necessary" when society is faced with the threat of terrorism. The limitation of human rights is one of the most disputed issues of interpretation. For instance, in accordance with some views, the rationale of human rights limitation should be reconsidered or at least reinterpreted due to the changed reality, affected by the threat and by fear. What may have seemed disproportional, unnecessary and unreasonable before 9/11 may now be deemed proportional, necessary and reasonable respectively. Under this approach, security demands prevail over human rights. Or, if one considers security as an aspect of human rights as discussed above, the need to protect the right to life prevails over protection of other human rights. The grounds for the limitations and derogations become broader.

[26] UN Doc. CCPR/C/21/Rev.1/Add.9, § 13.

[27] Scheinin and Vermeulen (2011) p. 51.

[28] *Ibid.* referring to A/HRC/13/37, § 17, and Etzioni. *Implications of Select New Technologies for Individual Rights and Public Safety*. In: Harvard Journal of Law and Technology. Vol. 15 (2002).

[29] See Council of Europe (2005).

This is especially relevant for aviation security, with a higher risk of terrorist attacks. For instance, in the USA, researchers note that the historical evolution of the law as it applies to airport screenings reveals a number of disturbing points concerning Fourth Amendment protections. A review of the cases illustrates that a person rarely wins a hearing and that searches are almost always found to be reasonable.[30]

In the EU, the ECtHR practice has formulated an approach to the "proportionality" and "necessity" issues. This will be considered in Chap. 5, but in general, they are common and can be applicable to any restrictions and any human rights.

In order to apply the limitation provisions to the security measures-versus-human rights dilemma, several steps should be taken that are similar to the interference test mentioned in Chap. 2. Similar to the right to privacy, an important source here is the case law developed around human rights, establishing and explaining many important principles and positions.

First of all, the exact impact of a concrete security measure on human rights should be analysed. This includes research on the security measure, its features, capabilities, operation modus etc. – i.e. how it actually works.

Then, it is necessary to indicate which rights are relevant with reference to this measure. In this step, the applicability of law protecting the rights should be ascertained. After that, the relation between the relevant rights and the security measure should be analysed in order to establish if there is interference. The rules on the limitations above will apply: if a security measure leads to limitation of human rights, this measure is allowed only, first, if the limitation is provided by law, and secondly, if the limitation complies with the requirements, i.e. is necessary in a democratic society in the interests of national security, public safety, public order, the rights and freedoms of others etc.

If this scheme is used, it becomes clear that some rights can be relevant, i.e. be protected by the law and at the same time interfered with by a security measure. In some cases the limitation is justified, i.e. the interference is substantiated, in some cases it is not, i.e. the right is violated.

3.3 Right to Life

The right to life is provided by Article 2 of ECHR: "Everyone's right to life shall be protected by law". As discussed, security itself may be a question of human rights too. It is aimed to secure people, to free them from certain harms, external and internal threats. It is obvious that people need to be protected from such harms in order to fulfil their right to life as such, and to conduct their lives as they wish. The right to life therefore has a special role in the human-rights-versus-security dispute discussed in this work. In contrast to other rights that run a risk of being *violated* by

[30] Kornblatt (2007) p. 403.

security measures, this right is supposed to be *protected* by them. Moreover, the states have both the right and the obligation to protect their citizens (see Sect. 4.6).

Within security undertakings, states must protect other human rights too. For instance, according to the Council of Europe "states are under the obligation to take the measures needed to protect the fundamental rights of everyone within their jurisdiction against terrorist acts".[31] However, as mentioned above, the right to life prevails over other rights and freedoms, including those considered in this work (privacy, non-discrimination, etc.). No derogations from this right can be made even in the case of emergency or war.

Aviation security organs thus may argue that they act in accordance with the principles of international air law and requirements of Article 2 of ECHR, fulfilling positive obligations of the state to safeguard the lives of people.

However, the security measures may be claimed to be inadequate, excessive or insufficient, in some cases resulting in failure to protect the right to life. Two dimensions can be distinguished. First, an air passenger can be injured in the course of a criminal act or terrorist attack. It can be claimed, then, that the security organs violated his right to life by failure to provide enough security and prevent the injury. Secondly, a passenger may suffer due to the actions of the security organs, for instance, in the course of an anti-terrorism or rescue operation, and then may claim excessive security.

As it will be discussed below, the aviation security measures may be divided into preventive security measures (protection against potential threats) and operative measures (response to acts of unlawful interference, i.e. reaction to realized threats). The nature of preventive measures differs from the nature of operative ones (the core idea of the former is to avoid "actual situations" at all) and the amount of possible damages to persons can hardly be compared. All the measures discussed in Special Part of this research are preventive security measures, with the purpose, roughly speaking, to search all the passengers in order to catch the one with the gun. Therefore, only the first dimension (failure to provide enough security and prevent the injury) can be applicable here. At the same time, the approaches and findings formulated by the Court with regard to the second dimension may be helpful for the discussion and will be considered as well.

With reference to the first type, in *Cesnulevicius v. Lithuania*[32] it was claimed that the state failed to comply with its positive obligations in order to prevent the death of a prisoner (§55). Here, the Court found a violation of Article 2 of the ECHR (§90). The Court emphasized that "persons in custody are in a vulnerable position and that the authorities are under a duty to protect them" (§84), and that the prisoner was "not [beaten] somewhere in forest or on the street at night, but in a State institution", thus whilst being in an apparently controlled environment" (§94). It was concluded that "the authorities did not respond adequately to the danger the person was

[31] See The Council of Europe Guidelines. *Human Rights and the fight against terrorism*, March 2005.

[32] No.13462/06, 10 Jan 2012.

in and thus did not fulfil their positive obligation to ensure that his right to life was upheld" (§89).

This case includes important considerations on ECHR Article 2. According to the Court, Article 2 "enjoins the State to take appropriate steps to safeguard the lives of those within its jurisdiction" (§82) and it "extends in appropriate circumstances to a positive obligation on the authorities to take preventive operational measures to protect an individual whose life is at risk from the criminal acts of another individual" (§83).

But such positive obligations are not unlimited. They arise only "where it has been established that the authorities knew or ought to have known at the time of the existence of a real and immediate risk to the life of an identified individual from the criminal acts of a third party and that they failed to take measures within the scope of their powers which, judged reasonably, might have been expected to avoid that risk".[33]

However, due to "the difficulties in policing modern societies, the unpredictability of human conduct and the operational choices which must be made in terms of priorities and resources", the Court tends "not to interpret Article 2 in such a way as to impose an impossible or disproportionate burden on authorities".[34] Accordingly, the Court stated that "not every claimed risk to life can entail a Convention requirement for the authorities to take operational measures to prevent that risk from materializing".[35]

These conclusions, as applied to this work, mean that state organs are obliged to take measures, including preventive measures, within the scope of their powers in order to prevent harm to air passengers and other persons. In the aviation security, with terrorism as the main threat,[36] the existence of a risk to the life of air passengers is known (however, it is arguable that this risk is "real and immediate") thus they take measures to avoid that risk, hence acting within Article 2 requirements.

However, this would be too simplified, since in the aviation security, there are the concepts of threat and risk assessment (discussed in more detail in Chap. 4) which are designed to determine the probability of a threat, the possible consequences of attacks, and the actions to be undertaken. The main idea is that the security measures must be threat-based and risk-managed. Equipped also with the modern technologies and methods, aviation security is designed to deal with the difficulties described above: unexpected risks, unpredictability of human conduct, etc., all actions should be proportionate.

Probably, however, if despite these endeavours the security should fail anyway, one might argue that the failure occurred due to the certain difficulties, and the positive obligation to prevent the risk will still be limited.

---

[33] See *Paul and Audrey Edwards v. the United Kingdom*, No. 46477/99, 14 Mar 2002, § 55; and *Bromiley v. the United Kingdom* (dec.), No. 33747/96, 23 Nov 1999, §83.

[34] *Bromiley v. the United Kingdom* (Dec.), No. 33747/96, 23 Nov 1999, §83.

[35] *Ibid.*

[36] Although the probability of this is in fact much lower than other risks; see Chap. 4.

With reference to the second type (where the measures taken by the authorities are excessive), there were cases when, for instance, the use of definite methods by authorities in actual situations such as anti-terrorist operations with hostages was questioned as being excessive and disproportionate under Article 2 of the ECHR.

The case of *Finogenov and others v. Russia*[37] concerned the hostage crisis in October 2002 in the Dubrovka theatre in Moscow and the decision to overcome the terrorists and liberate the hostages using gas. The applicants alleged, under Article 2 of the ECHR, that the authorities applied excessive force, which resulted in the death of hostages, and that the authorities failed to plan and conduct the rescue operation in such a way as to minimize the risks for the hostages (§3).

The Court determined that the situation was covered by Article 2 since the case was about the use of a dangerous substance by the authorities in conjunction with a rescue operation that resulted in the death of many of the victims that the authorities were trying to liberate and in mortal danger for many others (§203). Similar to the above-mentioned case, the Court considered that the authorities' positive obligations under Article 2 are not unqualified (§209). After examining whether the use of force was compatible with the requirements of Article 2, the Court held that there was no violation on account of the decision by the authorities to resolve the hostage crisis by force and to use the gas; however there was a violation of Article 2 on account of the inadequate planning and conduct of the rescue operation.

It can be seen that the Court distinguishes two aspects: first, the use of definite methods, secondly, the manner of carrying out these methods. Violation of Article 2 can refer to any or both of them.

In the course of preventive aviation security, if no direct victims, injuries or damages resulted, it is not so clear what can be questioned as "excessive" under Article 2. Therefore, it can hardly be stated that such measures, even if they have a strong negative effect on other human rights, may have a direct negative impact on the right to life. This concerns all the measures from the Special Part. The only, single preventive measure with such direct consequences known to me is shooting down of hijacked planes with the crew and passengers on board if it appears that terrorists intend to use them to attack targets on land, which may cause far more victims.

Whereas US pilots are trained to shoot down civilian aircraft in the event of any attack and in Russia this measure is provided by the law,[38] in Germany, the proposed measure was rejected by the Court, which found that the law infringed the right to life and human dignity and was unconstitutional and thus void.[39] In Poland, the measure was implemented in 2005, but in 2008, it was found unconstitutional and dismissed by the Court.[40] Obviously, this measure by the state, although having strong grounds, directly kills the passengers, deprives them of their right to life.

---

[37] *Finogenov and others v. Russia*, No. 18299/03 and 27311/03, 20 Dec 2011.

[38] Article 7(3) of Russian Federal law of 6 March 2006 N 35-FZ On counteraction against terrorism.

[39] §153, Judgment of the First Senate of 15 February 2006. Zitierung: BVerfG, 1 BvR 357/05, http://www.bverfg.de/entscheidungen/rs20060215_1bvr035705en.html

[40] First President of the Supreme Court, Judgment of 30 September 2008, K 44/07. Permissibility of shooting down a passenger aircraft in the event of a danger that it has been used for unlawful

Fortunately, cases of implementation of this measure in practice have not occurred. Hypothetically, if there were victims and the authorities were claimed to provide "excessive" security, again, it could be argued that the risk was worth it, although the court may judge differently when applying Article 2.

A more important conclusion for this work is the approach to the issue of excessiveness, which may be used not only with reference to Article 2, but to any other human right versus security undertakings. First, excessiveness may refer to both the form of particular measures and the manner of executing them. Secondly, the excessiveness issue depends greatly on concrete circumstances of a particular situation: e.g., the measure may be deemed to be non-excessive and thus proportionate and necessary despite the amount of victims and damages (in the case of other rights, different inconveniences such as loss of privacy) caused by this measure.

In aviation security, all these issues are supposed to be covered by the concepts of threat and risk assessment mentioned above: the security solutions should be proportionate to the probability of the threat and possible harms. The aim is also to ensure that the security resources are used effectively. But the key question is who decides? The decision making process is not democratic, and mainly unilateral decisions are taken by authorities. What is considered proportionate in the terms of threats, risks and harms may be disproportionate/excessive in the terms of human rights protection. Violations may also be considered "harms" – but, in contrast to the first harms, unintentional harms resulting from the security's actions rather than from the terrorists'.

As the result, in any case, the right to life – the obligation of the state to protect it – can serve as a good argument for the enhanced technologies and methods. With such an obvious aim and indisputable object to be protected, it may look like any security measures are permissible as long as they enhance security. But clearly, there should be reasonable limits. The impact of a security measure on passengers' human rights and freedoms, the considerations of proportionality, necessity and effectiveness – all in a broader sense than the concept of the aviation security usually implies – may serve contra arguments. And it is the government that has to prove that a disputable security measure has the right to exist. This will be further explored in the following chapters.

3.4 Right to Health

The right to health is a human right recognized in international human right law. At the same time, it is closely related to the right to life, since certain conditions of health are prerequisites for living at all.

acts, and where state security is threatened, http://trybunal.gov.pl/fileadmin/content/omowienia/K_44_07_GB.pdf.

The importance of this right is stressed in the above-mentioned provisions stipulating that certain other human rights may be limited for the protection of health. The Constitution of the World Health Organization (WHO) of 1946 defines health in its preamble as "a state of complete physical, mental and social well-being and not merely the absence of disease or infirmity". It also states that "the enjoyment of the highest attainable standard of health is one of the fundamental rights of every human being without distinction of race, religion, political belief, economic or social condition". Article 12 of ICESCR enshrines the right of everyone to the enjoyment of the highest attainable standard of physical and mental health.

Although the ECHR does not contain an article on the right to health, it has provisions to protect the right to life, security of the person, and other freedoms that affect the right to health. The CFREU provides in Article 35 that "A high level of human health protection shall be ensured in the definition and implementation of all the Union's policies and activities". Further, according to Article 11 of the European Social Charter[41] "With a view to ensuring the effective exercise of the right to protection of health, the Contracting Parties undertake, either directly or in co-operation with public or private organizations, to take appropriate measures designed inter alia: 1. to remove as far as possible the causes of ill-health; 2. to provide advisory and educational facilities for the promotion of health and the encouragement of individual responsibility in matters of health; 3. to prevent as far as possible epidemic, endemic and other diseases".

This right is frequently associated with access to health care, but actually it implies much more. It extends further, including a wide range of factors that can help one to lead a healthy life, such as safe drinking water and adequate sanitation, healthy working and environmental conditions, etc.[42] The realization of this right is thus dependent on the realization of economic, social and cultural rights and on the exercise of various freedoms. It is also directly related to peace and security. The right to health is therefore an interdependent right.

One of the most important factors for health is healthy environment. According to the WHO, almost one third of all illnesses are caused by environmental degradation.[43] Art.12(2)(b) of ICESCR obliges states to take steps necessary for the improvement of all aspects of environmental hygiene. Airports and aircraft are a part of daily life of thousands of people, thus the environment established there should satisfy to the requirements of normal environmental conditions. If a certain airport or airline security measure has an impact on passengers' health, then the provisions protecting this right may be violated.

Similar to the right to life, the right to health can hardly be subject to limitations due to security needs. Accordingly, every new or amended aviation security measure should be thoroughly tested on health effects before its use. Security practices dangerous to health should not be used even if they are highly effective.

---

[41] Adopted 18 October 1961, revised in 1966.

[42] United Nations (2008).

[43] WHO press release, 9 May 2002, http://www.who.int/mediacentre/news/releases/release36/en/index.html

Hypothetically, if a negative health effect of the already used security measure is proved and the devices are still in use, applications on violation may be submitted to respective authorities as violating Article 11 requirements to the states to remove the causes of ill-health and prevent diseases.

At the same time, reports on health issues originating from different experts may contradict each other, as in the case of body scanners (see Special Part). Apart from security needs, there are other different interests around health issues in civil aviation. The point is that flying as such is linked with cosmic radiation. Crew members, for instance, have restrictions on the number of hours they can fly. Accordingly, if the airline industry is really concerned about radiation, then scanners should not be the only potential cause discussed, "frequent flyers ought to be advised as to the health risks of their journeying…but I'm guessing that the commercial departments will resist all efforts to afford this issue the oxygen of publicity".[44] Thus, some actors may be interested to create uncertainty or silence around health issues.

The key problem is how to prove that the technology is safe or not. Definite requirements regarding certainty, objectiveness, independence of the assessment, with appropriate documentation should be ensured. Simple comparison with other daily life risks is not enough. The aggregate effect on the population, on frequent fliers, on vulnerable persons such as children, pregnant women, and long-time effects should be taken into account.

3.5 Right to Freedom of Movement

From the perspective of civil air transportation as such, the most important passenger right is the right to services provided by an airline, based on the fundamental right to freedom of movement and the contractual obligation which arises from selling a ticket. No matter the purpose for which you are travelling – visiting friends or family, on business, or just travelling for pleasure – you are in any case exercising your right to freedom of movement.

Analysis of international instruments allows concluding that the right to freedom of movement has three components: (1) the right to freedom of movement within a country, which includes the right to choose where to live within the country; (2) the right to leave any country, regardless of your citizenship; and (3) the right to enter a country of which you are a citizen.

For instance, Article 2 of Protocol No.4 to the ECHR[45] provides that "1. Everyone lawfully within the territory of a State shall, within that territory, have the right to liberty of movement and freedom to choose his residence. 2. Everyone shall be free to leave any country, including his own".

[44] Neeman. *Radiation in passenger screening: busting the myths*. In: Aviation Security International. Vol. 21 (2015) p. 26.

[45] Strasbourg, 16 Sept 1963.

Accordingly, the right to freedom of movement within human right law does not presume unlimited freedom of movement and limits this right by exclusion of the right to freely enter any countries where you are not a citizen. Clearly, a full and unlimited realization of this right implies a risk for security.

Of course, regional or national law may give preference to citizens/residents of particular states to enter certain states. For instance, the US Visa Waiver Program allows citizens of certain countries to travel to the USA without a visa for short stays. CFREU Article 45 provides that every citizen of the EU has the right to move and reside freely within the territory of the EU, and freedom of movement and residence may be granted to nationals of third countries legally resident in the territory of the EU. At the same time, EU Member States may restrict the freedom of movement of EU citizens on grounds of public policy or public security – but such restrictions must comply with the principle of proportionality and must be based exclusively on the citizen's personal conduct, representing a genuine, present and sufficiently serious threat.[46]

In general, however, entering foreign states is not free and is subject to various immigration and administrative requirements, including passport and visa regimes. In this context, these measures are quite understandable and substantiated.

In addition, limitations can be due to security needs too. For instance, there are different no-fly-zones, usually zones of military conflicts,[47] and different states may prohibit civil aircraft from flying to specific territories. For instance, in 2015, Russia and the UK suspended all flights to Egypt due to aviation security concerns in Egyptian airports in the aftermath of Metrojet's Airbus A321 tragedy.[48]

At the same time, the right to free movement in the scope mentioned above (i.e. already limited) can be limited too. Lawful grounds for limitation are contained in Sections 3-4 of Article 2 of Protocol No.4 to the ECHR:

> 3. No restrictions shall be placed on the exercise of these rights other than such as are in accordance with law and are necessary in a democratic society in the interests of national security or public safety, for the maintenance of ordre public, for the prevention of crime, for the protection of health or morals, or for the protection of the rights and freedoms of others. 4. The rights set forth in paragraph 1 may also be subject, in particular areas, to restrictions imposed in accordance with law and justified by the public interest in a democratic society.

Hence, two layers of limitations can be noted: the first is the limitation of the right to free movement in itself, regarding entrance into states of which you are not a citizen, and secondly, limitations to other components.

Civil aviation transports passengers both within and between the states, subject to limitations above. As a rule, it is presumed that a traveller must obtain all necessary prerequisites for legal travel beforehand. Although it is the task of immigration,

---

[46] Art.27 of Directive 2004/38/EC of the European Parliament and of the Council of 29 Apr 2004 on the right of citizens of the Union and their family members to move and reside freely within the territory of the Member States.

[47] See more details in Wikipedia: https://en.wikipedia.org/wiki/No-fly_zone

[48] Flight 31 Oct 2015 from Sharm el-Sheikh to St. Petersburg, 224 persons were killed.

police and other relevant organs to ensure that persons do not reside in the country or cross the borders illegally, aviation security control may contribute to detecting unlawful travellers.

In general, aviation security undertakings may prevent individuals from travelling both between and within states, constituting either a limitation on the already limited right of the person to enter a state where he/she is not a citizen or a limitation on the other aspects of freedom of movement respectively. Since reasons for the limitations are similar, they will be considered together.

In the aviation security field, the most "logical" exemption would be the national security needs and the protection of the rights of others, namely, the right to life. The limitations on the freedom of movement, accordingly, can be justified if they are necessary to protect these values, but similarly to other human rights, the obligation to prove that the measures are necessary lies with the state.

Another common lawful example of restriction of this right – overlapping with the one above – is in respect of persons charged with a criminal offence. It is believed that if a state, using different methods, undertakes to ensure that certain people are not able to leave a state (and possess a passport), such interference is justified by the public interest. For instance, in *Fedorov and Fedorova v. Russia*,[49] a restriction on leaving the place of residence lasting about 4 years and 3 months was found to be proportionate; consequently, no violation was found.

However there may appear different circumstances that make the Court find violation of Article 2 of Protocol No. 4 to the ECHR, as in the case of *Miażdżyk v. Poland*.[50] Here, the preventive measure was applied to the applicant for a period of 5 years and 2 months, with a pre-trial detention in one year (§34). But the Court, after assessing special features of the case, keeping in mind that "The restriction may be justified in a given case only if there are clear indications of a genuine public interest which outweigh the individual's right to freedom of movement" (§35), found that a fair balance between the demands of the general interest and the applicant's rights was not achieved (§41).

In *Luordo v. Italy*[51] the Court also found disproportionate an obligation imposed on the applicant for the duration of the bankruptcy proceedings (14 years and 8 months). This approach was followed in subsequent cases, where the duration of an obligation not to leave the territory of the respondent State varied between more than five years and more than ten years.[52]

Another type of infringement by applicant which led to violation of freedom of movement can be illustrated by *Stamose v. Bulgaria*.[53] The applicant infringed the terms and conditions of his US visa and was deported to Bulgaria. The Bulgarian border police imposed a two-year travel ban on the applicant, and directed the

---

[49] No. 31008/02, 13 Oct 2005.

[50] No. 23592/07, 24 Jan 2012.

[51] No. 32190/96, 17 Mar 2013, § 96.

[52] See respectively *Prescher v. Bulgaria*, No. 6767/04, 7 Jun 2011, § 47, and *Riener v. Bulgaria*, No. 46343/99, 23 May 2006, § 106.

[53] No. 29713/05, 27 Nov 2012.

competent authorities to seize his passport for breaching the US immigration laws. He alleged that the ban was unjustified and disproportionate, and that it prevented him from travelling to the US, where his mother and brother lived.

In §33, the Court observed that "the salient point is whether it was at all proportionate automatically to prohibit the applicant from travelling to any and every foreign country on account of his having committed a breach of the immigration laws of one particular country". In §34 the Court stated that it cannot consider "such a blanket and indiscriminate measure as being proportionate. The normal consequences of a serious breach of a country's immigration laws would be for the person concerned to be removed from that country and be prohibited (by the laws of that country) from re-entering its territory for a certain period of time". The Court stressed that individual circumstances of the person concerned should be taken into account (§36). Hence, the Court found violation of Article 2 of Protocol No.4 to the ECHR.

Therefore, when judging the applicability of this Article, the Court takes into account circumstances of the case, proportionality and necessity. However, all the above cases dealt with persons who have in different ways infringed definite rules, and the restrictions were applied post factum. But in aviation security, a practical possibility exists that restrictions may be used *before* any proved unlawful acts or infringements, thus, in respect not only of criminals, terrorists and potentially dangerous persons, but ordinary, innocent people as well.

With reference to aviation security, such restrictions may relate to: (i) existence of security requirements as such; (ii) particular aviation security measures, and (iii) immigration requirements if applied beyond immigration purposes and/or for the general security and aviation security purposes. Thus, there is overlapping between categories (ii) and (iii).

With reference to the first type of limitations, it is quite common that the possibility of travelling by air directly depends on going through the security controls. The general rule is that those denying to be screened are not allowed to proceed to the gates and will not fly.[54] Similarly, the airlines provide their right to refuse to carry such passengers in Contracts of Carriage.[55] Here, the needs to protect the public and other assets, obviously, substantiates that some control should be provided. How much control is sufficient, though, is yet another question – but this may relate to limitations of type two.

Another problem may emerge if a person does not refuse to be screened, but is nevertheless detained due to an "anomality" revealed by a security method. This can be a case of alarm by the body scanner's automatic detection system, by sniffing technology, by scanning of person's hand baggage, etc. In order to resolve the "anomality", the person may be subject to questions and further searches which take time and may ultimately result in missing a flight. As it will be discussed in the

[54] See, e.g. § 19 of Norwegian Regulation of 1 March 2011 No. 214 on the prevention of attacks against aviation security and Article 85(3) of Russian Air Code of 19 March 1997 N 60-FZ.

[55] See, e.g. British Airways. General Conditions of Carriage (§7a6). http://www.britishairways.com/en-no/information/legal/british-airways/general-conditions-of-carriage

Special Part, the cases of false positive detection are not so rare. At the same time, missing a flight relates more to economic consequences, issues of moral damages, etc. rather than the freedom of movement as such.

As for the second types of limitations, they may be caused by use of concrete aviation security measures, such as comparison of passenger data against watch lists, profiling, camera surveillance, traceability techniques, etc. They will be considered in the Special Part.

As mentioned above, in general, identity check/passport and visa control are connected with immigration and customs requirements in case of travelling between various states thus fall within the first layer of limitations mentioned above. However, today, these measures are often applied beyond travelling between states and/or for general security and aviation security reasons, thus, have potential to become limitation within the second layer too. Two dimensions can be noted.

First, immigration measures can be applied for domestic flights or flights between countries with no borders between them. For instance, in the Schengen system, despite its ultimate aim to abolish internal borders, third-country nationals must prove, if so required, that they have entered the territory lawfully, hold a valid visa, travel document or residence permit and satisfy entry conditions[56]; in other words, even when flying between Schengen member states or within a state. Failure to do so will lead to impossibility of travelling. But if the freedom of movement means the crossing of borders would be open to everyone, irrespective of their nationality, the scope of this right should encompass third-country nationals. Thus, these rules in fact limit the right to freedom of movement, thus, requires substantiation.

Secondly, the functions of the identity check/passport and visa control systems go far beyond immigration purposes and include security objectives. In the EU, for instance, along with abolishing of internal borders, new forms of control emerged, security measures which weaken the mobility of people.[57] They actively use and exchange personal data, create huge databases, etc. in order to identify, monitor and manage the movements across borders both inside and outside the EU.[58] This means increased opportunities for surveillance based on traceability, which raise concerns about freedom of movement. All these techniques fall within or overlap with aviation security measures and thus reflect limitations of type two above.

All these factors affect freedom of movement. It is argued that today, genuine free movement would only be possible inside a secured space with checkpoints and radars,[59] and that "aviation enables the free movement of individuals and goods. But

[56] Articles 19 to 22 of the Convention Implementing the Schengen Agreement (1990).

[57] See Atger. *The Abolition of Internal Border Checks in an Enlarged Schengen Area: Freedom of movement or a scattered web of security checks?* (2008) p. 4, 18.

[58] See Geyer. *Taking Stock: Databases and Systems of Information Exchange in the Area of Freedom, Security and Justice* (2008).

[59] Amicelli. *Surveillance and freedom of movement* In: Surveillance, fighting crime and violence (2012b) p. 270.

only that which can be monitored, traced, calculated, assessed and managed appears to be trustworthy".[60]

At the same time, civil aviation as such contributes to freedom of movement like no other means of transport, carrying people to other continents within hours, providing huge opportunities for travel which were impossible before aviation was born. Aviation security measures can also be considered to provide advantages for the right to free movement, since some people would not choose to fly if strict security was not provided.

Thus, while the need to choose between being subject to aviation security measures and not flying is probably substantiated by the security risks, limitations imposed by particular measures should be evaluated more thoroughly. The substantiation rules here are similar to the substantiation of privacy interference and can therefore be applied together.

3.6 Right to Equal Treatment and Non-discrimination

> We can ask today how it feels to be treated like a prisoner while being free: the randomness of Islamophobia at the airport is a case in point and the current 'war on terrorism' has managed to problematize and criminalize Muslims across the globe and treating all as guilty until proven innocent.[61]

In the aviation security context, the citation above, unfortunately, can be applicable not only to Muslims, but to broader groups of people: in the course of airport security, with all the methods and procedures involved, anyone can suddenly feel himself or herself being treated as guilty until proven innocent. But clearly, certain groups of people are more likely to get this feeling than others. The word that comes to mind in this regard is discrimination.

Non-discrimination and equality are fundamental human rights provided by Article 14 of ECHR, prohibiting any discrimination and providing that the enjoyment of other rights and freedoms shall be secured without discrimination on any ground such as sex, race, colour, language, religion, political or other opinion, national or social origin, association with a national minority, property, birth or other status. The list of the grounds is not exhaustive; other international instruments[62] include also belief, ethnic origin, genetic features, disability, age, sexual orientation. The formulation "other status" indicates that the application of this prohibition of discrimination is virtually unlimited.[63] However, in general, the grounds overlap with the categories referred to sensitive data by Article 8 of the DPD, which, as discussed, require additional protection.

[60] Leese (2013) p. 486.

[61] Bazian. *The 'Randomness' of Islamophobia at US Airports*, 3 September 2014, http://www.turkeyagenda.com/the-randomness-of-islamophobia-at-us-airports-1111.html

[62] Articles 2, 7 of UDHR, Article 26 ICCPR and Article 21 of CFREU.

[63] Vermeulen and Bellanova (2013) p. 306.

Article 1 §2 of Protocol no.12 to the ECHR[64] adds that no one shall be discriminated against by any public authority on any of the grounds. In comparison with ECHR Article 14, the prohibition of discrimination is not limited to enjoying others human rights and is extended to any legal right available in national law.[65] But the application of Protocol is limited, since not all signatories of ECHR signed the Protocol, e.g. the UK did not sign in order to avoid obligations not previously applicable to and accepted by the UK.[66]

Discrimination means any distinction, exclusion or restriction made on the basis of such grounds which has the effect or purpose of impairing or nullifying the recognition, enjoyment or exercise of human rights and fundamental freedoms. It is linked to the marginalization of specific population groups, making these groups more vulnerable.[67]

Equality of treatment is understood as steps that must be taken to prevent or remove differences in treatment with regard to other rights and freedoms. Thus, the right to non-discrimination and equality is interconnected and interdependent with other human rights and may be considered as their component: for example, all individuals are equal before the law, the rights to privacy and data protection are enjoyable by individuals regardless of race, origin, nationality, etc.

Thus, in addition to equality and non-discrimination as a fundamental right, it is possible to note a principle of equality and non-discrimination, which is an important principle in a democratic society and can be applicable in any sphere. This principle provides "rationale for resilience or resistance and a criterion for evaluating surveillance",[68] since the latter can have multiple negative effects on individuals or particular groups of people. Accordingly, equality is one of general legal principles discussed in Chap. 5, but due to overlap with equality as human right, discussed here.

With respect to aviation security, the principle of non-discrimination implies that any measure must not be discriminative on the basis of definite grounds mentioned above, and must not be applied in a manner which discriminates on these grounds. Accordingly, in the case of exercising security in the civil aviation, two dimensions of possible discrimination can be noted.

First, a security measure may be discriminatory in itself, on the whole. Technological advances and enhanced uses of personal data allow persons and groups to be classified in different ways[69] and also provide individual treatment.[70] Aviation security practices based on distinguishing persons and placing them in categories include, first, the above-mentioned black lists, watch lists, no-fly lists and

[64] Rome, 4 Nov 2000.

[65] Halstead. *Unlocking human rights* (2014) p. 298.

[66] *Ibid.*

[67] United Nations (2008).

[68] Wright et al. (2015) p. 288.

[69] Lyon (2007a) p. 5.

[70] Common examples are personalized advertising and different "smart" applications which work best when they know a lot about a person.

so on; profiling techniques, trusted traveller programmes (TTP). ID classifications at borders may also be discriminatory, e.g. the way they work in Israel "is not dissimilar from the ways in which they operated in South Africa under apartheid".[71]

Secondly, a security measure may be exercised in a discriminatory manner. For example, if a definite screening method such as a body scanner or hand search measure is applicable on a selective basis at the discretion of security personnel, a person may be selected to pass through it on the basis of specific criteria such as race, colour, sex, religion, ethnic or national origin (see more details in Special Part).

According to FRA, discrimination can be direct and indirect. Direct means that an individual is treated unfavourably in comparison to others in a similar situation, because of his/her particular characteristic, which falls under "sensitive" grounds mentioned above.[72] Indirect discrimination means that a neutral rule, criterion or practice affects a group in a significantly more negative way in comparison to others in a similar situation,[73] i.e. the grounds are beyond "sensitive" grounds.

The ECJ has an important decision regarding the principle of non-discrimination on the grounds of nationality in the protection of personal data within the context of EU citizenship. It held that the difference in treatment between Member State nationals and other EU citizens which arises during systematic processing of personal data relating only to the latter for the purposes of fighting crime, constitutes discrimination.[74] The same logic can be applied to the aviation security: the states cannot differ between different categories of citizens.

The ECtHR established applicable principles in its case law. First, there must be a difference in the treatment of persons in analogous, or relevantly similar, situations. Article 14 prohibits differences in treatment based on an identifiable, objective or personal characteristics, or "status", by which persons or groups of persons are distinguishable from one another.[75]

Further, such a difference of treatment is discriminatory if it has no objective and reasonable justification; in other words, if it does not pursue a legitimate aim or if there is not a reasonable relationship of proportionality between the means employed and the aim sought to be realized.[76]

Finally, the states enjoy a certain margin of appreciation in assessing whether and to what extent differences in otherwise similar situations justify a different treatment. The scope of this margin will vary according to the circumstances, the subject matter and its background.[77] For instance, in a case regarding discrimination on the grounds of nationality, the ECJ stated that different treatment may be justified only if it is based on objective considerations independent of the person's nationality

---

[71] Lyon (2004) p. 143 citing Zureik. *Constructing Palestine through surveillance practices*. In: British Journal of Middle Eastern Studies. Vol. 28 (2001).

[72] FRA. *Handbook on European non-discrimination law* (2011a) p. 22.

[73] *Ibid.* p. 29.

[74] *Huber v Bundesrepublic Deutschland*, C-524/06, 16 Dec 2008, §80.

[75] *Rangelov v. Germany*, No. 5123/07, 22 Mar 2012, §85.

[76] *Ibid.*§86.

[77] *Ibid.*§87.

and is proportionate to the legitimate aim.[78] As to the burden of proof, once the applicant has shown a difference in treatment, it is for the Government to show that it was justified.[79]

The ECtHR has been extremely strict in relation to discrimination based on race or ethnicity, stating: "no difference in treatment which is based exclusively or to a decisive extent on a person's ethnic origin is capable of being objectively justified in a contemporary democratic society built on the principles of pluralism and respect for different cultures".[80]

As it will be discussed in Special Part, in aviation security, as a rule, it is prohibited to exercise security measures in a discriminative manner, in particular, on the grounds of race, ethnicity, gender, national origin, or sexual orientation. Logically, if it is prohibited as such, no justifying grounds can be applied for making these ground criteria of security controls. However, in practice, these grounds are still used either illegally, or via exemptions: in the USA, for instance, profiling guidelines contain exemptions for certain activities at airports carried out by aviation security, immigration, and intelligence organs.[81]

In any way, similar to procedures described for the rights to privacy and the right to freedom of movement, if discrimination issues arise, first, it should be decided if the non-discrimination provisions are applicable to the concrete case, specifically, to the security measures or the manner they are practiced. Then, it should be evaluated whether there is interference to the right to non-discrimination. While it is obvious that the idea of both types is to distinguish persons according to specific grounds, it is up to the person to show that he/she is subject to a difference in treatment. The third step is to indicate whether different treatment has objective and reasonable justification, and it is up to the states to prove that.

3.7 Right to Freedom of Thought, Conscience and Religion

As indicated above, the religious grounds serve one of the main grounds for discrimination in the course of the aviation security measures and procedures. If practicing one's religion leads to almost automatic discrimination, in some sense this creates an overlay: in addition to the right to non-discrimination, the right to freedom of religion may be affected as well. Since the first dimension (discrimination) was discussed above, this section will focus on particular religious norms requiring of its practitioners definite behaviour or actions (or avoiding definite behaviour or actions), in particular during screening at the airports or in relation to any other aviation security measures.

---

[78] *Huber v Bundesrepublic Deutschland*, C-524/06, 16 Dec 2008, § 75.

[79] *Rangelov v. Germany*, §88.

[80] *Sejdić and Finci v. Bosnia and Herzegovina*, No. 27996/06 and 34836/06, 22 Dec 2009, §44.

[81] Quinlan. *Discrimination: the questionable effectiveness of screening based on race, religion, national origin or behavior*. In: Aviation Security International. Vol. 21 (2015) p. 37.

The right to freedom of thought, conscience and religion, including freedom to change his religion or belief and freedom, either alone or in community with others and in public or private, to manifest his religion or belief, in worship, teaching, practice and observance, is provided by Article 9 of ECHR.

As for the restrictions, the same article states in §2: "Freedom to manifest one's religion or beliefs shall be subject only to such limitations as are prescribed by law and are necessary in a democratic society in the interests of public safety, for the protection of public order, health or morals, or for the protection of the rights and freedoms of others".

Accordingly, if one party does not allow the other party to practice its religion, and there are no justifiable grounds, this will probably amount to violation. In the case of *Cyprus v. Turkey*,[82] the violations against ECHR Article 9 were quite "visible". The Court concluded that restricting access to places of worship as well as restricting their ability to travel outside their villages to attend religious ceremonies, failure to appoint priests to the area limited the religious life of the enclaved Greek-Cypriot population living in northern Cyprus all constituted a violation of Article 9 (§§241–244).

But what about religious rules that have an impact on a person's behaviour in public, non-religious places, such as airports or airplanes? Of course, the relation between the law and religious norms may be different in various cultures. The law may amount to religious norms and derive from them as in the case of Iran. But the law of vast majority of the modern states is more neutral; freedom to practice and manifest any religion is declared without any legal privileges (principle of secularism).

However, since some religions demand definite behaviour or practices from their adherents, conflicts between general law and order and religious rules may arise. There were cases when people fought for their religious beliefs to be allowed to be practised lawfully, such as wearing or not wearing hijab in public places. For example, the governments of Saudi Arabia and Iran oblige women to wear the hijab, while France, Turkey, Tunisia, Tajikistan have banned it in some public settings. In the USA and the UK, it is allowed to wear a hijab with a police uniform.[83]

In the ECtHR practice, a number of applications concerned allowance to wear hijab at schools and universities. Some of them were declared inadmissible holding that the measure had not been unreasonable.[84] In *Aktas v. France*,[85] the Court underlined the State's role as a neutral organizer of the exercise of various religions, faiths and beliefs. In other similar cases the Court found no violation.[86]

---

[82] *Cyprus v. Turkey*, No. 25781/94, 10 May 2001.

[83] Wikipedia. Hijab by country. http://en.wikipedia.org/wiki/Hijab#Hijab_by_country

[84] E.g. *Dahlab v. Switzerland*, No. 42393/98, declared inadmissible 15 Feb 2001.

[85] No. 43563/08, declared inadmissible 30 Jun 2009.

[86] *Leyla Sahin v. Turkey*, No. 44774/98, 10 Nov 2005, *Kervanci v. France*, No. 31645/04, 4 Dec 2008.

However, violation of Article 9 was found in *Ahmet Arslan and Others v. Turkey*[87] where the Court emphasized that in contrast to other cases, the case concerned punishment for the wearing of particular dress in public areas that were open to all, and not regulation of the wearing of religious symbols in public establishments, where religious neutrality might take precedence over the right to manifest one's religion.

In the course of security, in *El Morsli v. France*,[88] a Moroccan national married to a French man, was denied an entry visa to France, as she refused to remove her headscarf for an identity check by male personnel at the French consulate general. The Court declared the application inadmissible, holding that the identity check as part of the security measures of a consulate general served the legitimate aim of public safety and that requirement to remove the headscarf was very limited in time.

In the area of airport checks, some persons might face difficulties reconciling their philosophical or religious beliefs while being subjected to some of the security measures. For example, a procedure entailing a body image being viewed by a human screener may be very sensitive for many people. Muslim women, for instance, may be prevented from going through body scanners due to religious restrictions on modesty.[89] The Fiqh Council of North America (FCNA) stated that a general and public use of the scanners was "against the teachings of Islam, natural law and all religions and cultures that stand for decency and modesty".[90]

In 2010, the Dubai Airport authorities rejected the use of "naked" body scanners at the Emirates' airports as they violated ethical principles relevant to Islamic culture. In 2011, the Emirates considered installing the devices once again due to the technological developments, specifically, the option that blacks out sensitive parts of the body. But they decided not to install the scans anyway and to follow "the better practice, which is in line with religious and traditional values" instead.[91] But in this case, the key point is that the state here belongs to Islamic culture.

Taking into account the above-mentioned ECtHR practice, "questionable" devices like body scans constitute a part of the security measures served the legitimate aim of public safety and that the measure is limited in time, thus, if the same logic is followed, it can be argued that the case should be inadmissible or without violation (see *El Morsli v. France*). Moreover, in a research study considering aviation security versus basic fundamental human rights from an Islamic law perspective, it was concluded that aviation security provisions in general can be accommodated by the existing doctrines of the Islamic law.[92]

---

[87] No. 41135/98, 23 Feb 2010.

[88] No. 15585/06, declared inadmissible 04 Mar 2008.

[89] Peterson. *Emperor's New Scanner: Muslim Women at the Intersection of the First Amendment and Full-Body Scanners, The*. In: Hastings Women's LJ. Vol. 22 (2011).

[90] Fiqh Council of North America, Statement on the use of full body scanners for security at the airports and other places (Issued 9 February 2010). http://www.fiqhcouncil.org/node/4 (date accessed: 31 Jan 2012).

[91] Mazeina. *Full-body scanner gets a second look* (2011) http://www.thefreelibrary.com/Full-body+scanner+gets+a+second+look.-a0254431914

[92] Mustapha. *Aviation security provisions vs. basic fundamental human rights: an Islamic law perspective*. In: Aviation Security International. Vol. 21 (2015) p. 35.

However, the availability of alternative measures that could substantiate the "questionable" practices for those concerned may serve contra argument. For instance, in several cases,[93] applicants refused to perform military service for conscientious reasons and were convicted for that reason. They were willing to serve alternative service, but there were no respective laws in these countries. The Court found a violation of Article 9, taking into account that there existed effective alternatives capable of accommodating the competing interests involved in the overwhelming majority of European States. In the case of Armenia, it was taken into account that Mr. Bayatyan's conviction had happened at a time when Armenia had already pledged to introduce alternative service; in the case of Turkey, the Court invited Turkey to enact legislation concerning conscientious objectors and to introduce an alternative form of service.

Probably, taking into account the same logic, if alternative measures could be applied instead, but are not applied, non-violation of Article 9 in the case of some security measures is not so certain. Thus, it should be assessed whether people are provided with alternatives to scanning (e.g. the possibility of hand searches in private which would be acceptable to people with religious concerns), privacy safeguards of scanning equipment, etc., which all together may make it difficult to prove that scanners as such violate freedom of thought, conscience and religion.

However, since religious grounds can be used for discrimination in the course of the different aviation security measures, including the selection of persons to go through a scan, in these cases, substantiation discussed within the right to non-discrimination is needed (see above).

It should be also noted that both rights – to non-discrimination and to freedom of thought, conscience and religion – if violated, may have an impact on the right to free movement, as well. This illustrates the key idea of this chapter that all human rights are connected, and violation of one of them leads to violation of others, and for evaluation of proportionality of the aviation security regimes, they should be considered all together.

3.8 Rights of the Child

The rights of the child are stipulated in many international instruments such as ICESCR, the Convention on the Rights of the Child,[94] ICCPR. As a general rule, a child means every human being below the age of eighteen years.[95]

Due to the stage of their physical and mental development, in particular, their psychological, social or legal immaturity and less-developed awareness, children

---

[93] *Bayatyan v. Armenia,* No 23459/03, 7 July 2011, *Erçep v. Turkey,* No. 43965/04, 22 Nov 2011.

[94] Convention on the Rights of the Child, adopted and opened for signature, ratification and accession by General Assembly resolution 44/25 of 20 Nov 1989.

[95] *Ibid.* Art. 1.

are seen as a group "at the margin".[96] Thus, they are especially vulnerable and face particular challenges with reference to human rights. The core idea of human right law in respect of children is that children, as a vulnerable group, are afforded special protection, with additional guarantees. The guiding principle is "the best interest of the child".[97]

According to Article 24 of ICCPR, "Every child shall have, without any discrimination, the right to such measures of protection as are required by his status as a minor, on the part of his family, society and the state." Children shall have the right to such protection and care as is necessary for their well-being; in all actions relating to children, whether taken by public authorities or private institutions, courts of law, administrative authorities or legislative bodies, the child's best interests must be a primary consideration.[98]

Sector-specific instruments may include special provisions on protection of children. For example, since profiling children may have serious consequences for them throughout their life, and since they are unable to give their free, specific and informed consent, specific and appropriate measures for the protection of children are necessary in accordance with the UN Convention on the Rights of the Child.[99]

Both the EU GDPR and the "Police" Directive[100] stress that in addition to the right to data protection, other potentially affected fundamental rights include the rights of the child. According to the GDPR, children deserve specific protection of their personal data, as they may be less aware of risks, consequences, safeguards and their rights (recital 38). Article 8 sets out additional conditions for the lawfulness of the processing of children's personal data. The principle of transparency (providing information addressed to the public should be easily accessible and easy to understand), requires that with reference to children, information should be in such a clear and plain language that the child can easily understand (Recital 58). The GDPR also adopts "the right to be forgotten", which is particularly relevant, when the data subject has given their consent as a child and later wants to remove personal data (Recital 53). Further, although there are some exceptional circumstances when measures based on profiling by means of automated processing are allowed, such measure in any case should not concern a child (Recital 65). However, it is clear that some security measures are based on profiling, and information about the measures, which may include technical and special terms, may be difficult to understand and not clear enough.

---

[96] La Fors-Owczynik. *Minor protection or major injustice? – Children's rights and digital preventions directed at youth in the Dutch justice system.* In: Computer Law & Security Review. Vol. 31 (2015) p. 652.

[97] Convention on the Rights of the Child Art. 3.

[98] *Ibid.* Art. 3, CFREU Art. 24.

[99] Recommendation CM/Rec(2010)13 of the Committee of Ministers to member states on the protection of individuals with regard to automatic processing of personal data in the context of profiling adopted by the Committee of Ministers on 23 Nov 2010.

[100] See Chap. 2.

In the USA, despite the lack of general privacy and data protection law (see Chap. 2), children deserve special attention in this regard. For instance, in January 2013 the US Federal Trade Commission published rules interpreting the Children's Online Privacy Protection Act. Existing definitions were updated, e.g. "personal information" included Internet Protocol (IP) information and location information. These changes are designed to help to ensure that the personal information of children is not collected without parental permission. In Russia, there is a Federal Law on protection of children from information which may harm their health and development.[101]

Nevertheless, the EU report on human rights indicated that one of main challenges regarding children's rights is the lack of child-friendly services and systems.[102] This is very relevant for aviation security: different security measures and procedures are usually applicable to child passengers to the same degree as adults. I did not find any rules exempting children from particular aviation security measures, although these ideas are circulating.

For instance, in the UK, with reference to "naked scans", there was a discussion about making an exemption for children, since the scanners may violate the Protection of Children Act of 1978 by creating images or pseudo-images of nude children.[103] However, with new generation scans (see Chap. 6), no exemption is provided.

With regard to same problem in the USA, the concerns were raised over the viewing of their images, potential health effects, and safety when the child is separated from parents or guardians during the screening process. While TSA modified screening procedures for children under 12 years of age, it did not exclude them from body scanning or hand search.[104] The point is that with exemptions for children there may appear a risk that terrorists would then recruit children.

Another solution could be providing alternative measures for children, which would comply with the law. But the point is that alternative measures may be more expensive and take more time, thus, could be acceptable only if applied to a limited number of people. This logic can be noted for instance in the UK. Before 2013, no alternatives to scanners in the form of a private search were provided since it was considered to be disruptive to airport operations due to high volumes of expected searches required. But, experience showed the number of passengers refusing to be scanned was very low, thus, an alternative was provided.[105] The US TSA also reports

[101] Federal law of 29 December 2010 N 436-FZ On protection of children from information which may harm their health and development.

[102] State of democracy, human rights and the rule of law in Europe. Report by the Secretary General of the Council of Europe, Vienna, May 2014. SG (2014)1-FINAL, p. 69.

[103] Travis. *New scanners break child porn laws*, The Guardian, 4 January 2010, http://www.theguardian.com/politics/2010/jan/04/new-scanners-child-porn-laws.

[104] Elias. *Airport Body Scanners: The Role of Advanced Imaging Technology in Airline Passenger Screening* (2012) p. 5.

[105] UK Department for Transport. *Equality Impact Assessment on the use of security scanners at UK airports* (2013).

that since scanners were taken into use, 99% of passengers choose scanners instead of hand searches.[106]

This issue is similar to the possibility of providing alternatives on the grounds of religious beliefs discussed in section above. Apparently, the number of children is lower than the number of Muslims; thus, it could be easier to provide alternatives. However, taking into account that all the considered states – the USA, Russia and the UK – already provide alternatives in the form of hand searches universally, the unavailability of any other adequate alternatives (see Special Part), and the small number of people refusing scans, it can be problematic and may be unnecessary to establish other alternatives than those that currently exist.

As for the other aviation security measures, no alternatives are provided. In these cases, it must be considered whether the measures are necessary and no less intrusive alternatives are available, but these are consideration for all passengers in general (see Chap. 5 and Special Part).

In order to better protect the rights of child, instead of alternatives, may be preferable to concentrate on protecting all relevant human rights in the course of aviation security and minimizing the negative impact of aviation security measures in general. The point is that any negative effect may become more serious in terms of children, increasing aggregated harm to the population. Thus, in this regard, any interference requires more substantiation – compared with a situation in which only adults travelled by air. At the same time, aviation security is designed to protect travelling children along with other passengers, and this aim is also very important. Regardless, the less harm the better.

3.9 Concluding Remarks

The concept of human rights offers general principles and mechanisms of human rights protection and enforcement, as well as grounds for their restrictions and limitations. If applied to the civil aviation area, roughly speaking, the rights of air passengers – just as the rights of any human beings – are protected by law. But, some security measures may have impact on human rights, leading to their restrictions and limitations. The question that arises is whether or not the limitations are justified; if definite requirements are not satisfied, the respective human rights can be violated, and this is unacceptable. This has resulted in an aviation-security-versus-human-rights dilemma.

The right to life enjoys a special status. As discussed, it has priority in relation to other human rights. The risks of terrorism and crime in civil aviation and the positive obligation of the state to ensure the right to life are supposed to serve as substantiation for restriction and limitation of human rights in the course of security

[106] TSA. Advanced Imaging Technology. http://www.tsa.gov/traveler-information/advanced-imaging-technology (date accessed: 31 Jan 2015).

practices. For justification, the states may use a range of arguments, doctrines or constructions.

However, despite the fact that the right to life is supposed to be protected by security, it is possible that aviation security measures may be inadequate, excessive or insufficient, in some cases resulting in failure to protect the right to life. Hence, under certain circumstances, security is able to violate the right to life itself. While this does not make the right to life less important, this allows a more critical approach towards the aviation security measures, to contribute in the evaluation of aviation security features that may be excessive in relation to other human rights limitations.

A right close to the right to life is the right to health. It can hardly be subject to limitation due to security needs as well. If any aviation security measure has a negative health effect, the states must not use it or remove it immediately. Necessary prerequisites for evaluation of whether the measure is safe/unsafe are reliable, objective and documented reports on health effects conducted by independent experts. In the event a measure is unsafe, it does not matter how effective it is in terms of security, the impact it has on other rights, whether they are justified or not, since health concerns do not allow it to be used.

Regarding the rights to freedom of movement, to equal treatment and non-discrimination, and freedoms of thought, conscience and religion – different concerns deriving from aviation security measures can be indicated. Similarly to the right to privacy, the applicable law and the courts' practice suggest a common general scheme for analysis of the impact of security measures. Three steps include the following: establishing applicability of law protecting the right; establishing existence of interference; determining whether the interference is justified or there is violation. A common problem is interpreting and construing all the limitations properly. Difficulties may arise in substantiating the interference and implementing enough safeguards in practice.

Although positive obligations of the state to ensure the rights are not unlimited, it is usually up to the state to prove that a security measure having impact on the right(s) is justified and substantiated. At the same time, the states enjoy a certain margin of appreciation[107] in assessing whether and to what extent the questionable measure is justified, depending on concrete circumstances and situation, which make the value of human rights less absolute.

In order to evaluate the justification, a number of common criteria, especially the principles of legality and proportionality, can be used. With reference to the discussed human rights, courts formulated many of the important terms, principles and positions, such as approaches to proportionality, necessity, effectiveness and excessiveness, which allow us to make suggestions and draw tentative conclusions.

These approaches and principles will be further analysed in Chap. 5. Although the discussion is dedicated to privacy and data protection, since the approaches for evaluation and principles are general and common, they can be applied to other

[107] See more detail on margin of appreciation in Chap. 5.

human rights too. This will allow making common substantiation for particular aviation security measures in the Special Part.

As for the rights of the child, since the aviation security measures are, as a rule, applied universally to adults and children, if a measure raises human rights concerns, the consequences (possible limitations or violations) become more serious with respect to children due to their vulnerable status. Thus, any interference requires even more substantiation in addition to general substantiation.

In summary, the aviation-security-versus-human-rights dilemma is wider than the aviation-security-versus-privacy/data-protection dilemma and covers the latter. For the purposes of this research, it is important that the correlative "weight" of limitations and violations of *all* applicable rights by a security measure have an impact on the proportionality of the regime. The rule is: more concerns, both qualitatively and quantitatively, more substantiation and safeguards are required. At the same time, the measure's actual ability to protect the right to life should be also taken into account. However, some concerns, such as health, if proved, may be enough for non-use/removal of the measure, without the need for further justification and irrespective of the measure's effectiveness. This is the approach in the discussions below.

Chapter 4
Civil Aviation Security

4.1 Introduction

As mentioned in Chap. 1, in the aviation security vs. privacy debate, aviation security interests should be better understood and evaluated. This is the reason for the fairly detailed analysis of civil aviation security in this research.

In addition, a broader category, civil aviation, should be taken into account, since its status and role in the modern society has great influence on the measures to be taken for its protection, making civil aviation one of the most prominent areas of enhanced security measures.[1] As Chap. 6 will illustrate, aviation security measures become more advanced and sophisticated as a result of the most recent technological developments. It as though civil aviation is under constant pressure in terms of security, if compared, for instance, with other transport modes or public zones.

Thus, in addition to analysing the concept of aviation security and its legal regulation, it is important to address some other factors that influence the scope of security and, consequently, directly or indirectly, the aviation security vs. privacy dilemma. Specifically, there is a need to ascertain the reason why civil aviation has acquired such special status in terms of the need to be secured and how the aviation security regimes have developed into what we have today. This can be evaluated based on influential factors such as growth in aviation and historical developments in terms of both events and technological progress.

This chapter will start by discussing the term "aviation security" as well as the related and sometimes confusing term "safety". Following this, a presentation will be made of the phenomenon of civil aviation and its immense importance to the world economy, including some statistics and prognosis on its further development and growth. This allows preparing background for the discussion on the aviation security, including its quantitative and qualitative growth.

---

[1] Lyon. *Airports as data filters: converging surveillance systems after September 11th*. In: Journal of Information, Communication and Ethics in Society. Vol. 1 (2003a) pp. 13–20.

O.M. Enerstvedt, *Aviation Security, Privacy, Data Protection and Other Human Rights: Technologies and Legal Principles*, Law, Governance and Technology Series 37, DOI 10.1007/978-3-319-58139-2_4

The next section will analyse historical development of aviation security measures, especially those entailing concerns pertaining to privacy/data protection. Without aiming to provide a complete picture,[2] this presentation will help in understanding how, with the passage of time, more and more intrusive aviation security measures were introduced: from nothing more than general police control in the 1930s to the enhanced, sophisticated and high-tech security measures of today, that collect huge amounts of personal data. It will also give necessary illustrations of aviation security regulation and provide background for the selected measures in Chap. 6.

Obviously, responses to threats derive mainly from emerging threats – criminal tactics. As in any other area, the authorities who pursue criminals are always one step behind them. The history of aviation security may be divided into different stages according to different grounds. In this work, a number of key security incidents which caused serious changes in the aviation security regimes were selected as the threshold events marking four periods: (i) the earliest days of aviation – 1968 (the first "political" hijacking), (ii) 1968–11 September 2001 (9/11), (iii) aftermath of 9/11, and (iv) recent developments.

The next part is dedicated to the concept of aviation security – the rationale of aviation security decisions. The latter are based on risk assessment which is designed to determine how much security is needed in practice. Many factors are taken into account, such as assets, threats and risks, including probability and seriousness of harmful events, costs of protected assets and costs of security solutions, the security measure's benefits, etc.

Threats in aviation will be discussed first in general, followed by terrorist threats in particular. It is presumed that civil aviation is particularly vulnerable to terrorism, and even a minor explosion on board an aircraft almost always leads to disaster. Thus, the threats of terrorism usually serve as the main grounds for the enhanced aviation security measures. Moreover, the perceived terrorist threats on the national level are often directly linked to corresponding national aviation security decisions. This can be illustrated by the terrorist threats and respective aviation security regimes in the selected states. Thus, this section contains more detailed national "profiles".

Further, I will discuss legal principles of aviation security. There is a certain amount of overlapping of legal principles in this research. A more in-depth discussion of legal principles pertaining to privacy and data protection, and a review of more general principles applicable to aviation security and privacy follow in Chap. 5.

Finally, an overview of legal regulation of aviation security will be presented. As mentioned in Chap. 1, a comprehensive and detailed analysis of legal regulation of all three levels and all jurisdictions is beyond the aims of this work. Thus, the discussion is limited to an overview and some selected features that serve the aims of this research.

---

[2] A more detailed overview of the history is available in many books and articles, e.g. Price. *Practical aviation security: predicting and preventing future threats* (2013).

4.2 Security and Its Relation to Safety

Everybody wants to fly safely. It is often stated that it is statistically safer to fly than to use other transport modes. The European Aviation Safety Agency (EASA) states that air transport is one of the safest forms of travel.[3] However, this statement is only correct under certain conditions. The point is that in statistics, the impact of the numbers is very dependent on how one represents them. In the case of transport and safety, the numbers can be represented in terms of distance travelled, number of journeys or time of travel. The optimal and most favourable basis for the air transport industry will be the travel-per-kilometre basis, which usually reflects this means of transport as being the safest. However, if one would base the measurement on the number of journeys, air travel would be more dangerous than other transport modes.[4]

In any case, it has always been risky to fly. Obviously, with reference to aviation, air passengers are made safe or unsafe by different factors, including human errors (pilot error or inexperience, mistakes of dispatcher or other personnel, etc.), technical failures, hazardous weather, acts of unlawful interference, and military operations. Any of these factors may cause air crashes resulting in the loss of hundreds of lives and property damage worth huge amounts of money.

In order to minimize and prevent negative consequences deriving from all kinds of risks, special measures and undertakings are implemented. The terms safety and security are often confusing and the dividing line between them is blurred. In the Russian language, for instance, one word – безопасность, literally "without danger" – means both security and safety, so it is the context that may provide clarification. The same is true of the Norwegian language: "sikkerhet" means both safety and security, depending on the context. But even in other languages which do have two distinct words, the terms are very close.

The ICAO has the following definition of safety: "The state in which the possibility of harm to persons or property damage is reduced to, and maintained at or below, an acceptable level through a continuing process of hazard identification and risk management".[5] The Air Navigation Commission (ANC) defines aviation safety as "the state of freedom from unacceptable risk of injury to persons or damage to aircraft and property". In other words, acceptable safety means that all risk levels are kept below the risk limits, as low as reasonably practical (ALARP),[6] or that all hazard scenarios are kept in the acceptable region of the Tolerability matrix.[7] These definitions stress on the condition of being safe.

---

[3] European Aviation Safety Agency. EASA Annual Safety Review 2010.

[4] Brignell. *Number watch. The Risks of Travel* (2011) http://www.numberwatch.co.uk/risks_of_travel.htm

[5] § 1.3 of ICAO Doc 9735, Universal Safety Oversight Audit Programme Continuous Monitoring Manual, Third Edition, 2011.

[6] Aviation Glossary. http://aviationglossary.com/alarp-as-low-as-reasonably-practical/

[7] Aviation Glossary. http://aviationglossary.com/safety/

Security is defined by Annex 17 as "safeguarding civil aviation against acts of unlawful interference. This objective is achieved by a combination of measures and human and material resources".[8] The EU Regulation mainly repeats this, defining "aviation security" as the combination of measures and human and material resources intended to safeguard civil aviation against acts of unlawful interference that jeopardize the security of civil aviation.[9]

It is noteworthy that these definitions do not explicitly reflect the *result* of such undertakings – only the process. In contrast to this, Russian law interprets this term as "the state of the air traffic being safeguarded against an unlawful interference into the activity in the area of the aviation",[10] i.e. this definition stresses the condition of aviation as being *protected* and therefore puts the result in first place; in the second paragraph of the same article, the services that must ensure the air traffic security are stipulated.

However, for this research, aviation security measures that have impact on definite human rights apply, not the condition of protection; therefore, the term "aviation security" in this research will generally be used in conformity with Annex 17 and the EU definitions above.

In practice, aviation specialists make a distinction between aviation safety and aviation security based on factors of risk, which are usually classified into four types: human factors and performance, technical failures, fatal weather conditions, and criminal occurrences. Accordingly, aviation safety is a term encompassing the theory, investigation and categorization of flight failures (aviation accidents and incidents caused by the first three factors) and the prevention of such failures through regulation, education and training.

In contrast to this, aviation security exists to prevent unwanted deliberate, voluntary man-made harmful acts of unlawful interference against civil aviation or using civil aviation. In other words, safety aims at reducing the number of involuntary flight failures, security – to prevent wilful acts that aim at harming civil aviation.[11]

Nevertheless, safety and security are closely connected. For instance, all security equipment installed in the aircraft has to be certified by the aviation authority, taking into account safety terms: it must not jeopardize safety even if aimed at security. Hence, endeavours in the areas of both safety and security often come together, and security measures should first correspond to safety requirements and should secondly be deemed to have positive effects on safety.[12] According to IATA, "Safety is the promise the aviation industry makes to the billions of people who fly annually".[13]

---

[8] Chapter 1. Definitions.

[9] Article 3 of Regulation (EC) 300/2008.

[10] Air Code of 19 March 1997 N 60-FZ, Article 83. Translation into English by Garant database on Russian legislation (garant.ru).

[11] Nielsen. *Aviation security: legal framework and technical management*. In: International Journal of Private Law. Vol. 1 (2008) p. 82.

[12] Bjørnskau [et al.] *Transport Security and the Protection of Privacy. TØI report 914/2007* (2007) p. VIII.

[13] IATA. First Meeting of the RASG-MID Steering Committee (RSC/1) (Cairo, Egypt, 18–20 June 2012)
http://www.icao.int/MID/Documents/2012/rsc_1/docs/wp19.pdf

As a result, security can be considered one aspect of a safe industry.[14] Both safety and security are crucial elements in the aviation world. However, "security" is in focus in this research, since the analysed measures which have privacy/data protection concerns are aimed primarily at preventing man-made harms.

It should be also noted that I consider aviation security as a part of national security that encompasses state security, public security, environmental security, energy security, as well as other types of security.[15] This has direct relation to the security-versus-privacy dilemma, since the interests of national security serve as one of the lawful grounds for interference with human rights as discussed in Chaps. 2 and 3 (but in which other conditions apply).

4.3 Civil Aviation and Civil Aviation Security by Numbers

The most widely accepted date of the first flight in the history of aviation is 17 December 1903, when an aircraft was flown by the Wright brothers in the United States. The flight covered 120 ft (36 m) and lasted 12 s, but for the first time in history, a heavier-than-air machine had demonstrated powered and sustained flight under the complete control of the pilot.[16]

Air transportation developed rapidly: First scheduled commercial airline flight in 1914[17]; first non-stop flight across the Atlantic and first international scheduled services in 1919; first non-stop transatlantic passenger flights in 1928 – to name only a few milestones.[18] By the 1960s, with technological progress and the introduction of the 747 large-capacity aircraft, flying was taken out of the category of "adventure" and became a full-fledged practical means of transportation.[19]

The drivers and trends in future aviation include an increasing number of passengers, an increasing number of aircraft, higher capacity aircraft, increasing global competition, and increasing costs for security.[20]

Today, thanks to rapid economic growth and technology, there are aircraft that fly at over 40,000 ft, cruise at speeds almost twice the speed of sound and carry hundreds of passengers at one go. The largest passenger aircraft in the world today, Airbus A380 can seat up to 853 people in all-economy class configurations. The world's passenger aircraft fleet will total about 29,000 in 2030 (15,000 growth from 2010).[21]

---

[14] Everitt. *Safety and Security in Aviation – The Dividing Line is Blurred* (2000).

[15] See, e.g. Russian Federal law of 28 December 2010 N 390-FZ On security, Art. 1.

[16] Encyclopedia Britannica. Wright Brothers. http://www.britannica.com/EBchecked/topic/1071950/Wilbur-and-Orville-Wright

[17] IATA. New Year's Day 2014 marks 100 Years of Commercial Aviation. http://www.iata.org/pressroom/pr/Pages/2013-12-30-01.aspx

[18] Materials from Wikipedia.

[19] Steiner. *Aircraft evolution and airline growth.* In: Financial Analysts Journal (1967) p. 85.

[20] COPRA (2013).

[21] Airlines International, Special Report – Something Completely Different. 1 February 2011. http://airlines.iata.org/reports/special-report-something-completely-different

In recent years, air passengers have been provided with a wider range in destinations and suppliers, and with an increasing number of flight routes. Carriers fly to more cities and more remote areas, and it is often much cheaper to fly than to use other means of transportation. For instance, in the EU, since the creation of the European single aviation market in 1992, many customers of the low-cost carriers are either those who previously chose other modes of transport, or those who did do not travel at all; they only chose these carriers because of their low prices and would not otherwise have chosen an aircraft as means of transport.[22]

Some statistics: International scheduled air transport industry today is now more than 100 times larger than it was in 1945.[23] Moreover, the civil aviation market will apparently continue to grow. According to IATA, around 3.3 billion passengers were transported in 2014 and passenger numbers are expected to reach 7.3 billion by 2034.[24]

Today, the USA and the UK are leaders in international civil aviation market: the USA is the world's largest single market for both domestic passengers and international air travel, the UK is the second largest market of international air travel measured by number of passengers. However, a shift is expected in 2030, when China's economy is anticipated to overtake that of the USA. Other leading fast-growing markets, apart from the leaders – China and the USA – will be India, Indonesia and Brazil.[25] Russia is also one of the fastest growing markets for international air passenger traffic.

Clearly, since its appearance, the air transport industry has grown in terms of national and global assets, being one of the most vital and fastest paced economic forces in the global economy. For many states, the development of aviation, as well as air space usage in general, is one of the priorities in national policies, a matter of prestige. Aviation traditionally has symbolic importance for a state's reputation in the global arena.

Stimulating economic growth in all economies, aviation in many ways reflects the development of the key sectors of economy and technologies. It directly affects the employment of millions of aviation industry employers and indirectly affects many millions more.[26] Globally, aviation supports more than 58 million jobs and $2.4 trillion in economic activity; by 2034, these numbers are expected to reach 105 million and $6 trillion respectively.[27] For instance, in the USA, aviation supports some $669.5 billion of business, the equivalent of 4.9% of the GDP, and supports 9.3 million jobs[28]; in the UK – more than GBP 70 billion and 1.4 million jobs accordingly.[29]

---

[22] Macário [*et al.*] *The consequences of the growing European low-cost airline sector* (2007) p. 41.

[23] IATA official website. http://www.iata.org/about/Pages/history.aspx

[24] IATA Press Release, 16 Oct 2014. http://www.iata.org/pressroom/pr/pages/2014-10-16-01.aspx

[25] *Ibid.*

[26] Sweet *Aviation and airport security* (2004) p. 17.

[27] IATA Press Release, 16 Oct 2014.

[28] IATA Press Release, 23 Apr 2013. http://www.iata.org/pressroom/pr/Pages/2013-04-23-01.aspx

[29] IATA Statement. CAA's proposals do not address Heathrow cost issues, 30 April 2013.

Besides, even in the situation of economic uncertainty, the demand for connectivity remains strong,[30] and aviation is able to move people faster than any other form of transportation, which is essential to its economic viability.[31] Any interruption in flying, even temporary, is almost equivalent to a natural disaster, paralyzing the economy and the life of the country.[32]

Thus, both objectively and subjectively, the world depends on air travel, on continued and uninterrupted service.[33] However, along with enormous benefits, the industry requires many tasks and endeavours to keep it going. Air transport growth creates enormous opportunities but also presents a number of challenges.[34] One of them is aviation security. It is a common notion that the currently complex security system cannot be adapted to the enormous growth in passengers and cargo; it is a major restraint.[35]

Generally, all aviation occurrences can be divided into following categories: accidents, incidents, other occurrences and unknown occurrences. Aviation accident means, "an occurrence associated with the operation of an aircraft which takes place between the time any person boards the aircraft with the intention of flight and all such persons have disembarked, in which a person is fatally or seriously injured, the aircraft sustains damage or structural failure or the aircraft is missing or is completely inaccessible".[36] An occurrence "other than an accident, associated with the operation of an aircraft, which affects or could affect the safety of operations" is called incident.[37]

Consequently, other occurrences include safety occurrences that cannot be defined as accident or incident. Usually these cases involve aircraft being damaged (beyond repair) on the ground as a result of hurricanes, typhoons, sabotage, hangar fires etc.[38] Finally, unknown occurrences imply that insufficient information is available to determine the exact type of occurrence.[39]

However, the most important category for this research is a special occurrence category known as "security related" – criminal/security acts that result in accidents or incidents.[40] They can also be referred to as "acts of unlawful interference". Annex

[30] IATA Press Release, 6 Dec 2012. http://www.iata.org/pressroom/pr/Pages/2012-12-06-01.aspx

[31] Price (2013) p. 2.

[32] Bordunov (2007) p. 32.

[33] Sweet (2004) p. 6.

[34] Airlines International, Special Report – Growing Up. 1 April 2011. http://airlines.iata.org/reports/special-report-growing-up

[35] COPRA (2013).

[36] Convention on International Civil Aviation Annex 13 – Aircraft Accident and Incident Investigation.

[37] *Ibid.*

[38] ASN- Standards used on aviation-safety.net

[39] *Ibid.*

[40] Convention on International Civil Aviation Annex 13 – Aircraft Accident and Incident Investigation.

17 defines the latter as acts or attempted acts jeopardizing the safety of civil aviation, including but not limited to[41]:

- unlawful seizure of aircraft, or air piracy, aircraft hijacking, skyjacking, which are all synonyms, which means the forcible commandeering of an aircraft while in flight[42];
- destruction of an aircraft in service. Typically, it is explosive sabotage – bombings. The methods may be different. Saboteurs may put the bomb in the plane's baggage hold, carry onto the aircraft by a suicide bomber, or a passenger who disembarks at a subsequent transit point, leaving the device on board, or by a passenger who has no idea he is carrying a bomb. Other targets for bombings are aircraft on ground, airport facilities, e.g. arrival or departure halls, with usually high concentration of people, or nearby areas, where explosives may be put in a parked car.
- hostage-taking on board aircraft or on aerodromes;
- forcible intrusion on board an aircraft, at an airport or on the premises of an aeronautical facility;
- introduction on board an aircraft or at an airport of a weapon or hazardous device or material intended for criminal purposes;
- use of an aircraft in service for the purposes of causing death, serious bodily injury, or serious damage to property or the environment (aircraft as a weapon);
- communication of false information such as to jeopardize the safety of an aircraft in flight or on the ground, of passengers, crew, ground personnel or the general public, at an airport or on the premises of a civil aviation facility, e.g. bluff threats and threats from social media.

As mentioned above, aviation security focuses on terrorism, and the latter may refer to any types of acts of unlawful interference referred to above.

The term "terrorism" comes from the Latin verb *terrere* – "to frighten". This is the key word, since "the real targets of terrorism" are the billions who are not killed but who are terrorized by the killing.[43] According to a report, over the period 2008–2015, deaths from terrorism – globally and in all areas – have more than doubled (from 8450 to 17,958).[44] One can only imagine how many people were terrorized.

Terrorism is a difficult concept to define and there is no universally agreed and/or widely acceptable definition of terrorism. Although the UN adopted a number of treaties dealing with specific aspects of terrorism, none of the treaties contains a definition of "terrorism" or "terrorist acts".[45]

---

[41] Chapter 1 of Annex 17 to the Convention on International Civil Aviation, Security – Safeguarding International Civil Aviation Against Acts of Unlawful Interference. Ninth edition, March 2011 (Annex 17).

[42] Sweet (2004) p. 20.

[43] Schneier (2008) p. vii.

[44] Institute for Economics and Peace. *Global Peace Index 2015*, 2015.

[45] Gasser. *Acts of terror, "terrorism" and international humanitarian law.* In: International Review – Red Cross. Vol. 84 (2002). pp. 550–551.

According to the EU legislation, terrorist offences are those that may seriously damage a country or an international organization where committed with the aim of seriously intimidating a population, or unduly compelling a government or international organization to perform or abstain from performing any act, or seriously destabilizing or destroying the fundamental political, constitutional, economic or social structures of a country or an international organization: (a) attacks upon a person's life which may cause death; (b) attacks upon the physical integrity of a person; (c) kidnapping or hostage taking; (d) causing extensive destruction to a Government or public facility, a transport system, an infrastructure facility, including an information system, a fixed platform located on the continental shelf, a public place or private property likely to endanger human life or result in major economic loss; (e) seizure of aircraft, ships or other means of public or goods transport; etc.[46]

Clearly, terrorism is broader than acts of unlawful interference against civil aviation since it may include actions beyond civil aviation, and vice versa: some acts of unlawful interference are not terrorism.

The number of different occurrences in aviation is substantial. They may involve various damages to persons and/or the aircraft (in flight or on the ground), but may also involve damages to airport and infrastructure.[47] With reference to worldwide occurrence involving aircraft, the Aviation Safety Network's Database provides the following numbers (Table 4.1).

The point is that aviation security as a phenomenon was created due to the need to protect the aviation world from criminal/security-related occurrences – i.e. man-made threats and harms, which also may refer to acts of terrorism.[48]

Being a part of the aviation sector and at the same time relating to a broader general security sector, aviation security has become an industry in itself. Its growth in time, especially after the events of September 2001, is enormous, in terms of both scales and technologies involved.[49] Moreover, the enormous increase in passengers

Table 4.1 Aircraft safety occurrences worldwide (1921–2015)

Accidents	16,468
Criminal occurrences (excl. hijackings)	949
Hijackings	1067
Other occurrences	905
Unknown occurrences	556
Total	19,945

Aviation Safety Network (ASN) statistics: http://aviation-safety.net/database/ (last updated: 28 Dec 2015)

[46] Article 1 of Council Framework Decision of 13 June 2002 on combating terrorism (2002/475/JHA).

[47] These terms will be elaborated below.

[48] See further Sect. 4.5.4 – Threats.

[49] The overview of historical developments will be given below.

will have an important influence on the shape of aviation security in the future; airports will have to be able to process an increasing number of aircraft; higher capacity aircraft implies that more passengers will need to simultaneously pass security; etc.[50]

Today, the market is substantial, and it costs a lot to provide aviation security. The following statistics will help to give an overview.

Globally, airlines spend around $8.4 billion on security[51]; expenditures on airport security were $19.1 billion in 2011, and are expected to reach $45.43 billion by 2018.[52] Post September 2001, up to 32% of operational costs for major airports and 41% of staff are security related (compared to 5–8% pre 911).[53] This was also confirmed by Airports Council International (ACI) that reported that security operations at European airports account nowadays for about 35% of operating costs.[54] The USA spent $1.1 trillion on homeland security from 9/11 to 2011.[55] The TSA's annual budget is about $7 billion and it is requested even more.[56] As for the security equipment, global demand for it will increase 7% annually to $117 billion in 2016.[57]

Moreover, the aviation security market will grow despite the global economic downturn, due to an increase in global air travel, the persistent threat of terrorist attacks and new security legislation requiring airport operators to constantly update security procedures.[58] Accordingly, the costs for security will grow too due to the increase in the number of passengers that need to be checked, the increased costs of operating security systems, labour costs, etc.

At the same time, aviation security measures are being continually improved, and successfully: the number of terrorist attacks worldwide against aviation has

---

[50] For more details, see COPRA (2013).

[51] IATA Press release, 5 Mar 2013. http://www.iata.org/pressroom/pr/Pages/2013-03-05-02.aspx

[52] Security Director News. Global airport security expenditures expected to reach $45B by 2018. 23 April 2012.

http://www.securityinfowatch.com/news/10704628/report-global-airport-security-spending-to-increase-to-45b-by-2018

[53] Trembaczowski-Ryder. *The Economics of Security* (2010). Conference Transport Security Expo, London, September 2010.

[54] European Commission. *Impact assessment on the possible use of security scanners at EU airports* of 23 March 2011, §38.

[55] Israelsen. *Applying the Fourth Amendment's National-Security Exception to Airport Security and the TSA*. In: J. Air L. & Com. Vol. 78 (2013) p. 47.

[56] Sarkar. *The president's 2016 budget request: Transportation Security Administration*, 6 February 2015,http://www.fiercegovernmentit.com/story/presidents-2016-budget-request-transportation-security-administration/2015-02-06

[57] Miller. *Demand for security equipment projected to rise 7 percent a year through 2016*, Security Systems News, 20 May 2013, http://www.securitysystemsnews.com/article/demand-security-equipment-projected-rise-7-percent-year-through-2016

[58] Security Director News. Global airport security expenditures expected to reach $45B by 2018. 23 April 2012.

http://www.securityinfowatch.com/news/10704628/report-global-airport-security-spending-to-increase-to-45b-by-2018

decreased since the 1970s from about one or two per month to one to two per year.[59] The decrease in the total number of terrorist incidents and increase in prevented attacks can be explained mainly by an increase in the effectiveness of terrorism prevention by security and law enforcement authorities.[60]

Other statistics may help to visualize how many attempts at such acts fail, in many cases due to the endeavours of security. For instance, in 2009, at the Russian airports more than 4000 weapons were detected, 17,000 rounds of ammunition, 50 kg of explosives, 14 tons of flammable liquids, 5.5 tons of poisonous and other dangerous substances.[61] According to Chertok, the tendency is positive: the amount of seized explosives was estimated at hundreds of kilograms just a couple of years ago; in addition, not all the seized explosives were intended for terrorist purposes: a significant number was simply being transported.[62]

In the USA, according to TSA, the TSA confiscated approximately 50 million prohibited items over the past decade[63]; between 30 and 40 loaded firearms are still being detected every week.[64] Obviously, however, similar to the Russian case above, and taking into account quite liberal gun laws in the USA, the majority of the guns apparently belonged to ordinary people rather than terrorists.

Thus, it might be too early to celebrate success. Security experts agree that the threat of terrorism is still significant.[65] Despite fewer attacks and improved aviation security measures that have prevented many attacks, the international terrorists still have "the resources, sophistication and ruthlessness to find the weaknesses in global aviation security".[66] Aviation security faces new challenges every day due to emergence of new technologies, techniques, and ideas employed by terrorists, which are always one step ahead of those used by security officers. Accordingly, aviation security has to develop constantly too.

In summary, with substantial numbers in all areas and aspects, it is obvious that the aviation sector can be seen as much more than a means of transportation. It is a network on which global economic, social, and political spheres depend.[67] It has a direct and an indirect relationship with terrorism, commerce, and other areas. The numbers help to highlight the importance of the aviation industry as a whole and

---

[59] Aviation Security against Terrorist Threats – Conclusions of the conference of 31 October 2012, Nicosia,Cyprus.http://www.statewatch.org/news/2013/jan/eu-council-nicosia-aviation-security-16252-12.pdf

[60] *Ibid.*

[61] Aviation Explorer. http://www.aex.ru/docs/4/2010/7/25/1094/

[62] Chertok. *Aviation Security.* Aviation Security International Conference, St. Petersburg, 2010. http://global-port.ru/ru/avia2010articles

[63] The TSA's Administrator John Pistol's statement to the US House of Representatives' Committee on Homeland Security, 7 Feb 2012.

[64] Askew. *We need leadership from our leaders.* In: Aviation Security International. Vol. 21 (2015) p. 12

[65] See, e.g. Europol's EU 2012 Terrorism Situation and Trend (TESAT) report.

[66] Wilkinson and Jenkins. *Aviation terrorism and security* (1999) p. 156.

[67] Salter. *Imagining numbers: Risk, quantification, and aviation security.* In: Security dialogue. Vol. 39 (2008) p. 246.

aviation security as its essential part, as well as to predict their further development: both the industry and security will grow. The numbers also contribute to explain what makes the aviation so attractive for terrorists. This will be further analysed below.

4.4 Snapshots from History

4.4.1 The Earliest Days of Aviation – 1968

Civil aviation was not very old when the first aircraft was seized. Two incidents are commonly referred to as the first hijackings, both of them occurred in Peru: the first one in May 1930, when Peruvian revolutionaries seized a Pan American mail plane with the aim of dropping propaganda leaflets over Lima; nobody was injured. The second one occurred on 21 February 1931 in Peru. The pilot was surrounded by soldiers as he landed. He refused to comply with their orders and was detained. He was freed after the successful completion of the revolution was successfully.[68]

The first proven act of air sabotage occurred in the USA in October 1933, when Boeing 247 of United Airlines with 4 passengers and 3 crew members crashed near Chesterton, Indiana. Nobody survived, and the investigation concluded that the reason for the crash was a bomb containing nitro-glycerine as the probable explosive agent.

Although the background and purposes of criminals were different, the period after World War II until 1968 is often referred to as the "escape from persecutors phase" because many of the hijackings were in Eastern Europe and committed by people attempting to escape from the Soviet regime to seek asylum in the West,[69] or from Cuba to the USA, after Fidel Castro came to power in 1959. Conversely, some hijackers acted the opposite way, diverting American airlines to Cuba to support the regime. However, asylum hijackings are not pure terrorism.

As for the aviation security in this period, as long as there were no substantial threats to the public interests of the states, the problem of its ensuring had primarily a technical nature.[70] In the airports, there were no passenger or baggage screening requirements,[71] it was limited to routine police controlling.[72] Protection was mainly provided against such crimes as theft, pickpocketing, vandalism, and breaking and entering. Apparently, issues such as the aviation-security-versus-privacy dilemma did not yet exist.

---

[68] See, for example, Guinness World Records, First hijack of an aircraft, http://www.guinnessworldrecords.com/world-records/first-hijack-of-an-aircraft/

[69] Martin. *Air Piracy and Terrorism Directed against US Air Carriers* (1993) p. 2.

[70] Bordunov (2007) p. 36.

[71] Price (2013) p. 45.

[72] Hainmüller. *Why do Europeans fly safer? The politics of airport security in Europe and the US.* In: Terrorism and Political Violence. Vol. 15 (2003) p. 9.

4.4.2 1968–9/11

In the 1960s and early 1970s, the situation changed dramatically: civil aviation became a recognized target of international terrorism. Having both political and technological substantiations,[73] international terrorism is believed to have been born in this period.[74] The key event occurred on 23 July 1968, when terrorists from the Popular Front for the Liberation of Palestine hijacked an Israeli aircraft to Algiers. The terrorists demanded that Israel free a number of their colleagues imprisoned in Israel in return for the release of the passengers. For the first time, the terrorist act constituted a certain political statement; the choice of the target – an Israeli aircraft – was also significant.

This period contains the greatest number of hijackings worldwide; for instance, the total number from 1968 through 1972 was 326, meaning one every 5.6 days.[75] Not all were terrorist-related, however. For instance, while hijackings in Europe were principally a terrorist activity,[76] the main purpose of the Soviet hijackers was to escape from the USSR to one of the western countries.[77]

Up to the early 1980s, the general tactic of aviation terrorism was to hijack the aircraft, raise political demands, get the world's attention, and finally, after negotiation or an armed assault on the aircraft, free the hostages, i.e. to use the plane as a bargaining chip.[78] However, at the same time, the aircraft became the target itself. Statistics of aviation terrorism events from September 1970 until September 2001 shows a shift in terrorist tactics: decreasing numbers of hijackings and increasing numbers of sabotage or deliberate crashing.[79]

As a result, the 1980s and 1990s can be characterized as a period with fewer accidents, but a substantial number of victims per accident. The major event in the 1980s was the Lockerbie drama. On 21 December 1988, a bomb destroyed Pan American flight 103 over Lockerbie, Scotland, killing 270 people. Investigators found that the explosive had been concealed in a radio-cassette player.

Consequently, substantial changes to the aviation security regimes were made in this period. The key issue was that the threats from terrorists concerned first of all

---

[73] Jenkins. *Defense against terrorism*. In: Political Science Quarterly. Vol. 101 (1986) p. 774.

[74] Hoffman. *Terrorism trends and prospects*. In: Countering the new terrorism (1999) p. 16.

[75] Holden. *The contagiousness of aircraft hijacking*. In: The American Journal of Sociology. Vol. 91 (1986) p. 874.

[76] Poole. *Toward risk-based aviation security policy* (2008) p. 7.

[77] The first successful hijacking occurred on 15 October 1970, when Aeroflot flight 244 was hijacked to Turkey by father and son Brazinskas, killing a flight attendant and injuring three crew members. Later, the hijackers received American citizenship. Hijackings in the USSR up to 1973. http://hijacking.far.ru/2.html

[78] Avihai. *Between Two Septembers: From the bargaining chip of 6 September 1970 to a strategic agent of 11 September 2001*, 6 November 2006. http://www.ict.org.il/Article/948/Between%20Two%20Septembers%20From%20the%20bargaining%20chip%20of%20Sept%201970%20to%20a%20strategic%20agent%20of%20Sept%202001

[79] *Ibid.*

the public interests of the states. As a reaction, the international aviation community adopted several international conventions aimed at fighting against terrorism in the air (see Sect. 4.7.1).

National legislation of many states was updated accordingly, e.g. making hijacking an independent type of crime, as in the USSR in 1973.[80] The states also started to develop and introduce national security measures, established special aviation security departments at the airports and within the airlines. For instance, the formal history of the Soviet aviation security service can be referred to 1973, when in the Sheremetyevo International Airport, within the organization of transportation service, a screening group was established. In the UK, in 1974, in response to international terrorism, police officers were permanently deployed to Heathrow airport.[81]

The undertakings against hijackings and sabotage included such measures as manual searches of and interviews with passengers,[82] searches of carry-on baggage at the gates, walk-through or hand-held metal detectors (magnetometers) for passengers screening,[83] electronic screening of baggage, positive baggage match. These measures were used by many countries with varying levels of success, including the USA, Europe, and the USSR.

Other measures of this period included providing the crew with guns. The Soviet air company Aeroflot was almost the only airline in Europe that authorized crew members to carry personal firearms on board (Makarov pistol or PM, usually three items per crew). The crew members were trained on how to use the guns and were authorized to use them in case of violence connected with insubordination.[84] In addition, in the USSR, long before 9/11, from the 1990s, all cockpit doors were bulletproof and opening the doors while in flight was prohibited.

In the USA, other measures against bombings included the creation of the FAA Explosives Detection Canine Team Program, screening of portable computers and radios on flights from Europe and the Middle East,[85] a ban of certain hazardous materials. The airports started to locate lockers where they could be monitored.[86] In the 1970s, the Sky Marshal Service was created. Armed guards, dressed as typical passengers, were put on certain international flights to take action in the event of a terrorist action. In 1985, the Service was renamed the Federal Air Marshal Program and became a permanent part of the FAA workforce.

[80] Decree of the Presidium of the Supreme Council of the USSR of 3 January 1973. Bulletin of the Supreme Council of the USSR, 1973, N 1, art.3.

[81] Waldren. *Armed Police, The Police Use of Firearms since 1945* (2007) p. 224.

[82] For more details, see Hoofnagle. *Airport Searches* In: Encyclopedia of American civil liberties (2006).

[83] Wolff. *Screening technologies: an A to Z*. In: Aviation Security International. Vol. 21 (2015) p. 22.

[84] Hijackings. Legal aspects. http://hijacking.far.ru/1.html

[85] Manning. *United States' Response to International Air Safety, The*. In: J. Air L. & Com. Vol. 61 (1995) p. 526.

[86] Rumerman. *Aviation security*. In: Essays on the History of Flight (2003).

In the late 1990s, the first computer airline passenger profiling system was introduced in the USA – Computer Assisted Passenger Prescreening System (CAPPS). The Federal Aviation Administration Reauthorization Act (1996) required, among other things, passenger profiling.[87] The White House Commission on Aviation Safety and Security recommended that CAPPS should be used by all airlines.[88] This took effect in 1998.

It is clear that a number of security measures with privacy concerns appeared in this period, including camera surveillance, interviews with passengers, use of passenger personal data in general and the use of CAPPS and profiling in particular.

As mentioned in Chap. 1, a broader security-versus-privacy dilemma has been discussed since the mid-1970s. Clearly, in this period, technological advancements, the appearance of the Internet, surveillance, data mining and profiling techniques in general "contributed" to the dilemma.[89] In the 1990s, with the use of passenger personal data for security purposes, including profiling, a basis for specific aviation-security-versus-privacy dilemma was established.

4.4.3 *Aftermath of 9/11*

On 11 September 2001, terrorists hijacked four US airliners and crashed three of them into buildings and one into the ground, causing the death of thousands. There were other attacks as well, but 9/11 became the key event, the date which had consequences worldwide.

9/11 is often considered to be the first time that hijackers had used commercial airliners as weapons.[90] However, the first proved attempt was the hijacking of an Air France aircraft in December 1994 by Algerian terrorists who intended to crash the plane into the Eiffel Tower.[91] Another plot was the Project Bojinka in 1995 to blow up 11 planes simultaneously and crash a twelfth into CIA headquarters and thirteenth into the Pentagon.[92]

It is noteworthy that although many observers assumed that terrorist attacks and fatalities reached their maximum level just before 9/11, the maximum was in fact

---

[87] Dempsey. *Aviation Security: The Role of Law in the War Against Terrorism*. In: Columbia journal of transnational law. Vol. 41 (2003) p. 698.

[88] Poole (2008) pp. 9–10.

[89] See. e.g. Clarke. *Profiling: A hidden challenge to the regulation of data surveillance*. In: JL & Inf. Sci. Vol. 4 (1993); Turn *Information privacy issues for the 1990s* (1990); Bennett. *Computers, personal data, and theories of technology: Comparative approaches to privacy protection in the 1990s*. In: Science, Technology & Human Values. Vol. 16 (1991) etc.

[90] See, e.g. Robinson. *9/11: Threats About Airplanes As Weapons Prior To 9/11* (2011).

[91] Wilkinson and Jenkins (1999) p. 159.

[92] Thompson. *The terror timeline: year by year, day by day, minute by minute: a comprehensive chronicle of the road to 9/11—and America's response* (2004).

reached not in 2001, but in 1992.[93] Nevertheless, these tragic events are often used speculatively by both proponents and opponents of the enhanced security regimes. For instance, it is argued that 9/11 made a casus belli for the War on Terror (which resulted in the 2001 Afghanistan war), and an occasion to bring about increased militarization and police power, to gain a pretence for more and more governmental control of the public.[94]

These issues, as well as 9/11 advance-knowledge theories,[95] are beyond the scope of this work, but what is important is that the events of 9/11 highlighted the inadequacy of both laws and security systems[96] and led to stricter aviation security regimes all around the world. According to the researchers, the evidence showed that terrorists developed their tactics, whereas the authorities did not suggest any appropriate defensive or preventive measures.[97] In 1999, Wilkinson and Jenkins warned that despite all the statements to improve aviation security, a worse attack could happen.[98] Indeed, it happened in 2001.

The ICAO's Resolution A33–1 entitled "Declaration on misuse of civil aircraft as weapons of destruction and other terrorist acts involving civil aviation" was adopted. The ICAO also adopted Amendment 10 to Annex 17 on 7 December 2001 to address challenges posed by 9/11. Recommendations included: to lock flight deck doors for aircraft operated internationally and maintain measures on the ground; the States should actively share threat information; a comprehensive programme of a universal, regular, mandatory, systematic and harmonized aviation security audit (audit of States and airports) to be established by the ICAO.[99]

In the USA, the Aviation and Transportation Security Act gave the federal government direct responsibility for airport screening. The created Transportation Security Administration (TSA) became the principal agency responsible for security of all modes of transportation. It was charged with carrying out both the regulatory and operational responsibilities.

Additional security methods included the following. Each ticketed passenger had to display a valid government-issued ID before entering the security checkpoint and boarding a commercial flight. Parking and curbside access became limited. All individuals, vehicles and property entering into secured areas of an airport had to be screened. Only ticketed passengers, airport and airline employees, and certain approved law enforcement officials were allowed beyond the passenger screening checkpoints.

---

[93] Institute for Economics and Peace. *Global Terrorism Index 2012*, 2012.; La Free. *Countering Myths about Terrorism*. In: 38th Annual Theodore Standing Lecture, SUNY-Albany (2008).

[94] Materials from Wikipedia.

[95] Whether institutions or individuals other than the perpetrators had foreknowledge of the 11 attacks and should have taken action to try to prevent them.

[96] Acharya. *The Adequacy of Aviation Security Laws and Airport Security*. In: Central European Journal of International & Security Studies. Vol. 2 (2008) p. 107.

[97] Avihai (2006).

[98] Wilkinson and Jenkins (1999) p. 159.

[99] Abeyratne *Aviation Security Law* (2010) pp. 3–4.

Stricter procedures were implemented at checkpoints, including increased hand searches and random searches of carry-on-luggage. Limits were imposed on carry-on items (e.g. scissors or knives). All electronic items, such as laptops and mobile phones, could be subject to additional screening. More advanced metal detectors were installed. Travellers had to remove all metal objects prior to passing through the metal detectors in order to facilitate the screening process. Armed National Guardsmen were stationed at security checkpoints. All checked luggage on all commercial flights had to be screened for explosives, prohibited items and other dangerous materials, by either explosive detection machines, matching the bag to the passenger, hand searches, or the use of bomb-sniffing dogs.

For cases where a terrorist could potentially board an aircraft, additional measures were implemented: a massive expansion of the Federal Air Marshal Service,[100] the arming of trained pilots under the Federal Flight Deck Officer (FFDO) program, and the requirement that all cockpit doors be hardened and reinforced with heavier locks.[101] One of recommendations of a committee of airline industry executives and government officials after 9/11 was putting cameras on board aircraft, and in 2002, JetBlue was the first airline to install cameras in a passenger cabin.[102]

The event accelerated the further development of passenger profiling. It was a key recommendation of the 9/11 Commission that the government should assume responsibility of watch list matching from individual airlines. Thus, the use of a computer program that identified potentially suspicious travellers for bag searches and interviews was increased.[103] The CAPPS was redesigned and expanded to CAPPS II in 2003, entailing extended passenger checks. In 2004, it was replaced by a passenger-prescreening scheme Secure Flight, which compares passenger data against watch lists.

On 19 November 2001, the US implemented the Aviation and Transportation Security Act, requiring all airlines flying to or from the US to disclose to the US Bureau of Customs and Border Protection (CBP) and the Transportation Security Agency (TSA) personal data contained in PNR of air passengers. According to the Act, the transfers of passenger data must be completed before the plane takes off, or at the latest 15 min after departure. Not only the US Customs, but also all US federal agencies can have access to these data. From 5 March 2003 all international airlines had to provide the US government full electronic access to PNR; otherwise, fines and withdrawal of landing authorization can apply.

---

[100] Karber. *Re-constructing global aviation in an era of the civil aircraft as a weapon of destruction*. In: Harv. JL & Pub. Pol'y. Vol. 25 (2001). p. 794.

[101] The US Travel Association. *A Better Way: Building a World-Class System for Aviation Security*. 5 May 2011. https://www.ustravel.org/research/better-way-building-world-class-system-aviation-security

[102] Professional Pilots. *Cabin cameras set to keep watch in airliners*. 6 April 2002. http://www.pprune.org/rumours-news/49363-cabin-cameras-set-keep-watch-airliners.html

[103] Jenkins. *A primer on airport security*. In: Governance and Public Security New York: Campbell Public Affairs Institute (2002) p. 70.

The changes made to the aviation security regime were influenced by the security processes of Israeli's airline El Al, which by that time had the most intensive security screening in the world, operating under the paradigm that every individual is a guilty subject who must demonstrate their innocence,[104] used profiling and behavioural analysis.[105]

Apart from the USA, numerous countries, including Canada, China, the United Kingdom, France, Russia, Germany, India and Pakistan, also introduced new anti-terrorism legislation in the wake of the attacks, which, to a different extent, were copies of the American tactics.

In summary, as a result of 9/11, the use of measures which may constitute threats to privacy and data protection (passenger personal data transfer, profiling and behavioural analysis, camera surveillance, etc.) increased and spread throughout the world.

4.4.4 *Recent Developments*

Although 9/11 is still the first terrorist attack that is usually mentioned with regard to security developments, after this event, there were many other accidents and incidents that had an impact on aviation security regimes.

Along with enhanced security measures – not only in aviation, but also in other transport modes and crowded places – enforced in the aftermath of 9/11, terrorists' methods and tactics are being constantly developed too. The security regulators, as previously, have to catch up.

The list of events in recent history mentioned below is not exhaustive. It only includes, as illustrations, selected occurrences that are either "symbolic" for the civil aviation sector or, although not necessarily aviation-related, have had direct or indirect consequences for the aviation security regimes, with possible impact on the security- versus-privacy dilemma.

On 22 December 2001 the so-called "shoe bomber", an Islamic fundamentalist and self-proclaimed Al Qaeda operative Richard Colvin Reid from the UK attempted to blow up American Airlines Flight 63 from Paris to Miami using explosives hidden in the heels of his shoes. Passengers and crew overpowered him. As a result, a new shoe screening policy was implemented by many states to improve checks on passengers' footwear: travellers have to either remove their shoes for X-ray screening or screen the shoes in special devices without removing them.

The Madrid train bombings in March 2004, killing 192 people and wounding 2050, directed by an Al-Qaeda-inspired terrorist, led to the obligatory transfer of Advanced Passenger Information (API) by the airlines to the states in the EU.

On 24 August 2004, two female Chechen suicide bombers committed simultaneous terrorist attacks on two Russian passenger aircraft flown out of Domodedovo

[104] Laing. *Air Travel and Omnipresent Disaster* (2008) p. 707.

[105] See more detail in Sect. 6.4.4 – Profiling.

International Airport in Moscow. The attack killed 90 passengers and crew members. As a reaction, new screening technologies were implemented, including new explosive detection equipment; canine units; first body scanners were tested in the major Russian airports followed by permanent usage; new camera surveillance systems were put into practice. These enabled not only camera surveillance, but also recording and storage of the footage; the face-recognition programs were tested; special methods were employed by profiling experts, including interviews (psycho-emotional analysis) and express-analysers of speech, enabling analysis of a person's reaction, etc.[106]

Soon after the suicide bombers' attacks mentioned above, the Beslan school tragedy occurred in Russia in September 2004. The 3-day long ordeal during which 1100 people were held hostage ended in the deaths of at least 385. In the wake of Beslan, new laws on terrorism were implemented, the Air Code and other legal acts were amended; a new law on transport security was adopted,[107] the powers of law enforcement agencies, federal government and the President of Russia were expanded. The federal law On Counteraction Against Terrorism authorized the shooting down of hijacked planes if it appeared that terrorists intend to use them to attack the objects on the land.[108]

In the UK, additional anti-terror legislation was adopted following the London bombings in July 2005. Extra officers were deployed at airports to increase the use of stop and search powers; penalties for "encouraging" terrorism were increased.[109]

In August 2006, an alleged terrorist's attempt to smuggle liquid explosives on board US-bound aircraft was uncovered in the UK. This led to measures prohibiting passengers from carrying liquids, gels, and certain food items in excess of 100 ml onto aircraft all around the world.

Just a few days after car-bomb attacks in Glasgow and London in June 2007,[110] the EU announced a project containing anti-terrorism measures, including the creation of a European PNR system.

On 25 December 2009, at Northwest Airlines Flight 253 from Amsterdam to Detroit, the so-called "underwear bomber" Umar Farouk Abdulmutallab attempted to detonate plastic explosives hidden in his underwear. Consequently, those states that used body scanners accelerated their deployment (e.g. USA); a number of countries started to carry out their tests or introduced body scanners as a means of screening passengers. Another response to this event was the enhanced use of passenger hand search for screening of passengers.

---

[106] Soboleva. *Domodedovo promises to become the safest airport in Russia in the nearest future.* Finam.ru, 23 August 2005. http://www.finam.ru/analysis/newsitem149A300640/default.asp

[107] Federal law of 9 February 2007 N 16-FZ On transport security.

[108] Article 7(3) of Federal law of 6 March 2006 N 35-FZ On counteraction against terrorism.

[109] Walker. *Clamping down on terrorism in the United Kingdom.* In: Journal of International Criminal Justice. Vol. 4 (2006) p. 5.

[110] On 29 June 2007, in London, two car bombs were discovered and disabled before they could be detonated. The Glasgow International Airport attack, also using cars, occurred on 30 June 2007 and was substantially unsuccessful.

The Moscow Domodedovo airport suicide bombing of January 2011 killed 35 people. The bomb was detonated in the public area ahead of the checkpoints. As a reaction, the screening points were installed at the entrances to the airports terminals as well, i.e. screening not only passengers, but also everyone entering the buildings.

The worst incident in Russian/Soviet history of civil aviation happened on 31 October 2015, when Airbus A321 from Sharm el-Sheikh to St. Petersburg crashed after an inflight explosion, killing 224 persons on board. Very soon after this, there were series of shootings and bombings in Paris on 13 November 2015. Both events were connected to activities of ISIL.

As the result, Russia and the UK suspended all flights to Egypt until security is improved at its airports. It was reported that Moscow airports would strengthen aviation security measures[111]; the US DHS plan to expand security efforts at foreign international airports.[112] Similarly, the UK authorities plan to increase the staff of the security services and double the budget for security at foreign airports in order to assist other countries "to put in place the tightest security measures possible".[113] After the Paris bombing, procedures to accelerate adoption of the EU PNR system were also undertaken.

Thus, it is quite clear that in response to increasing threats, the regimes all around the world will change and will be stricter, greatly influenced by the US and similar regimes. Many other terrible events not related to aviation such as suicide bombings on the Moscow metro system in 2010, the attack of Anders Behring Breivik in 2011 in Norway, the Boston Marathon bombings in 2012, terrorist acts in Russia's Volgograd in 2013, the Paris bombings in January 2015, etc. have also led to some reactions and strengthening of national security regimes in general.

The overview of the historical events and consequent development of the aviation security measures demonstrates the reactive approach of the regulators and reactionary establishment of the regimes as such. It also provides insight into how the combination of the security measures, both newly developed and upgraded older measures, are gradually becoming stricter. Some of the forms, such as requiring passengers to take off shoes or installing additional metal detectors at the airport entrance, present less risks to human rights. Other measures, such as the increased use of hand searches and body scanners and increased use of personal data, imply more risks. The development can also be seen as an illustration of the rationale and logic of aviation security decisions, principles and regulation discussed below.

---

[111] Interfax. *Moscow airports strengthen inspection on departure and arrival after terrorist attacks in Paris*. 14 Nov 2015. http://www.interfax.ru/moscow/479279

[112] Markon. *Homeland Security plans to expand pre-clearance program at foreign airports*. Washingtonpost, 13 November 2015.

https://www.washingtonpost.com/news/federal-eye/wp/2014/11/13/homeland-security-plans-to-expand-pre-clearance-program-at-foreign-airports/

[113] Hope. *Britain to hire 2000 new spies at MI5, MI6 and GCHQ in wake of the Paris terror attacks*. The Telegraph, 16 Nov 2015, http://www.telegraph.co.uk/news/uknews/terrorism-in-the-uk/11997782/Paris-terror-attack-Britain-to-hire-2000-new-spies.html

4.5 Aviation Security Concept

4.5.1 Introduction

As mentioned above, the term "aviation security" in this research is used according to Annex 17 and the EU definition, implying the combination of measures and human and material resources. Nevertheless, what is important is that it also implies security decisions to determine exactly which measures should be deployed. The current common approach to aviation security stipulates that security measures must be threat-based and risk-managed.[114] The concept used to establish how much security is needed is called risk assessment.

Civil aviation has been subject to methodical and quantitative risk assessments since the 1960s.[115] In Annex 17, the risk assessment concept was first declared in 2006 in its 8th edition. Standard 3.1.3 of Annex 17 requires that each state "shall keep under constant review the level of threat to civil aviation within its territory, and establish and implement policies and procedures to adjust relevant elements of its national civil aviation security program, based on a security risk assessment carried out by the relevant national authorities".

ICAO's Doc 8973 provides guidance material on threat assessment and risk management methodologies. It specifies the methodologies of threat assessment and risk management, including analytical and semi-quantitative approaches based on numerical scores.[116] Similarly, European Civil Aviation Conference (ECAC) Aviation Security Handbook (restricted) describes the application of risk management to aviation security.

Although these processes and their results are as a rule not open to public, it is possible to determine general approaches. Aviation security decisions operate a number of distinct but interrelated concepts and factors: risk assessment, protected assets, threats, especially threat of terrorism in aviation, the security solution's benefits, in particular, effectiveness, etc. They will be considered below.

[114] See, e.g. Annex 17, § 3.1.3; European Civil Aviation Conference. ECAC Aviation Security Handbook (restricted), 2014.

[115] Stewart and Mueller. *A risk and cost-benefit assessment of United States aviation security measures*. In: Journal of Transportation Security. Vol. 1 (2008) p. 145.

[116] Tamasi. *Risk assessment techniques for civil aviation security*. In: Reliability Engineering & System Safety. Vol. 96 (2011) p. 893.

4.5.2 Risk Assessment

Generally, risk means "the statistical likelihood of a randomly exposed individual being adversely affected by some hazardous event".[117] In other words, it is a combination of the probability of occurrence and impact deriving from such an event, including all the possible consequences among different dimensions: casualties, delays, damage, etc.[118]

Risk assessment is a process aimed at determining the probability of a threat, the possible consequences of attacks, and the actions to be undertaken.[119] First, it is necessary to identify all the existing threats and to anticipate the new and emerging threat trends. At the same time, "the fear of harm should be proportional not merely to the gravity of the harm but also to the probability of the event".[120] The core principle in risk assessment is that while the risk of threats generally cannot be eliminated, enhancing protection from known or potential threats can reduce the risk, and the aviation security measures should be commensurate with the risk. Secondly, it is necessary to implement measures to counter that threat.[121]

Risk assessment can be divided into the following sub-tasks: (i) risk determination – the risk identification involving new risk and changes in the risk parameters; (ii) risk evaluation, consisting of risk aversion and risk acceptance; (iii) risk measurement – i.e. quantification, e.g. the number of accidents per unit of a system's output. Accordingly, the risk level is a quantitative value of the risk established in the risk assessment.[122] Aviation accidents are usually viewed in terms of fatal events with the system's output defined as the number of aircraft kilometres, passenger kilometres and/or aircraft departures over a given period.[123]

The quantitative risk assessment is divided in the followings steps: a threat assessment,[124] a vulnerability assessment,[125] and a criticality assessment.[126] Accordingly, the quantification can be done as follows:

Risk = Frequency × Consequence
Frequency = initiating event frequency × probability all safeguards fail

---

[117] Sage and White. *Methodologies for risk and hazard assessment: a survey and status report*. In: Systems, Man and Cybernetics, IEEE Transactions on. Vol. 10 (1980) p. 426.

[118] COPRA (2013).

[119] Tamasi (2011) p. 893.

[120] Ferguson *The ascent of money: A financial history of the world* (2008) p. 188.

[121] Alexander and Sochor. *Aerial piracy and aviation security* (1990) p. 21.

[122] COPRA (2013).

[123] Janic. *An assessment of risk and safety in civil aviation*. In: Journal of Air Transport Management. Vol. 6 (2000) p. 44.

[124] A threat assessment identifies and evaluates threats based on various factors, including capability and intentions as well as the potential lethality of an attack. Tamasi (2011) p. 893.

[125] A vulnerability assessment is a process that identifies weaknesses that may be exploited by terrorists and suggests options to eliminate or mitigate those weaknesses. Tamasi (2011) p. 893.

[126] A criticality assessment is a process designed to systematically identify and evaluate an organization's assets based on their values, the importance of its mission or function, the group of people at risk, or the significance of a structure. Tamasi (2011) p. 893.

Risk = Threat × Vulnerability × Criticality[127]

In addition, risk assessment contributes to an effective use of security resources[128]: theoretically, there should be an economic balance between the impact of risks and the cost of security solutions. The analysis of the Annual Losses Expected (ALE) permits one to ascertain the amount of economic resources necessary for countermeasures.[129]

Moreover, security needs to be delivered, but it also needs to be delivered in such a way that it does not place an unacceptable burden on the civil aviation business.[130] Experts admit that for aviation security, the main challenge is ensuring that the technologies are implemented in a way that is affordable.[131] There should also be a balance between the measures' costs and benefits.

For instance, according to TSA, since all the specific features of CCTV such as Video Analytics and other advanced capabilities[132] come at a price, thus, during development of the systems, airports "need to carefully weigh the benefits and costs, especially with regard to how these features contribute to established operational requirements, how they are to be implemented, and their downstream support requirements".[133] The producers of CCTV equipment advice that the airport should take into consideration the following three criteria when developing surveillance systems: functional purpose and objectives, specifications and the required parameters, and costs.[134]

At the same time, risk assessment incurs a number of challenges. There are uncertainties with application to the risk of terrorism, including false positives and false negatives, technological failure, reactive risk, displacement to organizational risks, too much precaution,[135] etc. In addition, there are challenges regarding uncertainties in quantifying the risks of terrorism (see below). All this makes security far from perfect.

Moreover, it is obvious that using quantification in different ways allows getting definite results for the indication of risks: for instance, as noted above, a per-

---

[127] Tamasi (2011) p. 893.

[128] Poole (2008) pp. 13–14.

[129] Tamasi (2011) p. 893.

[130] Atkins. *Online terrorism: Is the cyber threat real?* In: Aviation Security International. Vol. 18 (2012). p. 36.

[131] Bradley. *Airports Build a Case for Biometrics*, Aviationpros, 1 July 2015, http://www.aviationpros.com/article/12082409/airports-build-a-case-for-biometrics

[132] See Chap. 6.

[133] Transportation Security Administration *Recommended Security Guidelines for Airport Planning, Design and Construction* (2011) p. 124.

[134] See, e.g., Cmt-VC Company. Video surveillance systems. http://www.smt-vc.ru/catalog/sistemy-videonablyudeniya/

[135] Ericson. *Ten uncertainties of risk-management approaches to security*. In: Canadian Journal of Criminology and Criminal Justice/La Revue canadienne de criminologie et de justice pénale. Vol. 48 (2006).

kilometre basis will almost always indicate air transportation as the safest transport mode, while the number of journeys will prove the opposite.

Another problem is that from psychological aspect, people tend to overestimate and overreact to the risks of unlikely incidents that are quite rare, but well publicized, such as aviation incidents. Statistics shows that governments, e.g. the USA, spend more money per life saved on measures to prevent death from "rare" risks such as aviation incidents than on measures to prevent death from "ordinary" risks such as car incidents,[136] although statistically the latter have a far greater death rate. The difference lies in the threat perceptions: for example, following 9/11, the perception of the "rare" risk was enhanced by repeatedly televising the collapse of the twin towers.[137]

Wolff, applying the term Mathematical Expectations ("the probability of the occurrence of an event multiplied by the value associated with the event's occurrence") to terrorist attacks on aviation, argues that even if the probability of an event is low, if the result of the event occurring is substantial, such as the results of a successful attack, then the Mathematical Expectations might be higher than a different event with a higher probability but lesser consequences. Hence, the low probability, but high Mathematical Expectations should receive more attention.[138] This logic is understandable, the problem appears that this "attention" turns into enhanced security methods; the question appears how much "more attention" is enough. For instance, those interested in getting budgets will apparently always be for expensive security systems because more budget means more power; if, for example, the police are put in charge of security, the trade-offs they make will result in measures resembling a police state.[139]

The core question is who actually performs risk assessment and decides which measures are worth deploying. ICAO's Annex 17 cited above states: relevant national authorities – the states. The latter, obviously, employ security specialists and experts who provide assessments which are interesting and valuable for the state – i.e., in terms of which decision is more effective, more economical, etc. As a rule, the passengers and public in general are excluded from the discussion, under the pretext of secrecy for security needs, that security should be kept secured and secret, etc.

The problem is that the quantification of the security – as a rule, in secret – allows the governments, the companies, and the experts to express their concerns in numbers; this may lead to expansion of surveillance and control, to justification of decisions in the name of emergencies, crises, and exceptions.[140] As the result, objective risk as uncertainty can become an ideology,[141] "a governmental resource that can be

[136] Stewart and Mueller (2008) p. 145.

[137] Amicelli (2012a) p. 3.

[138] Wolff. *Are We Ignoring the "Risk" in Risk Based Screening?* In: Aviation Security International. Vol. 18 (2012) pp. 4–5.

[139] Schneier (2008) p. 77.

[140] Salter (2008) pp. 246–248.

[141] Rose [*et al.*] *Governmentality*. In: Annu. Rev. Law Soc. Sci. Vol. 2 (2006) p. 96.

deployed to create, obscure, or manage certain issues, institutions, or behaviours".[142] There is always a risk of misuse of statistics and numbers by politicians, bureaucrats or any other persons interested in generating alarm and panic. Accordingly, security can be governed in the name of risk, producing the justifications of risk factors and security threats.

At the same time, the cost of security solution includes not only economic resources, but also other risks,[143] including possible harms to passengers' rights and freedoms. It is thus argued that modern interpretations of risk, together with commercial pressure, technological innovation and cultural commitments lead to enormous surveillance, which both addresses risks and produces other risks[144] – to the rights to privacy, data protection and other human rights.

All this is especially applicable to the civil aviation sector, in which valuable or priceless assets should be protected and are often declared as targets of terrorism, thus receiving substantial funding in terms of aviation security. Thus, it is believed that in aviation, the most effective – but at the same often intrusive– technologies should be used. These issues will be discussed in more detail below.

4.5.3 Assets

The general goal of aviation security is the protection of assets. An asset is something you value and want to protect against all forms of threats. Anything of value can be an asset, but in the context of aviation security, the main assets are the lives of passengers, crew and people on the ground; aircraft; airports and airport infrastructure, goods, the continuation of business processes, etc.

Airports themselves are considered the infrastructure of the transport system. In addition, there are many components that make up the "airport infrastructure": runways, taxiways, aprons, terminal buildings, cargo buildings, maintenance buildings, administrative buildings, roadways, curb frontages, rental car areas, transit areas, taxi areas, limousine areas, fuelling systems, power systems, rescue units, air traffic control facilities, lighting, navigational aids, boarding devices, communications systems, security systems, parking lots, graphics, signage, landscaping, drainage, water supply, and sewer disposal.[145] Being a part of airport infrastructure, all these areas and components are subject to aviation security.

Security systems here play a "dual" role: on the one hand, they belong to airport infrastructure and should be protected along with other components; on the other hand, they are aimed at protecting all the assets from any attack. Thus, the protection of the security measures as such is an asset too, since it is a means by which to

---

142 Salter (2008) p. 248.

143 Schneier. *Beyond fear: Thinking sensibly about security in an uncertain world* (2003) pp. 14–15.

144 Lyon (2004) p. 137 and Salter (2008) p. 262.

145 Crockatt. *Airport infrastructure as an instrument for regional economic development* (2000). pp. 66–67.

ensure the continuation of security.[146] The protection of the latter is very important since criminals can try to disable security measures as to incapacitate them, e.g. cutting off cameras from the surveillance system, or altering the measure with the same goal, e.g. making cameras show recorded material to avoid detection.[147]

While it is quite clear that assets to be protected in civil aviation are priceless or of great value, other concepts and factors require more discussion.

4.5.4 *Threats*

As discussed, aviation security deals with man-made threats, and the relevant term is "acts of unlawful interference". Although the latter is broader than acts of terrorism, aviation security measures focus on preventing specifically this threat. It is the risk of terrorism that is the driver of most security measures. In this section, the following issues will be discussed: civil aviation as a target for terrorists, modern features of terrorism, "profiles" of the selected states in terms of terrorist threats, and other threats.

4.5.4.1 Civil Aviation as a Target for Terrorists

"Terrorists, criminals, and hostile nation-states have long viewed aviation as a target for attack and exploitation".[148] It is believed that no other industry has ever been under such a violent and sustained terrorist threat and other criminal attack as that endured by the aviation industry.[149] As discussed, the key point of terrorism is to *frighten* – not to achieve anything through the action in itself, but to provoke a reaction to it. Thus, blowing up a plane is not the real goal – only a tactic.[150] But what makes the civil aviation attractive to terrorists? Among the features, one might mention:

- Powerful fear factor
- Media attention
- Easy political/demographic selection
- Symbolism
- Economic driver

Already in the 1970s, a number of technological factors emerged affecting the terrorists' choice: live television as the "stage" for terrorist demands, increased air

[146] COPRA (2013).

[147] *Ibid.*

[148] DHS. National Strategy for Aviation Security. http://www.dhs.gov/publication/national-strategy-aviation-security

[149] Alexander and Sochor (1990) p. 21.

[150] Schneier (2008) p. 2.

travel, very sophisticated and advanced weapons.[151] Moreover, aviation provides a powerful symbolic target; it provides a unique multinational stage; it has enormous economic consequences for both the carrier and the targeted nation; it can create political embarrassment for the integrated targets; and it is a useful tool for revenge.[152] Some commentators also note the symbolic connection of civil aviation targets and other conflicts or interests, especially political, such as Middle Eastern geopolitical problems, ambitions of the USA as superpower etc.[153]

Every single terrorist act in aviation, even if failed, acquires worldwide attention. Air transport is very vulnerable to terrorist attacks, even a minor explosion on board the aircraft almost always leads to disaster, representing a huge threat for lives of passengers and crew, both in the air and on the ground. Thus, aviation provides a powerful fear factor. Hundreds of people can be on board aircraft at the same time, i.e. at a limited place with no way out before landing, while the airports typically are crowded with people every day.

The paradox is that attacks on other objects (other means of transport, different public places, schools, theatres, sport arenas, etc.) may attract the same worldwide attention, create as much fear and cause similar results in general. In comparison with attacks on aviation, however, they are much easier (and cheaper) to be carry out, mostly due to comparatively weak security measures.[154]

Moreover, if the entire transportation sector is taken into account, the amount of "aviation" attacks and victims thereof are not prevalent: e.g. between 9/11 and 2011, there were 75 terrorist attacks on aviation, resulting in 157 deaths, and nearly 2000 terrorist attacks on trains and buses, resulting in around 4000 fatalities.[155]

Thus, it can be argued that other targets are usually much more vulnerable. Despite this, aviation is assumed to have priorities in terms of security, acquiring the most enhanced security methods and technologies and corresponding security budgets. It is financed and equipped in this regard far better than other targets of terrorism.

Apparently, security will not win if terrorists simply switch to other targets. There are different ideas about this, including providing similar security standards for all transport modes[156]; in some countries, various objects are slowly getting more security. However, apparently, raising security levels for other means of transport to that of aviation would be unrealistic in the near future. Another question is whether

---

[151] Manning (1995) p. 508.

[152] Harrison. *International aviation and terrorism* (2009) p. 49.

[153] Salter (2008) p. 246.

[154] Cf. theatre hostage crisis in Moscow, 2002; school tragedy in Beslan, 2004; youth camp in Norway, 2011; Boston Marathon, 2012; Volgograd bus, 2013; different metro, cars and train attacks.

[155] Jenkins. *Aviation Security: After Four Decades, It's Time for a Fundamental Review* (2012) p. 1 giving statistics from the Mineta Transportation Institute's database of attacks on transportation targets.

[156] For instance, European airlines complain about different security costs in comparison with other transport modes.

this is needed at all, and whether the aviation sector really needs the security that is in place today.

Nevertheless, despite all the advanced measures, the aviation security experts state that operations against aviation are still relatively simple. Airlines move on a scheduled basis in predictable geographical locations. Thus, they provide easy political and demographic selection. Most importantly, the air transportation sector faces difficulties in providing adequate physical, personal, and operational security.[157] As will be shown in the next section, testing via undercover agents proves the vulnerability of the systems, in particular, the inability to detect fake bombs.

In addition, there is an opportunity, using the aircraft, to escape to a state granting the terrorists asylum.[158] It is also possible to use the misbalance in national aviation security systems and attack in one part of the world with a weaker security in order to damage a particular target, which can be better secured domestically, like bombing a Russian aircraft via placing a bomb while the aircraft is at an Egyptian airport.[159]

Arguably, there may be a psychological aspect similar to sport: terrorists may choose aviation particularly because of the challenges provided by a higher security level, since when successful, the satisfaction from the attack will be much greater. It may sound appealing to find a new gap in aviation security and carry out the attack using new methods that the security experts have not yet thought of. Finally, for the airports, the increased risk factors include large areas, constant movement, a continual influx of people, a large number of personnel, etc.

As a result, although it may be arguable whether the civil aviation is a "perfect" target for the terrorists, the security measures deployed are much more sophisticated and advanced in comparison with other objects.

4.5.4.2 Modern Terrorism

Another factor is that modern terrorism has been developing and using modern technological progress and other advances too. As can be seen from the historical perspective discussed above, certain types of terrorist attacks may be more relevant than others during a certain period, both due to security measures and terrorism's contagion effect[160] (cf. hijacking in 1970s and sabotage in 1980–1990s). Today, aviation security experts indicate the following threats as emerging: Improvised Explosive Devices (IED), Chemical, Biological, Radioactive (CBR) threats, cyber threats, ground-to-air and air-to-air missiles.[161]

---

[157] Sweet (2004) p. 15.

[158] Jenkins (1986) p. 778.

[159] Airbus A321, 31 October 2015.

[160] Tønnessen. Terrorisme mot luftfarten. Foredrag NFFs Ledersamling. Forsvarets forskningsinstitutt (FFI). 3 Feb 2011. http://nholt.no/getfile.php/Filer/pdf/NFF_Luftfart_FFI.pdf

[161] See Abeyratne (2010) p. 16 and COPRA (2013).

IED implies that the explosives, including home-made, may be hidden in such ordinary items as a calculator, hair dryer, fan, lamp, tape recorder, radio cassette player, electric shaver, coffee cans, thermos, etc.[162] They contain batteries, wires and electronic components to conceal the presence of IED's power sources, circuitry and switches. The techniques used include "clutter" (the mass of legitimate components serve to obscure the bomb parts), misrepresentation (the components may be seen, but look like something else), and "dispersal" (the components of the IED are separated; once in flight, the terrorist can either assemble and initiate the device himself or use other methods for unaccompanied devices; probably the most famous plot is the "liquid explosive" of 2006).[163]

Cyber-attacks in aviation usually target communications networks and information systems.[164] In January 2012, a Saudi hacker attacked El Al Israel Airlines' website; this did not cause catastrophic results, but demonstrated that even the most security conscious of airlines can be vulnerable to cyber-attacks.[165]

A missile is a self-propelled guided weapon system of various types. Anti-aircraft weapons range from huge, self-propelled or ship-mounted launchers to MANPADS (Man Portable Air Defence Systems). Laser beams directed in order to blind the pilot may also relate to this category. However, the aviation security undertakings analysed in this research can hardly have great influence on preventing missile threats.

Modern terrorism involves special features. First of all, there is globalization – incidents have clear international consequences. International terrorism can be considered as a specific form of terrorism, which means "terrorism involving citizens or the territory of more than one country".[166] Multinational groups operate and conduct attacks all around the world. The four major organizations: Taliban, Islamic State of Iraq and the Levant (ISIL), Boko Haram and Al Qaeda and its affiliates – were responsible for 63% of all deaths from terrorism in 2013.[167]

There is enhanced cross-border cooperation between violent extremist groups "both in terms of providing support for violent activities and in communications, to inform like-minded individuals about future actions and to inspire others".[168] Communication via e-mail, forums, websites, etc. allows global strategies to be developed.

Other features of modern terrorism include the existence of both amateurs "motivated by religious enmity, blind rage, or a mix of idiosyncratic motivations" and

---

[162] Information of National Antiterrorist Committee. Countering terrorism in the Russian Federation and personal security of citizens. 13 May 2013.

[163] Cochrane. *Dispersal: A means of infiltrating bomb onto aircraft*. In: Aviation Security International. Vol. 18 (2012). p. 39.

[164] See Sect. 2.2.2.3.

[165] Atkins (2012) pp. 34–35.

[166] Title 22 of the United States Code, Section 2656f(d).

[167] Institute for Economics and Peace (2015).

[168] Europol's EU 2012 Terrorism Situation and Trend (TESAT) report.

professional terrorists and their state sponsors[169]; extremists' ideology and intangible aims (e.g. Al Qaeda, which has formed its own ideology of "holy war" on the USA, American citizens, and Jews, and which is often cited as "a necessary evil"[170]).

There is also the use of suicide attacks and unpredictable and irrational tactics in general: terrorists are ready to kill and die for their aims (cf. "rational" terrorists from the past, who wanted to live and fight, whose aims were concrete and achievable, at least in their view).[171] There is an increased threat from small groups and lone actors, including those inspired by Al Qaeda's call for "individual jihad".[172]

Moreover, a major challenge is unexpected terrorists: a new type of terrorists whose radicalization is difficult or impossible to detect, since they come from the very inside of our societies. Terrorists' recruitment and training focuses on those who are isolated, independent, unstable, indoctrinated, convinced that they are the chosen ones.[173] A famous example is a blond American woman Colleen Renee LaRose, also known as JihadJane and Fatima LaRose, a convert to Islam who was charged with terrorism-related crimes. Recent acts of terrorism have demonstrated that an educated individual of any religion or skin colour may become a radicalized terrorist, like Norwegian Anders Behring Breivik.

In addition, the terrorists are aware of law enforcement authorities' tendency to suspect Islamic-looking people and therefore try to recruit persons who are less suspicious-looking. For instance, in Russia, terrorists using different methods of programming of the mind, prefer to convert ethnic Russians to Islam and then make them suicide bombers – "cannon fodder", fulfilling a double objective: killing Russians at the hands of Russians and raising less suspicion, thus making it easier to commit the acts.[174]

There may also be "insider threats", whereby a, presumably trusted and vetted individual can become the assailant. In civil aviation, many plots have been detected involving insiders, when airport or airline employees carried out or attempted to do harm either because of personal convictions or as a result of corruption.[175] Terrorists can recruit a pilot or another crewmember, or, although having nothing to do with terrorism, they may have the same intention: to destroy the aircraft, such as suicidal pilots. Statistically, the latter happens more often than suicidal terrorists commandeering aircraft and crashing them. Examples include incidents of Royal Air Maroc, SilkAir, EgyptAir and recently – Germanwings flight 9525 of 24 March 2015.[176]

---

169 Hoffman (1999) pp. 38–39.

170 Cochrane (2012) p. 39.

171 *Ibid.*

172 Europol's EU 2012 Terrorism Situation and Trend (TESAT) report.

173 Naudin. *A personal view*. In: Aviation Security International. Vol. 20 (2014) p. 44.

174 E.g. numerous terrorist acts in Volgograd, Russia, December 2013.

175 For overview, see Baum. *Metrojet Flight KGL9268*. ASI Special Report (2015b) http://www.asi-mag.com/metrojet-flight-kgl9268-a-special-report/

176 Baum. *Germanwings 9525: the challenge of suicidal pilots behind intrusion-proof cockpit doors*. ASI Special Report on Germanwings Flight 9525 (2015a).

Another great challenge is to anticipate an "unexpected" threat. New terror tactics have rarely been foreseen by the security authorities; they are often a combination of already-known tactics, or a further development of an old tactic.[177] One of the ideas is that terrorists of the future may choose exploitation of innocent travel circumstances and innocent items as well as new modes of operation. Some potential plots include the use of children, body cavity bombs, etc. Moreover, the components for such IEDs are often legally available (i.e. the Breivik case), and it is possible to get "instructions" for their manufacture from open sources such as the Internet.

All these factors, especially conflict and instability, aspects of modern technology, a pervasive ideology, and radicalization apparently will allow the terrorists to continue to exist.[178] But, when dealing with terrorism in terms of risk assessment, one problem encountered is how to determine the level of required precision to support the decisions,[179] since terrorist activity is very difficult to assess in terms of frequency and severity.[180] There are many uncertainties in quantifying the risks; although some statistical approaches for terrorist threat prediction exist, they rely on expert judgments from security and other experts, the uncertainties can still be great.[181]

How the concept of threats and risk works in practice can be illustrated by the consideration of the perceived threat of terrorism and corresponding security regimes in the selected states. Below, mode-detailed "profiles" of the selected states in this regard will be presented.

4.5.4.3 "Profiles" of the Selected States

The USA's 9/11 attacks constitute the largest incident recorded in the Global Terrorism Database and is commonly accepted culmination of terrorist attacks ever, not only against the civil aviation or the USA. Even before 9/11, about a third of all annual international terrorist attacks were directed against US targets, opposing "American values, culture, foreign policy and actions anywhere in the world".[182] However, after being number 1 in Global Terrorism Index (GTI)[183] in 2001,[184] in 2014, it was ranked as number 30 out of 124 countries in 2014.[185]

---

[177] Tønnessen (2011).

[178] HM Government. CONTEST: The United Kingdom's Strategy for Countering Terrorism. July 2011. §1.06.

[179] Tamasi (2011) pp. 893–894.

[180] Ericson. *Catastrophe risk, insurance and terrorism*. In: Economy and Society. Vol. 33 (2004).

[181] Stewart and Mueller (2008) p. 145.

[182] Wilkinson and Jenkins (1999) p. 148.

[183] GTI is an attempt to systematically rank the nations of the world according to terrorist activity.

[184] Institute for Economics and Peace (2012).

[185] Institute for Economics and Peace. *Global Terrorism Index 2014*, 2014.

The USA is a commonly accepted "superpower" – the collapse of the former Soviet Union in 1991 left the US as the sole remaining superpower.[186] Today, with the emergence of China, this "sole" status is arguable. Nevertheless, the field of security, both national and international, together with the officially announced war on terrorism, belong to the areas where the USA works hard not only to protect itself, but also to influence other parts of the world. The significance of the US aviation industry means that US rules and regulations frequently impact international carriers and airports in other states. The researchers note that ICAO rules all originate from the USA.[187] The US often becomes the first state to introduce new forms of surveillance and security measures in the field of air transportation.

The UK is a member of EU and an influential state in the aviation and security fields. Historically, with imperial background, it has always been pro-active with regard to world politics, international policies, national security and defence.

The most notable successful terrorist event were the London bombings of 7 July 2005, when attacks on the underground and bus service injured 784 people and killed 56. It is reported that between 2005 and 2014, 40 planned terrorist attacks were foiled.[188] In general, threats concern Northern Ireland-related terrorism and international terrorism, mainly due to the UK's support of the USA in its invasions.[189] Since the 9/11 attacks, the threat level in the UK has been assessed as "severe", meaning that a terrorist attack is "highly likely."[190] In 2015, the UK was ranked number 27 in the Global Terrorism Index.[191]

As the result, the UK developed US-style strategies to combat terrorism and was the first state to introduce them to the EU. According to the UK Department for Transport, more stringent measures are considered appropriate due to the level of terrorist threat.[192] Thus, it is often a "pioneer" when it comes to surveillance and security measures such as biometrics, surveillance cameras, databases, and DNA testing.[193] Not surprising, this concerns aviation security measures too.

Russia geographically belongs to both Europe and Asia and is a natural boundary and a natural bridge between Europe and Asia, with the largest territory in the world and enormous resources. It is the successor to the former Russian Empire and the

---

[186] Wilkinson and Jenkins (1999) p. 148.

[187] Naudin (2014) p. 44.

[188] Topping. *Theresa May claims 40 terror plots have been foiled since 7/7 attacks*, The Guardian, 24 Nov 2014, http://www.theguardian.com/politics/2014/nov/24/theresa-may-london-attacks-40-terror-plots-foiled

[189] Omand. *The terrorist threat to the UK in the post–9/11 decade*. In: Journal of Terrorism Research. Vol. 3 (2012) p. 8.

[190] HM Government. CONTEST: The United Kingdom's Strategy for Countering Terrorism. July 2011.

[191] Institute for Economics and Peace (2014).

[192] The UK Department for Transport. Better regulation for aviation security. Consultation document. July 2011.

[193] Foundation for Information Policy Research, Technology development and its effect on privacy & law enforcement, February 2004; House of Lords, Constitution Committee – Second Report. Surveillance: Citizens and the State, 2009.

former "superpower", the USSR, thereby having political ambitions too. It has always paid much attention to both the aviation/airspace industry and the national security.

Some of the world's most dramatic terrorist acts in recent decades took place in Russia. The worst incident in Russian/Soviet history of civil aviation occurred on 31 October 2015, when a bomb placed on board Airbus A321 from Sharm el-Sheikh to St. Petersburg exploded, killing 224 persons. Other acts of terrorism against civil aviation include suicide bombings on board aircraft 2004 and Domodedovo airport attack 2011. Beslan School Hostage Crisis of 2004 was ranked as number 3 among the worst attacks in 2002–2011[194]; there were many other terrorist attacks.

Terrorist threats are traditionally associated with the conflict between Russia and one of its regions, Chechnya. Today, an additional factor in terrorist risks in Russia is its participation in operations against Islamic State (ISIL) in Syria. In 2015, but prior to the A321 incident, Russia was ranked number 11 in the Global Terrorism Index.[195] Not surprisingly, advanced aviation security measures are used in Russia too.

Norway is a state outside the EU but a member of the EEU and thus subject to a part of the EU regulation. According to many ratings, Norway is one of the most comfortable states in the world to live in. In contrast to the USA, UK and Russia, Norway had been considered a country with little terrorism.[196] There were quite few aviation attacks in Norway. For example, six hijackings were registered from 1979 until 2004, resulting in almost no injuries; three of them were in connection with asylum seekers.

On 22 July 2011, the biggest act of terrorism in Norway occurred, when Anders Behring Breivik, a Norwegian radicalized right-wing extremist, carried out a double attack by bombing government buildings in the centre of Oslo followed by the mass shooting at a youth political camp on an island situated 30 km from Oslo; 77 people were killed. This attack dramatically changed the views on the risk level of terrorism; in 2011, Norway was ranked number 21 in Global Terrorism Index. Although in 2014 it was ranked 44th,[197] the Norwegian Police Security Service in its Annual Threat Assessment 2015 noted continued negative developments in the threat situation in Norway with reference to different forms of extremism.

The attack of 2011 also had an impact on security decisions. Although the Norwegian prime minister stated that "The Norwegian response to violence is more democracy, more openness and greater political participation",[198] allowing some security experts to declare a Norwegian alternative to the "American cycle of threat,

---

[194] Institute for Economics and Peace (2012).

[195] Institute for Economics and Peace (2014).

[196] For more details, see Fimreite [et al.] *After Oslo and Utøya: A shift in the balance between security and liberty in Norway?* In: Studies in Conflict & Terrorism. Vol. 36 (2013).

[197] Institute for Economics and Peace (2014).

[198] The Guardian. *Norway will not be intimidated by terror attacks, vows prime minister.* 27 July 2011, http://www.theguardian.com/world/2011/jul/27/norway-terror-attacks-prime-minister

fear, and overspending",[199] the attack influenced the approach to countering terrorism, strengthening counterterrorism laws and other legislation.[200]

This discussion reveals that the states that have global ambitions and high levels of terrorism threat (USA, UK and Russia) are "pioneers" in advanced aviation security measures and deploy all the selected measures (body scanners, CCTV, and PNR). Of course, since providing the measures is expensive, necessary resources should also be in place, so a high terrorism threat alone is not enough.

In contrast to that, in Norway, CCTV systems at airports are used extensively, but without the use of biometrics or behaviour analysis. However, Norwegian airports already concluded framework agreements with body scanner manufacturers, and there are some preparations for PNR system in Norway. Apparently, the latter will be established after adoption of the proposed EU PNR system.

As mentioned above, taking into account that ICAO's policies are based on those of the USA, and the endeavours of the USA and the UK to "export" their regimes internationally, this means that these measures will probably be used by many other countries all around the globe.

4.5.4.4 Other Threats

Despite a significant terrorist threat, aviation security is not just a counterterrorist operation and it is important to identify negative intent of whatever kind and wherever it can impact upon the safety and security of the aviation operations.[201]

Due to the sector-specific features, unlawful acts, apart from terrorist acts, ordinary crimes or offences may take specific forms, the most catastrophic being such incidents as suicidal pilots mentioned above. Another specific category is disruptive passenger incidents, which commonly involve intoxicated or otherwise impaired or disruptive persons who "disturb the good order or a discipline at an airport or on board the aircraft"[202] who may, even with no particular intent, cause a negative effect on safety.

There are many other types of criminal activity related to civil aviation, but without direct effect on safety which can be detected during aviation security processes, e.g. luggage theft, extortion, human and drug trafficking, facilitating illegal migrants,[203] etc. Another question is whether the aviation security undertakings introduced primarily to counter terrorism are necessary and proportionate.

---

[199] Schneier. *How the Norwegians Reacted to Terrorism* (2012a) https://www.schneier.com/blog/archives/2012/07/how_the_norwegi.html

[200] USA Department of State. Country Reports on Terrorism 2012 – Norway. 30 May 2013. http://www.refworld.org/docid/51a86e7918.html

[201] Baum (2015a).

[202] Chapter 1 of Annex 17.

[203] Baum (2015b).

4.5.5 *Benefits*

Aviation security decisions should also take into account security measures' benefits. The latter include consideration of suitability – mainly, this concerns effectiveness – the ability of the measure to fulfil the intended purpose, and necessity – is it necessary to deploy the measure, are there alternatives capable to provide same security, what will be in case of non-deployment? These factors are the key issues for determination of proportionality of aviation security measures from both aviation security and privacy/data protection perspectives.[204] But while the approach to effectiveness is common, necessity can be seen differently: apparently, for privacy, it is important to assess if there are other measures which are less intrusive or non-intrusive but which are capable of providing the same security.[205] In terms of aviation security, the matter of deploying the measure, non-deploying, or deploying alternatives entails evaluating various factors and costs.

In addition, there can be one more benefit that bridges the gap between aviation security and privacy/data protection. Security solutions may contribute to passenger experience, including privacy interests. For instance, searches by the machines are supposed to be less intrusive in comparison with manual full body searches or strip searches (however, certain problems remain, see Chap. 6). These common issues – effectiveness and passenger experience – will be analysed in more detail.

4.5.5.1 Effectiveness

Overall, in the aviation industry, it is believed that neither the best equipment in the world nor the most advanced technology is capable of providing 100% effective security. There is an understanding that 100% effective security is unobtainable and the concepts of threat and risk assessment are aimed at finding the most appropriate and most effective security measures.

The effectiveness of technology can be measured differently. First, it can be evaluated mathematically, i.e. quite objectively, using the cost-benefit analysis. The latter is designed to measure the cost associated with avoiding the risk and to determine whether such a cost is excessive.[206]

As mentioned above, there are huge budgets spent on aviation security. According to cost-benefit analysis, in the USA the annual cost per life saved ranges from \$64 million to \$600 million, which is greatly in excess of the regulatory safety goal – societal willingness to pay to save a life – of \$1–\$10 million per life saved. This means that \$300 billion spent by the US government on protection from terrorism since 2001 – total homeland security budget – fails a cost-benefit analysis.[207]

---

[204] See Chap. 5.

[205] *Ibid.*

[206] Stewart and Mueller (2008) p. 145.

[207] *Ibid.* p. 144.

However, this does not mean that every security measure is ineffective. There may be problems with specific measures, which are expensive but have low effectiveness. For instance, the annual cost of the Federal Air Marshal Service is $180 million per life saved, i.e. this measure fails a cost-benefit analysis, while the annual cost of hardening of cockpit doors is only $800,000 per life saved, showing that this is effective.[208]

In addition, the opportunity cost of these expenditures is considerable[209]: theoretically, if the amounts paid for an "ineffective" measure could be spent on different more cost-effective security measures, saving more lives, this would mean that deploying the former measure actually reduces security.

Apparently, however, not all security measures can be evaluated like this, due to lack of information, deployment secrets, etc. The problem that arises is that factors which could measure aviation security effectiveness – the quantification of impact of successful attacks, the frequency of attacks, and the detection of prohibited items – are currently not technically feasible.[210]

However, improved testing, evaluation, and performance measurement can enhance effectiveness.[211] Testing implies that before implementing a measure, before purchasing the equipment and installing it in the airport, the technology should be thoroughly tried out in conditions close to "real". Accordingly, the results of testing should be evaluated. This practice is quite common, however; the GAO found that American TSA, while testing certain technologies, did not account for all factors affecting the systems and did not consistently evaluate the effectiveness before adopting them.[212] This is likely a concern of security organs in other states too.

Performance measurement uses indirect, measurable variables: throughput of a screening point, percentage of a total sent to secondary screening, number of prohibited items detected, technical sensitivity of detection equipment, etc.[213] For instance, for screening, a number of aspects such as detection capability and speed, user-friendliness, mobility and overall life-cycle, cost will ultimately determine the degree of success of the various technologies.[214] In this regard, GAO reports that TSA did not consistently establish performance measures that fully reflect the measure's goals. This concerned for example, a passenger prescreening program Secure Flight, where one of the goals was to accurately identify passengers on various

---

[208] *Ibid.* p. 143.

[209] *Ibid.*

[210] Salter (2008) p. 258.

[211] GAO *Aviation Security: Improved Testing, Evaluation, and Performance Measurement Could Enhance Effectiveness* (2015a).

[212] *Ibid.*

[213] Salter (2008) p. 258.

[214] Neeman. *Effectively screening people: Pat down searchers, wands, archways and portals*. In: Aviation Security International. Vol. 19 (2013) p. 15.

watch lists, but TSA did not have measures to assess the extent of system matching errors.[215]

The effectiveness of technology can be measured in terms of its ability to accurately detect threats while minimizing false alarms, i.e. constituting a trade-off between detection and false alarms. However, while false alarm rates can be easily measured in operational settings, it may be problematic to measure detection rates because of uncertainty over what may have gone undetected,[216] e.g. the number of passengers that are actual matches to watch lists are missed by Secure Flight, CCTV, etc.

The report from the Freedonia Group shows that false alarms will apparently continue to be a critical issue for the security industry.[217] It can only be imagined how many innocent people will be inconvenienced in this regard. Clearly, any minor error may cause negative consequences for a passenger being processed, having to go through additional security checks and ultimately being denied access to boarding.

In some cases, high failure rates may lead to cancellation of the measure. For instance, iris scanners installed in the UK airports for facilitating passenger identification were removed in 2012 when it turned out that every tenth passenger was rejected by scanners by technological error.[218] In other cases, despite the failure rates, measures are in operation or are planned to be used (see Special Part).

Failure to distinguish dangerous object, apart from being due to incorrect instrument reading, can be due to incorrect human judgment.[219] In reality, all technology ultimately relies on human operators; judgment of humans may involve the examination of body scanner images, resolution of metal detector alarms, physically searching baggage, monitoring CCTV, and so on.

It is very difficult for the security officers, staring for hours at monitors, to stay alert when processing thousands of passengers a day and to be able to recognize, for example, the IED among everyday electronic devices. For instance, in the USA, the screeners missed a loaded 40 calibre gun in carry-on luggage; it is reported that federal agencies conduct covert tests of airport security by carrying concealed guns, bomb parts or knives, and the failure rate in detection of these items is up to 70%.[220] Apparently, a part of this is due to human error. As for the CCTV, studies show that watching only two monitors, the operator misses 45% of the data after 10 min, and 95% of the data after 22 min.[221]

---

215 GAO (2015a).

216 Elias (2012) p. 9.

217 Miller (2013).

218 Aviaport Digest. *Airports disclaim slow scanners*. 17 February 2012, http://www.aviaport.ru/digest/2012/02/17/229881.html

219 Vegh. *Magnetic Resonance Technology: a replacement for existing airport screening system?* In: Aviation Security International. Vol. 18 (2012) p. 20.

220 Quinn. *Man boards plane at IAH with loaded gun in carry-on*, ABC News, 17 December 2010. http://abc13.com/archive/7848683/

221 Houlis. *Visibly efficient: the future of integrated CCTV* (2010) http://www.ifsecglobal.com/visibly-efficient-the-future-of-integrated-cctv/

Another problem is the finite period for security procedures; security officers may not have the required expertise to intercept the data and may have only minimal training before being posted, thus, all the advantages of the technologies may be neglected. The key solution, according to security experts, is excellent security training and constant vigilance: a combination of knowledge and alertness.[222]

A number of commentators go further and express scepticism and doubt that *all* the enhanced technologies and novelties will work. For example, according to Naudin, the protection of civil aviation consists first of all of intelligence work, while "screening as currently implemented in airports, is merely Brownian motion.".[223]

Jenkins argues that "The most significant improvements in security may have come through pilots becoming non-passive and passengers becoming aggressive towards terrorism".[224] There were some examples of hijackers being neutralized on board by heroic actions of the crew and fellow passengers, such as during the accident of 29 June 2012 in China.[225] Of course, both pro- and contra- views may be speculative.

Another interesting method by which to test the effectiveness of the measures is the use of undercover agents. This may also permit identifying the system's weakest points. For instance, in 2013, an agent with a fake bomb managed to pass through a metal detector and hand search in the USA.[226] According to GAO's report of 2015, in 67 out of 70 tests at American airports, during screening, inspectors were unable to infiltrate prohibited items, including dummy IEDs, indicating a 95% failure rate.[227]

As discussed above, the process of effectiveness determination can hardly be democratic. It is the states and security organs that decide which security measures to deploy, and in what amount; the general public is excluded from the discussion. This is understandable, but as the result of such secrecy and lack of transparency, it becomes very difficult to determine objectively whether a security measure is effective or not, and, as a result, to provide an objective proportionality test (see Chaps. 5 and 7).

The researchers argue that people should insist that the effectiveness of security measures be proved and that it must be the state's obligation to prove effectiveness of the measures.[228] The Article 29 Data Protection Working Party suggests the evi-

[222] Cochrane (2012) p. 41.

[223] Naudin (2014) p. 44.

[224] Jenkins (2002) p. 69.

[225] Six men boarded in a Tianjin Airlines flight and, pretending to be disabled and hiding weapons in their crutches, tried to hijack the aircraft. See: Baum. *Terror in Bulgaria, Hijacking in Xinjiang: concern in Reykjavik, but joy in London*. In: Aviation Security International. Vol. 18 (2012b) p. 1.

[226] AbcNews. TSA Agent Gets Past Newark Airport Security With Hidden Fake Bomb, 8 March 2013,http://abcnews.go.com/blogs/headlines/2013/03/tsa-agent-gets-past-two-security-layers-with-hidden-fake-bomb/

[227] Baum (2015b).

[228] Waldron. *Security and Liberty: The Image of Balance*. In: Journal of Political Philosophy. Vol. 11 (2003) p. 210.

dence test: the needs of aviation security should be substantiated by relevant and sufficient reasons, the proposed measures must be based on evidence-based research, statistics, forecasts etc., justifying the measure, only this way the measure can be proportionate.[229] However, these reasonable suggestions may be difficult to realize, not only due to secrecy, but also to the unavailability of adequate means by which to measure aviation security effectiveness and other difficulties as noted above.

4.5.5.2 Improving Passenger Experience

A positive development is that almost all aviation security stakeholders representing international organizations – the aviation industry and security providers, states, national law enforcement and security agencies – take part, to various extents, in the security-versus-privacy debate, proposing standards, requirements, guidelines, expressing positions, etc. All seem to be very concerned that the passenger's experience should be improved. Evaluating and improving passenger experience, which includes evaluation of the impact of aviation security measures on passengers, is proposed to be one of the principles, priorities, drivers and trends in future aviation security. For instance, one of the primary goals of the SmartS project developed by IATA in cooperation with governments is improving the passenger experience by, among other things, "replacing intrusive and time consuming screening with more pleasant technology solutions".[230]

The EU research project COPRA, which took into account opinions of many different aviation security stakeholders, substantiated the need for quicker process time of checks, privacy concerns, and the demand for safe, comfortable and less intrusive checks among drivers and trends in future aviation security. It suggested that future aviation security concepts should, among other things, consider the effect of security measures for relevant stakeholders (including social and ethical aspects of security measures), be seamless, quick, comfortable, acceptable and safe for all stakeholders.[231] "Comfortable" here means a quick flow and no intrusiveness, user friendliness and good service; "safe" means minimizing the impact on health, environment, privacy and damages to goods. It can be noted that privacy is both part of comfortable (as a part of "not intrusive", both physically and digitally) as well as safe, where it is mentioned explicitly, as this entails the information of each stakeholder is not misused and is protected against misuse by thirds.[232]

Accordingly, evaluation of passenger experience, including impact of aviation security measures on passengers, should be taken into account before any security measure is adopted. It should be an integral part of any assessment of new technologies, processes or systems, along with such factors as security effectiveness, opera-

---

229 Article 29 Data Protection Working Party (2014a).

230 IATA Press release, 5 Mar 2013. http://www.iata.org/pressroom/pr/Pages/2013-03-05-02.aspx

231 COPRA (2013).

232 *Ibid.*

tional efficiency, cost-effectiveness, and legal considerations.[233] Of course, as will be further analysed in this work, certain improvements in one area can lead to negative trends in other areas. However, "an overall better passenger screening process should be an attempt to optimise and improve upon as many fields as possible while acknowledging some necessary trade-offs and perhaps minimising such necessary negative trends".[234]

An appropriate process for the evaluation of impact of aviation security technologies and procedures on passengers' privacy and data protection is the Privacy Impact Assessment (PIA). Establishing and enforcing this strategy as a part of the aviation security proportionality principle – evaluating the security measures' impact on passengers, in particular, on their rights to privacy/data protection – may directly contribute to the security versus privacy dilemma.

4.5.6 Conclusion

For the purposes of this research, the connection between the threats, risks, and advanced security measures leading to privacy concerns is particularly important. These components can be seen in two dimensions.

First, obviously, the more clearly perceived terrorist threats are, the easier it is to justify a particular advanced security measure. Thus, the concept of threats and risks is generally used to explain the need for balancing and compromising human rights in the name of security, to substantiate and justify the use of enhanced measures and techniques.

In general, the formula can be presented as follows: (i) threats exist, in particular, terrorism, which jeopardizes the security of civil aviation, (ii) it is necessary to safeguard people from these threats by adopting definite aviation security measures, (iii) the risks of the negative consequences of the threats, especially the amount of possible victims and damages, should justify the security measures, even if the latter cause privacy and data protection concerns.

At the same time, the impact is not one-sided. As will be further analysed in this work, privacy/human rights risks may lead to cancellation or amendment of a measure with serious human right concerns, even if a measure is promising in security terms.

Secondly, in order to evaluate the risks and establish how much security is needed, a number of concepts and factors are used. As a result, when planning to implement a measure or a technology, aviation security providers use the proportionality principle extensively and have to balance threats and risks; costs of assets and costs of harms; impact of risks and the cost of security solutions; the costs of security solutions and their benefits; the benefits and threats.

---

233 Neeman (2013) p. 12.

234 *Ibid.*

Ideally, the benefits derived from the security solutions should be proportionate to all these factors: threats, costs, etc. Another question is how the security decisions are made, who takes the decisions, and what interests lie behind the decisions. As discussed, this is not a democratic process, and it is the security organs who decide about most of the factors above, often in secret, making it difficult to objectively evaluate such factors as risks, costs and effectiveness. Moreover, aviation security equipment manufacturers represent a powerful industry and usually have a lobby in the state organs and are able to influence decisions. Thus, although aviation security implies extensive usage of the proportionality principle, it is a question of whether today's security decisions are proportionate to all the factors indicated above.

The proportionality principle seen from a privacy/data protection point of view will be thoroughly discussed in Chap. 5 and then applied in Chap. 7 to evaluate whether the benefits of the selected aviation security measures are proportionate to the privacy/data protection/other human rights risks. In general, however, a number of aviation security factors taken into account for evaluation of proportionality of security solutions overlap with factors used within the proportionality principle seen from a privacy/data protection perspective.

Specifically, taking into account the perceived enhanced level of threats in civil aviation, the aviation security measure should be suitable and necessary. In addition, if a measure can improve the passenger's experience, including privacy, it can be considered as one of the security solution's benefits. Similarly, privacy/data protection risks can be considered as part of the security solutions' costs. However, while both the security solution's benefits and privacy/data protection risks are important, benefits and other aviation security aspects are the most important.

4.6 Principles of Aviation Security

The notion of principles for this research will be elaborated in Chap. 5. In short, principles of aviation security present axioms which establish the basis for aviation security regimes and have basis in law, thus, they are legal principles.

A number of general legal principles – legality, proportionality and non-discrimination apply in aviation security as in any other sector. They are discussed in other parts of this work.[235] Here, more specific aviation security principles will be analysed.

In contrast to legal principles of privacy and data protection in this research, which should be established by international or EU law (by virtue of the fact that the EU has acquired regulative influence in the data protection field worldwide), the aviation security principles in this research must have a basis in international law.

The selected aviation security principles belong to general principles of international air law.[236] In addition to international legal sources, these principles may have

[235] Chapter 5; Sect. 4.5 and Chap. 5; Chap. 3 respectively.

[236] Bordunov (2007) p. 39.

basis in the national law too. For instance, the most important principle of international air law – the principle of state sovereignty over the airspace – is only recognized by the Chicago Convention, since it is already enshrined in the national legislation of many countries. The same can be said about the principle of ensuring the safety and security of civil aviation: it is a basic principle of international law, while national air laws are based on the mandatory requirements to ensure the safety of civil aviation.[237]

In general, the principles of air law fulfil the following tasks: (i) they unite the norms of international aviation security law provided by various institutions into a single legal system; (ii) when new issues require legal regulation, the principles serve the criteria whether the newly adopted rules are lawful; (iii) they help in assessing the actions of operators of aircraft flying over the territory of the state[238]; (iv) they serve as criteria of legality of norms of bilateral agreements.[239]

The following principles of air law play important role in aviation security: the principle of ensuring safety and security of civil aviation, the principle of state sovereignty over the airspace, and the principle of freedom of flights outside the national territory.

The principle of ensuring the safety and security of international civil aviation is the core principle. It includes, first, the sovereign right of the state to adopt technical, operational, organizational, informational and air navigation measures ensuring safety of flights, and secondly, the duty of the state to maintain a high level of security, to adopt measures protecting from acts of unlawful interference against civil aviation.[240]

The 1944 Chicago Convention states that its main objective is to develop international civil aviation "in a safe and orderly manner" (Preamble). The Charter of the United Nations (1945), although it does not contain provisions dealing specifically with the security of civil aviation, is one of the most salutary international legal documents in the area of civil aviation security,[241] and the principles of security are established in Articles 1(2), 2(4), 33, 55, 56. For example, Article 1(2) provides the principle of equal rights and self-determination of peoples. On the national level, the states must provide appropriate measures according to this principle, such as regulating airworthiness of the aircraft, establishing technical requirements and standards for airports, defining the procedure for investigating aviation accidents and incidents, providing aviation security, etc.

The principle of ensuring the security of civil aviation constitutes the basis of aviation security regimes: the states have both the right and obligation to ensure

---

[237] *Ibid.* p. 40.

[238] E.g. if, instead of following the prescribed route, foreign aircraft follow a different route, this can be considered a violation of the principle of security of international civil aviation. A flight of a foreign aircraft without the permission of the sovereign state can be considered as invasion and violation of the principle of sovereignty over airspace.

[239] Bordunov (2007) p. 39.

[240] *Ibid.* p. 46.

[241] Abeyratne (2010) p. 212.

aviation security, i.e. establish various procedures and requirements. The wording "ensuring security" means that security measures should be effective. Apparently, this implies the usage of modern technologies, tactics and strategies, a part of which may have privacy and data protection concerns. As a result, it can be said that this principle in many ways "creates" the security-versus-privacy dilemma.

The question that arises is how far the states may go in these measures, both quantitatively, i.e. on which territories the state can deploy the measures and exercise its jurisdiction, and qualitatively: exactly which aviation security measures should be deployed. The other principles may clarify these issues.

The Chicago Convention rests on the notion that states are sovereign over their land and airspace.[242] The principle of state sovereignty over the airspace, among other things, constitutes the legal basis for the national security of the state; it protects its sovereign rights and legitimate interests from illegal attacks from the outside.[243]

Another principle is the principle of freedom of flights outside the national territory. It usually applies to the airspace over the exclusive economic zone, the high seas or over international territories.[244] However, the freedom is not absolute: the flights must comply with the applicable rules.

These two principles are closely connected with the jurisdiction (for the security purposes, it usually implies judicial and administrative powers to solve cases of unlawful interference on board the aircraft). Jurisdiction can be determined according to principles established by international,[245] regional[246] and national law.[247] The following areas of jurisdictions can be distinguished: (i) territorial jurisdiction: the state has sovereignty over the airspace above its territory, (ii) national: applicability of the law of the state registration of the aircraft, and (iii) mixed, combining to a definite extent both of these principles.[248]

---

[242] See Chicago Convention Article 1 – Sovereignty and Article 2 – Territory.

[243] Bordunov (2007) p. 40.

[244] Territories which do not belong to a particular state have special, mainly international status (Antarctica etc.).

[245] Some rules can be found in international conventions such as the Convention for the Unification of certain rules relating to international carriage by air (Warsaw Convention, 1929), Convention on Damage Caused by Foreign Aircraft to Third Parties on the Surface (Rome Convention, 1952), Tokyo Convention (1963), Hague Convention (1970), Montreal Convention (1971) discussed in Chap. 2.

[246] E.g. Article 9 of Council Framework Decision of 13 June 2002 on combating terrorism (2002/475/JHA) provides that Member States must establish jurisdiction over terrorist offences where the offence is committed on board a vessel flying its flag or an aircraft registered there, and other rules of jurisdiction issues.

[247] E.g. Article 1 of Russian Air Code of 19 March 1997 N 60-FZ provides that 1.The Russian Federation shall enjoy a complete and an exclusive sovereignty with respect to the air space of the Russian Federation. 2. The air space of the Russian Federation is the air space over the territory of the Russian Federation, including the air space over the inland waters and over the territorial sea.

[248] See Chicago Convention Article 2 – Territory, Article 12 on High Seas read in conjunction with Annexes 2, 6, 11, 12 of the Chicago Convention.

As a general rule, national jurisdiction – the law of the state where the aircraft is registered – has jurisdiction over it when it flies in the high seas or over international territories; aircraft landed on foreign territory falls under the jurisdiction of the state of landing. For aircraft in the airspace of a foreign country, jurisdiction of the state or the state registration of the aircraft can be applied.

At the same time, the territorial sovereignty and jurisdiction issues are relevant in cases if personal data necessary for the security are stored in a foreign jurisdiction. Obviously, access to that data may require legal authority. While in current international treaties interpretations are fragmented, it is proposed that access can be seen as an extraterritorial application of jurisdiction to enforce, thus, in order to avoid breaching the other state's sovereignty, such activities should have explicit legal grounds.[249]

In general, the state has the right and obligation to ensure aviation security on its territory and in its airspace as well as on board all aircraft registered in this state. The second issue – qualitative – is more complicated. The proportionality principle discussed above with reference to security decisions is the most applicable.

At the same time, as will be discussed below, some states, such as the USA, pressure other states to adopt a level of security similar to their own. Moreover, as noted above, terrorists may exploit security gaps in foreign airports which are "weaker" in terms of security, serving an argument for "exporting" aviation security by such states as the USA and UK. Thus, in this sense, the principle of ensuring the safety and security of international civil aviation can be seen more as an obligation rather than a right, and ambitions and influences of some particular states are being spread beyond their state jurisdictions.

4.7 Aviation Security Regulation

4.7.1 International

As soon as the first aircraft crossed the sovereign state borders in the air, it became clear that aviation by its very nature is an international sphere. Since aircraft fly from one national jurisdiction to another, this sphere required international attention and appropriate international regulation. The need arose to develop international air law.

International air law can be defined as "a subsystem of the system of international law, which is aimed at legal support of international air transport and the maintenance of safe, effective and economical international air services".[250] Consequently, aviation security law is an essential part of air law, but with its specific features: e.g. it is closely integrated with security law, criminal law, etc.

---

[249] See more details: Osula. *Transborder access and territorial sovereignty*. In: Computer Law & Security Review (2015).

[250] Bordunov (2007) p. 15.

International harmonization of regulations is cited as one of the drivers and trends in future aviation.[251] Since many of modern aviation security issues have global character, it is believed that these issues require global cooperation between countries, agencies and all industry stakeholders, while "sovereignty will need to be left at the door".[252] Thus, the role of international aviation security regulation is becoming more and more important.

Multilateral conventions constitute the primary source of air law,[253] including aviation security law. They include a special group of conventions on acts of unlawful interference: the 1963 Tokyo Convention,[254] the 1970 Hague Convention (or the "Hijacking Convention"),[255] the 1971 Montreal Convention (or the "Sabotage Convention")[256] with Montreal Supplementary Protocol 1988,[257] Montreal Convention 1991,[258] and finally, 2010 Beijing Convention and Protocol[259] adopted as a reaction to new forms of threats: chemical, biological, radiological, and nuclear (CBRN) materials and the use of aircraft as a weapon.

These conventions constitute the globally recognized basis for the fight against different acts of unlawful interference in civil aviation, but deal mainly with responses to acts of unlawful interference rather than preventive issues: they obliged the states to categorize hijacking, sabotage and other specific acts of unlawful interference against aviation as crimes and provided necessary definitions, to establish jurisdiction over offences and crimes, dealt with extradition and prosecution issues. The conventions also served as a basis for the development of Annex 17 (see below) and its amendments. Nevertheless, it is impossible in practice to make the countries follow the extradition and prosecution requirements. The commentators argue that the conventions did not provide for and guarantee the trial of an offender, and did not specify adequate punitive measures.[260]

The most important international convention on civil aviation in general is the Chicago Convention of 1944.[261] It established International Civil Aviation

[251] COPRA (2013).

[252] Brooks. *A new approach for a new breed of threat?* In: Aviation Security International. Vol. 21 (2015) p. 31.

[253] Diederiks-Verschoor. *An introduction to air law* (2006) p. 4.

[254] Convention on Offenses and Certain Other Acts Committed on Board Aircraft of 14 September 1963.

[255] Convention for the Suppression of Unlawful Seizure of Aircraft of 16 December 1970.

[256] Convention for the Suppression of Unlawful Acts against the Safety of Civil Aviation of 23 September 1971.

[257] Protocol for the Suppression of Unlawful Acts of Violence at Airports serving International Civil Aviation of 24 February 1988 (Montreal Protocol).

[258] Convention on the Marking of Plastic Explosives for the Purpose of Detection of 1 March 1991.

[259] Convention on the Suppression of Unlawful Acts Relating to International Civil Aviation of 10 September 2010 (Beijing Convention) and the 2010 Beijing Protocol to the 1971 Hague Convention on the Suppression of Unlawful Seizure of Aircraft.

[260] Manning (1995) p. 514.

[261] Chicago Convention, 7 December 1944, ICAO Doc 7300/6. The Convention is now in its ninth edition.

Organization (ICAO) – a universal intergovernmental organization charged with coordinating and regulating international air travel. The ICAO consists of government representatives of its member states, which currently number 191.[262] Being the international forum for states and stakeholders including regional organizations and regional civil aviation bodies, the ICAO is unique in its capability to facilitate the processes of establishing effective cooperation and partnerships.[263]

With regard to aviation security, the ICAO's work is carried out in three areas: policy initiatives, audits of the organization's contracting states, and assistance to states that are unable to address serious security deficiencies highlighted by audits. The ICAO obliges member states to ratify the international conventions on security, assist in the development of relevant conventions, annexes, standards and recommendations, and to enact appropriate supporting legislation. It is believed that the states should co-operate with each other and follow the internationally established principles and standards. All members, in addition, must oversee the ICAO policy's implementation.

As for Chicago Convention, "[A]lthough no explicit mention was made of the security of international civil aviation since such acts were unknown at that time, several States made reference to the significance of the Convention to security and safety of air travel".[264]

The Preamble to the Convention and some provisions indicate clearly that security of civil aviation is one of its main objectives. The Preamble states that "the future development of international civil aviation can greatly help to create and preserve friendship and understanding among the nations and peoples of the world, yet its abuse can become a threat to the general security". The Preamble also states that one of its aims is to develop the international civil aviation "in a safe and orderly manner".

According to Article 44, the aims and objectives of the ICAO are to develop the principles and techniques of international air navigation and to foster the planning and development of international air transport so as to, inter alia, meet the needs of the peoples of the world for safe, regular, efficient and economical air transport; promote safety of flight in international air navigation.

However, the Convention does not contain any details on the actions states are required to perform with regard to security. In order to provide such details, and to harmonize regulations and their application, ICAO adopts additional legal instruments.

The first category is Standards and Recommended Practices (SARPs), which can be considered "auxiliary" sources of international air law.[265] The key SARP is Annex 17 to Chicago Convention – Security (Annex 17). First adopted by the ICAO in 1974, it was revised and updated several times in response to the emergence of new threats and trends. This document establishes the foundations of an interna-

[262] ICAO. Member States. http://www.icao.int/about-icao/Pages/member-states.aspx

[263] Benjamin *ICAO and Civil Aviation Challenges* (2009).

[264] Abeyratne (2010) p. 210.

[265] Bordunov (2007) p. 57.

tional aviation security program as well as minimum aviation security standards expected from all member states.

The legal nature of standards and recommended practices is different. In relation to standards, the ICAO provides that states must conform to them in accordance to the Convention[266] thus they are mandatory. Long practice of application of the standards shows that they became established customs, and the states do assign the ICAO standards a mandatory character.[267] However, since ICAO has no enforcement powers over its members, provision is made for contracting states to opt out of regulations that they find unacceptable,[268] with the obligation to report deviations: "in the event of impossibility of compliance, notification to the Council is compulsory under Article 38 of the Convention.".[269]

Recommended practice is defined by Annex 17 as "Any specification… the uniform application of which is recognized as desirable in the interests of safety, regularity or efficiency of international air navigation, and to which Contracting States will endeavour to conform in accordance with the Convention". Thus, in comparison with standards, they are recommendations by nature. The same refers to ICAO's guidance materials and guidelines combining aviation security and privacy issues such as Guidelines on PNR.[270]

An example of guidance material and supplementing Annex 17 is the Aviation Security Manual (Doc 8973 – Restricted).[271] It "provides detailed procedures and guidance on aspects of aviation security and is intended to assist States in the implementation of their respective national civil aviation security programmes required by the specifications in the Annexes to the Convention on International Civil Aviation".[272] In other words, it provides "guidance material on achieving civil aviation security objectives through application of SARPs".[273] Its five volumes include national organization and administration; recruitment, selection and training; airport security; preventive security measures; and crisis management and response in connection with acts of unlawful interference.

Some soft law in aviation security derives from various non-governmental industry organizations such as IATA or ACI.

IATA is an international trade body created by a group of airlines. "IATA was founded... to be a voice of the civil aviation industry relative to many aspects of aviation issues".[274] Similarly, ACI is a trade body for airports, accordingly, "the voice of the airports".[275] Today, IATA and ACI's security initiatives include recogni-

---

[266] Annex 17 p. (ix).

[267] Bordunov (2007) p. 57.

[268] Wilkinson and Jenkins (1999) p. 85.

[269] Annex 17 p. (ix).

[270] ICAO Document 9944 – Guidelines on Passenger Name Record (PNR) data of 2010.

[271] Seventh edition, 2009.

[272] Annex 17 p. (viii).

[273] *Ibid.* § 2.1.4.

[274] Harrison (2009) p. 29.

[275] ACI Policy and Recommended Practices Handbook. Seventh edition. November 2009. Page i.

tion of equivalence (One-stop security) – a process avoiding a second security screening for transfer passengers, SmartS, Security Management Systems (SEMS), Facilitation and Passenger Data and other projects.[276]

All these projects are aimed at harmonization and application of common standards in aviation security globally and require joint efforts of many different stakeholders, but problems remain. For instance, one-stop security concept is already partly realized, e.g. within the EU and, unilaterally, for flights from the USA to Europe; between Canada and the USA. However, currently, in other regions, precious time, efforts and resources are being wasted by re-screening transfer passengers.[277]

Overall, global harmonization of aviation security regulation is a challenge. The problem is poor cooperation between governments and the fact that the ICAO's instruments have no enforcement mechanisms. All ICAO member states are different, thus it is not surprising that actual implementation of security policies and enacting of supporting legislation can differ greatly. Audits of member states which are conducted periodically by ICAO detect deficiencies and non-compliance with the ICAO's standards.[278] In addition, the implementation of the ICAO's recommendations is quite expensive, and not all the states are able to enforce them. Substantial differences in technical, economic, and threat environments of the ICAO member states result in chaotic application of what were ostensibly to be universal security standards.[279]

Clearly, it is much more difficult to solve privacy and data protection concerns (which are quite common) in differing regimes and without common security standards. This will be further explored in the Special Part.

4.7.2 EU

In contrast to global endeavours on harmonization which suffer a number of problems, more successfully harmonized practices can be noted in the EU. Before 9/11, aviation security was addressed essentially on a national level. In 2002, the EU adopted its first common rules in the field of civil aviation security.[280] Since then, the EU legislation on aviation security has developed. This regulation was revised substantially and updated in 2008, and then overhauled by a new framework– Regulation

---

[276] IATA. Aviation security and facilitation. http://www.iata.org/whatwedo/security/Pages/index.aspx

[277] IATA. One-stop Security. http://www.iata.org/whatwedo/security/Pages/one-stop.aspx

[278] E.g. The Ministry of Transport of RF. The ICAO Commission completed the inspection on the organization and oversight of aviation security in the Russian Federation. 18 March 2011. http://www.mintrans.ru/news/detail.php?ELEMENT_ID=15869&phrase_id=1289896

[279] Harrison (2009) p. 4.

[280] Framework Regulation (EC) No 2320/2002 of the European Parliament and of the Council of 16 December 2002.

No 300/2008.[281] The new Regulation simplified, harmonized and clarified the existing rules.

The common standards are supplemented by detailed implementation measures, most notably through Regulation No 185/2010[282] and Commission Regulation (EC) No 272/2009.[283] The EU regulatory security regime applies, in addition to the EU Member States, to a number of states that are not EU members, but are members of EFTA[284] and EEA (Switzerland, Norway, Liechtenstein and Iceland).

The Chicago Convention was used as the basis for EU's security regime. The latter covered most of the ICAO's standards on aviation security. The key objective of Regulation No 300/2008 is to provide a "common interpretation of Annex 17" within Europe. Another instrument which is often considered the actual basis for the Regulation was ECAC's Document 30.[285] In contrast to both Annex 17 and Document 30, the EU regulation has a binding effect.

According to Regulation No 300/2008, each EU Member State must designate a single appropriate authority to be responsible for the implementation of the common basic standards. In addition, it must devise and implement several security programmes. The first one is a national civil aviation security programme[286] that would define responsibilities for the implementation of the common basic standards and describe the measures required by operators and entities for this purpose. The second one is national quality control program to monitor compliance with this regulation and with the country's national civil aviation security program.

Each operator and entity must devise and implement one of the following programs: (i) airport security program to set out the methods and procedures to be used by the airport operator to ensure compliance with this regulation and with the country's national civil aviation security program; (ii) air carrier security program to set out the methods and procedures to be used by the air carrier to ensure compliance with this regulation and with the country's national civil aviation security program; or (iii) entity security program to set out the methods and procedures to be used by the entity to ensure compliance with this regulation and with the country's national civil aviation security program.

---

[281] Regulation (EC) No 300/2008 of the European Parliament and of the Council of 11 March 2008 on common rules in the field of civil aviation security.

[282] Commission Regulation (EC) No 185/2010 laying down measures for the implementation of the common basic standards on aviation security.

[283] Commission Regulation (EC) No 272/2009 supplementing the common basic standards on civil aviation security laid down in the Annex to Regulation (EC) No 300/2008.

[284] The European Free Trade Organization was established in 1960 as a trade-bloc alternative for European countries who were unable or unwilling to join the, at the time, EEC, the predecessor of the contemporary EU.

[285] European Civil Aviation Conference (ECAC) is established by the ICAO and the Council of Europe. The main security instrument is Document 30 – ECAC policy statement in the field of aviation facilitation (11th Edition, December 2009). Similar to Annex 17, it establishes the foundations of an EU aviation security programme as well as minimum aviation security standards expected from all Member States.

[286] Article 10.

The EU legislation provides a list of allowed methods of passenger screening (see Chap. 6). However, along with the idea of having common policies in the EU, it is still obvious that Member States differ in terms of the levels of threat risk, technical development, economic situation and other factors. Thus the EU framework provides two possibilities for Member States to practice measures which differ from the commonly accepted and recommended.

First, the EU countries are allowed to apply more stringent measures than the common basic standards, following a risk assessment and providing that those measures are "relevant, objective, non-discriminatory, and proportional to the risk being addressed".[287] EU countries must inform the Commission of these measures and the Commission will then communicate this information to the other EU countries.

Secondly, Member States may exercise the right to conduct trials of new technical processes or methods for a maximum period of 30 months.[288] A "trial" in the sense of the EU aviation security legislation is conducted when a Member State agrees with the Commission that it will use a particular means or method not recognized under the terms of the legislation to replace one of the recognized security controls, for a limited period. The term does not, in the legal sense, apply when a Member State or entity is conducting an evaluation of a new security control deployed in addition to one or more of those already covered by the legislation.[289]

For example, despite the absence of common EU regulations on scanners before 2011, the actual use of the machines started on the EU territory much earlier either by exercising the members' right to apply security measures that are more stringent than existing EU requirements or by exercising their right to conduct trials. In these cases, the Member States decided if and under what conditions to use scanners.

The options to apply more stringent measures may be considered both positive and negative. Positive because they may take into account the individual countries' needs. They may be helpful in transferring from the historically developed reactive approach in security towards a risk-based approach, since the latter implies more national autonomy in applying security measures.[290] The negative factor is that these options made a genuine One-stop security concept impossible, creating non-uniform operating conditions for the Member States, therefore not benefiting passengers.[291] It entailed the possibility of some states to abuse the regime, to enact measures which do not satisfy the common standards, as it is unclear which actions may be put against a state which applies measures not satisfying the requirements of being relevant, objective, proportional to the risk etc.

To summarize: in the EU, a better harmonization in comparison with the international level can be noted. The EU presses hard to ensure that its Member States have

---

[287] Regulation No 300/2008 Article 6.

[288] Commission Regulation (EU) No 185/2010, Chapter 12.8.

[289] Report from the Commission to the European Parliament and the Council 2010. Annual Report on the Implementation of Regulation (EC) No. 300/2008 on Common Rules in the Field of Civil Aviation Security/*Com/2011/0649 Final.

[290] Poole (2008) p. 8.

[291] European Parliament, Resolution on aviation security of 6 July 2011 (2010/2154(INI).

same level of security at the airports, with common privacy/data protection guarantees. This concerns, for instance, body scanners, as it will be discussed in the Special Part. However, due to the differences between Member States, and gaps in privacy/ data protection regulation related to particular aviation security technologies (see Chap. 2), some challenges remain.

4.7.3 National

"Civil aviation is one of the most regulated industries in the global economy, internationally through the ICAO, and nationally through individual states' agencies…"[292] Despite the needs and trends of global regulation, today, concrete rules on aviation security regimes have to be adopted and enforced by the national regimes.

All the considered states are ICAO members. Specific relations occur between the international aviation security law and domestic law. First, it is common that national law gives priority to international treaties to which the state is a party over national law.[293] Secondly, in the legal system of each state there are national laws for adopting and implementing international rules into the national legal system. The need of integration of the international and domestic law is also due to the nature of many norms of the Chicago Convention giving references to the national law of states.[294]

However, as discussed, there are no enforcement mechanisms for the international standards. On the one hand, this creates difficulties and differences between member states. On the other hand, the actual implementation belongs to individual states under the "host state" doctrine and each state is free to flesh out the respective details.[295] Thus, the government of each state has a paramount role to play in its own area of responsibility. This constitutes "regulatory duopoly"[296] where the relationship between international and national regulations is an important aspect.

In addition, the EU/EEA Member States – here, the UK and Norway – are obliged to follow the common European rules on security, which are based on the ICAO standards. Again, despite the fact that these are obligatory, differences may occur both in implementation and in detailed legislation.

On the national level, there has been a long-running debate on who must be responsible for aviation security (including its funding): the states, the airports, or perhaps the airlines?

As discussed above, the ultimate target of terrorism is not an airline or an airport – it is the state, its government and its policies, and aviation security measures

---

[292] Harrison (2009) p. 4.

[293] See e.g. Article 3 of Russian Air Code of 19 March 1997 N 60-FZ.

[294] Bordunov (2007) p. 32.

[295] Harrison (2009) p. 4.

[296] *Ibid.*

constitute a part of national security, counter terrorism strategy of the states, hence, the latter must be responsible for aviation security and its funding.[297]

However, this does not mean that aviation security must be provided by state bodies alone. As concluded at the ICAO Conference in 2002, aviation security is the responsibility of states; those states that outsource aviation security programs must then ensure adequate governmental control and supervision.[298] Annex 17 provides that the state must "ensure that the appropriate authority arranges for the supporting resources and facilities required by the aviation security services to be available at each airport serving international civil aviation".[299]

Accordingly, government responsibility for aviation security is the predominant model today. ICAO's Annex 17 specifies that each member state must designate and specify to ICAO an appropriate authority within its administration to be responsible for the development, implementation and maintenance of the national civil aviation security programme.[300] The state must require the appropriate authority to define and allocate tasks and coordinate activities between the departments, agencies and other organizations of the state, airport and aircraft operators and other entities concerned with or responsible for the implementation of various aspects of the security program.[301] The state must also establish a national aviation security committee for the purpose of coordinating security activities between the departments, agencies and other organizations of the state, airport and aircraft operators and other entities concerned with or responsible for the implementation of various aspects of the security programme.[302]

Therefore, when fulfilling tasks of aviation security in practice, governments delegate the concrete powers to specific agencies: transportation and aviation authorities, Interior ministries, police and other law enforcement and security organs, security intelligence agencies, military and defence organs, immigration organs and customs, specific departments at airports and within airlines, etc. The concrete structure, combination, status and responsibilities differ in particular states, and may constitute a complicated structure. For instance, simplified, Russian system of aviation security can be illustrated like this.

The development and implementation of the state policy in aviation security in Russia is fulfilled by the Federal Air Transport Agency (Rosaviatcia). However, this agency is only a part of the Federal Aviation Security System of Russia (see Fig. 4.1). Thus, actual aviation security activities are performed by Rosaviatcia in cooperation with other transportation and law enforcement organs, in particularly the National Anti-Terrorist Committee, the Ministry of Transportation and Federal Transportation Oversight Agency (Rostransnadzor). The relevant agencies include the Ministry of

---

[297] See, e.g. ACI position brief. November 2010. http://www.aci.aero/Media/aci/file/Position%20Briefs/2010/position%20brief_SECURITY%202010.pdf

[298] Abeyratne (2010) p. 4.

[299] §3.1.8.

[300] §3.1.2.

[301] §3.1.4.

[302] §3.1.5.

the Interior, Federal Security Service (FSB), General Prosecution, Federal Customs Service, Ministry of Defence, as well as other interested federal organs of the executive branch. In addition, there are some special organs, such as the Interregional Air Transportation Administration, which coordinates the activities of regional agencies and other organizations. Airlines and airports have special aviation security departments too. In addition, there are other minor actors, such as various committees, training and certification centres etc.

In the UK, relevant bodies include Civil Aviation Authority, Home Office, the Department for Transport (DfT), the UK Border Agency, and UK Visas and Immigration. In addition, they work closely with other organs such as the Centre for the Protection of National Infrastructure, the British Transport Police and the UK Security Service, often known as MI5. The Home Office is the leading UK government department for counter-terrorism and police, immigration and passports etc. One of its offices is the Office for Security and Counter-Terrorism (OSCT). Home Office coordinates the activities of many agencies and government departments. It also has direct responsibility for certain aspects of the counter-terrorist strategy.

In the USA, the Department of Homeland Security (DHS) is equivalent to the Interior ministries of other countries. One of its agencies is a special organ responsible for transport security – the Transportation Security Administration (TSA). Another agency of the DHS is the Bureau of Customs and Border Protection (CBP). Prior to November 2001, it was the Federal Aviation Administration (FAA) that controlled aviation security in the USA. Today, FAA's major responsibilities include civil aviation safety, air traffic control and navigation, etc. Other relevant organs

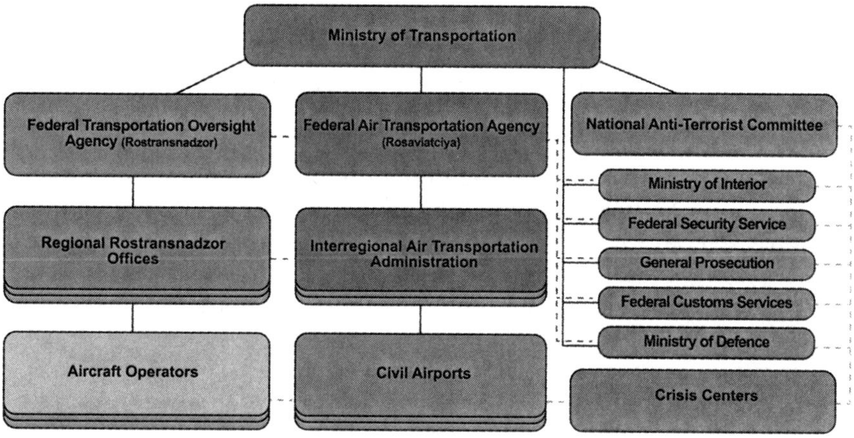

Fig. 4.1 Aviation security system of Russia (Based on Yegorova. *Aviation security in Russia: 40 years of evolution*. In: Aviation Security International. Vol. 19 (2013) p. 20, www.asi-mag.com)

include police and security intelligence agencies (e.g. FBI, CIA, and NSA[303]), the US Department of Defense. Information sharing agencies and programmes include fusion centres, the Terrorism Liaison Officer Program, InfraGard.[304]

In relation to the USA, simply mentioning the security organs is not enough. As mentioned above, the USA has obtained substantial influence in the aviation security field, and it is argued that all ICAO rules originate from the USA, strengthened by the facts that 40% of world air traffic is North American, 70% is operated by allied English-speaking countries and industrial aeronautic supremacy of the North American manufacturers.[305] Accordingly, many of the above-mentioned aviation security concepts, decisions, views and so on originate from the USA, in particular, from DHS/TSA.

In Norway, the Ministry of Transportation and Communication (MoTC) have the overall national responsibility for the civil aviation sector. One of its departments is the Department of Civil Aviation, Postal services and Telecommunications. The Civil Aviation Authority (Luftfarttilsynet) is an independent civil aviation agency which is directly governed by and reports to the MoTC. Luftfarttilsynet is responsible for introducing and adapting national and international legislation and provides the detailed regulations in the form of Regulations for Civil Aviation based on the EU and Norwegian legislation. It has a section for security and publishes the National Aviation Security Programme. An important part is information, communication of updates on aviation security.

The actual providers of airport security are airport operators. Avinor AS can be considered a monopolist as it operates the vast majority airports in Norway. Avinor was established as a limited company, wholly owned by the state, in 2003. The ownership is administered by the Ministry of Transport and Communications. Other relevant organs include the Ministry of Defence, Ministry of Justice and the Police, and the Norwegian Police Security Service (Politiets sikkerhetstjeneste, PST).

Overall, identifying and understanding structures of state aviation security is quite problematic due to flexibility and variations of tasks and missions of the agencies, relations between them, and the diverse nature of aviation systems in different nations.[306] Ideally, all of them are dedicated to effective security, dissemination of information and knowledge, developing strategies against threats. If seen from a data protection perspective, complexity of the systems means that in the course of aviation security measures, various organs may take part in personal data collection, analysis and sharing, increasing the risks of misuse, unlimited data retention, etc. (see Chap. 7).

One of the consequences of the complicated structures of relevant organs is a complex system of national regulations applicable to aviation security. Apart from aviation security regulation as such, more general transport security regulation

303 Federal Bureau of Investigation, Central Intelligence Agency, and National Security Agency respectively.

304 For the details, see Price (2013), Chap. 4 – The Role of Government in Aviation Security.

305 Naudin (2014) p. 44.

306 Price (2013) p. 169.

applies. In addition, as mentioned above, aviation security forms a part of national security. Measures taken to ensure national security include anti-terrorism legislation, using counterintelligence services or secret police to protect the nation from internal threats, etc. which all may be relevant to civil aviation as well. Moreover, as mentioned above, aviation security focuses on terrorists, making anti-terrorist regulation the core objective. Beyond this, regulations governing other relevant areas, bodies, etc. apply, including general criminal law, border and immigration, police, law establishing and giving powers to different state organs, etc.

National aviation security regulation in many ways reflects the state's general security policies, perceived level of terrorist threats (see Sect. 4.5.4.3) and general surveillance level (Sect. 2.3.3). All these specific national factors ultimately result in the aviation security regimes and the amount of the used aviation security measures. A summary of the main national regulations serving as the basis for these regimes/measures can be seen in Table 4.2.

Common trends in national aviation security regulation include, first of all, the fact that, in line with international and regional regulation, aviation security is becoming more and more regulated. In general, it can be noted that those states that are concerned with security, where terrorism risks are deemed to be severe, that have a high level of general surveillance are "pioneers" in the adoption of advanced aviation security measures as well. As discussed, airports in the USA, UK and Russia use many of intrusive aviation security measures, with regulation serving their legal basis.

For instance, as for aviation security laws, national UK laws often lay down more stringent measures in the UK than those of the EU, as permitted under Article 6 of Regulation No 300/2008, e.g., the use of body scanners before they were generally permitted by the EU regulation. Another example is the use of PNR within the UK border control programs from 2004, i.e., before the EU PNR system is adopted. The UK also is a leader in CCTV use. In addition, the researchers note that the UK has more anti-terrorist legislation than almost any other developed democracy.[307]

Norwegian regulators seem to be very careful with introducing new measures and apply only those that are officially confirmed by the EU regulations, but, as mentioned above, the surveillance level is growing and the implementation of body scanners and PNR is likely only a matter of time.

A comprehensive analysis of all these regulations is beyond the scope of this work. However, in the Special Part, selected regulations regarding selected measures will be used for analysis on an illustrative basis. After that, some important general findings will be mentioned in Chap. 8.

But in general, the majority of aviation security laws are dedicated to stopping prohibited articles and persons from getting on board aircraft. In addition, all states developed substantial anti-terrorism laws that apply to stop, search, question and detain, including in the airports. In contrast to that, regulation of other practices

[307] Report on the Anti-terrorism, Crime and Security Bill 2001 (2001-02 HC 351), §1.

Table 4.2 Aviation security and other relevant regulation in the selected states

	UK	Norway	USA	Russia
Main aviation security/ transport security regulation	Aviation Security Act 1982, the Airports Act 1986, the Aviation and Maritime Security Act 1990, and Aviation (Offences) Act 2003	Aviation Act 1993, Regulation on the prevention of attacks against aviation security 2011. In addition, substantial part of Norwegian security legislation is connected to the country's participation in the Schengen Area cooperation[a].	Aviation and Transportation Security Act of 2001; the TSA issues and administers Transportation Security Regulations which are codified in U.S. Code Title 49, § 44903 – Air transportation security[b]	Air Code 1997, Federal laws On Security 2010, On Transport Security 2007, etc.
Main anti-terrorism/ criminal/ police regulation	Criminal Law Act 1967, Criminal Law Act 1977, Police Act 1997, etc., Counter-Terrorism and Security Act 2015, Terrorism Prevention and Investigation Measures Act 2011, Counter-Terrorism Act 2008, Terrorism Act 2006, Prevention of Terrorism Act 2005, Anti-terrorism, Crime and Security Act (2001), Terrorism Act (2000), Criminal Justice Act 1998, etc.	General Civil Criminal Code 2005, Criminal Procedure Act 1981, Police Act 1995, Law on preventive security service 1998, etc.	18 U.S. Code Chapter 113B – Terrorism; PATRIOT Act 2001, Homeland Security Act 2002, Support Anti-Terrorism by Fostering Effective Technologies Act (SAFETY Act) 2002, Border Protection, Anti-terrorism, and Illegal Immigration Control Act 2005, REAL ID Act 2005, etc.	Criminal Code 1996, Criminal Procedure Code 2001, Federal law On police 2011, Federal law On counteraction against terrorism 2006, Presidential decree On measures against terrorism 2006, Presidential decree On the Procedure for establishing levels of terrorist threat, with additional measures to ensure security of the individual, society and the state 2012, etc.

The lists are not exhaustive and include selected items for illustration purposes

[a]Schengen Agreement of 14 June 1985. Norway has been associated with the Schengen Agreement since 19 December 1996. The agreement on Norway's association with the implementation, application and development of the Schengen *acquis*, as based on the Council Decision 1999/439/EC of 17 May 1999, was signed by Norway and the EU on 18 May 1999

[b]The US Code contains the general and permanent laws of the USA, organized into titles based on subject matter

such as camera surveillance is lacking.[308] Thus, regulation of different aviation security segments and measures is unequal.

A related problem is that in aviation security, changes in technology occur much faster than regulators can address those changes. Thus, legislation on aviation security measures traditionally lags behind the actual situation, and it is common that the measures are established by decisions of executive agencies such as the Ministry of Transport or Home Office in the form of orders, resolutions, guidelines, etc.

On the one hand, in general, the use of such acts rather than laws may have positive effects on security. Indeed, the former are more dynamic and operative in reacting to the constantly changing situation: security measures should be able to be implemented faster than through slow and bureaucratic regulatory process.[309] On the other hand, it may have negative effects in the long term: the legislation, which presumes certain requirements and procedures for making valid laws, is "replaced" by decisions of executive bodies, compromising the legality of the regimes (see Chaps. 5 and 7) and the concept of separation of powers, creating the risk of bureaucratic abuse of power, etc.

Furthermore, the law must work. In some states, problems may arise in terms of proper implementation of laws. In Russia, for instance, this is a general problem, spreading beyond aviation security. Negative factors in this regard may derive from the economic situation, low salaries of security inspectors etc. In the USA, salaries of TSA's officers are deemed to be the lowest in the federal government.[310]

Consequently, there may be anomalies with fatal consequences. For instance, in 2004, two Chechen suicide bombers carried out attacks on two aircraft departing from Moscow Domodedovo Airport (2004), using bribes.[311] It is often said that you can install the best available devices, but since it is a human who controls them, there is always the chance that they can be misused.

The key question for this work is the aviation security-privacy perspective. As noted in Chap. 2, Norway and the UK are subject to the EU common rules governing both aviation security and data protection. In addition, national regulation may include technology-specific laws. In particular, with reference to body scanners, CCTV and some other technologies, there are respective Codes of Practices and other soft laws in the UK and in the USA.

In Norway, as noted, there is a tendency to adopt technology-neutral laws. At the same time, CCTV at airports are used with certain limitations (without facial recognition or behaviour analysis, for instance), body scanners/PNR systems are not yet used – and if used, they will have to correspond to the EU standards.

---

[308] Meads. *CCTV Data: is it fit for purpose?* In: Aviation Security International. Vol. 21 (2015) p. 18.

[309] Brooks (2015) p. 31.

[310] Naylor. *TSA Officers Among Lowest Paid Of Federal Workers*, NPR, 26 January 2010, http://www.npr.org/templates/story/story.php?storyId=122948752

[311] RIAnovosti. *Terrorist attacks of 24 August 2004 in Tu-154 and Tu-134. Reference.* 24 August 2009, http://ria.ru/society/20090824/182146689.html

In Russia, where the privacy/data protection regulation is technology-neutral, the aviation security regulators tend to rely on general data protection law. Thus, they tend to repeat or refer to the latter, as in the rules about the Russian PNR system.

In Chap. 7, it will be analysed whether selected existing provisions regarding scanners, CCTV and PNR are capable of ensuring compliance with privacy/data protection principles.

4.8 Concluding Remarks

Quantitative growth in aviation in numbers, costs, etc., historical developments, the rationale and logic of aviation security decisions, overview of principles of aviation security and legal regulation contribute to explain the aviation security regimes we have today. They also have direct or indirect impact on the aviation security-privacy dilemma.

The civil aviation sector cannot be seen only as a means of transportation. It has key importance for the global and national economies, driving development of many other sectors. Aviation is a matter of prestige for many states. It grows irrespective of any economic crisis and it will continue to grow. Accordingly, aviation security as its essential part grows too, in terms of scales, costs, and technologies involved, and will continue to grow, and, as other parts of this work show, not without privacy concerns.

The overview of the historical events and consequent development of the aviation security measures permits us to note that the aviation security focuses on terrorists, and the majority of measures, including body scanners and PNR are introduced under the motto of combatting terrorism and crime. While some of the events are very explicit and are well connected to consequences in the form of concrete measures, e.g. the "underwear bomber" of 2009 and the global spread of body scanners, other measures, such as more and more sophisticated camera surveillance and PNR systems, are a progressive reaction to a longer line of different events and terrorists' developing tactics.

Consideration of the concept of aviation security, including risk assessment, assets, threats and benefits, reveals that serious and actual threats in civil aviation serve as a good argument for enhanced security measures. As discussed, the threats of terrorism in civil aviation are highlighted, and the security measures deployed there are much more sophisticated and advanced in comparison with other objects. The concept of risk in civil aviation security, on the one hand, is aimed to serve as a tool for security decisions to determine the most appropriate and most effective security measures for protection of assets. On the other hand, the security sceptics note, and not without grounds, that this concept can be used for enhanced surveillance, political decisions and other interests.

In general, the concept of risk assessment does not include evaluation of risks for privacy/data protection before adopting a security measure. Nevertheless, today, regulators are beginning to take into account privacy, for instance, through Privacy

Impact Assessment procedures. At the same time, the security decisions are often made in secret rather than democratically, making difficult, if not impossible, any public knowledge and review of such issues to ascertain actual risks and effectiveness of the measures. All this has direct impact on the proportionality, transparency and other principle discussed in Chaps. 5 and 7.

Nonetheless, the states, in particular security organs, are the driving force. Via risk assessment, they decide which measures to use, and adopt respective regulations. Along with the growing use of various technologies, aviation security is becoming more and more regulated, not only nationally, but internationally and regionally too. For all ICAO members, ICAO instruments are valid; for EU/EEA members, the EU regulation is applicable.

Aviation security, as discussed, is an international area by its nature. Endeavours to harmonize aviation security legislation globally can be considered as a positive development not only for aviation security as such, but also for the aviation security-privacy dilemma, since this can contribute to common approaches, including privacy/data protection. It can be helpful to legally establish common universal principles of data protection for definite technologies.

Although there are endeavours of to harmonize aviation security legislation, the lack of enforcement mechanisms and divergences between states slow down these processes. National aviation security practices differ greatly. More detailed regulation issues will be considered in Chap. 7 within the legality principle. In general, however, the regulation problems may constitute an unfavourable factor for privacy and data protection in aviation security: it is much more difficult to solve privacy and data protection concerns in complicated and differing systems, without universal standards and with internal problems.

As this chapter reveals, aviation security measures are necessary in the contemporary world: aviation security must be delivered. Most people understand this need. Another question entirely is whether those measures that are used are proportionate between aviation security and privacy/data protection concerns. This will be explored below.

Chapter 5
Legal Principles of Privacy and Data Protection

5.1 Introduction

As discussed in Chap. 4, the states are obliged to ensure the security of civil aviation and have the right to establish security procedures that they believe are necessary. However, this does not mean that security agencies are free to do anything. In addition to the aviation security considerations discussed in Chap. 4, such as threats and risks, costs and benefits, etc., there are other mechanisms to review the operations of aviation security agencies, for the purposes of this research – in the terms of privacy and data protection. The principles of privacy and data protection can serve as such mechanisms.

Since technological development is rapid, and legislation is not, and since there is a lack of laws on every specific aviation security technology, it may be more fruitful, instead of analysing existing regulations with reference to every aviation security measure, to apply privacy and data protection principles on a case-to-case basis. The point is that principles provide a normative rationale for judging the acceptability of surveillance, on the basis of which opposition or adaptation may take place.[1] They express the essence of privacy/data protection values and constitute the basis of the applicable regulation, allowing identification of exact privacy/data protection concerns that arise as a result of concrete aviation security measures and how these can be dealt with.

Legal principles are important in other dimensions too. As some researchers suggest, in the world of the increasingly global and interactive market, increasing connectivity among jurisdictions, legal professions and doctrinal disciplines, there should be more emphasis placed on legal principles as a tool in modern legal training.[2] When assessing aviation security measures as a complex and global network,

---

[1] Wright et al. (2015) p. 288.

[2] Cankorel. *Cognitive Classification of Legal Principles: A New Approach to International Legal Training*. In: Ankara Law Review. Vol. 5 (2008) p. 154.

O.M. Enerstvedt, *Aviation Security, Privacy, Data Protection and Other Human Rights: Technologies and Legal Principles*, Law, Governance and Technology Series 37, DOI 10.1007/978-3-319-58139-2_5

with illustrations from a number of selected states, it would be too complicated and insufficient to take into account national privacy/data protection law separately for each measure.

Thus, for research such as the present study, which entails various disciplines, jurisdictions and languages, the analysis of applicable legal principles has a pragmatic goal: it is intended to suggest an effective universal (common, general) model for approaches to the aviation security-versus-privacy dilemma, for any jurisdiction and any legal regime.

Accordingly, this chapter is dedicated to the core legal principles of privacy and data protection which are the key elements for this research, since they will constitute a practical tool for evaluation of concrete aviation security measures in the Special Part.

As I mentioned in Chap. 1, in the search of concrete principles and determination of what the legal principles are, I was faced with a number of methodological difficulties and endeavoured to formulate my own approach to principles. The main condition is that principles for this research are *legal* principles. In contrast to *political* principles, these legal principles must be contained in the law. In addition, they can be formulated and applied by courts.

5.2 What Are Principles in This Research?

In general, legal principles constitute the original, defining ideas, positions, attitudes that make up the moral and organizational basis for the appearance, development and operation of the law; they express the essence of law: what the law consists of, what it should be focused on and concentrated in its development.[3] It is common to make a distinction between legal rules and legal principles; however, there are many approaches to formulating what the principles are.

Some philosophers, in particular Dworkin, believe that the principles play a crucial role in the law.[4] Others "attack" principles, for instance by treating them as binding upon judges or as summaries of judges when the latter are forced to go beyond binding standards.[5] They claim that via Dworkin's interpretation, rules in effect are dropped "out of the picture, leaving the field to principles".[6]

Nevertheless, as Alexy notes, two main positions can be differed: (1) principles express the idea of optimization; they are optimization commands, and it is this feature that differs them from the rules, (2) the optimization thesis is wrong.[7] But after exploring different views, Alexy concludes that no explanation of what

[3] Matuzov. *The theory of state and law: lectures*. (2001) pp. 83–86.

[4] Raz. *Legal principles and the limits of law*. In: Yale Law Journal (1972) p. 824 referring to Dworkin. *The model of rules*. In: The University of Chicago Law Review (1967).

[5] Dworkin (1967) p. 31.

[6] Alexander and Kress. *Against legal principles*. In: Iowa L. Rev. Vol. 82 (1996) p. 739.

[7] Alexy. *On the structure of legal principles*. In: Ratio Juris. Vol. 13 (2000) p. 294.

principles are "is more promising than a construction of them as obligations to be optimized, where this corresponds to obligations to optimize. This construction seems to be the best expression of the idea of an ideal "ought" "and of ideal validity".[8]

I have stated repeatedly that principles are essential for this work. I will focus on explaining the essence and role of principles rather than debating whether principles as such are important or not.

There are a number of differences between rules and principles. First of all, the traditional approach indicates first of all the formal binding effect (legal rules have them, but not the principles). However, today, the principles also may be expressed in written norms[9] and have binding effect, thus, legal principles are sources of law. Like other laws, they can be enacted or repealed by legislatures and administrative authorities and can become legally binding via establishment by the courts.[10]

For instance, with reference to data protection principles, although they are primarily abstractions, they simultaneously "have a normative force of their own".[11] This force is achieved in three ways: (i) they are expressly incorporated in data protection laws as rules[12] automatically giving them legal force, (ii) they are guiding standards in interest-balancing processes, and (iii) they shape drafting of new data protection legislation.[13]

Accordingly, due to their connection to overriding values and aims, legal principles may have very practical force, including normative force: they may serve as sources for emerging legal norms; they may contribute to organize and coordinate arguments around a legal question[14] as well as serve as guiding standards in interest-balancing processes (in our case, security versus privacy).

The second difference relates to the logical character and content. Principles are abstractions and imply more general, fundamental norms of a legal order (norms of general application that do not take into account specific legal facts[15]), while legal rules imply concrete provisions.[16] Accordingly, if the facts of a case are such that the conditions of application of a valid legal rule have been met, then the rule must be applied; if they have not been met, then the rule cannot contribute to the case.[17] In

---

[8] *Ibid.* p. 304.

[9] Black. *Regulatory conversations*. In: Journal of Law and Society. Vol. 29 (2002) p. 172.

[10] Raz (1972) p. 848.

[11] Bygrave (2014) p. 145.

[12] See DPD Articles 6, 7, 10–12, 16–17, 22–23; OECD Guidelines on the Protection of Privacy and Transborder Flows of Personal Data (1980) Paragraph 14; the Council of Europe Convention for the Protection of Individuals with regard to Automatic Processing of Personal Data (1981).

[13] Bygrave (2014) p. 145.

[14] Malt. *Sameirettslige Prinsipper* In: Selskap, kontrakt, konkurs og rettskilder (2010) p. 183.

[15] Muñiz. *Legal Principles and Legal Theory*. In: Ratio Juris. Vol. 10 (1997).

[16] Petersen. *Customary Law without Custom-Rules, Principles, and the Role of State Practice in International Norm Creation*. In: Am. U. Int'l L. Rev. Vol. 23 (2007). p. 287.

[17] Perry. *Two models of legal principles*. In: Iowa L. Rev. Vol. 82 (1996) p. 788.

contrast to that, while principles can be unable to solve a concrete case, they may assist in practical reasoning, about what ought to be done.[18]

As for the content, principles are often indistinguishable from various values, interests, rights, policies and goals (i.e. have value-oriented content), while rules mainly specify concrete actions for particular circumstances (action-oriented).[19]

This nature of principles presents both advantages and disadvantages. The abstract nature of principles facilitates legal analysis: an expert in these abstractions analyses legal problems more efficiently than a novice who takes concrete and inefficient steps.[20] This is another advantage of using principles as the basis of evaluation in this research.

However, the principles are formulated broadly and not specifically, thus, their legal effect may be uncertain. For proper application in practice, in order to be applied to specific cases with correct results, principles often require further interpretations and clarifications, and the approaches of different states and actors here may differ greatly.

Another result of the principles' general nature is that in the case of a conflict between principles and rules, the latter will often prevail as a more specific norm.[21] An exception is the proportionality principle, which will often defeat a rule. Nevertheless, the most important consequence for the terms of this research is that on the one hand, it is always possible to make exceptions and deviate from principles by a concrete legal rule. For instance, the principles can be restricted due to the aviation security needs. With regard to data protection, there are proposals to overcome this nature and adopt legislation to clarify the application of some key principles and include data protection principles in one comprehensive legal framework.[22] These initiatives may be helpful in the sense that they could solve some uncertainties and unclear issues, but they are not yet realized.

On the other hand, "principles can justify rules, but not vice versa".[23] If concrete legal norms contain exceptions and restrictions to privacy and data protection due to aviation security needs, in this situation, the principles are aimed to ensure that the restrictions are as limited as possible. Thus, principles in this case can be used to find a balance between security and privacy.[24]

Thirdly, principles are less prone to remain unchanged for a long time, because they are formulated in a general way, and because principles are starting points which presuppose the possibility of exceptions. In contrast to that, legal rules

[18] *Ibid.* p. 787.

[19] *Ibid.* p. 788.

[20] Cankorel (2008) p. 156.

[21] Alexy and Rivers. *A theory of constitutional rights* (2010) pp. 83–84.

[22] Article 29 Data Protection Working Party and the Working Party on Police and Justice. *The Future of Privacy. Joint contribution of the Article 29 Data Protection Working Party and the Working Party on Police and Justice to the Consultation of the European Commission on the legal framework for the fundamental right to protection of personal data* (2009) p. 6.

[23] Perry (1996) p. 788.

[24] Aquilina (2010) p. 140.

usually correspond to a certain historical period. For instance, the data protection principles are likely to remain in the future revisions of the data protection laws. Despite the challenges provoked by new technologies and globalization, it is argued that the main principles of data protection are still valid.[25] At the same time, a negative consequence of this nature is that data protection principles do not fully address the surveillance issues, in particular those connected with new technologies and information systems.[26] As mentioned in Chap. 2, there are currently endeavours to modernize existing data protection law, and this concerns principles too. The aim is to keep the core essence of the principles, but update them according to new realities. Thus when discussing concrete data protection principles, I will take into account both existing rules and proposed.

Fourthly, many of the legal principles are transferrable in the sense that they transcend jurisdictional, professional and doctrinal boundaries.[27] This is particularly important for this work, which deals with numerous jurisdictions, levels and spheres (privacy and security) and at the same time needs to find proper connections between them. In this sense, common principles may help to unite all these pieces.

Finally, due to the connection with overriding values and aims, and because they are prone to internal organization within their abstract system, legal principles are teachable,[28] and application of principles provides room for dynamic and adaptive law.

All the above-mentioned features of legal principles determine the roles of principles of privacy/data protection and the tasks they fulfil. First, the principles of privacy/data protection unite legal rules, which may have different legal status and may be provided by various institutions, in a single legal system. They will allow determining "axioms" of privacy and data protection, in particular, common privacy and data protection standards and requirements for various aviation security techniques which can be applied universally.

Secondly, when new issues require legal regulation, the principles provide a good starting point. On the one hand, they can assist in developing regulation, or at least best practices and guidelines for the use of aviation security technologies, to provide information to regulators about how rights to privacy and data protection can be ensured. It is of course preferable that legal regulation – based on principles – should appear before any new aviation security procedure or amendment to existing procedure takes place. Accordingly, along with other considerations such as the aviation security measure's costs and benefits (see Chap. 4), effects on other human rights (Chap. 3), the principles of privacy/data protection should be taken into account before the measure is implemented. On the other hand, principles can

[25] Article 29 Data Protection Working Party and the Working Party on Police and Justice (2009) p. 6.

[26] Raab. *Surveillance: Extending the limits of privacy impact assessment* In: Privacy impact assessment (2012b) p. 379.

[27] Cankorel (2008) p. 154.

[28] *Ibid.* p. 156.

assist to evaluate if the existing legal regulation is adequate and sufficient, e.g. are all principles taken into account?

As discussed, in reality, the law usually goes behind the technology. Of course, as mentioned above, privacy/data protection principles may also face difficulties when addressing new technologies, but their advantage is that they can be applied in a broader, a more flexible way, taking into account the essence, the nature of the values they imply. As the result, in the situation of lack of regulation or non-regulation, principles can be applied to existing aviation security measures directly. Moreover, in the situation of lack of regulation or non-regulation, they may contribute to mechanisms of self-regulation, i.e. serve as guidelines for manufacturers, integrators, distributors, security experts, operators of aviation security technologies and other relevant actors.

Thirdly, principles may help in assessing the actions of these different stakeholders, especially those who are usually not subject to specific legal regulation such as data protection rules: security providers, security experts and so on.

Fourthly, principles may help educate the public about how privacy challenges associated with the use of these technologies are being addressed.

What is especially important for this research is that principles may help in evaluating a concrete aviation security measure at any stage of its development and lifetime. Hence, a practical task of the principles in this research is to serve as tests of particular security measures in the Special Part: taking into account technical characteristics and operation of the measure, they allow determining which concrete privacy/data protection concerns arise in the course of these concrete measures and, taking into account other factors such as effectiveness and other human rights concerns, evaluate whether the concerns are justified or not. Ultimately, therefore, principles can be used to evaluate whether aviation security regimes are lawful, necessary and proportionate. Accordingly, the principles will be a necessary tool for this research.

With reference to a number of issues, the use of the principles approach is already accepted formally. For instance, with regard to passenger data transfer from the EU, data protection principles are the key to determining whether the country requesting data can be considered as providing an adequate level of data protection. Sufficient compliance with data protection principles and procedural/enforcement requirements constitutes a minimum requirement for protection to be considered adequate.[29]

The main groups of principles discussed in this research are:

1. General principles
2. Principles of aviation security discussed in Chap. 4
3. Privacy principles
4. Data protection principles

[29] See Article 29 Data Protection Working Party. *Working Document. Transfers of personal data to third countries: Applying Articles 25 and 26 of the EU data protection directive* (1998).

Obviously, as will be discussed below, there exist overlays between all of these categories and there are also overlays between different principles within one group. Some of the selected data protection principles can hardly be considered as individual and/or wholly independent and separate from other principles. These overlays, as well as close relations between principles, indicate that the principles are interdependent, indivisible and interrelated in varying degrees.

It should be noted that concrete privacy principles in this chapter will be considered within either general or data protection principles. The point is, as discussed in Chap. 2, that the EU case law has formulated approaches on criteria of ECHR Art.8(2) (in accordance with the law, necessity, etc.). All these elements fall within or have close connection with either general or data protection principles.

Thus, below, two groups of principles will be discussed in more detail: general and data protection principles, with indication of possible connections and relations between each other. The lists of the discussed principles are not deemed to be exhaustive and present selected items, which, in my opinion, are critical for the security versus privacy dilemma.

5.3 General Principles

The first group comprises international general principles that are "a manifestation of international law".[30] International law here is understood as general or common international law, which is customary law valid for all states belonging to the international community.[31] Accordingly, these principles can also be defined as "generally recognized norms of international law of the most common character".[32] These principles are mandatory and contain obligations *erga omnes*, i.e., obligations of each and every member of the international community. They are stipulated in international legal instruments in different contexts, often repeated by national laws.

For instance, international general principles include such principles as *pacta sunt servanda*, the principle of state sovereignty, equality of states, legality, proportionality, etc. They are universal and valid in any sphere, from application to any human right to the basics of international air law: these principles will apply no matter where the aircraft is flying. These principles are applied by courts when determining the lawfulness of legislative and administrative measures of the states.

In this research, three of international general principles play important role for the aviation security versus privacy dilemma: legality, proportionality, equality and non-discrimination. While the latter was discussed in Chap. 3 together with the right to equality and non-discrimination, legality and proportionality will be analysed in more detail in this section.

---

[30] Voigt. *The Role of General Principles in International Law and their Relationship to Treaty Law*. In: Retfærd. Vol. 31 (2008). p. 5.

[31] Kelsen. *Principles of international law* (1952) p. 19.

[32] Bordunov (2007) p. 39.

5.3.1 *Legality*

5.3.1.1 Introduction

In general, aviation security measures may fit into a democratic, rule-of-law society if such systems comply with the fundamental legal requirements of such societies.[33] As mentioned above, the principle of legality is paramount in democratic society and it is a basis of any other principles. It is enshrined in the majority of international conventions and treaties as well as national laws.

The legality principle is closely related to the concepts of legal formalism and the rule of law. It implies that states should be governed by law. For instance, law enforcement agencies must observe the rule of law, whereby all acts of law enforcement agencies have to comply with the law and such acts cannot be carried out outside the limits and boundaries delineated by the law.[34] Accordingly, the law must be clear, ascertainable and no law must be given retroactive effect.[35] Hence, decision-makers must apply legal rules that have been declared beforehand, and not alter the legal situation retrospectively by discretionary departures from established law.[36]

In Norwegian law, for instance, the principle of legality ("legalitetsprinsippet") enjoys constitutional status; it is an unwritten norm requiring that actions by public authorities that infringe significantly on the rights and freedoms of private citizens (e.g. police surveillance), be authorised by statute.[37] Moreover, "no one may be sentenced except according to law, or be punished except after a court judgment. Everyone has the right to be presumed innocent until proved guilty according to law".[38]

With reference to the aviation security, the principle of legality applies directly. For instance, Russian laws stipulate that legality is a principle of transport security,[39] general security,[40] as well as a principle of anti-terrorism activities.[41] Clearly, aviation security decision-makers must apply legal rules that have been declared beforehand; all acts of security agencies must comply with the law, and the decisions establishing security measures or techniques as well as their practical implementation should have basis in law. However, as mentioned in Chaps. 1 and 4, aviation security is characterized by a high amount of restricted information. This has direct effect on openness and transparency regarding legal information about aviation security.

---

[33] Wright et al. (2015) p. 290.

[34] Aquilina (2010) p. 141.

[35] Norwegian Constitution § 97.

[36] Wikipedia. Principle of legality. http://en.wikipedia.org/wiki/Principle_of_legality

[37] Bygrave (2001) p. 333.

[38] Norwegian Constitution § 96.

[39] RF Federal law of 9 February 2007 N 16-FZ On transport security, Art. 3(1).

[40] RF Federal law of 28 December 2010 N 390-FZ On security, Art. 2(2).

[41] RF Federal law of 6 March 2006 N 35-FZ On counteraction against terrorism, Art. 2.

In the privacy context, the term "in accordance with the law" (lawfulness) and "legitimate aim" are used as the conditions for limitation of the privacy right (Article 8(2) ECHR).[42] In data protection field, it is common to indicate the principle of lawful processing.

Below, important issues for this research will be analysed in more detail: (1) information openness and transparency (overlapping with the requirements of accessibility, foreseeability and specific, precise and clearly defined law within "in accordance with the law" and also with data protection transparency principle), (2) in accordance with the law, (3) legitimate aim, and (4) lawful processing.

5.3.1.2 Information Openness and Transparency

In general, *publicatio legis* principle implies that ideally, laws and other legal sources should be communicated to the citizens and/or published so that citizens can familiarize themselves with the laws and rules relating to them.

A closely related term is the transparency principle. As mentioned in Chap. 3, it requires that information provided to the public should be easily accessible and easy to understand; in other words, the information must have two attributes: clarity and accessibility. The states should be transparent about their actions, in particular, about the use and scope of surveillance techniques and powers.[43]

In general, transparency principle includes (i) transparency of legal rules on aviation security practices, and (ii) transparency of the security practices. The latter is mandated to various degrees by legal rules, but it is distinct from the former type of transparency. In this section, the first type of transparency will be addressed. The second type will be mentioned in Sect. 7.2.2 where applicable in connection with legal transparency, since it is important for understanding the context and a broader picture. It will then be addressed in Sect. 8.2.4.3 in association with the notion of transparency in general – an important emerging legal principle.

The absence of legal information makes it impossible to fully realize individual rights, and it is the duty of the state to ensure its availability. One of the most important factors of availability of legal information is its quality: information should be scientific, objective, reliable, specific, complete, adequate, relevant, timeliness, optimal, accurate, and concise.[44]

Thus, in addition to a "negative" meaning of the legality principle – that the state powers must have a legal basis (see below), the legality principle also implies some "positive" obligations of the state – to ensure the availability of legal information. The importance of the latter cannot be underestimated since it may serve as a

---

[42] See Chap. 2.

[43] International Principles on the Application of Human Rights to Communications Surveillance, signed by 258 non-governmental organizations. Final version 10 July 2013. https://en.necessaryandproportionate.org/text

[44] Kovaleva (2007) p. 7.

safeguard for the rule of law, contributing to predictability, awareness, open political debate, etc.

Making legal information publicly available allows persons to become aware of the impact on their privacy and data protection rights as well as other human rights and gives them instructions concerning where to lodge a complaint in the event they believe their rights are violated, and provides them with the opportunity to redress. Moreover, transparency plays a positive role in public acceptance of aviation security technology: for instance, pursuant to a study from the UK, presenting passengers with information on body scanners results in positive increases in their acceptance of this technology.[45] The importance of this principle can be compared with the importance of the proportionality principle: "If we can only remember two concepts as regards the legislation on privacy, it needs to be these two [proportionality and transparency]".[46]

The researchers also use the terms (i) deliberation and (ii) awareness and communication. The former means that when new measures are being considered, or when existing schemes are being expanded, the deliberative and democratic process should be as open, consultative and fair as possible, while awareness and communication provide a platform for debate and change. They are already practiced by for instance DPAs.[47]

At the same time, according to categories of access, information can be divided into open (public) and restricted. Normally, the states issue Freedom of Information Acts (FOIA) entailing responsibility on the part of the state to provide public access to information held by public authorities.[48] In general, the intent of FOIA is to avoid unnecessary secrecy.[49] The core idea is that legal information, if not already available in open sources such as Internet, can be obtained by citizens via FOIA request.

FOIA provide exemptions which authorize government agencies to withhold information deemed to be restricted. In the USA, this pertains to information relating to national defence or foreign policy, information that is exempt under other laws, trade secrets and confidential business information, law enforcement records or information, etc.[50] Russian law provides a broad formulation, "restricted information".[51] Analysis of other laws allows us to infer that this includes state secrets[52] and confidential information such as personal data, information about

[45] Mitchener-Nissen [et al.] *Public attitudes to airport security: The case of whole body scanners.* In: Security Journal (2011).

[46] Poullet (2009) p. 224.

[47] Wright et al. (2015) p. 288.

[48] In the selected states, FOIA were enacted in the USA in 1966, in UK in 2000, in Norway in 2006, in Russia in 2009.

[49] UK ICO. What is FOIA? https://ico.org.uk/for-organisations/guide-to-freedom-of-information/what-is-the-foi-act/

[50] Title 5 of the US Code, section 552.

[51] Art.20(4) of Russian Federal law of 9 February 2009 N 8-FZ On providing access to information about the activities of government bodies and local authorities.

[52] See e.g. Russian Federal law of 21 July 1993 N 5485–1 On state secrets.

investigations and legal proceedings, service secrets, professional secrets, trade secrets, details of the invention.[53] Information about "the special means, techniques and tactics of the measures to combat terrorism as well as the composition of their participants"[54] is also confidential.

Thus, not all so-called public information can be made publicly available. As mentioned in Chap. 4, aviation security decision-making is not a democratic process, and as Chap. 6 will discuss, the details on the specific technologies and their *modus operandi* cannot be fully debated in a public forum.

In the case law in the area of combating terrorism, the ECtHR admitted that "Democratic societies nowadays find themselves threatened by highly sophisticated forms of espionage and by terrorism, with the result that the State must be able, in order effectively to counter such threats, to undertake the secret surveillance of subversive elements operating within its jurisdiction".[55] The Court thus recognizes that intelligence – secret services – may legitimately exist.

The problem is that when invoking "security needs", security organs may apply broader restrictions than necessary, creating unnecessary secrecy. Thus, certain limits apply. In particular, pursuant to case law, secret surveillance is justified only in so far as it is strictly necessary for safeguarding the democratic institution: the Court, being aware of the danger inherent in secret surveillance measures "of undermining or even destroying democracy on the ground of defending it, affirms that the Contracting States may not, in the name of the struggle against espionage and terrorism, adopt whatever measures they deem appropriate".[56]

For instance, with regard to secret surveillance databases, in *Shimovolos v. Russia*,[57] the case concerned the registration of a human rights activist in a secret surveillance security database and the tracking of his movement as well as related arrest. The database in which his name had been registered had been created on the basis of a ministerial order which had not been published and was not accessible to the public. Therefore, people could not know why individuals were registered in it, what type of information was included and for how long, how it was stored and used or who had control over it.[58] Violation of Article 8 thus took place.[59] This means that despite security needs, some transparency still requires.

In *Khan v. the United Kingdom*, the Court found that it was not in accordance with law that the law did not exist and the rules were not directly publicly accessible[60] – i.e. it was presumed the rules could be open to the public.

[53] Presidential decree of 6 March 1997 N 188 On approval of the List of confidential information (Russia).

[54] Federal law of 6 March 2006 N 35-FZ On counteraction against terrorism, Article 2 (10).

[55] *Klass and others v. Germany*, No. 5029/71, 6 Sept 1978, §42.

[56] *Ibid.* §49.

[57] *Shimovolos v. Russia*, No. 30194/09, 21 June 2011.

[58] *Ibid.* §69.

[59] *Ibid.* §71.

[60] *Khan v. the United Kingdom*, No. 35394/97, 12 May 2000, §27.

Thus, air passengers should have information about the aviation security technologies/methods used as well as their consequences. Of course, it would be impossible to provide all the details of the technology capabilities, but general information to the extent that it does not jeopardize the security should be provided. This idea is stipulated in the Norwegian Personal Data Act 2000, for instance: the controller must inform a person of "the security measures implemented in connection with the processing insofar as such access does not prejudice security".[61] But, a problem emerges in determining what information jeopardizes security and what does not. Similar to any security decisions, it is not a democratic process, and it is the security organ who decides. Thus, there may always be a tendency to keep secret as much information as possible under the pretext of ensuring the safeguarding of security.

Accordingly, in practice, secrecy, either necessary or unnecessary, creates a problem for realization of the principle of transparency. Additionally, there may appear problems with such requirements of "in accordance with the law" criteria as accessibility of law, foreseeability of interference; precise, specific and clearly defined law (see below).

5.3.1.3 In Accordance with the Law

The criteria "in accordance with law" implies, first, that any limitation, any interference must have a legal basis ("some basis in domestic law").[62] No security measure that interferes with human rights can be adopted in the absence of an existing publicly available legislative act. For instance, the ECtHR held that the use of a covert listening device by the UK authorities was not in accordance with law because there was no statutory system to regulate the use of such devices. The latter was governed by Home Office Guidelines which were neither legally binding nor directly publicly accessible.[63] Two aspects can be seen here: first, it is not enough to follow a voluntary code of practice, internal guidelines, etc., and any security measure should be subject to legal regulations. Secondly, the formulation "directly publicly accessible" implies that the Court considered that the respective rules could refer to open rather than restricted information.

One of the main problems in aviation security is that changes in technology occur much faster than regulators can address those changes. Thus, as discussed in Chap. 4, it is quite common that a measure is established by a decision of executive body in the form of order, guidelines, etc. Of course, the latter may present some advantages such as more timely reaction to technological changes. Nevertheless, if an aviation security measure is not specifically authorized by law and is regulated only by administrative practice, it can hardly be legitimate.

---

[61] Section 18 (b).

[62] *Kruslin* v. *France*, No. 11801/85, 24 April 1990, §27, *Kopp* v. *Switzerland*, 13/1997/797/1000, 25 March 1998, §55.

[63] *Khan v. the United Kingdom*, No. 35394/97, 12 May 2000, §27.

However, the mere presence of law does not make the regime automatically "in accordance with the law", since the law itself must have the quality of law. This implies a number of requirements.

First, the law in question should not contravene other laws, in particular superior laws. For instance, the ECJ established that even provisions with a basis in European law may yet lack the qualities of law by contravening common European human rights standards and thereby may be contrary to common European law.[64] Examples include ECJ decisions finding that the EU Data Retention Directive[65] and the EU-USA "Safe Harbour" agreement on transferring personal data to the USA[66] both were in violation of the EU law.

Specifically, the Court ruled that the Data Retention Directive, which concerned all persons and required the retention of all means of electronic communication, exceeded the limits imposed by compliance with the principle of proportionality and was invalid.[67] The Court found that the "Safe Harbour" agreement was invalid,[68] despite the fact that the European Commission had earlier decided that the Safe Harbour regime was adequate. Not only the Snowden revelations of mass surveillance in the USA were grounds for the case: even before the revelations, various reports showed that the entire US data privacy framework was inadequate.[69]

Another example derives from the ECtHR, which found the discretionary powers of stop and search in the UK Terrorism Act 2000 to be a breach of ECHR Article 8, notwithstanding that it was provided for by a statutory measure, on the ground that the provision of Article 8 stipulating "in accordance with the law" was not met, in particular, due to wide framed powers of the police officer, which were "neither sufficiently circumscribed nor subject to adequate legal safeguards against abuse".[70]

This means that some other conditions must be satisfied too. Logically, if a measure in aviation security is authorized by law, but does not meet requirements of privacy and data protection law, and does not provide adequate safeguards, then the measure can be argued to be not in accordance with law.

Secondly, the law should be adequately accessible to the person concerned, and it should ensure that any interference is reasonably foreseeable to the person

---

[64] C-402/05 P and C-415/05, *Yassin Abdullah Kadi and Al Barakaat International Foundation v Council and Commission*, 3 September 2008, §269.

[65] Directive 2006/24/EC of the European Parliament and of the Council of 15 March 2006 on the retention of data generated or processed in connection with the provision of publicly available electronic communications services or of public communications networks.

[66] Commission Decision 2000/520/EC of 26 July 2000 pursuant to Directive 95/46/EC of the European Parliament and of the Council on the adequacy of the protection provided by the safe harbour privacy principles and frequently asked, related questions issued by the US Department of Commerce.

[67] Joined Cases C-293/12 and C-594/12, *Digital Rights Ireland Ltd. v. Minister for Communications, etc.*, 8 April 2014, §§69 and 71. For comprehensive analysis of the decision, see Graver and Harborg. *Datalagring og menneskerettighetene* (2015).

[68] C-362/14, *Maximillian Schrems v. Data Protection Commissioner*, 6 Oct 2015, § 106.

[69] See Sect. 2.3.3.3.

[70] *Gillan and Quinton v. United Kingdom*, No. 4158/05, 12 Jan 2010, §87.

concerned, that is, formulated with sufficient precision to enable the individual to regulate his conduct.[71] As mentioned above, this overlaps with the requirements of transparency (see above).

The requirements of accessibility and foreseeability constitute an essential legal protection against arbitrariness when fundamental rights are being limited.[72] The ECtHR has held that protection against arbitrariness is even more important as regards surveillance measures,[73] which may be very relevant for aviation security.

At first sight, the accessibility requirement is quite clear: ordinary citizens must have access to the law. However, the question is whether such access to the law is provided in the event, for example, that a specific document is contained only in paid databases, or has restricted access only (see above), or otherwise the access to it is limited due to one of "accessibility barriers" such as language availability, secrecy, etc.[74] Of course, the most relevant limitation in aviation security is restricting information due to security needs (see above).

The foreseeability requirement can be problematic as well. The ECtHR acknowledged that the requirements of the Convention, with regard to foreseeability, "cannot be exactly the same in the special context of interception of communications for the purposes of police investigations as they are where the object of the relevant law is to place restrictions on the conduct of individuals. In particular, the requirement of foreseeability cannot mean that an individual should be enabled to foresee when the authorities are likely to intercept his communications so that he can adapt his conduct accordingly".[75]

Nevertheless, the Court said that in a system applicable to citizens generally, "the law must be sufficiently clear in its terms to give… an adequate indication as to the circumstances in which and the conditions on which the public authorities are empowered to resort to this kind of secret and potentially dangerous interference with private life".[76] Accordingly, since aviation security measures are applicable to citizens generally rather than in the context of police investigation, the requirement of foreseeability is valid.

Thirdly, a law must be specific, precise and clearly defined. These requirements derive from the Human Rights Committee[77] comments on wire-tapping[78] and from §8 of General Comment 16,[79] which specifies that even in cases of lawful interference

[71] *Ibid.* §76. *S. and Marper v. the United Kingdom*, No. 30562/04 and 30566/04, 4 Dec 2008, §§ 95 and 96.

[72] *Liberty and Others v. United Kingdom*, No. 58243/00, 1 July 2008, §69.

[73] *Klass and others v. Germany*, No. 5029/71, 6 September 1978.

[74] See Bing. *Rettslige kommunikasjonsprosesser: bidrag til en generell teori* (1982) pp. 234–235.

[75] *Malone v. The United Kingdom*, No. 8691/79, 2 Aug 1984, §67.

[76] *Leander v. Sweden*, No. 9248/81, 26 March 1987, §51.

[77] The UN Human Rights Committee is the body of independent experts who monitor implementation of the ICCPR by its State parties.

[78] ACLU. *Privacy Rights in the Digital Age*. March 2014. https://www.aclu.org/sites/default/files/assets/jus14-report-iccpr-web-rel1.pdf

[79] General Comment 16, issued 23 March 1988 (UN Doc A/43/40, 181–183; UN Doc CCPR/C/21/Add.6; UN Doc HRI/GEN/1/Rev 1, 21–23).

with the right to privacy, "relevant legislation must specify in detail the precise circumstances in which such interferences may be permitted". This approach can also be found in case law.

For instance, according to the ECtHR, in cases of telephone tapping, in order to make a surveillance measure lawful, national laws should define:

- categories of people liable to have their communications monitored;
- nature of the offences which may give rise to an interception order;
- limits on the duration of such monitoring;
- procedure to be followed for examining, using and storing the data obtained;
- precautions to be taken when communicating the data to other parties;
- circumstances in which data obtained may or must be erased or the tapes destroyed.[80]

However, these six requirements cannot be applicable generally, and should be applied on a case-by-case basis: e.g. the ECtHR distinguishes a greater degree of interference with privacy (such as wiretapping and surveillance of telecommunications) and a lower one (such as GPS surveillance), accordingly, the requirements for the first case will be stricter.[81]

The required details on the measure relate to such data protection principles as minimality, purpose limitation, etc. Thus, explicit laws aimed at specifying the measure's correspondence to these principles can be considered as a pre-condition to a measure being lawful. However, as mentioned above, different restrictions for security needs may apply.

"In accordance with law" also implies that the domestic law must contain a measure of legal protection against arbitrary interferences by public authorities with the human rights,[82] covering judicial dimension of this principle. In the context of privacy/data protection, "domestic law should provide appropriate sanctions and remedies in cases of breach of the provisions of domestic law"[83] enforcing the principles of privacy and data protection.

As noted above, it is quite unrealistic that an individual can challenge a security measure as such. If a person refuses to undergo a particular aviation security measure, he/she can be either subject to an alternative measure (if any) or be denied the right to fly. Of course, a person may exercise various protest actions, such as passive radical action (not flying at all); wearing sunglasses or hats in order not to be recognized by face recognition, destroying CCTV cameras, giving false personal data

---

[80] *Huvig v. France*, No. 11105/84, 24 April 1990, §34 and *Kruslin* v. *France*, No. 11801/85, 24 April 1990, §35. *Uzun v. Germany*, No. 35623/05, 2 September 2010, §65.

[81] Increasing Resilience in Surveillance Societies (IRISS) (2013) p. 34.

[82] *Gillan and Quinton v. United Kingdom*, No. 4158/05, 12 Jan 2010, §77, *Ollson* v. *Sweden*, No. 10465/83, 24 March 1988, §61.

[83] §7 of Recommendation CM/Rec(2010)13 of the Committee of Ministers to member states on the protection of individuals with regard to automatic processing of personal data in the context of profiling adopted by the Committee of Ministers on 23 Nov 2010.

when booking. But, if wearing sunglasses, he/she may be targeted by security personnel and thereby be subject to even more measures; in the two latter cases, criminal liability may arise.

Typically, such protests and individual actions against the system are not very effective. The point is that those who carry out the security measures in practice have no interest or power in changing the system – they are not the ones who control the security system.[84] As Schneier notes, "The only way to change security is to step outside the system and negotiate with people in charge".[85] Helpful societal measures may include opinions and actions of opinion leaders, celebrities, activists (e.g. Edward Snowden's case), collective actions via specific groups, e.g. privacy advocates, demonstrations, counter-surveillance (watching the watchers), public opinion, etc.[86] As mentioned in Chap. 1, one of the most effective and practical roles in these processes is that played by data protection authorities, human rights organizations, privacy advocates etc., that is, groups who in some cases can influence the regimes.

In this section, judicial/administrative remedy refers to particular remedies for a person who believes he/she suffered in the result of existing procedures. In particular, if privacy/data protection invading technology is used by a security organ. Even if the measures are used according to the law and in compliance with the limitations allowed by human rights law, this organ's actions can be challenged before a judge or some other independent and impartial mechanism.[87]

Judicial protection is closely connected to the data subject influence principle (see below). For instance, data subjects should have a method available to them to hold data collectors accountable for following the data protection principles (DPD Articles 22–23).

This leads to a question: how to provide this legal protection against arbitrary interferences, where should air passengers apply if they believe that their privacy/data protection rights are violated in the course of security measures? This depends greatly on the national systems, which are supposed to establish concrete procedures, but typically, the following available mechanisms for privacy/data protection issues can be relevant: (i) airport or airport security department, airline, entity or law enforcement agency personnel who actually perform aviation security tasks (in data protection terms – controllers or processors); a usual method will be lodging of a complaint; (ii) national data protection authorities or other relevant supervisory organs; and (iii) courts. Chap. 7 will provide more details.

---

[84] Schneier (2008) p. 73.

[85] *Ibid.* p. 74.

[86] See Wright et al. (2015) p. 289.

[87] Aquilina (2010) p. 142.

5.3.1.4 Legitimate Aim

The "legitimate aim" criterion is connected to both the general legality principle and the data protection's purpose limitation principle. In general, surveillance can be permitted by laws only if it contributes to achieve a legitimate aim that corresponds to a predominantly important legal interest that is necessary in a democratic society.[88] The aviation security measure must be in pursuit of one of the aims set out under ECHR Article 8(2).

As discussed in Chap. 4, aviation security measures are carried out to safeguard civil aviation against acts of unlawful interference – more broadly, in the interests of national security, hence falling within Article 8(2). Thus it seems it is quite easy for the aviation security measures to satisfy the "legitimate aim" criteria. The problem is that such objective may be very wide in scope. Since it is the state that identifies the objective(s) of the interference, it can usually make "a plausible case in support of the interference".[89]

It may occur that the reason given by the state is not the "real" reason motivating the interference, with other reasons and motivations behind, but in most cases, the court will still accept that states are acting for legitimate purpose.[90] The main idea is that terrorist offences and serious crimes constitute threats to other fundamental rights, especially the right to life; thus, their prevention, detection, investigation and prosecution are also legitimate in order to protect the rights and freedoms of others.

In this case, the overlapping data protection principle – purpose limitation – can contribute to defining "legitimate aim" since it requires that the goal should be set clearly and be purpose specific (see below).

5.3.1.5 Lawful Processing

Furthermore, the principle of fair and lawful processing is a primary principle for data protection law stipulated in DPD Art. 6(1)(a). "Lawful processing" can be defined from the discussion above: in general, this means that any processing should have legal grounds and should be carried out according to the law, i.e. satisfying all other data protection requirements as well. DPD Article 7 stipulates criteria for making a data processing legitimate: if the controller pursues a legitimate aim (§§e-f); if the data subject has given his unambiguous consent (§a).[91] In the UK Information Commissioner's guide to data protection, one of the conditions for

[88] International Principles on the Application of Human Rights to Communications Surveillance, signed by 258 non-governmental organizations. Final version 10 July 2013. https://en.necessaryandproportionate.org/text

[89] Kilkelly (2001) p. 30.

[90] *Ibid.*

[91] Gutwirth [et al.] *Deliverable D1: legal, social, economic and ethical conceptualisations of privacy and data protection* (2011) p. 27.

lawful processing of data is to "make sure you do not do anything unlawful with the data"[92] implying that other data protection principles should be respected.

Moreover, if national data protection legislation is absent, some endeavours should be made by the states to establish procedures, develop laws or rules for protection of personal data process by security measure in question.[93] This means that requirements to the law concern not only the laws on aviation security measures, but the laws on specific data protection as well.

5.3.1.6 Concluding Remarks

It is clear that the legality principle should be applied on a case-by-case basis. But in general, pursuant to the legality principle, aviation security measures should serve legitimate aims, have a legal basis in law which respects other laws and is accessible to the citizens, who must be able to foresee the consequences with regard to privacy interference, the law must contain judicial protection against arbitrary interferences, and processing of personal data should be lawful. However, due to the specific nature of aviation security, with a substantial volume of restricted information and the possibility of restriction due to security needs, some particular requirements can be difficult if not impossible to fulfil. At the same time, as indicated above, any exceptions and restrictions must be provided by law, be necessary and proportionate.

5.3.2 Proportionality Principle

5.3.2.1 Introduction

The proportionality principle is particularly important for this research and its results, since it is the key instrument for analysing aviation security versus privacy dilemma. As mentioned in the Introduction, a close term which is used extensively with respect to the dilemma and sometimes is confusing with proportionality is "balancing". Here, a more detailed analysis of these terms will be presented.

In addition, the proportionality principle is relevant for privacy and data protection and for aviation security. As discussed in Chap. 4, security solutions should be proportionate to the probability of the threat and possible harms, costs and benefits, etc. Obviously, the concrete set of factors to be taken for assessing proportionality within the aviation security will differ from those used within privacy/data protection; however, there are some overlapping factors.

---

[92] ICO. Guide to data protection. https://ico.org.uk/for-organisations/guide-to-data-protection/principle-1-fair-and-lawful/

[93] See, e.g. ICAO PNR Guidelines, §2.12.

While effectiveness was discussed in Chap. 4, here, the proportionality principle in privacy/data protection will be developed in more detail below to indicate how this principle will be further used in the Special Part.

5.3.2.2 Balancing

With reference to "security versus privacy", a common suggestion is the idea of finding the balance between these values, in other words, to balance privacy (or, broadly, liberty) against security using the so-called trade-off model. According to the latter, it is not possible to increase one of the values without decreasing the other one: "respecting civil liberties often has real costs in the form of reduced security" and vice versa.[94] The key issue here is optimization: to choose the point that maximizes the joint benefits of security and liberty.[95]

The term "balance" is undefined in these theories. But generally, this word has a number of meanings, with some of them applicable here. The first one is "a state of equilibrium or equipoise; equal distribution of weight, amount, etc".[96] In the philosophical context, balance is used to mean "a point between two opposite forces that is desirable over purely one state or the other".[97] Aristotle and other philosophers called this point "the golden mean" (i.e. the middle ground between extremes) and applied it to ethics or politics.

Another meaning of "balance" is an "instrument for comparing the weights of two bodies, usually for scientific purposes, to determine the difference in mass (or weight)".[98] As Waldron states, balance "has connotations of quantity and precision".[99] Apparently, this can be relevant to trade-offs between security and privacy or any other conflicting interests. Consequently, within these two meanings of the term "balance", balancing is aimed to find the golden mean for the privacy and security, and at the same time, can "measure" both ends (although the problem of measuring intangible values remains).

The balancing idea can be found in the ECtHR case law, stating that "inherent in the whole of the Convention [ECHR] is a search for a fair balance between the demands of the general interest of the community and the requirements of the protection of the individual's fundamental rights".[100] Therefore, the rights of the individual can be balanced with the interests of the state.[101]

---

[94] Posner and Vermeule (2007) p. 32. See also Vermeule. *Security and Liberty: Critiques of the Tradeoff Thesis*. In: Harvard Law School Public Law & Legal Theory Working Paper Series (2011).

[95] *Ibid.* pp. 26–27.

[96] Dictionary.com. http://dictionary.reference.com/browse/balance

[97] Wikipedia. http://en.wikipedia.org/wiki/Balance_(metaphysics)

[98] Encyclopedia Britannica. http://www.britannica.com/search?query=balance

[99] Waldron (2003) p. 192.

[100] *Soering v. the United Kingdom*, No. 14038/88, 7 July 1989, §89.

[101] Kilkelly (2001) p. 31.

In aviation security, balancing is important too. First, it can be stipulated by law, as in Russian Law on transport security, where "balancing the interests of the individual, society and the state" is one of basic principles of transport security.[102] The issues of threats and risks and effectiveness of aviation security discussed in Chap. 4 are generally used to explain the need for balancing and trade-offs with compromising the human rights in the name of security.

In addition, since the threat to security constantly grows, the balance is supposed to be changed accordingly. For instance, terrorist incidents in the 1970s had maximum death tolls of about a dozen; attacks in the 1980s and 1990s – hundreds; 9/11 – thousands; today, weapons of mass destruction may kill hundreds of thousands – therefore, "As risks change, we who care about civil liberties need to realign balances between security and freedom. It is a wrenching, odious task, but we liberals need to learn from 9/11 just as much as the FBI does".[103]

It looks like a change in the scale of the threats, especially after 9/11, justifies a change in the scheme of civil liberties. Accordingly, the weight of security is growing while the weight of privacy is sinking, leading to a more vertical new balance line. This process is often described as "striking a new balance between liberty and security",[104] leading to compromises in the name of security. Hence, any limitations or violations of human rights and freedoms that may have seemed disproportionate, unnecessary and unreasonable before 9/11 may now be deemed proportional, necessary and reasonable respectively.

The concept of the balance and trade-off model is subject to criticism first of all because it justifies such compromises, with the right to privacy being one of the first "victims of a balancing approach".[105] Two types of critics can be distinguished: internal and external.[106]

"Internal" critics object to the way in which trade-offs are evaluated and judged, claiming that the balance is always skewed in favour of security, at the expense of the right to privacy, leading to a systematic erosion of civil liberties in the name of security.[107] Some argue that accepting that the weight of some human rights depends on circumstance makes human rights principles inconstant, and the process of balancing can possibly "swallow up the rights".[108] The ECtHR issued a reminder that although measures which interfere with privacy may be designed to protect democracy, they should not destroy it in the process.[109]

---

[102] RF Federal law of 9 February 2007 N 16-FZ On transport security, Art. 3(2).

[103] Waldron (2003) p. 192 citing Kristoff. *Liberal reality check: we must look anew at freedom vs. security*. In: Pittsburgh Post-Gazette (2002) p. A9.

[104] *Ibid.*

[105] Scheinin and Vermeulen (2011) p. 49.

[106] Amicelli (2012a) p. 13.

[107] *Ibid.*

[108] Cali. *Balancing Human Rights? Methodological Problems with Weights, Scales and Proportions*. In: Human Rights Quarterly. Vol. 29 (2007) p. 253.

[109] *Malone v. The United Kingdom*, No. 8691/79, 2 Aug 1984, §82.

But, it is noteworthy that even juridical review can be of little help, since it runs the risk of accepting too many compromises in the name of balancing: "If such a court in normal times lets itself be strongly "pulled" into the approach of balancing, it may be unable to break loose of the frame when the day comes when it should".[110] It is also claimed that the process of balancing "tangible values against intangible harms" presents the debate from the view that privacy is "a mere abstraction, a luxury with little concrete value".[111] Internal critics, nevertheless, offer no objection to the security-privacy balance and often acknowledge that trade-offs are necessary.[112]

"External" critics may deny first the whole trade-off model. But, such denial may be discarded here: apparently, no matter how you describe or interpret the security and privacy dilemma, it is not possible today to avoid balancing and, accordingly, the concepts of balance and trade-off. Any stakeholders may take part in the balancing process. Trade-offs in practice at the airports can be seen from multiple perspectives. If the trade-off is evaluated from the air passengers' view, in exchange to the ability to fly, they have to perform certain actions, including going through the security checks and complying with security rules. From the view of the aviation industry, commercial airlines, the trade-off is made for the possibility of conducting business operations. However, if the security measures place an unacceptable burden and may substantially infringe on business, they might be unacceptable. Regulation of security procedures and their implementation have certainly invoked some trade-offs, but it is up to the decision-makers to establish the final balance, where the interests and convenience of passengers, for example, may be considered as less important than the security effectiveness issues. Thus, in this research, it is presumed that balancing and trade-offs do exist.

Secondly, external critics deny that the issues of security and privacy are best thought of in terms of balancing or trading off of one for the other.[113] For instance, within aviation security, threats, risks, effectiveness and other factors are used to balance aviation security needs with human rights and compromise the latter in the name of security. The problem is that it is difficult to define the trade-offs sufficiently enough to get a clear sense of how much security is bought for how much cost: it is impossible to say with confidence that e.g. this particular aviation security measure reduces by X percent the chance of a bomb getting on board; instead, the avoidance of terrorist attacks is treated as an absolute goal.[114] Clearly, the aviation security agencies' interests and motivations are not same as those of the population as a whole: e.g. their biggest interest is not in finding the public's preferences about where to find the balance, but in avoiding a terrorist attack: for instance, if an attack

---

[110] Scheinin. *Terrorism and the "Pull" of Balancing in the name of security* In: Law and Security: facing the dilemmas (2009) p. 57.

[111] Spencer (2002) p. 519.

[112] Amicelli (2012a) p. 13.

[113] *Ibid.*

[114] Pillar. *The Price of Aviation Security*. The National Interest. 22 November 2010. http://nationalinterest.org/node/4463

occurs, few will defend the responsible aviation security agency that it was operating according to public privacy preferences; instead, the agency will be blamed that it failed to do its job to determine what was necessary to stop the attack.[115] Thus, the result of any balancing is always definite.

Clearly, the security versus privacy dilemma is "much more complex, nuanced and subtle than suggested by the trade-off model".[116] According to this critical view, the processes of finding a "balance", trading civil liberties for security are not only unnecessary, but may also be dangerous. The critics urge caution about giving up civil liberties/privacy for the sake of security and refer to Franklin's famous: "they that can give up essential liberty to obtain a little temporary safety deserve neither liberty nor safety".[117] Schneier, expressing scepticism towards all security measures (apart from those above that cause no negative impact on privacy) and calling them "security theatre," adds that "We are not trading privacy for security; we are giving up privacy and getting no security in return".[118]

Accordingly, the privacy-security dialectic approach was formulated, seeing a positive relationship between privacy and security: "Security and privacy are not opposite ends of a seesaw; you don't have to accept less of one to get more of the other;"[119] "Liberty cannot be enjoyed without security, and security is not worth enjoying without liberty;"[120] liberty requires security without intrusion, i.e. security plus privacy, and there is no security without privacy[121]; privacy can be protected without cost to security,[122] etc. As Raab notes, this approach may strengthen the management of security and keep it within a society that values equality and freedom.[123] For this approach, the use of the proportionality principle discussed below can be helpful.

5.3.2.3 Proportionality Principle, in Particular in Privacy/Data Protection

The general principle of proportionality is a worldwide legal principle, a key organizing principle of legal thought.[124] It is found in both the common and civil law, bringing them together to a global uniform jus commune; it is found before national

[115] Pillar (2010).

[116] Amicelli (2012a) p. 14.

[117] Franklin. *Memoirs of the life and writings of Benjamin Franklin* (1818). Available also in: Benjamin Franklin, Historical Review of Pennsylvania, 1759? John Bartlett (1820–1905). Familiar Quotations, 10th ed. 1919.

[118] Schneier (2008) p. 9.

[119] *Ibid.* p. 69.

[120] Vermeule (2011) p. 26.

[121] Schneier (2008) p. 70.

[122] Solove (2011) p. 3.

[123] Raab. *Privacy as a Security Value*. In: Jon Bing: En Hyllest/A Tribute (2014) p. 58.

[124] Engle. *The History of the General Principle of Proportionality: An Overview* (7 July 2009). In: Dartmouth Law Journal, Vol. X-1 (2012) p. 3.

and transnational courts.[125] It is enshrined in international/EU/national legal instruments with reference to different areas.[126]

In Europe, the test of proportionality is the most acknowledged method of legal evaluation of conflicts of fundamental rights and legitimate interests, such as privacy and security.[127] It is relatively closely tied to people's concrete non-legal world experiences: it only requires a comparison of two countervailing factors.[128]

At the same time, the need of such "comparison" reveals that there is a close connection between the balancing and proportionality principle, and in the ultimate scenario, the proportionality principle in practice can easily turn to balancing. This allowed some researchers to differ "weak" and "strong" proportionality tests, with the weak one amounting to balancing privacy and security, presuming that the one per se weakens the other and excluding the possibility that both interests can co-exist together.[129]

In contrast to that, the key point within a strong proportionality test – hereinafter referred in this work to as "proportionality principle" – is that the simple fact that technologies are available, affordable or are believed to get public approval does not constitute sufficient reason for them to be used.[130] As mentioned in Chap. 1, in aviation security, there is a tendency that if new technologies enhance security, they should be used as soon as they are available, but from a privacy/data protection perspective, any contribution to security in itself is not a justification: the availability of means should not justify the end.[131]

The key idea is that the limitations on the rights of individuals by technology should be proportionate to the expected security benefits. In other words, the restriction on privacy/data protection should fulfil its purpose, and there has to be concrete evidence that there is genuine threat to public security and this is an appropriate way of responding.[132]

Even though a security measure contributes to security, such a measure may be unacceptable if it harms privacy to an excessive degree. Thus, instead of balancing, where it is accepted that privacy is outweighed by security, it is important to answer how much harm to a human right would be tolerable in a democratic state in which human rights are protected.[133] It is proposed that the language of rights – not the

---

[125] *Ibid.*

[126] E.g. it applies under public international law, private international law, public and private domestic law. With applicability to the EU institutions, the principle of proportionality is laid down in Article 5 of the Treaty on EU.

[127] Wright et al. (2015) p. 288.

[128] Cankorel (2008) p. 180.

[129] Gutwirth et al. (2011) p. 27.

[130] Information Commissioner's Office (2014) and Lyon (2004) p. 142.

[131] European Data Protection Supervisor. *Opinion on the Communication from the Commission on the global approach to transfers of Passenger Name Record (PNR) data to third countries* (2010).

[132] Colvin. *The Human Rights Act and CCTV*, CCTV User Group, 12 October 2007, https://www.cctvusergroup.com/art.php?art=39

[133] Gutwirth et al. (2011) p. 27.

language of balance should be used.[134] In other words, first, a presumptive weight to the right to privacy should be given, and then, the limitations to the right in a narrow manner should be construed.[135]

The point, as discussed in Chap. 3, is that human rights law provides lawful grounds for derogations, restrictions and limitations if they serve definite purposes: to protect national security, public order, public health or morals or the rights and freedoms of others. Having these existing norms – *lex lata* – as a point of departure, in respect of *lex ferenda,* it can be noted that from being only *subject* to balancing and the trade-off model, human rights should move to another dimension. Specifically, they should regulate the balancing process as such, via applying the test of limitations and the proportionality test in that process.[136]

The CFREU Article 52 provides explicitly that "[s]ubject to the principle of proportionality, limitations may be made only if they are necessary and genuinely meet objectives of general interest recognized by the Union or the need to protect the rights and freedoms of others". This refers to criteria and legitimate aim discussed above, as well as necessity, which will be discussed below. However, a mechanism for the application of the proportionality principle is missing.

In contrast to privacy, processing of personal data is not framed as an interference with a right (cf. ECHR Article 8(2)). DPD Articles 6(c) ("relevance", "not excessive"), 7, 8 and 13 ("necessary") manifest the proportionality principle. Altogether, this means that even when the aims of data processing are legitimate according to Article 7, they will only be legitimate if the data collection and processing correspond to the needs for realizing a specified purpose and are proportional, i.e., necessary, adequate, relevant and not excessive; moreover, if the specified purpose can be reached without personal data processing or by processing less personal data, than the processing can be considered as disproportionate and thus illegitimate.[137] However, the DPD, although establishing these requirements, does not contain any further explanations on what "proportional", "necessary", and other terms imply.

Thus, it is difficult to assess criteria of proportionality objectively. In addition, the methods and criteria of the proportionality test vary from jurisdiction to jurisdiction, from court to court and from case to case.[138] In this research, criteria will be addressed taking into account the views of the ECtHR and the ECJ, as well as data protection authorities and research literature.

The case law, in particular the practice of the ECJ[139] and the ECtHR, increasingly recognize the proportionality principle for both privacy and data protection, for the latter, as a core data protection principle in its own right.[140] The strict methodology

[134] Walker (2006).

[135] Gallagher (2002) p. 288.

[136] Scheinin and Vermeulen (2011) p. 50.

[137] Gutwirth et al. (2011) p. 28.

[138] *Ibid.* p. 26.

[139] C-465/00, *Rechnungshof* decision, 20 May 2003, § 91.

[140] See, for example, Joined Cases C-92/09 and C-93/09 *Volker und Markus Schecke GbR and Hartmut Eifert v. Land Hessen*, 9 Nov 2010.

of the proportionality test is used when the courts make decisions on the justifiability of concrete cases of restricting fundamental rights, such as the application of surveillance measures. If the legitimacy of surveillance is questioned, the dispute in most cases is resolved by courts, applying the test of proportionality.[141]

The case law has developed specific criteria to assess whether a surveillance measure can be considered proportionate. For the purposes of this research, the criteria can be combined into several groups.

First, the proportionality principle requires that the grounds for the security measure's interference must be sufficiently important to limit the right,[142] in other words, the interference must meet "a pressing social need".[143] Accordingly, the aviation security measure must be suitable (appropriate, relevant, adequate) to fulfil the specific legitimate aim. Hence, two important issues should be considered in combination: legitimate aim and the measure's suitability.

As for legitimate aim, as mentioned above, it should be set clearly and be purpose-specific. In particular, it should specify the specific harm that a measure is intended to prevent/detect. Thus, purpose limitation principle should be applied, since understanding of the purpose helps to determine additional criteria for the measure to be proportionate: the categories or types of data needed, the type of processing, the quality of the data, etc.[144]

Determination of suitability of aviation security measures depends greatly on the factors discussed in Chap. 4, where security decisions are aimed to determine appropriate and most effective security measures taking into account threats, risks, possible harms, costs, benefits, etc. Possible harm if the measure is not provided at all can also be taken into account. The key element for fulfilling the legitimate aim is effectiveness, which implies that the measure concerned must be effective in achieving its specific purpose.

The second criterion is necessity – the measure concerned must be necessary to realizing these goals.[145] In some cases, a more stringent standard is applied – in order to be necessary, the measure should be indispensable. According to the ECJ, in order for security measure be proportionate, it has to be demonstrated that other less intrusive methods are not available.[146] If there is a choice between several appropriate measures, recourse must be used to the least onerous.[147] Accordingly,

---

[141] Wright et al. (2015) p. 288.

[142] Fordham and De La Mare. *Identifying the Principles of Proportionality.* In: Understanding Human Rights Principles, London: Justice and Hart Publishing (2001).

[143] Spielmann. *Allowing the Right Margin the European Court of Human Rights and the National Margin of Appreciation Doctrine: Waiver or Subsidiarity of European Review?* In: Centre for European Legal Studies, Working Paper Series (2012) p. 22.

[144] Article 29 Data Protection Working Party (2014a).

[145] Case C-524/06, *Heinz Huber v. Bundesrepublik Deutschland*, 16 December 2008, § 66.

[146] Joined Cases C-92/09 and C-93/09 *Volker und Markus Schecke GbR and Hartmut Eifert v. Land Hessen*, 9 Nov 2010, §§ 81 and 86.

[147] Case T-13/99, *Pfizer Animal Health SA* v. *Council of the European Union*, 11 Sept 2002, §12.

taking into account reasonable costs, if there are other aviation security measures which are less intrusive or non-intrusive and capable of providing equal security, i.e. to achieve same purpose with the same result, this least harmful choice should be used.

Therefore, before adopting any new measure, existing measures and alternatives should be assessed, including effectiveness issues and evidence on why existing measures are not sufficient; it should be explained whether other measures were found to be more or less privacy-intrusive; if those which were found to be less intrusive were rejected, this should be justified.[148] As discussed in Chap. 4, it is possible to increase security at no cost to privacy, e.g. via reinforcing cockpit doors, in-flight security officers, and training crew and passengers. The problem is that neutral measures probably cannot substitute all other measures which are used or are planned to be used.

Thirdly, as formulated in case law, the measures "should not exceed the limits of what is appropriate and necessary in order to attain the legitimate objectives".[149] This is criteria of non-excessiveness: the measure must not go further than is necessary to realize the goals.[150] As discussed in Chap. 3, excessiveness may refer to both the form of particular measures and the manner of executing them. It depends greatly on concrete circumstances of a particular situation: the measure may be deemed to be non-excessive despite inconveniences such as loss of privacy caused by this measure.

But if a measure is nevertheless used, its scope must be limited to the absolute minimum. For instance, such elements as the number of people affected by the aviation security measure, the amount of personal data collected, how long it is retained, who has access to it, may often be excessive,[151] and enhancing these options must be done only with due cause. In order to minimize the excessiveness, different safeguards can be helpful and should be used, e.g. organizational or technical means to limit the scope of the measure. Here, an overlap with the minimality principle of data protection discussed below can be noted.

Simply said, in respect to every separate aviation security measure, it is necessary to answer the questions: how good the measure is to actually provide security and to justify privacy concerns, and whether everything possible is done to minimize the latter. The relationships can be illustrated like this:

As shown in Fig. 5.1, privacy risks here should strive towards the minimum, while security advantages should strive towards the maximum, ideally, with zero privacy risks and maximum security benefits indicated as (1). The diagonal blue line indicates rough borders between "proportionate" and "disproportionate" zones. If privacy risks are high, but security benefits are low, the measure cannot be considered proportionate (2). In addition, certain limitation to allowed maximum risks to privacy should be determined too (red line). Apparently, (3) and (4) can be seen as

[148] Article 29 Data Protection Working Party (2014a).

[149] *Pfizer Animal Health SA* v. *Council of the European Union*, §12.

[150] Fordham and De La Mare (2001).

[151] Article 29 Data Protection Working Party (2014a).

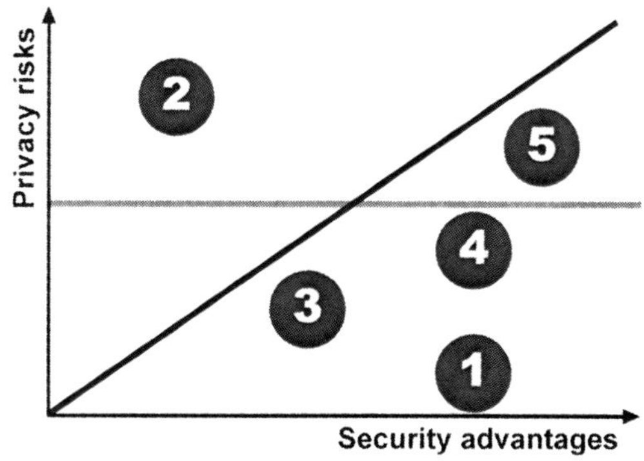

Fig. 5.1 Proportionality principle applied to privacy and security

proportionate, if privacy risks are lower than security benefits and do not go beyond the red line. If risks exceed this allowed maximum (5), no matter which security benefits can be expected, this measure cannot be considered appropriate. An absurd example is naked passengers on board.

The problem that emerges is that it is impossible to determine the risks of privacy and security advantages or the golden middle point mathematically and place them on the figure exactly. While some extreme examples are quite easy to judge, such as metal detectors or cockpit doors without privacy concerns, many measures appear somewhere in the middle.

Thus, within this approach, assessment of what is proportionate must be ascertained on a case-by-case basis, and many different factors should be taken into account. In particular, all circumstances of the case, including the context and nature of the interference, scope and duration of the measure, the type of information collected (sensitive or not), relevant authorities permitting, carrying out and supervising the measures, and remedy provided by the national law,[152] provision of safeguards, etc. For example, the privacy considerations in terms of context are very different for CCTV cameras in public zones or in toilets at the airport or aircraft.

Broadly, the need to analyse many different factors means that any possible concerns related to other privacy and data protection principles – and even more broadly, to other applicable human rights discussed in Chap. 3 should be taken into account when assessing proportionality of the measure. This approach confirms the idea that the all the principles and all the human rights should be considered together, since concerns associated with any of them may have influence on the aggregated impact of the security measure.

It is commonly accepted that it is to the state's obligation to justify the interference. On the one hand, the more severe the interference and the more intrusive the technology, the more approvals and the stronger the reasons required to justify it.[153]

[152] See, e.g. *Klass and others v. Germany*, No. 5029/71, 6 Sept 1978, §50, *Malone v. the United Kingdom*, §§66–68, *Liberty and other organisations v. the United Kingdom*, No. 58234/00, 1 July 2008, §§93–94.

[153] Kilkelly (2001) p. 32 and Aquilina (2010) p. 14.

Since aviation security measures are mainly preventive measures, and are carried out without a concrete and individualized law enforcement context, this requires even more substantiation of the measure. On the other hand, the more severe the issue and more substantial the harm which society may be exposed to, the more interference may be justified.[154]

In aviation security, as discussed in Chap. 4, the risks of terrorism and crime and the consequent harms are considered as enhanced, thus, the non-deployment of security measures would be deemed inadequate, and the aviation security measures seem to be easier to justify than those seeking to protect, for instance, morals.[155] At the same time, if the technologies imply serious harms to privacy, they require justification.

Thus, in aviation security, despite the possible harms, justification is required, in particular, if impact on privacy is serious, and it is the state that should prove the suitability of the aviation security measures, their non-excessiveness, necessity, as well as correspondence to other criteria.

However, in determining whether a particular measure is compatible with ECHR Article 8 the state, when identifying their pressing social need and the level of interference when pursing a legitimate aim, has a certain degree of discretion, or a margin of appreciation. It means a freedom to act, or the latitude of defence or error which the Strasburg organs will allow to national bodies before it is prepared to declare a national derogation from the ECHR, or restriction or limitation upon a right guaranteed by the ECHR, to constitute a violation of the ECHR.[156] According to the Court, margin of appreciation must be derived from "a just balance between the protection of the general interest of the community and the respect due to fundamental human rights while attaching particular importance to the latter".[157]

According to this doctrine, domestic authorities are best to settle a dispute.[158] Since security falls within a valid aim within Article 8(2), a margin of appreciation granted to the state is wide.[159] Nevertheless, the states do not get an unlimited power of appreciation and it is the ECtHR who gives the final ruling on whether interference can be justified under Article 8 (2). But, according to some researchers, if distinguishing between weak and strong proportionality tests discussed above, the

---

[154] Article 29 Data Protection Working Party (2014a).

[155] Kilkelly (2001) p. 32.

[156] Yourow. *The margin of appreciation doctrine in the dynamics of European human rights jurisprudence* (1996) p. 13.

[157] Case *"Relating to certain aspects of the laws on the use of languages in education in Belgium" v. Belgium*, No 1474/62; 1677/62; 1691/62; 1769/63; 1994/63; 2126/64, 23 July 1968, §5. See also e.g. *Lingens* v. *Austria*, No. 9815/82, 8 July 1986, §39.

[158] Spielmann (2012) p. 2.

[159] *Ibid.* pp. 14, 16. For example, in *Leader v. Sweden*, the court did not find violation of Article 8 because the applicant was regarded as a national security risk. *Leander v. Sweden*, No. 9248/81, 26 March 1987, §59.

Court acknowledges the need to take effective measures against crime and terrorism, but applies a weak version of the proportionality test or avoids it.[160]

The margin of appreciation doctrine has a number of other weaknesses: a degree of vagueness, a risk of manipulation of the identified factors and parameters, and lack of legal certainty.[161] Despite the core of the legal balancing between privacy and security lies in the proportionality and "necessary in a democratic society" tests (and the parameter of proportionality offers more legal certainty[162]), in many privacy cases the courts put the emphasis on the legality test.[163] This means that if the aviation security measure has a basis in law, this will be more valued than the proportionality/necessity considerations.

Consequently, it is argued that the proportionality principle is more a mediatory principle, which is "effective in questioning means of action, but silent on the ends of the action".[164] Thus it helps to judge, but can hardly provide a clear and concrete decision.

At the same time, with reference to recent ECJ jurisprudence in the data privacy field, the court is increasingly applying the proportionality principle in a stringent way. Examples include the *Schrems* case, where the ECJ found that the "Safe Harbour" agreement on transferring personal data to the USA is invalid.[165] In particular, it was noted that the US authorities were able to access and process personal data transferred from the EU to the USA under Safe Harbour beyond what was strictly necessary and proportionate to the protection of national security (§90).

In *Google Spain*, by virtue of the principle of proportionality, it was held that the operator of a search engine is responsible for the processing that it carries out of personal information which appears on web pages published by third parties.[166] In *SABAM*, the ECJ found that imposing a general filtering obligation on Internet service provider does not respect the requirement of proportionality.[167] All these rulings resulted in concrete decisions, with direct consequences for the respective situations or regimes.

[160] Gutwirth et al. (2011) p. 26.

[161] Lester. *The European Court of Human Rights after 50 years*. In: European Human Rights Law Review. Vol. 4 (2009). p. 474, Spielmann (2012) p. 28.

[162] Arai-Takahashi. *The Margin of Appreciation Doctrine and the Principle of Proportionality in the Jurisprudence of the ECHR* (2002) p. 14.

[163] Gutwirth et al. (2011) p. 27.

[164] Cali (2007) p. 270.

[165] Case C-362/14, *Maximillian Schrems v. Data Protection Commissioner*, 6 Oct 2015, § 106.

[166] Case C-131/12, *Google Spain SL, Google Inc. V. Agencia Española de Protección de Datos, Mario Costeja González*, 13 May 2014, §§ 63 and 3.

[167] Case C-70/10, *Scarlet Extended SA v. Société belge des auteurs, compositeurs et éditeurs SCRL (SABAM)*, 24 November 2011, § 53.

5.3.2.4 Concluding Remarks

The proportionality principle is the key principle for this research. Moreover, it seems to be increasing in importance: it is proposed, for example, that the methodology of proportionality principle should be used at the level of planning, introducing or increasing aviation security measures.[168] Regulatory or self-regulatory measures should be taken in order to encourage or oblige the interested parties, formally and substantially to apply this methodology.[169]

Within the proportionality principle, the general rule is: if an aviation security measure implies an intrusive act that interferes with the rights to privacy and data protection, decisions about it must be made by weighing the benefit sought to be achieved against the harm that would be caused to the rights. The compromise has to be proportionate to the threat perceived and the benefits of the use of that concrete aviation security measure.

As discussed, although taking into account passenger experience is increasing, it is still not so common for the aviation security to evaluate possible harms to privacy and data protection *before* enforcing a particular aviation security measure. Frequently they are considered post-factum.

Accordingly, two processes which both use proportionality tests can be noted: (i) evaluating by security providers which security solutions to implement, according to threats and risks, costs and benefits, and (ii) evaluating whether the privacy/data protection risks are proportionate to the benefits of the measure, which, by the way, takes into account all the conclusions of the first one.

As a rule, these are two separate processes, with different aims. They do not necessary run parallel – the second one may be conducted after the first one, sometimes with a period of time between the two. For instance, the installation of body scanners in some airports took place much earlier than the decision-makers started to participate in public discussions about privacy concerns, e.g. in the USA and the UK. However, in some cases, privacy concerns raised *before* the implementation of a measure may cancel or postpone it until the concerns are solved.[170]

Another problem is that the first process is carried out by the relevant national authorities in secret; as a rule, the processes of threats and risks assessments and their results are not open to public. Thus, they are outside the scope of democratic debate – a proper, democratically accountable way of making aviation security decisions is missing.

Taking into account all these factors, in general, it can be concluded that what is proportionate for the aviation security is not always proportionate for privacy/data protection. At the same time, the overlap exists, since the suitability (effectiveness) and necessity may have direct effect on both aviation security and privacy/data protection considerations. Moreover, aviation security, as discussed, slowly but surely

[168] Wright et al. (2015) p. 288.

[169] *Ibid.*

[170] As in the case of body scanners in Norway, which testing was cancelled due to privacy concerns in 2007. However, in 2015, framework contracts for scanners were concluded, see Chap. 6.

takes into account privacy/data protection considerations both as the cost of security solutions and as benefits – the means to improve passenger experience. In some cases, privacy risks may freeze or cancel even promising in security terms measure. In other cases, enhancing privacy may contribute to passenger experience.

Accordingly, the proportionality principle is indeed the key principle for this research, and will be extremely helpful for the security-versus-privacy discussion. In the Special Part, to evaluate the proportionality of a concrete aviation security measure in privacy/data protection terms, the proportionality principle will be applied using the following scheme:

1. Suitability – is the aviation security measure suitable to fulfil the specific legitimate aim? The key element here is effectiveness, which implies that the measure concerned must be effective in achieving its intended purpose.
2. Necessity – is the measure in question necessary to realizing these goals, or are there other aviation security measures which are less intrusive but capable of achieving same result?
3. Non-excessiveness – does the measure go further than is necessary to realize the goals?

In addition, in order to assess whether a particular measure is proportionate, it is necessary to evaluate many other factors, including the severity of the interference, the sensitivity of the practice or data collected, and other contextual factors. Above all, it is necessary to assess the measure's compliance with other legal principles of privacy and data protection, its impact of other human rights, in order to determine the aggregated impact and compare it with the measure's security benefits. This will be done in the Special Part.

5.4 Principles of Data Protection

Similar to other legal principles discussed in this chapter, principles of data protection comprise the essence, nature and roles of principles discussed above and are contained in law. Data protection principles are also often called "standards", "requirements" and can be defined as "the basic principles applied by data privacy laws to the processing of personal data".[171] The core idea is that all the principles must be satisfied: complying with one of them does not exempt you from complying with the others; if the principles are understood, the whole data protection law will fall into place.[172] Accordingly, if all of them are applied to aviation security measures, they will allow to establish whether collection, storage, share, transfer, analysis or any other type of processing of personal data for aviation security purposes correspond to the requirements of data protection law.

---

[171] Bygrave (2014) p. 145.

[172] Information Commissioner's Office. The Eight Data Protection Principles. http://www.belb.org.uk/downloads/foi_data_principles.pdf

The principles of data protection are international – they are formulated in the DPD, Convention 108, OECD guidelines, etc. However, the principles contained in the EU law are the most influential internationally – they have legal weight beyond the EU – and there are jurisprudence and academic works clarifying them.[173] It is clear that they may be effective outside the EU jurisdiction repeated by laws of non-EU states, where data protection law was greatly influenced by the EU regime (see Chap. 2). In the aviation security and privacy context, these principles are fully or partly repeated by e.g. ICAO in its recommendations concerning PNR usage by the states, by EU and national data protection authorities giving guidelines on the use of biometrics, body scanners, CCTV, profiling, etc.

However, although principles of data protection are international, they are not global thus they lack the universal status of the international general principles, e.g. they do not apply on the territories which do not have data protection law as such.

For instance, while the USA assigns data protection principles a significant role in its regulatory regime, the principles are formulated differently (there are similar categories known as "the principles of fair information practices"[174]) and are applied less extensively than data protection principles in Europe. Some of the principles are reflected in the Privacy Act[175] and other sector-specific laws, however, these laws contain many exceptions so that important provisions are not applicable to the use of personal data on counterterrorism purposes, by law enforcement and intelligence agencies.[176] While general protection of commercial data in the USA is lacking, the principles of fair information practices provide a good starting point for governmental use of commercial databases.[177] Despite certain similarities in comparison with the EU data protection principles, it cannot be stated that the latter apply in the USA.

Moreover, there are variations in formulation and application of concrete data protection principles from instrument to instrument, from state to state, etc. which may also contribute to non-harmonization and divergences. For instance, there are differences between the provisions of the DPD, Convention 108, and provisions of more specific instruments with different legal status providing further details on concrete data processing technology or practice. Since the principles expressed in the DPD are "the most widely implemented privacy principles globally"[178] and keeping in mind the leading role of the DPD, this research will use the latter as the main source of the data protection principles.

Another issue is that even when using the DPD as the point of departure, there are divergences in interpretations and formulations of particular principles. Bygrave, for instance, indicates the following "core principles of data privacy law":

---

[173] Bygrave (2014) p. 146.

[174] See Dempsey and Flint. *Commercial data and national security*. In: Geo. Wash. L. Rev. Vol. 72 (2004) p. 1489.

[175] Privacy Act of 1974, 5 U.S.C. § 552a (2000).

[176] Dempsey and Flint (2004) p. 1489.

[177] *Ibid.* p. 1495.

[178] Greenleaf. *Five years of the APEC Privacy Framework: Failure or promise?* In: Computer Law & Security Report. Vol. 25 (2009). p. 32.

1. Fair and lawful processing
2. Proportionality
3. Minimality
4. Purpose limitation
5. Data subject influence
6. Data quality
7. Data security
8. Sensitivity[179]

Schedule 1 to the UK Data Protection Act lists the following eight principles:

1. Fairly and lawfully processed
2. Processed for limited purposes and not in any manner incompatible with those purposes
3. Adequate, relevant and not excessive
4. Accurate and where necessary, up to date
5. Not kept for longer than is necessary
6. Processed in line with the data subject's rights
7. Security
8. Personal information shall not be transferred to countries outside the EEA without adequate protection[180]

Taking into account the concepts of the Privacy Act 1974, the US DHS developed a set of Fair Information Practice Principles[181]:

1. Principle of Transparency
2. Principle of Individual Participation
3. Principle of Purpose Specification
4. Principle of Data Minimization
5. Principle of Use Limitation
6. Principle of Data Quality and Integrity
7. Principle of Security
8. Principle of Accountability and Auditing

Article 5 of Russian Law on Personal Data provides the following seven principles of personal data processing:

1. legality of goals, faithful and diligent conduct when processing personal data;
2. compliance with purposes determined and declared at the time of collection of personal data to process personal data exclusively within the scope of the authorities granted to them;

---

[179] Bygrave (2014).

[180] UK Information Commissioner's Office. Data protection principles. https://ico.org.uk/for-organisations/guide-to-data-protection/data-protection-principles/.

[181] Department of Homeland Security. *Privacy Impact Assessment Update for TSA Advanced Imaging Technology*, 18 December 2015, p.3.

3. inadmissibility to integrate information databases created for different purposes;
4. inadmissibility to process personal data irrelevant to the purposes declared at the time of collection
5. compliance of scope and character of personal data to be processed and methods of processing with intended purposes of such data processing;
6. reliability of personal data, adequacy of personal data for processing purposes;
7. personal data shall be stored in a way that allows verification of the identity of the individual concerned only to the extent necessary for processing purposes. Personal data shall be destroyed upon achieving the set goals as well as when such goals cease to be relevant.

Other regulators, entities and researchers provide their own lists of principles. This fact does not allow making any exhaustive, generally accepted "official" list of all the principles and their names. On the one hand, the principles' components are quite similar. On the other hand, individual principles of data protection are not completely independent and there is great degree of overlap between them. In addition, they are not equal in the sense that they have different "weight". For instance, the purpose limitation principle is particularly important, since every aspect of processing – and any other principles – should be seen in the light of the purpose(s) of processing.

Thus, I have devised my own list based on the available materials:

1. Fair and lawful processing (discussed above within general principles)
2. Proportionality (discussed above)
3. Purpose limitation (a very broad and important principle)
4. Minimality (overlap with 3, but the emphasis is on ensuring limitation of personal data at the stage of data collection)
5. Non-retention of data beyond a certain period of time (overlap with 3 and 6, but in aviation security context is usually discussed separately, due to importance of limiting storage periods)
6. Data quality (common principle)
7. Data security (common principle)
8. Data subject influence (common principle)[182]

As for the issues of sensitivity and transborder transfer, due to some overlap, I will discuss them within principles 3 and 4. Thus, this research is supposed to cover all the important data protection principles that are usually indicated by different sources.

At the same time, in the aviation security field, exceptions and restrictions may apply to data protection principles. Similarly to possible limitation of the right to

[182] The principle of accountability is often argued to be a separate data privacy principle. Nevertheless, I consider it as an important emerging principle and will discuss it towards the end of this work in Chap. 8. The same applies to a more general transparency principle as indicated in Sect. 5.3.1.2 (broader than the one I discuss above within the principle of legality – legal transparency - and data subjects' right to be informed about data processing, see below).

privacy (see Chap. 2), the states – by way of a legislative measure – may be permitted not to apply data protection principles if such a restriction constitutes a necessary and proportionate measure in a democratic society to safeguard public security, public safety, prevention, investigation, detection and prosecution of criminal offences, protecting the data subject or the rights and freedoms of others, etc.[183]

Thus, data protection principles can be compromised, but essential conditions are that exceptions and restrictions must be provided by law, be necessary and proportionate – i.e. similar to requirements applicable to substantiate interference to the right to privacy. Thus, for the analysis and substantiation of privacy/data protection limitations, legal principles of privacy and data protection will be used in a complex all together. Concrete restrictions as well as issues of their substantiation will be considered in Chap. 7 with respect to concrete aviation security measures.

5.4.1 Purpose Limitation

As discussed in the Legality section, in relation to privacy, a connected term which is applied by the courts in the interference context is "legitimate aim". Due to its broadness, applying purpose limitation principles along with "legitimate aim" can be helpful to narrow the scope of legitimate purposes.

The purpose limitation principle is one of the fundamental principles of data protection, also known as "purpose specification" and "finality". Personal data must be collected for specified, explicit and legitimate purposes and not further processed in any way that is incompatible with those purposes (DPD Article 6(1)(b)). The notion "legitimate aim" was discussed above; "specified" and "explicit" purposes mean that the purposes cannot be unlimited and must be defined clearly and unambiguously, taking into account also the transparency and notification requirements, as well as prior checking mechanisms.[184]

The problem is that function creep – data collected for one purpose is used for another – is a particular concern for any smart technology, including those used in aviation security. In 2006, the presenters of IBM's "Smart Surveillance Solution" stated: "There is a lot of video captured and stored, and often the value of the video is unknown until well after the time of capture. Stored video is potentially valuable later".[185] This indicates very well how little awareness exists among technologists, security and business people about considering the possible negative effects.[186]

---

[183] See, e.g. § 6 of Recommendation CM/Rec(2010)13 of the Committee of Ministers to member states on the protection of individuals with regard to automatic processing of personal data in the context of profiling adopted by the Committee of Ministers on 23 Nov 2010, GDPR Art.23 and Recital 73.

[184] Article 29 Data Protection Working Party (2004).

[185] Wright [et al.] *Sorting out smart surveillance*. In: Computer Law & Security Review. Vol. 26 (2010) p. 349.

[186] *Ibid.*

Consequently, aviation security, which extensively uses data initially collected for purposes other than security purposes – in particular, personal data deriving from the commercial sector – clearly contravenes the purpose limitation principle. However, the regulator may use different exceptions and derogations, primarily due to the security needs, to substantiate such use, e.g. by establishing international PNR agreements, see Special Part.

Yet, while the aviation security measures, as discussed, may easily fall within the "legitimate aim" criteria in relation to the data processing, the purpose limitation principle also implies some additional requirements, which can make the process of justification more complicated. As can be seen in the very term 'purpose limitation principle', the key word is limitation; accordingly, a number of other elements connected to purpose can be limited too.

For instance, the data should not be further processed for other purposes; purpose refers to *defined* purposes and not actual purposes, implying that the controller must have decided that personal data will be used for security purposes; how the data will be handled for security purposes if the data were initially collected for other, e.g. purely commercial purposes. In addition, with reference to both privacy and data protection, other criteria such as lawfulness and proportionality will serve as a limitation in order not to let the state go with new security measures as far as it wishes.

This principle generally implies that every aspect of how personal data are processed should be viewed in the light of purposes of the processing. This relates, for instance, to other data protection principles such as data quality and non-retention of data beyond a certain period of time discussed below.[187] Thus, clear aims and specific purposes may also contribute towards limiting the categories or types of data needed, the amount of people affected, the type of processing, the duration of storage, the quality of the data required, etc. This will mean better compliance with privacy and data protection rules. For aviation security, a logical idea could be that data collected pertaining to innocent people should not amount to data collected on targeted, suspected, or dangerous individuals. Biometric and sensitive data should be subject to additional rules: each part of the process from collecting, processing and storing such data must be taken into account to ensure that each element of the data processing is fully justified.[188]

An important criterion is compatibility. As Article 29 Working Party notes, further processing for a different purpose does not necessarily mean that it is automatically incompatible; this needs to be assessed on a case-by-case basis.[189] Working Party distinguishes formal or substantive assessment. The former compares the purposes that were initially provided by the data controller with any further uses to find out whether these uses were covered. Substantive assessment is more flexible, pragmatic and effective, since it takes into account the way the purposes are (or should

[187] See, e.g. Norwegian Personal Data Act of 2000, §11.

[188] Article 29 Data Protection Working Party (2014a).

[189] Article 29 Data Protection Working Party. Opinion 03/2013 on purpose limitation, Brussels, 2 April 2013 (2013a), p. 21.

be) understood, depending on the context.[190] In order to assess compatibility, the following factors can be helpful:

- the relationship between the purposes for which the data have been collected and the purposes of further processing
- the context in which the data have been collected and the reasonable expectations of the data subjects as to their further use
- the nature of the data and the impact of the further processing on the data subjects
- the safeguards applied by the controller to ensure fair processing and to prevent any undue impact on the data subjects[191]

This principle also implies that personal data should not be disclosed, made available or otherwise used for purposes other than those specified, except with the consent of the data subject or by the authority of law (DPD Article 6). Of course, disclosure can be of different types, e.g. domestic sharing and across borders. With regard to domestic sharing, limitation concerns first persons within the organization and secondly sharing with other organs. Both, in principle, should not involve persons or entities that are unrelated to the aviation security activities. Ideally, the persons who have access to personal data should be restricted to minimum and, should there be a disclosure of personal data by such officials, the applicable punishments should be established.[192]

This is especially relevant for situations when not all personal data to which aviation security organs needs access are processed by them, and they may need access to data originally collected by other organizations for other purposes, such as PNR collected by airlines. To safeguard the data, access to the system and the number of people who have access to data should be limited, as well as the length of time that the data is stored, and compliance with other data protection principles should be ensured. For example, the ICAO PNR Guidelines require that the state should ensure that every state authority having access to personal data ensures the appropriate level of data protection.[193]

It can be seen that this principle overlaps with and may contribute to compliance with other principles, both general and data protection principles. This makes it one of the most important for the researched area, with the circulation of personal data not only among multiple entities, but transborder as well.

---

[190] Article 29 Data Protection Working Party (2013a), p. 21.

[191] For more detail, see Article 29 Data Protection Working Party (2013a), pp. 23–27.

[192] Aquilina (2010) p. 142.

[193] §2.12.

5.4.2 Minimality

As discussed in Chap. 2, the biggest challenges to data protection today are technological developments, globalization and Big Data. In aviation security, state organs, airlines, airports and other entities, via different technologies, are capable of collecting huge amounts of different types of personal data from multiple sources (see Chap. 6), which are then processed for security purposes and other uses.

These huge opportunities create advantages for aviation security, but challenges for data protection. The minimality principle – also known as "data minimization" – can contribute towards limiting data collection. It is stipulated in DPD and Convention 108 (Article 5(c)), which both are aimed at ensuring minimality at the stage of collection: personal data must be adequate, relevant and not excessive in relation to the purposes for which they are collected and/or further processed (DPD Article 6(1)(c)). It requires that the minimum amount of personal data to be collected and processed, limiting this amount to what is absolutely necessary to achieve the purpose for its collection and processing.[194]

In simple words, keeping in mind the purposes of processing, the data controller should identify the minimum amount of personal data needed to properly fulfil these purposes. Accordingly, data can be collected only in that necessary minimum amount – not more.[195]

This principle is aimed at preventing or reducing, to the greatest possible degree, the processing of personal data: in fact, many purposes can be actually achieved without recourse to personal data, or by using really anonymous data, even though they may initially seem to require the use of personal information.[196]

It can be seen that this principle is dependent on the purpose limitation principle. It is also very close to the proportionality principles, since for determination of how much data is needed, various interests – data protection interests and other interests – should be considered. It also relates to the next principle below – non-retention of data beyond a certain period of time – which is also aimed at minimization, but minimization of aspects such as data storage.

Moreover, minimality contributes to data security too: the rule that the minimum amount of data should be kept for the minimum amount of time is supposed to reduce the likelihood of data being leaked, lost or misused.[197]

For instance, a good practical illustration can be found in the employment field, where the employer's access to a part of employee's personal data (namely, speed, fuel consumption and travel times) can be limited via programming. This is supposed to protect the employee's personal data, which will not be open to the

---

[194] Bygrave (2014) pp. 151–152.

[195] UK Information Commissioner's Office. Data protection principles. https://ico.org.uk/for-organisations/guide-to-data-protection/data-protection-principles/

[196] Article 29 Data Protection Working Party (2004).

[197] Gilbert. *Dilemmas of privacy and surveillance: chaljenges of technological change* (2007) p. 26.

employer, but can still exist in the system and be useful for the employee to evaluate his driving habits.[198] Similarly, some information about air passengers, although kept in airlines' systems, could be secured by limiting authorization for access only to passengers, thus, contributing to the minimality principle and other principles.

A problem which arises is how to classify all the potentially collected data: which are needed and which are excessive. Apparently, with current risk-based and intelligence-led aviation security (see Chap. 6), security organs may consider that any pieces of information may be useful for predicting threats, thus, may be unwilling to limit them at the stage of collection.

A special treatment in the terms of minimality deserves sensitive data (see Chap. 2), which collection at the initial stage must be justified by stricter requirements. In order to assess the sensitivity of data, the context of the processing should be taken into account.[199] But in every respective case, such practices should be checked to ensure that they satisfy legal requirements. In addition to the applicability of the exemptions, other principles should be applied as well: e.g. whether the collection and usage of this data is justified and really needed, or whether collection is unnecessary and excessive.

5.4.3 Non-retention of Data Beyond a Certain Period of Time

According to DPD Article 6(1)(e), personal data can be kept for no longer than is necessary for the purposes for which the data were collected or for which they are further processed. This constitutes the principle of non-retention of data beyond a certain period of time (for convenience, "data retention principle"): data should be stored only for such periods as stipulated by law and should be deleted after the data is no longer relevant.

This principle is closely related to the principles of proportionality, minimality, purpose limitation and data quality. First, data should not be kept unnecessarily; these data must be deleted as soon as they are no longer necessary for the specific purpose.[200] Secondly, deletion of personal data when no longer needed is supposed to reduce the risk of the data becoming inaccurate, outdated or irrelevant.[201] Moreover, similar to the data minimality principle, keeping data for the minimum amount of time is supposed to enhance data security, reducing chances of data misuse.

The data retention period is commonly an essential item in the analysis of a concrete aviation security practice or regime, e.g. storage of PNR data, storage of

---

[198] Schartum. *Rettslige aspekter ved feltteknologi i arbeidslivet* (2013) p. 73.

[199] Article 29 Data Protection Working Party *Opinion 3/2012 on developments in biometric technologies* (2012b).

[200] European Committee on Legal Co-operation (2003).

[201] UK Information Commissioner's Office. Data protection principles. https://ico.org.uk/for-organisations/guide-to-data-protection/data-protection-principles/

CCTV images, etc. In particular, the data retention issues cause concerns if personal data on all passengers – including innocent people – are stored rather than only on suspects, targeted individuals and the like.

Case law around ECHR also clearly confirms that storage of personal data for longer than is necessary contradicts the "necessary in a democratic society" test.[202] Thus, excessive data retention also means interference with privacy.[203] Typically, a concrete practice often violates all the related privacy and data protection principles (see above). As the Article 29 Working Party notes, failure to specify the purpose, in terms of data protection, may lead to a breach of the minimality and data retention principles; in terms of privacy, this will mean that the legitimate aim criterion suffers, leading in turn to proportionality concerns.[204] For instance, the lack of defined purposes, disproportionality and lack of a clear link with the purpose for personal data retention are applicable pursuant to Article 29 Working Party, to PNR.[205]

Accordingly, it is essential to stipulate rules on data retention and re-evaluate such retention periodically to ensure compliance with a person's right to a private life and valid data protection law,[206] making this principle important per se and also an important condition for ensuring other principles discussed in this chapter.

5.4.4 Data Quality

The data quality principle requires that personal data must be relevant, accurate, timely, complete, kept up to date (DPD Article 6(1)(c,d)). Multiple terms are used, and in different legal instruments, there can be variation in terminology, scope and stringency of monitoring requirements.[207] But in general, two aspects of data quality can be noted: the first concerns validity – data should be valid with respect to what it describes, and secondly, data should be relevant and complete regarding the purposes of processing.[208]

The data quality principle is dependent on the principle of purpose limitation and is close to the minimality principle requiring that personal data should be adequate. Ensuring the accuracy of personal data is supposed to contribute to data minimality.

Ideally, according to this principle, persons responsible for the compilation of files or those responsible for keeping them have an obligation to conduct regular checks on the accuracy and relevance of the data recorded and to ensure that they are kept as complete as possible in order to avoid errors of omission as well as to

---

202 See *S and Marper v. the United Kingdom,* No. 30562/04 and 30566/04, 4 Dec 2008.

203 Article 29 Data Protection Working Party (2014a).

204 *Ibid.*

205 *Ibid.*

206 *Ibid.*

207 See further Bygrave (2014) pp. 163–164.

208 Bygrave (2014) p. 163.

ensure that they are regularly updated or kept current when the information contained in a file is used and for as long as it is being processed.[209]

However, in practice, persons in some cases have to contact the data collectors to rectify/delete various inaccuracies in their data. This can be done via realizing another data protection principle – data subjects' influence. This will be discussed below, but in general, it may be difficult for a person to rectify/delete inaccuracies. Thus, the realization of the principle of data quality is dependent on the provision of particular measures to be undertaken by the data controller to ensure that data is valid, accurate, relevant, etc.

5.4.5 Data Security

Personal data should be processed in a secure mode and protected by security safeguards against such risks as accidental or unlawful destruction, loss, alteration, disclosure, access, as well as any other unlawful forms of processing.[210] This is principle of data security. Clearly, there is no "one size fits all" solution to data security universal for all spheres and all contexts, thus, appropriate security measures should be adopted on a case-to-case basis.[211]

As discussed in Sect. 2.2.2.3, a part of aviation security is information security which is aimed to protect the respective systems against attacks. The latter can have very serious dramatic consequences. This substantiates enhanced requirements to information security. Consequently, since undertakings on data security of personal data often come parallel with broader undertakings on information security, the requirements may be enhanced too.

It should be also noted that in the considered jurisdictions, the application of this principle – or the approach to security requirements – differs greatly between Russia and the EU/USA.

In the EU/USA, the laws generally indicate that the methods of data protection must be reasonable and sufficient, leaving the implementation of these principles to the controller, who will take full responsibility if the measures taken are insufficient. Thus, general technical standards are not provided. For example, Article 17(1) of DPD provides that data controllers and processors must implement two types of appropriate security measures: technical and organizational; Recital 46 stipulates that these types of measures should be taken both at the time of the design of the processing system and at the time of the processing itself. In the EU, there is a considerable disparity in the security requirements of Member States: while some

[209] The United Nations General Assembly Guidelines for the Regulation of Computerized Personal Data Files.

[210] Art. 17(1) of DPD, Art. 11 of OECD Privacy Framework 2013.

[211] UK Information Commissioner's Office. Data protection principles. https://ico.org.uk/for-organisations/guide-to-data-protection/data-protection-principles/

specify more details, in the UK, for example, the requirements simply repeat those of the DPD.[212]

Data security requirements in Russia are very comprehensive and detailed. It is even argued that this law is aimed largely at stipulating technical requirements to personal data processing – so complicated that they are analogous to protection of state secrets – rather than protection of data subjects.[213]

Data controllers must provide technical measures according to the security levels determined by the RF Government.[214] The choice of the means of protection of personal data made by the controller in accordance with the regulations adopted by the FSB and the Federal Service for Technical and Export Control of the RF (FSTEC). In practice, concrete methods and techniques appear to be excessive and expensive: expenses for security equipment (which must be produced by companies licensed by the FSTEC and the FSB) constitute up to 200% of the annual turnover and then 10–15% of the cost for annual maintenance.[215] But in reality, personal data in Russia are usually stolen through bribery of responsible employees rather than by breaking the security systems, so all these requirements may be devoid of meaning and effectiveness.

Nevertheless, the aim of all the regimes is uniform – to keep personal data secure. Particular aviation security technologies dealing with personal data imply concrete security undertakings and are very similar across jurisdictions, as will be discussed in Special Part.

5.4.6 Data Subject Influence

It is not only the controller who ensures that the data are kept secure, accurate and so on. The data subjects also have their self-interest at stake and thus should have the right to participate. This is another fundamental principle of data protection – data subject influence (sometimes also called a principle of cooperation).

This principle refers to a number of rules providing various rights of individuals. According to Bygrave, the relevant rules include (i) the rules to make people aware of data-processing activities in general, (ii) to be aware of basic details of the processing of data on themselves (including the rules on collecting data directly from data subjects, rules to orient data subjects directly about certain information on processing, and rules prohibiting processing without the consent of data subject), (iii) the right to gain access to data kept on them, and (iv) to object to others' processing of data on themselves (which is close to the rules prohibiting some types of

[212] Esayas (2015) p. 16.

[213] Chernova (2013).

[214] Resolution of Government of 1 November 2012 N 1119 On approval of requirements for the protection of personal data during their processing in information systems of personal data.

[215] Modern Telecommunications Russia. *The Council of Federation adopted Personal Data Law*, 21 July 2011. http://www.telecomru.ru/article/?id=6067

processing without data subject's consent) and to demand that the data be rectified or erased.[216]

The UK ICO provides the following list of relevant rights:

1. a right of access to a copy of the information comprised in their personal data;
2. a right to object to processing that is likely to cause or is causing damage or distress;
3. a right to prevent processing for direct marketing;
4. a right to object to decisions being taken by automated means;
5. a right in certain circumstances to have inaccurate personal data rectified, blocked, erased or destroyed; and
6. a right to claim compensation for damages caused.[217]

As can be seen, the principle of data subject influence relates to a wide category of rules, but they overlap greatly. Below, this principle will be discussed taking into account the rules indicated by Bygrave.

The DPD contains duties of information and orientation of the data subject: first, general information should be provided to public, and it should be easily accessible and easy to understand. This relates to the more general principle discussed above within the legality principle, specifically, the principle of *publicatio legis*/transparency. As noted, security needs may limit access to legal information if the latter falls within a restricted category.

Data subjects should be informed of the data processed on them, purposes of such processing and the identity of who is collecting their data. Information should be also given about the measures for protection of the data, whether there is any data share, the identity of the officer collecting and responsible for the data, procedures available for redress and contact information for persons to whom to address questions or concerns about data retention (Articles 10–11).

In general, the data subject can object at any time to the processing of data relating to him on compelling legitimate grounds relating to his particular situation.[218] In addition, the ability to object is connected to the rules prohibiting certain data processing without the data subject's consent[219] (DPD Articles 7–8). The latter means "any freely given specific and informed indication of his wishes by which the data subject signifies his agreement to personal data relating to him being processed" (DPD Article 2(h)). It should be given unambiguously (DPD Articles 7(a)) and explicitly ((DPD Articles 8(2)). The "unambiguous" means that, as a general rule, data subjects should give consent to the terms and conditions of the processing of his/her personal data, and their actions should leave no doubt that they have given

[216] Further see Bygrave (2014) pp. 158–163.

[217] UK Information Commissioner's Office. Data protection principles. https://ico.org.uk/for-organisations/guide-to-data-protection/data-protection-principles/

[218] Article 29 Data Protection Working Party (2004).

[219] Bygrave (2014) p. 160.

consent.[220] "Explicit" consent is a more stringent requirement – there must be specific request for permission from the data subject followed by specific reply.[221]

The problems which arise concern ensuring the criteria informed, specific (or explicit) and freely given consent. In principal, free consent is difficult to realize in reality: in data processing, data subjects are almost always the weakest party, making consent a formality.[222] The point is that aviation security procedures fundamentally differ from free choice. Passengers and the security organs are not equal in powers, and passengers cannot object to screening or personal data collection unless they accept not to fly at all.

Moreover, the exceptions from the duty to obtain consent are quite widespread and include such categories as processing being necessary for the performance of a contract or for the performance of a task carried out in the public interest, in order to protect the vital interests of the data subject, etc. (DPD Article 7). Clearly, the interests of aviation security may fall under any of the exceptions and thereby justify most instances of processing.

As a result, genuine, freely given and informed consent in aviation security can hardly be ensured or required either, and in practice, people are not able to consent or reject airport surveillance.[223]

Further, data subjects should be allowed to access their data and make corrections to any inaccurate data (DPD Article 12, Article 8 of the CFREU). This allows a data subject to have an indirect right of access to carry out checks over public security files to ensure that the file contents are accurate. It is common that stipulating the rights of data subjects by national data protection law are supported by corresponding provisions on data controller's obligations to provide these rights to data subjects.

For instance, the Russian Personal Data Law stipulates the rights of the data subject to obtain information related to the processing of his/her personal data, to access it, to redress breaches of personal data processing, to correct, block or destroy personal data (Article 14) as well as respective duty of data controller to provide these opportunities to data subject (Articles 20 and 21).

It is also essential that the agencies concerned – i.e. data controllers – must have guidelines or procedures on access to and sharing of personal data in order to ensure this obligation to provide ensure the data subjects' rights to access, rectification and deletion.[224] However, in practice, there appear difficulties with the duties of access/rectification: it is very often that data controllers have no established procedures and mechanisms thus are not capable to provide these rights. Notable examples are unsuccessful attempts to get air passenger personal records from authorities and airlines, which will be discussed in Special Part.

---

[220] *Ibid.*

[221] *Ibid.*

[222] Gutwirth et al. (2011) p. 27.

[223] Gilbert (2007) p. 33.

[224] Aquilina (2010) p. 142.

5.5 Concluding Remarks to Chap. 5 and to General Part

The principles, as discussed, are the key elements of this research. They express the essence of the considered values – aviation security and privacy/data protection – and constitute the basis of the applicable regulation, allowing us to identify the privacy/data protection concerns that arise from concrete aviation security measures and how they can be dealt with.

The key principle is the proportionality principle. It has two roles: first, it is valuable as such, as one of general legal principles of which the core essence is the comparison of different values. Thus, it can be used as an instrument for finding proportionality between a number of values: threats and risks, costs and benefits, the need to interference and the need to protect a human right, ultimately – between aviation security and privacy/data protection.

Secondly, it is valuable since it unites different factors and issues discussed in General Part of this research into one "puzzle" allowing analysing aviation security versus privacy dilemma from multiple corners.

In particular, aviation security issues discussed in Chap. 4 have impact and/or define aviation security. Similarly, privacy/data protection concepts, regulation (Chap. 2) and principles (Chap. 5) define privacy/data protection. At the same time, as discussed above, security decisions and principles of aviation security discussed in Chap. 4, privacy and data protection principles discussed in Chap. 5, as well as all other relevant human rights discussed in Chap. 3 have impact on the proportionality of the aviation security regimes.

It can be seen that some factors perform a double task. For instance, with reference to the section on aviation security, the rationale of aviation security decisions and aviation security principles explains, on the one hand, the logic of regulators adopting and using different aviation security techniques and methods. On the other hand, the security regulators use the proportionality principles to find a balance between threats and risks, costs and benefits. Cost and benefits may include costs in the form of privacy risks and benefits, if a measure contributes to passenger experience, for instance, respects or even enhances privacy. These issues, as well as issues of effectiveness of aviation security measures, overlap with the use of proportionality principle from privacy/data protection perspective.

Altogether, these factors permit the establishment of the approach to the aviation security and privacy dilemma that will be used in the Special Part to evaluate the impact on privacy/data protection and the proportionality of the selected aviation security measures. Other relevant human rights will be evaluated additionally to better assess proportionality of the regimes.

Accordingly, the impact of aviation security measures on passenger rights will be assessed using the following scheme. First, the applicability of privacy/data protection to the selected aviation security measures should be analysed using their technological and operational characteristics and knowledge from Chaps. 2 and 3.

If positive, it should be established whether there are relevant limitations, concerns or problems. If positive, the next step is to consider whether the limitations, concerns or problems can be justified. For these two steps, legal principles of privacy/data protection indicated in Chap. 5 will be used, taking into account knowledge from Chap. 4 and other chapters. Other human rights discussed in Chap. 3 will be evaluated similarly, but on a limited basis, in the amount necessary to indicate additional problems, if any.

Clearly, proportionality is not a fixed condition and can range from "proportionate" in the best case to "non-proportionate" in the worse. The first condition is perfect, implying that there are either no concerns, or the concerns are outweighed by the security needs. Apparently, practical results from evaluation of particular measures will differ, entailing that each must be placed somewhere between the poles. At the same time, however, it will be possible to evaluate an aggregated effect of all the considered measures on all privacy and data protection principles as well as on all human rights. Consequently, it will allow – with certain limitations – evaluation of contemporary aviation security regimes as such.

Part II
Special Part

Chapter 6
Aviation Security Technologies

6.1 Introduction

This chapter will start with an overview of types of aviation security measures as well as their locations – airport, aircraft and beyond. This will allow us to pinpoint the selected security measures more precisely and to provide a broader picture of aviation security measures in general.

Next, I will discuss a number of contemporary aviation security technologies, including body scanners, CCTV and PNR systems. Focus will be laid on their technical and operational characteristics. As mentioned in Chap. 1, they will be discussed taking into account their possible impact on individuals' rights. The technologies' effectiveness issues, that is, their benefits for security and the ability to protect the right to life will be evaluated too, since they are critical for justification of the measures' negative impact on passengers' other rights.

In addition to privacy and data protection, other human rights' applicability will be considered as well. Since the issues of the freedom of movement and the right to non-discrimination are quite similar for all the technologies, they will be analysed in separate sections. With reference to scanners, another human right to be discussed is the right to health. With regard to the freedom of thought, conscience and religion, religion can be grounds for discrimination, thus, any "discriminatory" measure with may have impact on this freedom indirectly. This will mainly be discussed within the discrimination issues; other aspects will be mentioned where applicable. As for the rights of the child, in general, aviation security measures are applicable to children in the same manner as to adults. But as discussed in Chap. 3, any negative impact on adults' rights may become more serious in respect of children.

Focusing on current aviation security regimes, I will offer relevant facts about the development of the measures, specifically to illustrate two parallel, sometimes overlapping, but quite controversial tendencies: (i) endeavours to make the technologies

O.M. Enerstvedt, *Aviation Security, Privacy, Data Protection and Other Human Rights: Technologies and Legal Principles*, Law, Governance and Technology Series 37, DOI 10.1007/978-3-319-58139-2_6

more privacy- and data protection-friendly by providing different safeguards, including automation, and (ii) technological developments making the measures more sophisticated and "smart", with the growing use of personal data and enhanced opportunities for personal data processing, automated analysis, etc.

In this research, as discussed above, three main aviation security measures are selected. However, other techniques may constitute an essential part of them, or otherwise be connected, and at the same time can constitute separate aviation security measures. Specifically, kin conjunction with CCTV, it is biometrics, in particular, facial recognition, and behaviour analysis; with reference to selection of passenger for body scanners, CCTV and PNR – profiling. I will mention these techniques as well.

This chapter will have two "conclusions". First, a discussion on a broader perspective involves modern technological trends in aviation security that to different extents relate to the selected measures. They include intelligence-led, pro-active and risk-based approach, global information sharing, randomness in aviation security and combination of different technologies. Of particular interest are emerging trusted traveller programs (TTP). All these trends may have positive effect on the measures' effectiveness, but may imply additional impact on privacy and other human rights.

The conclusion concerns a more specific perspective: the selected measures' effectiveness, i.e. the ability to protect the right to life, their application to the rights to privacy/data protection, and other human rights applicability and concerns. More detailed privacy/data protection concerns and issues of the legality and proportionality of the regimes, common for the applicable human rights, will be discussed in Chap. 7.

6.2 Types of Aviation Security Measures

If taking into account the political aspect of civil aviation and attacks on it mentioned above, it can be argued that aviation security is "an accidental by-product of national security",[1] and that it depends greatly on foreign policy – for instance, Middle Eastern geopolitical problems must be solved to prevent terrorism in aviation security.[2] This pertains especially to the USA's position as a superpower. Following this logic, in Russia, the Chechen/Caucasian problem should be first solved: indeed, many acts of terrorism in Russia have Chechen/Caucasian causal "roots".

While these issues lie beyond the scope of this research, another prominent "solution" to aviation insecurity is the securitization of the entire aviation system.[3] However, the answer to the question how much security is enough may be quite

[1] Salter (2008) p. 246.

[2] St. John. *The Politics of Aviation Security* In: Aviation terrorism and security (1999) p. 46.

[3] Salter (2008) p. 246.

speculative; while state organs may constantly argue that the security is still weak and request more funds, some researchers call this "security theatre".[4] In addition, some events may cause disproportional "securitization". For instance, 9/11 led to an enormous increase in state investment in security technologies; expenditures covered, among other things, passenger inspection and screening; information management technology and intelligence, etc.[5] However, the purpose of this section is to present modern aviation security measures as objectively as possible, no matter whether it is viewed as "aviation security" or "security theatre".

As discussed in Chap. 4, the ultimate aim of the risks assessment is to protect the aviation system, and to do it in the most effective way. In practice, this means determining which concrete aviation security measures should be used and in what amount.

All the security measures may be classified according to different grounds. Annex 17 distinguishes (i) "Preventive security measures"[6] from (ii) "Management of response to acts of unlawful interference",[7] and as noted, this research focuses on preventive measures.

It is also believed that the security undertakings must cover each and every level and stage at which the risk that such acts may occur exists. All measures can in general be classified according to specific security areas/segments or assets to be protected, including but are not limited to the following: measures relating to airport security, aircraft security, in-flight security, measures relating to passengers and their cabin baggage, checked baggage, special screening procedures/special categories of passengers,[8] cargo, mail and other goods.[9]

The key feature of an international airport is that it is a national and international border, it is both "a continuation of urban space and a connection to the globe".[10] Thus, in addition to aviation security, border security is applicable. It should be noted that customs and border protection measures are usually formally excluded from the aviation security and are the task of border security – government control authority, customs and immigration, as in the case of the EU and Russia. In the USA, customs and border protection is included in the layers of aviation security.[11] However, even if they are regarded as separate entities, as discussed in Chap. 4, customs and immigration authorities are a part of aviation security systems, and there is overlapping due to similarity of tasks and methods used.

---

[4] Schneier (2008) p. 69.

[5] Amicelli (2012a) p. 3.

[6] Chapter 4.

[7] Chapter 5.

[8] May include diplomats, persons exempted from screening, persons in custody and under administrative control, airline crew, airport staff and other non-passenger, monitoring performance of security equipment, carriage of weapons, travel of armed personnel, etc.

[9] E.g. Annex 17; Annex to Regulation (EC) No 300/2008.

[10] Salter (2008) p. 256.

[11] E.g. the US TSA's "20 Layers of Security". http://www.tsa.gov/about-tsa/layers-security (date accessed: 27 Feb 2013).

Airport security is a term which is often used as a synonym for aviation security, since air travel security is primarily based in airports. The importance of the airport security measures cannot be underestimated: "while still absolutely vital, [they] are at best a last line of defense".[12] Airport security thus amounts to aviation security but mainly with reference to the ground. Airport security includes perimeter security, airside security, airport personnel identification, airport planning requirements, access control, screening of persons other than passengers and items carried, examination of vehicles, surveillance, patrols and other physical controls.

Taking into consideration all the airport infrastructure components mentioned above in Sect. 4.5.3 (assets), it may be impossible to draw the line between the airport and non-airport areas such as parking or train stations. However, the measures which are aimed at protecting the airport complex together with such connected objects are associated with perimeter security.[13] The latter involves physical security (fences, barriers, doors and locks), security patrols, camera monitoring, radars and many other different technologies such as thermal imaging, a twin beam technology,[14] indoor and outdoor electronic perimeter intrusion detection systems, etc. An essential premise for this research is that camera surveillance may start far before the airport entrance, since cameras cover related locations as well, including private areas, if any.

Further, for the purposes of security, the airport area is divided into airside ("the movement area of an airport, adjacent terrain and buildings or portions thereof, access to which is restricted") and landside ("those parts of an airport, adjacent terrain and buildings or portions thereof that are not airside").[15]

The prime concern is the airside, access to which is strictly controlled, including the passenger terminal, where the passenger security screenings and checks are carried out. There is also division into "sterile" and "public" zones based on identification cards and authorized entry. Measures of access control include different access control systems, readers, controllers, software solutions, etc. They may include biometrics. Needless to say, all airport areas are thoroughly equipped with camera surveillance systems.

As stated earlier, this research focuses on three aviation security measures: body scanners, camera surveillance and PNR systems. In order to locate the selected measures in a standard airport, I developed Fig. 6.1 which is a basic "map" of these as well as some other general measures. Although it is not intended to provide a comprehensive picture due to, among other things, local divergences and differences even between airports within one state, it does allow us to see major common lines.

---

[12] ACI position brief. November 2010. http://www.aci.aero/Media/aci/file/Position%20Briefs/2010/position%20brief_SECURITY%202010.pdf

[13] See more in Barry and Mazel. *Airport Perimeter Security: Where we've been, Where we are, and Where we're going* (2008) p. 990.

[14] A wireless barrier beams which provide a warning of intrusion.

[15] Regulation (EC) No 300/2008 Article 3. Cf. Annex 17: Airside – "the movement area of an airport, adjacent terrain and buildings or portions thereof, access to which is controlled", page 1–1.

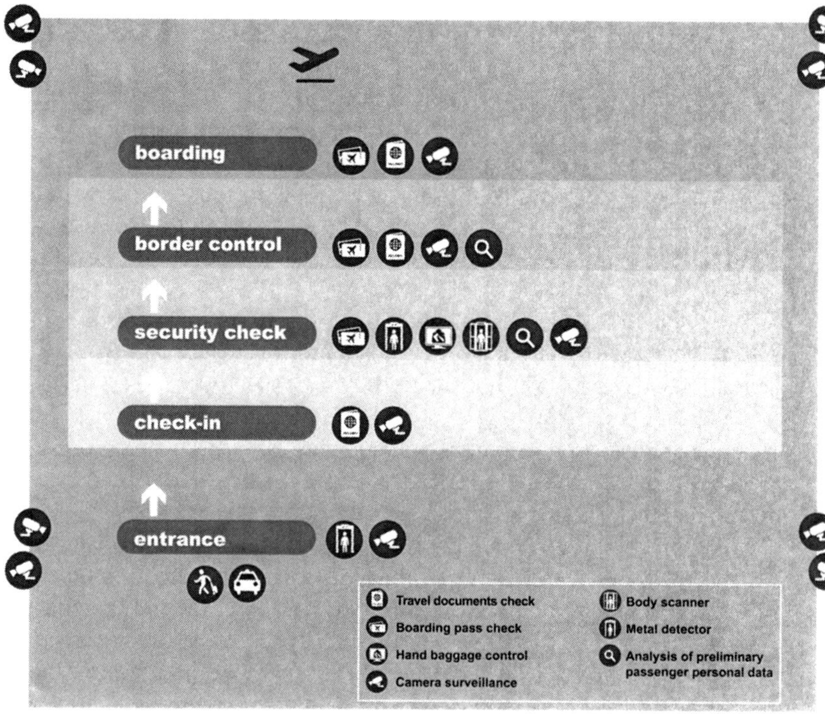

Fig. 6.1 Aviation security measures in a standard airport

Passenger screening is considered the most visible and the most personally intrusive component of airport security system. First of all, screening implies application of rules prohibiting or restricting the carrying of certain items. Secondly, it involves procedures aimed at controlling and preventing passengers from carrying those items: removing shoes, coats, belts, etc., physical inspection of passengers using manual methods or technologies.

For instance, according to the EU legislation, a list of permitted methods of passenger screening includes the following procedures: (a) hand search; (b) walk-through metal detection equipment; (c) hand-held metal detection equipment; (d) explosive detection dogs; (e) explosive trace detection equipment; (f) security scanners which do not use ionising radiation.[16] In Russia, the regulation describes very broadly technologies and special equipment used for inspection at airports: technologies and special equipment used for inspection at airports include stationary

[16]Annex to Commission Regulation (EC) No 272/2009 of 2 April 2009 supplementing the common basic standards on civil aviation security laid down in the Annex to Regulation (EC) No 300/2008 of the European Parliament and of the Council, as amended by Commission Regulation (EU) No 1141/2011 of 10 November 2011 amending Regulation (EC) No 272/2009 supplementing the common basic standards on civil aviation security as regards the use of security scanners at EU airports.

roentgen television introscopes and metal detectors, portable (hand) metal detectors, roentgen image scanners, the systems of introvision in the terahertz range of the electromagnetic spectrum, and other means certified in accordance with established procedure as well as sniffer dogs of canine units.[17]

But, although aviation security regulation focuses on screening rules and procedures, passenger-related measures go far beyond this. Other basic forms are camera surveillance, with cameras installed throughout the airport and in zones nearby; analysis of passenger personal data, which implies use of data collected beforehand and from different sources such as governmental watch lists, databases of different law enforcement and intelligence bodies, data from airlines, frequent flyers lists, etc.; profiling, including subtypes such as behaviour analysis, computerized profiling of personal data; use of biometrics, etc. In addition, as will be discussed below, these measures can be used in combination. All this broadens the scope of measures relating to passengers substantially, going far beyond security checkpoints.

Further, despite all these undertakings on the ground, clearing a passenger and allowing him/her on board is not the equivalent of the absence of threats. Thus, aviation security measures continue on board the aircraft.

The terms aircraft security and in-flight security both refer to the procedures and infrastructure designed to avoid security problems aboard the aircraft. Sometimes aircraft security is used to cover both dimensions; however, the EU law makes a distinction between these two terms. Aircraft security is related to the procedures carried out while the aircraft is still on the ground: for instance, before departure, an aircraft is subjected to an aircraft security check or aircraft security search in order to ensure that no prohibited articles are present on board.[18]

In-flight security refers to the procedures relevant while the aircraft is in the air. The prime aim of security is to prevent a terrorist from boarding in the first place. There are also difficulties in avoiding external threats while in the air; therefore, the security measures employed while the aircraft is in the air are relatively limited. The applicable measures are illustrated in Fig. 6.2.

In general, unauthorized persons are not allowed to enter the cockpit during a flight, and there are some rules governing how to deal with potentially disruptive passengers. Measures may include a hardened cockpit door, training of flight crew, cabin staff, and passengers, security officers carrying weapons,[19] providing crew with weapons; cockpit door surveillance, cabin surveillance and cargo surveillance systems which allow pilots to watch all these areas of aircraft on cockpit screens.

Such measures as hardened cockpit door, training crew and passengers, and in-flight security officers can be considered quite effective. However, some of them can become a threat if they fall into the "wrong hands". For example, the weapons can be used by "insiders" – a pilot, crew member or security officer recruited by

[17] Art. 10 of Order of the Ministry of Transport of 25 July 2007 N 104 On approval of rules for pre- and post-flight inspections.

[18] §3 of Annex to Regulation (EC) No 300/2008.

[19] E.g. Federal Air Marshal Service, including Law Enforcement Officers on some flights in the USA.

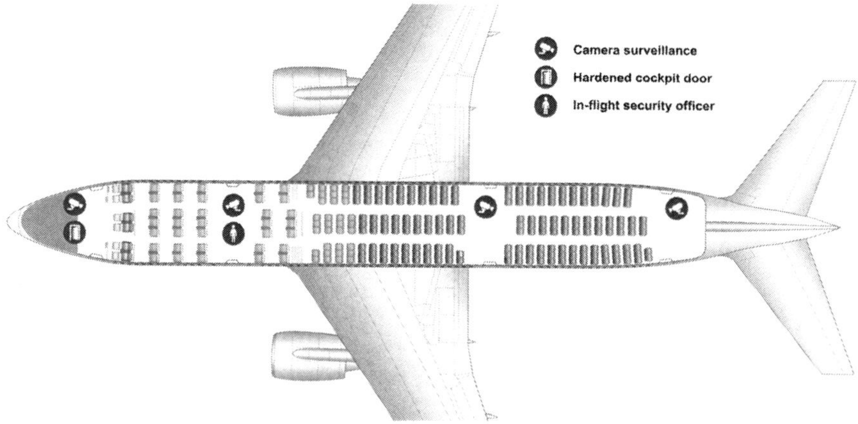

Fig. 6.2 Aviation security measures in standard aircraft

terrorists, or the enhanced cockpit door may serve as a barrier to crew and passengers attempting to overpower suicidal pilots who lock themselves inside the cockpit.[20] The advantage of these measures is they do not involve human rights concerns.[21] In contrast to that, the use of cabin surveillance systems, for example, raises questions, especially if the cameras are hidden.

Overall, as can be seen in Figs. 6.1 and 6.2, the majority of measures which have privacy/data protection concerns are associated with the airports, and a smaller part to on board the aircraft. However, the scope of aviation security measures is beyond these physical locations, as shown in Fig. 6.3.

Accordingly, in addition to airports, aircraft, and airport infrastructure, which, as discussed, can refer to multiple zones and objects, aviation security measures may also apply to other public zones, and actually to anywhere in general civil society.

In particular, this concerns personal data, the collection, sharing and circulating of which can be difficult if not impossible to detect and control, at least by air passengers. Namely, some security measures may start already at the moment when a person is sitting at home and booking a ticket via Internet. During this process, personal data which he/she fills out in the airline's website comes to the airline's computer reservation systems immediately. Then these data may be stored, processed, transferred or otherwise used for security purposes not only in connection with particular flight, but beyond this, and for years. The latter can be the case if a person's name appears on a watch list (even by error). Accordingly, it can be presumed that security measures do not cease when a person leaves the airport, and may take place even if he/she did not fly at all after reserving a flight (see more detail in the section on PNR below). Thus, aviation security measures are not limited to definite times and places.

[20] E.g. Germanwings flight 9525 of 24 March 2015. Baum (2015a).

[21] Schneier (2008) p. 69.

Fig. 6.3 Aviation security measures scope

6.3 Body Scanners

6.3.1 Introduction

State-of-the-art imaging technologies, in other words, body scanners, are one of the most "visible" airport security solutions and at the same time one of the most controversial issues of contention for both security professionals and civil rights advocates.

Similar to any other relatively new security measures, they are being introduced in the airports throughout the world under the banner of combatting terrorism. Taking into account the nature of recent attacks and the emergence of new technologies that enable the relatively easy manufacture of non-metallic weapons,[22] body scanners are designed to detect such objects concealed in clothing.

Traditional measures for detecting hidden objects are metal detectors and hand search. Accordingly, detectors are limited to detecting metal items. Hand searches may differ, but the most effective is a full search of all body areas, or a strip-search. Even if done in private, such searches raise privacy concerns, require additional staff and cause delays for passengers. On the one hand, privacy concerns prohibit direct touching of private body areas as part of the search. On the other hand, it is

[22] Neeman (2013) p. 13.

relatively easy to hide threats in sensitive body parts.[23] Hand searches as alternative to body scanners and respective problematic issues will be further discussed in Chap. 7, but in general, the scanner's advantages are quickness and no need of physical touching. The choice between hand search and body scanners is thus a trade-off between security effectiveness and privacy concerns.[24]

It is noteworthy that as present, instead of the term "body scanner" or "whole-body scanner", more neutral and technical terms are used, like "security scanner" in the EU, "advanced imaging technology" (AIT) in the USA, "radio scan portal" in Russia. Clearly, this is done in order to present, via language, the scanning process as automatic, anonymized, universal, neutral, routine, and professional, downplaying the core essence – inspection of the naked body.[25] At the same time, the machines are also often derisively referred to as "naked scanners", "nudeoscopes" and "digital strip searchers" – in particular, by privacy advocates.

The first body scanner was created by Dr. Steven W Smith in the USA in 1992. It was an ultra-low-dose Secure 1000 backscatter X-ray scanner. It is believed that the first airport in the world to implement body scanners was the Schiphol in the Netherlands in 2006, but the trials of the technologies started much earlier. According to some information, the USA started trials not later than in 2002,[26] Russia – in the middle of 1990s.[27]

But the hour of body scanners came apparently after 25 December 2009, when the so-called "underwear bomber" attempted to detonate plastic explosives hidden in his underwear. Then, many other states committed to installing the devices as well. The UK, for example, installed the machines immediately; Schiphol purchased additional units, etc.

Detection performance lies in the scanner's ability to detect any prohibited object that the person screened may be carrying on the body or in the clothing. Several technologies are used in order to identify such hidden objects. They include those based on natural thermal radiation (heat from the body), but the most commercially available security scanners use the following technologies: backscatter X-ray,[28] active millimetre wave (MMW), and X-ray transmission imaging technology.

The early commonly used backscatter and MMW devices produced and showed to a human operator extremely detailed images of the screened persons as if they were naked. Thus the reaction of the general public, privacy organizations and other

---

[23] *Ibid.*

[24] *Ibid.*

[25] Tirosh and Birnhack (2013) p. 1274.

[26] For example, Post-gazette.com in March 2002 informed that new X-ray scanners were tested in the government technical center in New Jersey and were tested on volunteers at several airports.

[27] Federal press, 16 Sep 2011. http://fedpress.ru/federal/econom/train/id_250058.html (date accessed: 3 February 2012).

[28] Backscatter (low X-ray) technology is based on the X-ray Compton scattering effect. It detects the radiation reflected from the object to form a two-dimensional image, revealing details of the surface of the body and a high resolution image of any external objects. Passengers must stand between two large boxes and raise their arms above their head.

interested persons and agencies was appropriate: protests and scandals. As a reaction, amendments were made to present images visible to operator in more privacy-friendly forms: blurring faces and intimate parts, using mannequin figures, etc. (see more detail below).

As for the health risks, the difference between the mentioned types of scans is the use (by backscatter and X-ray transmission) and non-use (by MMW) of ionising radiation. The latter will be discussed below in more detail.

As for backscatter, according to a number of reports, there are no health risks, as a person receives less radiation from X-ray scan than from 2 min of airline flight. For instance, the UK Health Protection Agency concluded that the dose received from being scanned is far below the allowed levels in the UK and does not constitute any unacceptable risks to health.[29]

According to other opinions, it is much more. For example, Peter Rez from Arizona State University (USA) discovered that the radiation dose was often higher than the manufacturers claimed. He said the probability of dying from radiation from a body scanner and that of being killed in a terror attack are both about one in 30 million, making body scanners redundant.[30] According to a report from ProPublica, "Research suggests that anywhere from six to 100 U.S. airline passengers each year could get cancer from the machines".[31]

Potential harms may exist to the frequent traveller, children, persons with specific medical conditions such as pregnancy, chronic health conditions, and to security and airport personnel who are repeatedly exposed to radiation. There are fears that the frequent use of scans may cause DNA damage in frequent fliers, which may lead to cancers[32] as well as increase doses of cumulative radiation of population.

Transmission scanners deliver a radiation dose that is 20 to 85 times higher than the backscatter scanners thus raise severe health concerns. However, it is also noted that "whilst those figures taken out of context may seem high, when compared to cosmic radiation exposure on the average long-haul flight, they are not so shocking".[33]

Nevertheless, transmission scanners are not used for passenger screening. Backscatters were prohibited in the EU in 2012. The last UK backscatter was removed in 2012, the US in mid-2013. In Russia, backscatter scanners were used in Domodedovo in 2007, but were substituted by MMW machines. The reasons for refusal from backscatters were privacy and health issues rather than the security

[29] UK Department for Transport *Guidance. Security scanners implementation information* (2015b).

[30] Daily Mail Reporter. *Airport body scanners are 'just as likely to kill you as terrorist bombs'*. 19 November 2010. http://www.dailymail.co.uk/travel/article-1331185/Airport-body-scanners-just-likely-kill-terrorist-bombs.html

[31] Grabell. *U.S. Government Glossed Over Cancer Concerns As It Rolled Out Airport X-Ray*, ProPublica, 1 November 2011. http://www.propublica.org/article/u.s.-government-glossed-over-cancer-concerns-as-it-rolled-out-airport-x-ray

[32] Mick. *Despite Allegations of Fraud, TSA Lets Scanner Maker Keep Nearly $300M*, Dailytech, 18 January 2013. http://www.dailytech.com/Despite+Allegations+of+Fraud+TSA+Lets+Scanner+Maker+Keep+Nearly+300M/article29691.htm

[33] Neeman (2015) p. 26.

issues. Security experts claim that high resolution backscatter systems are more effective in comparison with lower spatial resolution MMW.[34]

According to the official US TSA's web-site, there are about 740 MMW units at nearly 160 airports.[35] By November 2015, the list of UK airports with scanners included 16 airports.[36] In Russia, body scans are installed at several major airports, including Moscow's Sheremetyevo, Domodedovo, and Vnukovo; Koltsovo in Ekaterinburg, Sochi and Norilsk. The list of airports currently deploying scanners is constantly growing and includes also airports in Canada, France, the Netherlands, South Korea, Nigeria, etc.[37] In Norway, the machines were tested, framework contracts with scanners' producers were signed, and as soon as financial issues are resolved, the scanners will be in use.[38]

This section analyses MMW scanners which are used in the selected states. X-ray transmission imaging technology will also be presented as an example of possible future technologies. For the extensive discussion of older generation scanners versus privacy and data protection, see my earlier work.[39] It should be noted that due to some uncertainly about the MMW technology, part of the discussion is dedicated to finding out the truth about the scanners, and as a result of this uncertainty, the subsequent legal discussion becomes partly "speculative".

6.3.2 Millimetre Wave

6.3.2.1 General

Millimetre wave (MMW) technology has existed since 1986. It was first developed for the clothing industry, where it was used for taking measurements. The technology has since been used for several years as a security resource.[40] There are two

[34] Huey. *When Politics Meets the Hard Realities: Cold Fact, Speculation and Surrealism at the Checkpoint*, 19 September 2012, http://www.noquarterusa.net/blog/72258/when-politics-meets-the-hard-realities-cold-fact-speculation-and-surrealism-at-the-checkpoint/

[35] TSA. Advanced Imaging Technology. http://www.tsa.gov/traveler-information/advanced-imaging-technology (date accessed: 31 Jan 2015).

[36] The UK Secretary of State. Direction to certain aerodrome managers under the Aviation Security Act 1982 relating to security scanners 2015. https://www.gov.uk/government/uploads/system/uploads/attachment_data/file/397617/Consolidated_Scanners_Direction_Redacted.pdf

[37] There are several "lists of airports with body scanners" available on the web, but none of them can be considered as full and up to date, as the situation changes constantly. An example of such lists can be found at http://www.flyertalk.com/forum/practical-travel-safety-issues/1138014-complete-list-airports-whole-body-imaging-advanced-imaging-technology-scanner.html

[38] Interview with Avinor representative, 22 Oct 2015.

[39] Mironenko. *Body scanners versus privacy and data protection*. In: Computer Law & Security Review. Vol. 27 (2011).

[40] National Coordinator for Counterterrorism (Netherlands). Frequently Asked Questions Security Scan. How long has this technology existed? https://english.nctv.nl/themes_en/themes-a-z/Security_and_civil_aviation/Security_Scan/faq-14.aspx

types of nonionizing millimetre radio wavelengths: passive[41] and active. It is the latter which is used on the airports and will be further discussed.

In contrast to backscatter, MMW do not expose passengers to ionising radiation.[42] A form of electromagnetic radiation is used, which do not alter atoms in molecules by removing electrons, and which lie in the spectral region between radio waves and infrared to obtain images. Clothing and other organic materials are translucent in extremely high radio frequency bands (short wavelength). The millimetre wave is transmitted simultaneously from two antennas rotating around the body. The wave energy reflected back from the body or other objects on the body is used to construct a high resolution three-dimensional image and thus locate any items of concern.[43] For screening, passengers must step in a special booth, where the radio waves are emitted.

Today, two basic types of operation can be differed:

Type A: Security scanner with automatic threat detection and indication of the location of detected objects on a stick figure.
Type B: Security scanner with a human reviewer to analyse the image and to indicate the location of detected objects.[44]

Manufacturers of MMW scanners currently used include L3 Communications, Rohde & Schwarz, Smiths Group. In this section, devices of L3 will be used as illustration.

Type A models are represented by L3's Provision 2 and Provision ATD. They are quite similar with regard to operation and privacy safeguards and are qualified in both the USA and the EU. According to both the DfT (UK) and TSA (USA), only Type A is in operation in these states. Type B model is SafeView (ProVision). Among the considered states, it is used in Russia.

The images are displayed on the monitor as 3-D black and white silhouettes. Faces and intimate zones on presented images are blurred. The images are viewable only by an officer sitting in a remote and secure room, so that screeners viewing an image are not able to see the person being scanned.

The scanners can be configured for analytics software known as Automated Target Recognition (ATR) – i.e. the equipment uses a special program for automatic detection of weapons and other prohibited items. For instance, this option is used in Russia. If an anomaly is detected, the information is transferred to the security officer at the checkpoint. Optional "image-on-demand" mode enables operator to request an image analysis according to the situation. In case of the network configuration and if several scanners are used at the same time, the operator does not know which of the scanners generates the image, providing additional level of

[41] Passive millimetre makes an image from natural radiation from the body, with a rough body image and a clearer image of external objects.

[42] Health issues will be discussed below.

[43] Neeman (2015) p. 24.

[44] ECAC. Security Scanners. Update 18.09.2015. https://www.ecac-ceac.org/cep

confidentiality. It is reported that the images are not saved after analysis.[45] Another option is that images can be routed to same-gender screeners for analysis.

In contrast to Type B, Type A is claimed to be image-free and ATR is used as default. "Anomalies" on passengers' bodies are detected automatically without human intervention. ATR interprets the scan data and identifies areas where items may be concealed, i.e. areas of the body which should receive a targeted hand-search.[46] The displays show only a generic outline with no body features visible that is identical for all passengers. The software highlights the locations of potential concealed threats on the generic outline. If the machine does not detect any threat, it will merely display a green "OK" without outline.[47] Due to all these features, it is stated that privacy settings can be compared with those of metal detectors, because both systems do not reproduce the image of the inspected persons.[48]

At the same time, some of these statements, in particular, that Type A has no image, and that neither type involves quality images or saves images, contradict to information from other sources. Since these issues may be important for the results of this research,[49] they require additional consideration.

First of all, despite Type A does not show an image on the monitor, this does not mean that images do not exist. As cited above "ATR software interprets the scan data;" a question is what these "scan data" can be apart from images. The idea that they are still images was confirmed to me in a number of conversations with both scanners' producers and users – airports. Otherwise, there would be nothing to analyse. Thus, the images are probably only presented as non-images.

Another fact that may prove this view is that Type A can be configured in two modes: Standard 1 and Standard 2.[50] The difference between them is how sensitive they are and how much false alarms are allowed. Standard 2 is more accurate than Standard 1.[51] In addition, some models allow changing configuration depending on a person entering the machines. This can be done both manually and automatically. For instance, it is possible to program to use Standard 2 for all passengers heading to particular flight. The selection can be done via scanning boarding pass before entering the machines: if a person is going to this flight, the machine will automatically use Standard 2. Apparently, these operations would be impossible without the

[45] Response to the author from Moscow Domodedovo airport dated 7 Feb 2012, confirmed again on 12 Mar 2015.

[46] UK Department for Transport (2015a).

[47] TSA. Advanced Imaging Technology. How it works. http://www.tsa.gov/ait-how-it-works (date accessed: 31 Jan 2015).

[48] Borlas Secuirty Systems. Provision ATD. http://borlassecurity.ru/equipment_provision.html

[49] E.g. the storage of images may have importance for analysis of various data protection principles such as purpose limitation and minimality.

[50] This classification is used in ECAC's Security Scanners. Update 18 Sep 2015. https://www.ecac-ceac.org/cep

[51] Data on actual sensitivity is graded confidential and is unavailable to the public. Interview with Avinor representative, 22 Oct 2015.

machine creating images of different quality. Thus, in this research, it is presumed that images exist for both types.

The next question is the format and quality of images. According to manufacturers of body scanners, the images can be presented in a variety of formats: more or less detailed, with or without modesty adjustments, with or without a facial image, etc. depending on the legal requirements.[52] The key word is "presented". Same idea can be found in one of FAQ about scanners, where the answer to the question: "Will I be naked in the image?" is "No, not in any image *that will be seen* by a human" (emphasis added).[53] In informal conversations, the experts also admitted that the high-resolution images are still being captured.[54] Thus, it may be the case that images are only presented as non-images or as mannequin figures, and hypothetically, quality images exist.

At the same time, security organs assure that in case of finding an "anomality", a person will be subject to private search. This order implies that there is no need to analyse high-resolution images by a human operator. If these images are inside the machines and are never viewed and analysed by humans and are irrevocably deleted once the scanning is over, this constitutes a protective mechanism. However, it is a question whether this can be trusted (see discussion on the storage below). Additionally, any "anomality", even if it is falsely positive, will lead to a manual search, which can also be privacy intrusive.

The question of storage or non-storage of images is one of the most controversial. All the official statements of manufacturers and aviation security and other agencies assure that the images cannot be stored and are removed once the passenger has passed control and security officers are convinced that he is clean. The DHS/TSA stress that the storage option is not included in the scanners software, and there is no need to perform the disabling of the storage function;[55] accordingly, since the image is not stored, it cannot be used for any other purpose or shared with anyone. The UK Code of Practice requires that after the scanning analysis, "all data relating to the individual must be destroyed, made irretrievable and incapable of being copied or sent".[56] But can this be trusted?

From a technical point of view, the machines can be designed so that they store images locally, in a central database, or do not store at all.[57] The latter means that images cannot be transferred, copied, or printed, and are permanently removed after security personnel review them. However, manufacturers also admit that, even if scanners do not routinely store images, they can be adjusted to do so, for development

[52] Summary of 1st Body Scanner Task Force meeting, 12 December 2008. Annex IV to European Commission. *Impact Assessment on the possible use of security scanners at EU airports* of 23 March 2011.

[53] Barden. *Airport body scanners… all your questions answered.* 20 May 2015. http://www.aph.com/community/holidays/airport-body-scanners-all-your-questions-answered/

[54] Conversations of the author with employees of Adani and Avinor in the period 2010–2015.

[55] Department of Homeland Security (2015) p 2.

[56] UK Department for Transport (2015a).

[57] Tirosh and Birnhack (2013) p. 1274.

research, etc.[58] In conversation with the author, L-3 representatives stated that the scans "do not save images".[59] Seller of L-3 devices in Russia, Borlas Security Systems, refused to confirm or comment on this issue due to secrecy reasons.[60] In an informal conversation with the author, however, other experts admitted that the high-resolution images mentioned above are still being captured and stored by these "changed" devices. There are a number of factors which may also contribute to the dispute.

For instance, the UK Impact Assessment stipulates that in order "to ensure that personal data continues to be protected, the Government has decided to require airports to undertake twice yearly checks of scanner equipment to ensure that data cannot be saved, copied or transmitted".[61] The question which arises is that, if the storage option is not included, what is the sense of checking?

It is quite logical that the security organs actually need to retain the images as evidence, should they actually find a real terrorist, and probably also for testing and training purposes. Moreover, the machines would need the capability to save images for later inspection to find out what went wrong with the scans if there were a successful terrorist attack.

According to some sources, the images produced by the machines can be stored in computer memory and used as evidence of presence of prohibited items,[62] and multi-level access codes ensure that all the data in the computer can be used only by those who have permission to do so.

In the EU Commission's Report of 2010 it is stated that there is no need for images to be captured or stored for future use such as, for example, evidence in a Court case since it would be the discovery of an actual forbidden object on the person, and not the discovery of an image on a machine, that would be the basis for prosecuting a person.

But who can decide for the Court how to treat such cases in the future? In the event the person later states that the object did not belong to him and appeared during the arrest, important evidence which could prove the opposite may be missed. Thus the Commission's conclusion can be changed by the Courts' practice any time.

Supporting this point of view, the EU's Impact Assessment of 2011, states that "In principle, storage and retrieval of images created by the security scanner should not be possible once a person has been cleared as not carrying any threat items. Only if an individual is stopped for carrying such a prohibited article may an image be retained as evidence until the passenger is ultimately cleared or denied access to

---

[58] Summary of 1st Body Scanner Task Force meeting, 12 December 2008. Annex IV to European Commission. *Impact Assessment on the possible use of security scanners at EU airports* of 23 March 2011.

[59] Interview with L-3 representatives in September 2010.

[60] Written correspondence with Borlas representative in January 2012.

[61] UK Department for Transport (2013)

[62] Reznichenko. *Inspection equipment – weapons of anti-terror* (2006) http://www.cnews.ru/reviews/free/transport2006/articles/antiterror/

the security restricted area and eventually the aircraft".[63] This means that for some period of time, the images can be stored.

To the contrary, TSA assures that after considering the privacy issues of the storage feature and evaluating potential uses of the images for training, investigations, or possible prosecution of persons caught with prohibited items, it was decided not to include data storage capabilities at all.[64]

In the past, with reference to older generation scanners, the TSA stated they could not save images either, and that storage option was disabled. Moreover, those statements contradicted the TSA's own Procurement Specs which specifically required that the machines have the ability to record and transmit images, even if those features might be initially turned off on delivery.[65]

With reference to old "naked" devices, lots of images appeared on the Internet and the EPIC obtained evidence that the scanners had functions permitting them to store and transfer the images (for example, via USB or protocol TCP/IP). According to a letter dated 15 April 2010 from TSA to EPIC, the TSA confirmed that it had about 2000 body scanner photos from devices that the TSA said earlier "could not store or record images". FOIA documents obtained by EPIC revealed that TSA could disable the body scanner's privacy settings, the images could be stored on the machines, and the scanners ran on a standard operating system making them vulnerable to outside security threats.[66] In addition, the TSA documents indicated that the software (namely, Type B) still may be capable of storing and transmitting unfiltered images of naked airline travellers.[67]

Today, TSA admits that initial versions were manufactured with storage functions that TSA required manufacturers to disable, but as mentioned above, states that current versions do not store the images. However, after it was proved that previous statements were apparently lies, it may be not so easy to believe that current statements are true. Thus, it is unknown whether the risk of possible abuse has been eliminated or may nevertheless emerge. These uncertain issues will be further discussed within data protection principles in the next chapter.

Further, the issues which are important for this research will be discussed: whether the MMW scanners help the security and respect individual's rights.

---

[63] European Commission (2011), §136.

[64] Department of Homeland Security (2011).

[65] Procurement specification for whole body imager devices for checkpoint operations, U.S. Department of Homeland Security, TSA. Final, Version 1.02, 23 Sep 2008.

[66] EPIC, Suspension of Body Scanner Program, http://epic.org/privacy/body_scanners/episc_v_dhs_suspension_of_body.html (Aug. 22, 2014) (date accessed: 2 Feb 2015).

[67] EPIC Alert, Volume 18.16, 17 Aug 2011. https://epic.org/alert/epic_alert_1816.html

6.3.2.2 Effectiveness

Apparently, criteria and parameters for scanners' security effectiveness exist and scanners undergo certification testing.[68] In the USA, in addition, the cost effectiveness of scanners was evaluated in terms of lives saved and limitation of damage to the economy. However, this is not public information (see Chap. 4). Other available materials show different, even opposite views.

Proponents – mainly manufacturers and regulators – argue that body scanners are effective as they are capable of identifying both metallic and non-metallic objects. According to L3, all its scanners detect "a broad variety of targets which include the following objects and materials (metallic and non-metallic): weapons, standard and home-made explosives (sheet and bulk), liquids, gels, plastics, powders, metals, ceramics, and other solids, etc."[69] Recital 4 of Commission Regulation states that security scanners are an effective method for passenger screening.[70] The continued use of the scanners by many states also suggests that they believe that the scanners are a useful tool.

Another popular pro-argument is timing. Scanners are supposed to improve passenger flow by performing screening at a very acceptable speed. In contrast to manual searches requiring 2–3 min, MMW scanning takes 1.5 s, processing 200–300 people per hour.[71] Another positive factor of ATR usage is deemed to be minimizing or even eliminating the human factor in the inspection, thus reduction or elimination of the possibility of corruption, i.e. "agreement/collusion" between terrorists and the security officer at the check point.[72]

Opponents argue that scans are inadequate, ineffective, inconvenient and inaccurate, with a too high false alarm rate, ultimately being a waste of time and money.[73] These arguments will be considered in more detail.

Firstly, neither MMW technology nor backscatters can detect explosives carried inside the body. The problem is that explosives can be inserted in the rectum or vagina, or swallowed, or surgically inserted into the body, e.g. into silicone breast implants.[74] The first known attempt involving the use of the internal method was July 2009, when Al-Qaeda suicide bomber Abdullah Hassan Tali al-Asiri inserted around half a kilo of explosive and a detonator into his rectum and blew himself up

---

[68] Elias (2012) p. 9.

[69] L3, ProVision ATD Fact Sheet. http://www.sds.l-3com.com/advancedimaging/provision-at.htm

[70] European Commission (2011), §135.

[71] L3, ProVision ATD Fact Sheet. http://www.sds.l-3com.com/advancedimaging/provision-at.htm

[72] For more about corruption at the check points, see Jones. *Checkpoint security: gateways, airports and the architecture of security* In: Technologies of Insecurity (2008) p. 97.

[73] Schneier. *A Waste of Money and Time*, The New York Times, 17 October 2012 (2012b) http://www.nytimes.com/roomfordebate/2010/11/22/do-body-scanners-make-us-safer/a-waste-of-money-and-time

[74] For more details, see Koffler. *The latest Al-Qaeda threat: Body bomb*. In: Aviation Security International. Vol. 17 (2011); Bunker. *Body cavity bombs: Fantasy or reality?* In: Aviation Security International. Vol. 20 (2014).

in the presence of the Prince Nayef of Saudi Arabia, killing himself but only causing minor injuries to the Prince. The second major incident was the previously mentioned attempt by the "underwear bomber". Al-Asiri's target was not the aircraft, but he took two flights and passed through two airport security screening systems to reach his intended target.

The "underwear bomber" boarded the flight at Schiphol airport, which already had 15 body scanners. The scanner's manufacturers admitted that they would not have detected the underwear bomb because it was in a light powdered form and the detonator was hidden in a body cavity. Even if the radiation exposure were to be increased significantly, it would be difficult if not impossible to detect plastic explosive carried internally.[75]

Secondly, if taking into account objects on the body, as security experts state, MMW is a low resolution imaging system that has inherent limitations and gaps in coverage and detection. It cannot identify well-concealed IED, IED components or a weapon on a passenger. An example is an incident when a TSA agent went through scan with a gun.[76] In addition, some areas of a person's body could not be detected with security scanners at all.[77] As noted above, security experts claim that the prohibited backscatter systems were more effective in comparison with MMW. And certainly, transmission scans are more effective.

Thirdly, according to the Association of Independent Aviation Security Professionals (AIASP), the scientists and inventors involved in the design of the systems as well as years of experience confirm that ATR, when applied to the already poor resolution of MMW scan further degrades its detection capability.[78] Two factors play a role here: quality of images and human involvement.

Indeed, from a security perspective, it is more effective to have both ATR and a high resolution body image that provides screeners with more information to assess the probability of a threat carried on the passenger's body.[79] The detection was more reliable when the image related directly to the person being screened rather than just giving an indication on a generic outline.[80] But, privacy concerns over "naked"

---

[75] McNeal. *Security Scanners in Comparative Perspective*. In: Health Matrix. Vol. 22 (2012) p. 475.

[76] Stinchfield. *TSA Source: Armed Agent Slips Past DFW Body Scanner*, NBC, 18 February 2011, http://www.nbcdfw.com/news/local/TSA-Agent-Slips-Through-DFW-Body-Scanner-With-a-Gun-116497568.html

[77] Summary of the 2nd Task Force Meeting on Security Scanners, 14 September 2010. Annex IV to European Commission. *Impact Assessment on the possible use of security scanners at EU airports* of 23 March 2011.

[78] Association of Independent Aviation Security Professionals. *Vulnerability of Automated Millimeter Wave Imaging Technology: An Unintended Consequence of "One Size Fits All" Checkpoint Screening* (2011).

[79] Neeman (2013) p. 14.

[80] Summary of 1st Body Scanner Task Force meeting, 12 December 2008. Annex IV to European Commission. *Impact Assessment on the possible use of security scanners at EU airports* of 23 March 2011.

images have led to the use of Types A and B – at the expense of security effectiveness.[81]

Some security experts believe that the technology is not advanced enough to remove detection by the human eye, and an algorithm can hardly replace a human being looking at an image.[82] In this regard, Type B has security advantages compared with Type A, since it provides added visibility for small-size threats and allows operators to detect smaller and more difficult objects, while the capability of Type A to actually detect threats is limited.

However, human factors may have a negative side too. For Type B, which relies on human operators rather than ATR, an extremely finite period of time means evidence might be missed. The operators may not have the required technical expertise to intercept the data and may have only minimal training before being posted, thus, all the advantages of the technologies may be neglected.[83]

In summary, the ATR function which is supposed to enhance privacy is provided at the cost of a considerable reduction of the scanners effectiveness.

Fourthly, there appears to be a problem in that the devices fail to distinguish between prohibited items and other objects. ATR only identifies an anomaly, but cannot interpret its nature: is it a gun, IED, an electronic device, or an implant? The technology fails to differentiate between such items, causing false alarms. The percentage of false positives may be high, increasing the need for manual searches to find out what the issue is, despite the fact that the governments claim that scanners reduce such a need.

According to a number of sources, the alarm rate on scanners was originally over 70% in various tests in Europe.[84] In an experiment, an observer counted true false positives (an alarm where there is no anomaly present) of around 45% (almost half the passengers inspected).[85] It is noted that an alarm rate of over 20 % severely compromises the trust in the technology and reduces the effectiveness of the whole system.[86] Since false alarms are currently impossible to exclude, many innocent people may be bothered.

Finally, scanners are quite expensive. The price of a basic scanner ranges between EUR 100,000 and EUR 200,000 per item, excluding training, installation, and maintenance costs. According to cost-benefit analysis made by academic researchers, for the USA, more than one attack every 2 years would need to originate from

---

[81] Grynpas. *Advanced Imaging technology: privacy and health concerns*. In: Aviation Security International. Vol. 18 (2012) p. 39.

[82] Richardson. *TSA plan to replace humans with software at airport scanners generates criticism*, Secuirty Director News, August 2011 (2011b), http://www.securitydirectornews.com/?p=article&id=sd201107R4VP0J; Huey *"Live Free or Die": The Unintended Consequences of "One Size Fits All" Aviation Security Screening*, 5 March 2011 (2011a), http://www.noquarterusa.net/blog/57130/"live-free-or-die"-the-unintended-consequences-of-"one-size-fits-all"-aviation-security-screening/

[83] Cochrane (2012) p. 41.

[84] Huey (2011a) and Association of Independent Aviation Security Professionals (2011).

[85] Association of Independent Aviation Security Professionals (2011).

[86] *Ibid.*

US airports for scans to pass a cost-benefit analysis, however, the attack probability needs to exceed 160–330 percent per year to be 90% certain that scans are cost-benefit.[87] This means that absence of documented prevented attacks may show that costs exceed the benefits.

Theoretically, if the amounts paid for the scanners could be spent on different security measures providing more security, this would mean that deploying scanners actually reduces security.

In summary, it is not only technology's weaknesses, but privacy and health concerns that may reduce the machines' effectiveness in locating dangerous items. But do the safeguards really protect privacy and data protection? These issues will be discussed below using tools from Chap. 2.

6.3.2.3 Data Protection Issues

For the data protection regime to apply, as discussed in Sect. 2.2.2, two conditions must be satisfied: data should be personal and should be processed; in general, images may fall within personal data.

The first question which arises is whether the images produced by the devices amount to personal data. The EU Commission states that "the image shall not be linked to any data concerning the screened person and his/her identity shall be kept anonymous".[88] Anonymity implies that it is impossible to link a face, or one's identity, with an image produced by the scanner. In this case, images can fall within the category non-personal data, thus, will not be subject to data protection.

As mentioned in Chap. 2, four "bricks" are important for assessment of "personal" data. While it is quite clear that scanners' images satisfy the conditions "any information," "relating to", and "natural person", the requirement of "an identified or identifiable" should be analysed.

Three cases can be distinguished: images visible to operators for Types A and B and quality format images inside the machines. As mentioned above, Type A presents images as non-images; Type B settings use "stick figures" and blur the face and intimate zones of the person screened. However, despite undertakings in both cases, the data provided by the scanner will still allow for the indirect, if not direct, identification of the data subject.

For example, apart from dangerous items, images allow one to identify distinguishing features of an individual that may identify the person. The body scanning process is capable of revealing very sensitive areas of a person's private life, medical

[87] The analysis considered threat probability, risk reduction, losses, and security costs. Stewart and Mueller. *Cost-benefit analysis of advanced imaging technology full body scanners for airline passenger security screening*. In: Journal of homeland security and emergency management. Vol. 8 (2011) p. 15.

[88] Commission Implementing Regulation (EU) No 1147/2011 of 11 November 2011 amending Regulation (EU) No 185/2010 implementing the common basic standards on civil aviation security as regards the use of security scanners at EU airports.

aids and conditions, such as intentionally concealed physical features (for instance of transsexuals; breast implants; bras with gel pads) or medical information (evidence of a mastectomy, colostomy, prostheses, diapers, menstrual pads, urinal bag, piercings etc.), which people prefer not to be revealed and which may cause embarrassment to the persons being scanned. In any of these cases, if the scanners detect these objects, the passenger is required to undergo a further hand search. Thus, a relation between the image and a particular person can be established – that is, identification is possible.

There are also doubts that in case of Type B, the reviewer in a remote location would be unable to discover who was going through the machines if he really wanted to do so. For example, the operators, like any other human beings, gossip and can easily communicate to each other about a celebrity being scanned[89] or any other person with "remarkable" characteristics. For instance, it was reported that security officials at Lagos airport hurried over to the line in order to catch a glimpse of some of the passengers entering the machine and then immediately returned to view the naked images in order to match the faces of their favourites with the images.[90]

In the USA, in 2012, there were complaints filed by several women stating that TSA agents intentionally targeted them to undergo screening in a body scanner to view their bodies. One woman said that a female agent sent her through a body scanner three times acting on a request from male agents who were in a separate room viewing the scans.[91] This and similar stories strongly indicate that abuses leading to identification are possible.

Furthermore, as explained by the European Data Protection Supervisor (EDPS),[92] identification is possible since the image will be used to decide whether the passenger can access security restricted areas.[93] Therefore, the image itself might still be indirectly related to the data subject, especially if the screener cannot determine whether or not the passenger is carrying prohibited articles or if the passenger is not cleared.[94] This is relevant to both Types. In addition, in 2011, the TSA documents obtained by EPIC indicated that passengers passing through the machines were identified by gender and assigned a unique identification number.[95]

---

[89] Tirosh and Birnhack (2013) p. 1283.

[90] Eze *Now Showing at MMIA: Nude Images of Passengers*. (2010).

[91] Zetter. *Female Passengers Say They Were Targeted for TSA Body Scanners*. Wired, 4 February 2012, http://www.wired.com/2012/02/female-body-scans/.

[92] Established by the European Parliament and European Council. Regulation (EC) 45/2001 of 18 Dec 2000 on the protection of individuals with regard to the processing of personal data by the institutions and bodies of the Community and on the free movement of such data. EDPS is an independent supervisory authority that regulates the processing of data.

[93] EDPS letter to Commission Vice-president Sim Kallas of 1 July 2010.

[94] EDPS comments on the draft proposals for a Commission Regulation amending Regulation (EC) No 272/2009 and for a Commission implementing Regulation amending Regulation (EC) No 185/2010 on common basic standards on civil aviation security as regards the use of security scanners at EU airports.

[95] EPIC Alert, Volume 18.16, 17 Aug 2011. https://epic.org/alert/epic_alert_1816.html

Moreover, as mentioned above, a boarding pass bearing the person's personal data can be scanned prior entering the machine in order to determine standard of screening. This process also provides some kind of identification, a relation between a particular person registered for a particular flight and the image.

Another common issue is traceability. As discussed in Chap. 2, in contrast to other transport modes such as metro or buses, the airlines carry *identified* air passengers. Passengers identify themselves with a boarding pass and a passport as they enter the security area of the airport. They are not again identified at the scanner, but their anonymity is momentary and can be easily reversed.[96]

Finally, high-dimension images presenting persons' bodies in detail, obviously, if taken outside the machines, allow identification, thus, constitute personal data. As discussed above, if they are kept inside, this is a strong safeguard; but it is a question if this can be trusted.

Thus, the argument that people cannot be identifiable or identified by images is quite weak. Therefore it may be said that the images generated by the scans fall within the category personal data, and it does not matter if the images are quality, are presented as mannequin figures or as no image. In any case, a person may be identified. Moreover, the images of any type can directly (quality images) or indirectly (other images) reveal racial or ethnic origin, religious beliefs, and data concerning health and sex life, thus, deal with sensitive data.

The next question is whether the use of body scanners constitutes the processing of personal data. It is often stated that if images are not stored, then the use of a body scanner would not constitute the processing of personal data. But, as mentioned in Chap. 2, "processing" is *any* operation performed upon personal data. Thus, even if the images are not recorded and available only during the time the individual is being screened, only the fact of image short analysis is enough. Thus, scanning passengers falls within "processing," and in this research, data protection rights are considered as applicable to both types.

6.3.2.4 Privacy Issues

As for privacy, for many cultures, wearing of clothing is not merely limited to keeping warm or impairing visibility of the sexual organs. People use clothes according to traditional concepts of decency and modesty, covering extremely personal parts of the human body.[97] Moreover, exposing body parts that must be covered is criminal offence in some countries.[98]

Therefore, the objective of the body scans automatically moves into a very intimate and personal sphere. Tirosh and Birnhack point to dress/undress dissonance caused by the use of scanners: on the one hand, the state expects people to conceal

---

[96] Tirosh and Birnhack (2013) p. 1283.

[97] FRA. *The use of body scanners: 10 questions and answers* (2010b).

[98] Tirosh and Birnhack (2013) p. 1289.

their bodies via anti-nudity laws, on the other hand, it requires them to subject themselves to visual inspection via scanners, which virtually undresses persons.[99]

In general, body (or strip) searches carried out for security reasons will normally constitute interference with respect to private life. Such measures have not been found to be incompatible with ECHR Article 8 in certain environments such as prisons: frequent strip searches were found to be required by the exceptional security requirements of the prison.[100] This was also justified by lack of physical contact and the presence of a third officer to avoid abuse.

Clearly, strip searches are more acceptable in prisons rather than at airports, where in addition, the commercial and time pressures of the airport environment apply. But in order to indicate whether strip searches of ordinary passengers fall within exceptional requirement or not, other circumstances should be taken into account, including how the searches are fulfilled: if by humans – how exactly, if by scanners – with what kind of safeguards.

With regard to body scanners, the search is done via producing images that reveal individuals' naked bodies. As discussed above, the images apparently exist and are used for analysis. In this case, inspection of a passenger's body via scanner while providing security screening constitutes entry into personal space, an intrusion into privacy. However, the seriousness of the impact will differ depending on the used safeguards. As discussed above, there is a difference if the quality images remain inside the machines or are seen by human operators. It should also be noted that if a person opts out from scanning, entry into personal space will still occur via the alternative – namely, inspection of the passenger's body by hand search, which can also be considered intrusive.

For scanners, relevant types of privacy (see Chap. 2) will be first of all physical or bodily privacy. Secondly, as described above, scans reveal very sensitive areas of a person's private life, medical aids and conditions. This includes information about behaviour such as augmentation surgeries or medical related practices. Thus, privacy of behaviour and action is relevant. Although an overlap with bodily privacy is clear, the two types are separate since it is the activities revealed by the images which people wish to conceal rather than the bodies or images themselves.[101]

As a result, body scanners may have an impact on members of particular groups including disabled people, transgendered people, older people, children, women and religious groups, all of whom can experience additional negative effects on privacy. Here, normal/abnormal dissonance can be noted: on the one hand, the state signals that all citizens are equal before the law; on the other hand, body scanners amplify physical differences via detecting bodily anomalies.[102] Although the technology is not meant to discriminate between bodies and people, its technological design and its use produce differential treatment.[103] Thus, for the specific groups,

[99] *Ibid.* p. 1267.

[100] See *McFeeley v. the UK*, No. 8317/78, 15 May 1980.

[101] Finn et al. (2013) p. 12.

[102] Tirosh and Birnhack (2013) p. 1268.

[103] *Ibid.*

the right to privacy is allocated unequally.[104] When scanners detect an "anomality", the passenger is required to undergo a further search. This can be done in private rooms, but in any way, for such people, screening by a body scanner means that they have discuss their private issues and to disclose personal conditions to security staff in order to explain why a body scanner alarm has been triggered.[105]

An illustration is the situation with a transgender woman who complained that she was detained and harassed at Orlando airport security checkpoint after a body scanner detected an "anomaly" – male organ – on her. According to her, she disclosed her reality immediately, but officials searched her for 40 min, "fully disassembled" her luggage and put her in an empty room with an officer holding the door; an agent told her to "get back in the machine as a man or it was going to be a problem;" and she missed her flight.[106]

With reference to informational privacy, the interference to privacy is constituted by first the capture – production of the image – and then the analysis of the image.[107] Similarly to data protection applicability, the production of the image alone is enough for interference. Although it is argued that since images are only an outcome of the search,[108] and are fruits of the violation rather than the actual violation,[109] if considering privacy types, images constitute a separate type of privacy interference.

In summary, the following types of privacy apply: physical (privacy of the person, or bodily), interactional (privacy of behaviour and action) and informational (privacy of data and image).

6.3.2.5 Other Human Rights Issues

With reference to other applicable human rights, the first one to be mentioned is the right to health. The states and producers assure that the MMW device is completely safe to health. For instance, a statement by the International Commission on Non-Ionizing Radiation Protection reveals that, while the ICNIRP recommends limiting human exposure to non-ionizing radiation, exposure to humans from body scanners currently in use is about 1/10th of its recommended guidelines for the general

---

104 *Ibid.* p. 1299.

105 Australian Department of Infrustructure and Transport. *The use of body scanners for aviation security screening in Australia: Privacy Impact Assessment*, 2012. Page 24.

106 Queally. *Transgender Woman Says TSA Detained, Humiliated Her Over Body 'Anomaly.'* Los Angeles Times, 22 Sept 2015, http://www.latimes.com/nation/nationnow/la-na-nn-tsa-transgender-20150921-story.html

107 Mountfield. *Advice in the matter of the human rights and equality implications of the introduction of full body scanners at airports*. In: Matrix Chambers. Vol. 8 (2010). p. 6 and European Commission (2011), §§133–134.

108 Tirosh and Birnhack (2013) p. 1300.

109 Taylor. *Bending Broken Rules: The Fourth Amendment Implications of Full-Body Scanners in Preflight Screening*. In: Richmond Journal of Law and Technology. Vol. 17 (2010) p. 15.

public.[110] The amount of radiation emitted by MMW scans is claimed to be many times lower than that emitted by other commercial radio frequency devices such as mobile phones, wireless handsets and other standard household devices[111] and is exceedingly small compared with exposure to cosmic radiation during air travel.[112]

A problem is that health effects are so small that they are impossible to verify or exclude empirically.[113] MMW scans presumably do not have the potential for cancer-causing DNA damage and it may be that the risks are truly extremely low. However, while the short-term effect seems to be negligible, the long-term effects – which are likely to be very low – are still uncertain.[114] At the same time, even if there are effects, it would be very difficult to prove that cancer results from MMW, since people are exposed to background radiation every day and the health effects can take years to appear.

Currently, despite a degree of uncertainty, as long as the MMW scans are officially considered "safe", the grounds to allege violation of the right to health may be too weak. Nevertheless, as security experts note, this uncertainty is used by opponents of scanners who argue that people, in particular pregnant women and children, should not be unnecessarily screened.[115] But if health risks were treated seriously, the debate should be addressed to radiation from flying as such rather than from security screening.[116] Today, passengers who have fears about health may always opt out from scanning to hand search, although to avoid radiation, a better option would be not flying at all, as well as not using mobile phones and other devices.

Freedom of movement and the right to equality will be discussed in separate sections below (Sects. 6.6 and 6.7). As for the freedom of thought, conscience and religion, this could be applicable in cases where people opt out from scanning on religious grounds but no alternative to scanners is provided. In all the considered jurisdictions using scans, all people irrespective of religion or any other factors are allowed to choose a hand search instead of scanning. Taking into account that options for hand searches in private usually exist, it can be concluded that it would be difficult to prove that scanners as such violate freedom of thought, conscience and religion. At the same time, religious grounds can be used for discrimination in the course of scanning, thus, the issues that will be discussed in Section 6.7 may apply too.

---

[110] Elias (2012) p. 8 citing International Commission on Non-Ionizing Radiation Protection, *ICNIRP Statement—Health Issues Associated With Millimeter Wave Whole Body Imaging Technology,* Health Physics, 102(1), pp. 81–82, January 2012.

[111] UK Department for Transport (2015a) and L3, SafeView Fact Sheet. http://www.sds.l-3com.com/advancedimaging/safeview.htm

[112] Auvinen. *Ionising Radiation: Myths & Realities,* Body Search 2015 Conference Program, http://network.icom.museum/fileadmin/user_upload/minisites/icms/pdfs/Body_Search_SPEAKER_PROGRAMME.pdf

[113] *Ibid.*

[114] Neeman (2015) p. 25.

[115] *Ibid.* p. 26.

[116] *Ibid.*

6.3.3 Transmission X-Ray

Several other technologies have been developed, for example X-ray transmission imaging technology (transmission scanners), which can produce images compared to medical X-ray images. They can see inside body cavities and underneath skin and thereby detect explosives carried inside the body, e.g. swallowed, rectally or surgically inserted IEDs, also known as body bombs. Such items are undetectable by backscatter, MMW or a manual search. Examples of images produced by the technology can be seen in Fig. 6.4.

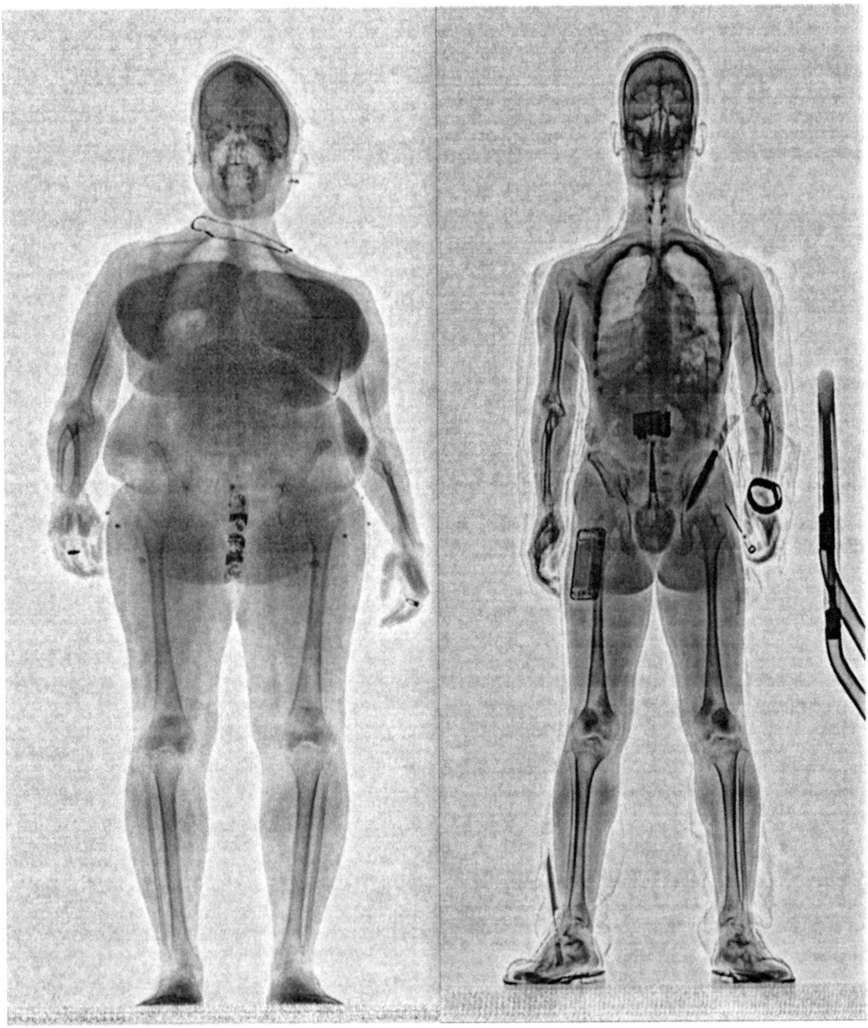

Fig. 6.4 Explosive in rectum; ceramic knife, metal knife, phone, scissors and screw driver (Pictures were provided to the author by Adani, producer of X-Ray Inspection System ConPass)

The technology is currently used in prisons, diamond mines, and in some states by customs authorities for analysing passengers suspected of internal carries. For instance, in Russia, the devices are called "scanners for detection of objects hidden inside the human body" and are used by the Federal Custom Service to detect internal concealments by the Federal Custom Service.[117] Similarly, they are used by Customs in the Netherlands at Schiphol airport,[118] Oslo, London, etc. Normally, the authorities must have serious cause for concern to use these devices on passengers.[119]

The use of this technology could prove to be highly effective in routine screening of passengers. It is health concerns regarding high radiation doses mentioned above that prevents the scanners from being employed broadly. This illustrates that full aviation security is impossible without putting passengers' health at risk. Here, a relevant dilemma is the right to life/security versus the right to health. As discussed in Chap. 3, no security benefits can substantiate health risks.

Images of transmission scanners amount to personal data and, similarly to quality "naked" images produced by MMW, include sensitive data. The latter are collected to a greater extent, since the devices in addition show what is *inside* the body, meaning more potential for information concerning health and other intimate details. Accordingly, privacy issues similar to those of MMW, but potentially more serious, may arise.

However, similarly to manufactures of other devices, producers of transmission scanners declare that privacy concerns are solved. Specifically, the software automatically blurs intimate zones on images visible to the operator. Examples can be seen in Fig. 6.5.

It is argued that, like the backscatter X-ray, the actual X-ray dosage is lower than background radiation in typical flights, so there is no strong scientific foundation for such concerns,[120] thus, transmission X-ray solutions are totally safe.[121]

Another solution could be to make the technology available at screening checkpoints, but to use it only when there are specific suspicions concerning a given passenger caused by their behaviour or as a result of intelligence.[122] Security experts believe that such deployment is long overdue, and every checkpoint should be able

[117] Aviaport Digest. *15 Russian airports will accommodate scanners to curb drug trafficking*. 13 November 2014, http://www.aviaport.ru/digest/2014/11/13/314589.html

[118] Schiphol. Security Scan at Amsterdam Airport Schiphol. http://www.schiphol.nl/Travellers/AtSchiphol/CheckinControl/SecurityChecksUponDeparture/SecurityScan.ht

[119] Koffler (2011) p. 46.

[120] Neeman. *Effectively screening people: Pat down searchers, wands, archways and portals*. In: Aviation Security International. Vol. 19 (2013) p. 15.

[121] Sosenko. *Transmission X-ray: Advantages & Successes*, Body Search 2015 Conference Program, http://network.icom.museum/fileadmin/user_upload/minisites/icms/pdfs/Body_Search_SPEAKER_PROGRAMME.pdf

[122] Neeman (2015) p. 26.

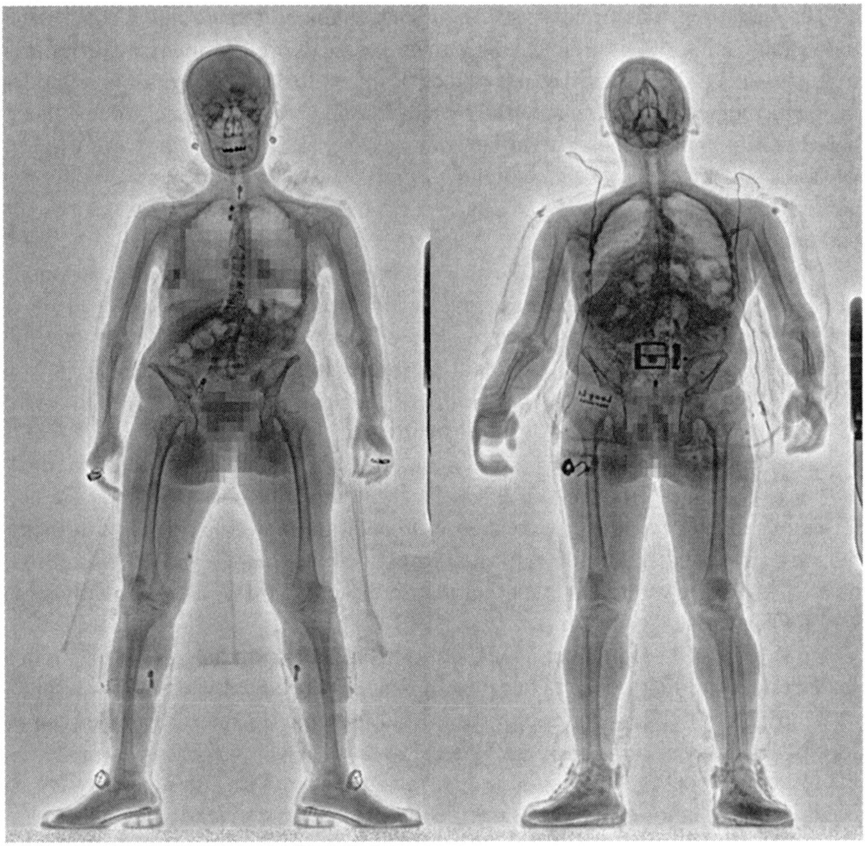

Fig. 6.5 Privacy undertakings (Pictures were provided to the author by Adani, producer of X-Ray Inspection System ConPass)

to determine whether or not a person behaving abnormally is transporting prohibited items as internal carries or surgical implants.[123]

This means that the use of scanners would require serious grounds similar to those currently used by Customs. Here, a proportionality issue similar to proportionality issues of other techniques in this research can be applied. The difference is that proportionality of the measure will be based on proved effectiveness, but the measure will not necessarily be used.

[123] Baum. *Searching bodies: the civil liberties dilemma*. In: Aviation Security International. Vol. 21 (2015c) p. 1.

6.3.4 Concluding Remarks

As discussed, the current effectiveness of scans, failing to detect dangerous items and confusing them with other objects, is quite questionable. In general, the decisions to use MMW Type A instead of Type B, MMW instead of backscatter and in particular instead of "naked" scans (i.e. operator viewing quality images) are done in order to reduce health and privacy risks, but for the cost of security. With reference to transmission scanners, despite their effectiveness and advantages in comparison with backscatter or MMW, health issues prevent their use for passenger screening, underscoring again that no security benefits can substantiate health risks.

Nevertheless, since the questions of existence of images, their quality and storage are controversial, it cannot be excluded that "naked" images are simply masked and exist somewhere inside the machines of both MMW types. Blurring, using mannequin figures or no image at all can be examples of "masking" privacy/data protection infringements rather than protecting privacy. On the one hand, indeed, the quality images have great value to security authorities, and the fact that they were given up may be controversial. On the other hand, if high-resolution images are solely inside the machines and are used for analysis only by computer rather than by humans, this may contribute towards safeguarding privacy. Can one be certain, however, that they are irrevocably deleted after scanning? As discussed, this is uncertain.

Another problematic issue is that in current conditions, scanners can only indicate "anomalies", and a hand search, which is performed by a human, is required to clarify them. Some definite groups of people incur increased risk to trigger scan alarms in this regard. In this dimension, a search by a "robot" – without senses and unable to talk would probably provide more privacy, but it is a question of future developments.

There are also some promising MMW projects, such as Flat-Panel body scanners,[124] an idea to create a MMW system that would take up less space and would not require passengers to stop and stand still; integrating the scans devices into airport building structures, different objects of the checkpoints, etc. Passengers will not even know at which place and which moment they are inspected. There is also advanced whole body imaging technology based on Terahertz radiation and other developments.[125] However, the issue is whether new technologies may make scanning process more passenger-friendly or, on the contrary, only further exacerbate privacy/data protection concerns by for example making the process unnoticeable.

Today, the process of scanning has impact on physical, behavioural and informational privacy, and involves processing of personal data. As discussed, the question of image quality or storage has little effect on applicability of both rights; however, it may have influence on the impact seriousness. As for the other human rights, as

[124] Neeman (2015) p. 24.

[125] Neeman (2013) p. 15.

long as the official reports state that the MMW devices are safe, the grounds to allege violation of the right to health may be too weak. The rights to free movement and to equality and non-discrimination will be discussed below.

6.4 Camera Surveillance

6.4.1 Introduction

One aviation security measure that is deemed effective and at the same time is associated with permanent surveillance of people and raises privacy and data protection concerns is the use of camera surveillance.

The enhanced risk factors at airports put special demands on the technical solutions for security, including requirements to camera surveillance, making this technology one of the most important airport security measures. Moreover, as mentioned in the General Part, cameras can be used in the air too, namely, if cockpit is equipped with devices showing what happens on board the aircraft. As the result, people can be under extensive camera surveillance 24/7 both on the ground and in the air.

CCTV started in the 1960s, but began to grow in the 1990s.[126] Over the past years, there is constant and increasing installation of CCTV cameras all around the globe. The world's four largest regional markets of camera surveillance equipment are China, the USA, Japan and the UK.[127] For instance, in the UK, as the British Security Industry Association reported in July 2013, there were up to 5.9 million CCTV cameras, approximately 1 for every 11 citizens.[128] The 2012 Olympic Games in London accelerated and reflected the booming CCTV usage.[129]

Camera surveillance in the USA is booming. Similarly, there are more and more cameras in Norway, and the number of cameras is growing. Oslo was repeatedly ranked as the second city in Europe – after London – in terms of the number of cameras used.[130] In Russia, camera surveillance is growing too; however, the total number of cameras may be less than in the UK or USA. For instance, there are "only" 140 thousand CCTV cameras in Moscow.[131]

---

[126] Wagenaar. *Zooming in on 'heterotopia': CCTV-operator practices at Schiphol Airport.* In: Information Polity. Vol. 17 (2012). p. 9.

[127] The SecurityLion. *New Surveillance Code of Practice "will not slow UK CCTV market growth".* 21 August 2013, https://thesecuritylion.wordpress.com/2013/08/21/new-surveillance-code-of-practice-will-not-slow-uk-cctv-market-growth/

[128] Annereau. *Regulating CCTV use in the UK.* March 2014, http://united-kingdom.taylorwessing.com/globaldatahub/article_regulating_cctv.html

[129] Courtney. *The great Olympics zoom boom [CCTV services].* In: Engineering & Technology. Vol. 7 (2012). p. 60.

[130] NRK. *Overvåkingen av Oslo øker, men kriminaliteten går ikke ned.* 12 November 2011, http://www.nrk.no/ostlandssendingen/_-overvakning-gir-ikke-resultater-1.7869062

[131] RiaNovosti. *70% of crimes in Moscow are disclosed using CCTV systems.* 24 March 2015, http://ria.ru/moscow/20150324/1054179260.html

But of course, it is not only a question of the quantity of cameras, but of quality too. Although it is quite common for a non-expert to associate the words "CCTV" with a number of cameras controlled manually by operators, and such cameras are still in operation, this is already outdated technology. Without added technologies and intelligence, they are limited and not always efficient: studies show that watching only 2 monitors, after 10 min an operator misses 45% of the data, after 22 min – 95% of the data.[132] Thus, today, cameras are often equipped with sophisticated analytical software (see below).

There are different, often contradictory views, evaluations and studies concerning how effective CCTV can be. According to proponents, namely, the producers and the security organs, camera surveillance is one of the most effective forms of security, able to provide the highest awareness in the control area and provide reliable information about events, with the widest range of opportunities:

- Continuous monitoring of the protected area at any time of the day;
- Motion detection in controlled areas;
- Accumulation of information on the event;
- Automatic recognition and comparison of recorded data with the data from the databases, etc.[133]

Opponents argue that CCTV is ineffective while increasing surveillance,[134] with a poor record for crime prevention.[135] They argue that all available studies fail to provide evidence that surveillance brings significant benefits, or that any of its benefits outweigh the limitations on privacy.[136]

Some privacy advocates go further and argue that justified interference with the right to privacy "in the interests of national security, public safety or the economic well-being of the country, for the prevention of disorder or crime, for the protection of health or morals, or for the protection of the rights and freedoms of others" does not apply to CCTV in public places, including transport facilities, and that total camera surveillance could be justified only if the crimes were a way of life of most citizens. Accordingly, CCTV can be used only where they are absolutely necessary for national security: on strategic defence enterprises, intelligence centres and similar institutions.[137] However, such views are radical rather than common.

Of course, the best assessment of the effectiveness would be statistics on prevented or disclosed crimes. However, available figures are very controversial. For

---

[132] Houlis (2010).

[133] Cmt-VC Company. Video surveillance systems. http://www.smt-vc.ru/catalog/sistemy-video nablyudeniya/

[134] Medwecki. *CCTV: ensuring effective surveillance at airports*. In: Aviation Security International. Vol. 19 (2013) p. 34.

[135] E.g. EPIC. *EPIC Framework for Protecting Privacy & Civil Liberties If CCTV Systems Are Contemplated* (2008) and ACLU. *What's Wrong With Public Video Surveillance?* 2002, https://www.aclu.org/technology-and-liberty/whats-wrong-public-video-surveillance

[136] Gilbert (2007) p. 34.

[137] Puzanov. *Video camera in the law?* In: EZH-Jurist. Vol. 8 (2012). p. 10.

instance, it was reported that about 70% of all crimes in Moscow were disclosed using city CCTV systems,[138] while in the USA, the Metropolitan Police revealed that only 3% of crimes are solved using CCTV and only one crime was solved per 1000 cameras.[139] While many researchers present empirical studies showing that CCTV systems have little effect on crime prevention,[140] security experts and CCTV manufacturers claim the opposite.[141]

Nevertheless, the negative evaluations are usually based on the evaluation of CCTV as a whole, without taking into account specific contexts and locations. One cannot deny the whole technology if it proves ineffective in some definite places, but works well in others. It is believed that installation of CCTV in connection with minor offences should be avoided,[142] CCTV surveillance can be appropriate in certain restrictively defined situations, at specific, high-profile public places that are potential terrorist targets.[143] And, of course, as with any technology, it would be unwise to rely on it alone. The evidence also suggests that CCTV works most effectively when combined with other preventative measures.[144] Thus, it is important to consider the CCTV effectiveness issues with regard to a particular situation and environment, in our case, airports and aircraft (see below).

Along with other areas, CCTV within the aviation industry is booming and constitutes an integral part of aviation security. Airports of the considered states have hundreds and thousands of CCTV cameras in the airports, especially in the major cities. For instance, only in 2009, TSA provided USD $70 million in funding to install CCTV installation in 20 airports.[145] In 2012, the Minneapolis International Airport started to install 1800 new high definition digital cameras, incorporating video analytics and facial recognition technology,[146] and the same was done in other airports.

---

[138] RiaNovosti (2015).

[139] Privacy International (2011) p. 822.

[140] Welsh and Farringtom. *Crime prevention effects of closed circuit television: a systematic review.* Home Office Research Study 252, August 2002, http://www.popcenter.org/Responses/video_surveillance/PDFs/Welsh%26Farrington_2002.pdf; Armitage. *To CCTV or not to CCTV.* In: A review of current research into the effectiveness of CCTV systems in reducing crime. Londen: Nacro (2002).; EPIC (2008); Larsen. *Setting the watch: privacy and the ethics of CCTV surveillance,* Hart Publishing (2011).

[141] See any web-site of any CCTV provider.

[142] Article 29 Data Protection Working Party (2004).

[143] See e.g. ACLU (2002), Larsen (2011), Ratcliffe *Video surveillance of public places* (2006) p. 21.

[144] Agustina. *The impact of CCTV on fundamental rights and crime prevention strategies: The case of the Catalan Control Commission of Video surveillance Devices.* In: Computer Law & Security Review. Vol. 27 (2011). p. 169.

[145] Elliott. *Smile! The TSA is taping you — and here's what you need to know about it.* 21 October 2009, http://elliott.org/blog/smile-the-tsa-is-taping-you-and-heres-what-you-need-to-know-about-it/

[146] Richardson. *Minneapolis airport to spend $20m on new surveillance system,* Secuirty Director News, May 2012, http://www.securitydirectornews.com/public-sector/minneapolis-airport-spend-20m-new-surveillance-system

In 2002, American JetBlue became the first airline to get approval from the US Federal Aviation Administration to install video cameras on all its aircraft.[147] Since then, many other airlines have been using cameras on board aircraft. Commercial aircraft video surveillance systems market is estimated to be USD $127.13 million in 2014, and is expected to grow and reach $173.83 million by 2020.[148] The same tendencies – both quantitative and qualitative growth – can be noted in other states.

A commonly used term for camera surveillance systems is Closed Circuit Television (CCTV), however, technical advances have made "CCTV" outdated and misleading. Strictly speaking, CCTV is a technology system of surveillance by cameras that consists of video cameras connected in a closed circuit television. Images are sent to a central television monitor or recorded in a control room.[149] Today, camera surveillance can include various equipment and systems. Most modern surveillance systems use networked, digital cameras rather than "closed-circuit".[150]

Despite the inaccuracies, the term CCTV is generally used today for any camera surveillance and for convenience will be used in this research synonymously with camera surveillance. These terms will mean as broad category of relevant equipment as possible and will include any camera surveillance systems or any other systems for recording, viewing or any processing of visual images for surveillance purposes, as well as any systems connected with these systems.

This section will first describe the use of camera surveillance technology in the airports, on board aircraft, focusing on those features which have influence on applicability of privacy/data protection rights. Special attention will be given to the CCTV's use of facial recognition and behaviour analysis, as presenting specific risks for privacy and data protection.

It should be noted that in general, cameras are able to use sound in combination with images, which also can falls in the category of personal data. With regard to the airports, the aviation security organs and manufacturers assure that this option is not used and would be useless due the noise factor. Some statements regarding CCTV systems on board aircraft ensure that sound is not used either, however, available equipment for such systems shows that sound can potentially be used. Thus, this issue will be discussed only with reference to in-flight CCTV.

---

147 Blank. *Surveillance Cameras Set To Keep Watch In Airliners*. New York Times, 6 April 2002, http://www.nytimes.com/2002/04/06/business/surveillance-cameras-set-to-keep-watch-in-airliners.html

148 PRNewswire. *Research and Markets*. 20 April 2015, http://www.prnewswire.com/news-releases/global-commercial-aircraft-video-surveillance-systems-cockpit-door-surveillance-systems-cabin-surveillance-system-ground-maneuvering-camera-system-flight-safety-camera-systems-market-2014-2020-300068367

149 Carli (2008) p. 4.

150 Closed system allows watching retrospectively, while open system refers to images on the camera as viewed in real time.

6.4.2 Camera Surveillance at Airports

6.4.2.1 General

CCTV monitoring and analytics belongs to physical airport security in the airports which may cover both internal and perimeter security and monitor most airport operations. This includes surveillance of various zones and objects, including perimeter fence areas, aircraft, cargo, baggage, storage areas, etc., but due to the purposes of this research, surveillance of people will be in the focus. This can be done in such areas as terminals and public areas, arrival/departure, immigration and customs control, security checkpoints, landside passenger pickup and drop-off, concourse and gate area access points.[151]

All these areas require surveillance according to its own unique security challenges. As mentioned in Sect. 6.2, CCTV systems may be present at many different zones at airports and beyond, and ultimately, should provide as much monitoring and awareness as possible. Moreover, the purposes of CCTV are quite comprehensive, including:

- Operational Purposes (traffic flow and management, passenger flow, etc.)
- Safety Purposes (public safety, protection of staff and passengers from assault, etc.)
- Security Purposes (passenger movement, enforcement of security regulations, etc.)
- Law Enforcement Purposes (border / government agency enforcement, prevention of crime, detection of crime, apprehension and prosecution of offenders, evidential / court proceedings).[152]

As mentioned in Chap. 2, in terms of applicability of data protection, it is important to determine whether surveillance cameras deal with personal data and processing; for privacy, the key issues are the context and types of privacy. Thus, CCTV technical capabilities will be discussed, primarily with regard to these issues.

In the airports, CCTV cameras include fixed cameras and Pan-Tilt-Zoom cameras. The latter are capable of remote directional and zoom control, including by automated means. Huge technical opportunities are provided by the use of Internet protocol (IP) and wireless technologies.

IP cameras are network-ready imaging appliances that digitize the video signal and simplify the network infrastructure.[153] It is believed that IP video systems enhance camera surveillance solutions and lead to increased efficiency, ensuring interconnections over virtually any distance: data can be stored and accessed at any time via wired or wireless connections anywhere in the world and by multiple

[151] See Price (2013) p. 213 and Medwecki (2013) p. 35.

[152] Cardiff Airport. *CCTV Code of Practice* (2013).

[153] Transportation Security Administration (2011) p. 120.

users.[154] Unlike an analogue system there is no limit to the amount of components allowed, IP systems enable integration with other systems, including intruder alarm systems, access control and intruder detection, fire systems, help points, as well as the option for remote monitoring of additional sites.[155] The latest trend is the use of networked Pan-Tilt-Zoom camera, which allows auto tracking.

Advances in digitalisation and miniaturisation increase the ability and capacity to capture images and information.[156] The evidential quality images are one of the fundamental requirements of a camera surveillance system.[157] The image resolution is one of the most important CCTV's performance factors,[158] due to the need to see the smallest details of video, such as people's faces, different objects, numbers of cars, etc. Overall, it is of paramount importance that airport CCTV is capable of identifying individuals. For instance, footage can be used to document and resolve such cases as thefts, improper hand search, etc.

Different technical solutions are available and used in the airports. Modern monitors provide high-resolution, multi-image, multi-screen displays that allow observers and operators to view images with maximum optical information.[159] Megapixel cameras have a significant advantage over analogue cameras; an operator can zoom in on specific targets; megapixel images can provide excellent evidence.[160]

However, with megapixel cameras, the farther you zoom into the picture, the higher the loss of detail, making the picture blurred. According to the producers of another technology – high definition camera with multifocal sensor system, – it provides highest detail resolution at every distance due to its lens and sensor concept.[161] The question that arises is whether images can make natural persons identifiable (see below). In addition, since it may have impact on privacy and seriousness of data protection concerns, similarly to the scanners, the question of storage issues should be discussed in more detail.

A common idea is that hundreds of thousands of people pass through airports daily, and you cannot handle all alarm events in real time.[162] Obviously, CCTV are designed to record the data. For instance, it is reported that Digital Video Recorders for CCTV surveillance systems offer hard disk drive capacities for up to ten weeks

---

[154] Houlis (2010).

[155] Siemens. CCTV: Analogue or IP? http://www.ssp-cctv.com/ssp-cctv/cctv_analogueorip.php, 2015.

[156] Home Office (2013).

[157] Siemens. CCTV: Analogue or IP? http://www.ssp-cctv.com/ssp-cctv/cctv_analogueorip.php, 2015.

[158] Price (2013) p. 214.

[159] Batt. *The use of video surveillance in today's airports*. In: Aviation Security International. Vol. 19 (2013) pp. 18–19.

[160] Medwecki (2013) p. 36.

[161] Dallmeier. Dallmeier CCTV / IP video surveillance solutions for airports. http://www.dallmeier.com/en/solutions/markets/airport/airport.html

[162] Vasilyev. *Modern airport: integrated system of video surveillance*. In: Security Systems. Vol. 6 (2010).

of recording time.[163] The TSA stipulates that a video stream can be compressed and stored in digital format in several forms.[164] The technologies provide a wide range of opportunities for these tasks, including transmission and recording of video and audio signals from any number of cameras in broadcast quality via an Ethernet/IP network.[165]

Thus, it is presumed that images can be recorded, and there are data archives. The actual technical capabilities of various devices may differ, but ultimately, it is up to the controller to establish the limits: the questions are what footage must be stored, if any at all, for what period of time, and at what quality level. These are operational, not technical, requirements.[166] These issues will be discussed in more detail in Chap. 7.

Moreover, qualitative development in camera surveillance has resulted in the "intelligent CCTV" concept. Using IP and wireless technologies, CCTV can be integrated with intelligence gathering, behaviour analysis, biometric identifiers, etc. The key term is Video Analytics, also known as Intelligent Video Analytics or Automated Video Content Analysis.

Video Analytics can broadly be defined as the extraction of meaningful and relevant information from digital video.[167] Video Analytics presents a form of intelligent video in which decision rules, implemented in software algorithms, apply specific behaviours and/or changes derived from the video signal to detect, classify, track and recognize the activities.[168]

In other words, it uses computers to monitor CCTV images, and applies analytical methods for automated detection of certain events from CCTV cameras. The key notions here are "decision rules" and automation. The scope of the former can be very wide, since rules can cover anything from representation of the law to "internal rules", e.g. definition of different "anomalies" situations and events such as suspicious behaviour. While normal CCTV relies on human beings to monitor and deal with potential attacks, Video Analytics can detect events automatically, classify them, track, recognize, and highlight the objects on the screen, and alert security guards.[169]

Besides the software, one more essential element for any "smart" CCTV system is the hardware creating images. Obviously, the quality of images is important too, since poor quality may affect the results.[170] Thus, development of technologies are

163 PACT The Privacy & Security – Research Paper Series: 2012 49.

164 Transportation Security Administration (2011) p. 126.

165 Dallmeier. Dallmeier CCTV / IP video surveillance solutions for airports.

166 Transportation Security Administration (2011) p. 128.

167 Gagvani. *Introduction to video analytics*. EETimes, 22 August 2008, http://www.eetimes.com/document.asp?doc_id=1273834

168 Regazzoni [et al.] *Video analytics for surveillance: Theory and practice [from the guest editors]*. In: Signal Processing Magazine, IEEE. Vol. 27 (2010) p. 16.

169 Batt (2013) pp. 18–19.

170 Meads (2015) p. 18.

making advances towards mitigating any possible problems related to such factors as light, luminance, colour, etc.

Application of Video Analytics is quite wide and may include two basic types of tasks:

(i) techniques of identification/authentication, such as biometrical techniques (facial and gait recognition, iris scanning),
(ii) recognition of certain pre-defined and pre-classified events or behavioural patterns: intruder detection, slip and fall detection, graffiti detection, theft detection, abandoned object detection, e.g. an isolated bag which does not move for a pre-set period of time,[171] suspicious behaviour analysis and tracking, etc.[172]

The key difference between applications (i) and (ii) is that the first one is more active, since it requires "sample" from the person or object to analyse and evidently compare with other samples or databases, the latter is more passive and is aimed at detecting. At the same time, the first one offers more objective data, in contrast to the second one which use some embedded rules implying that criteria may be less objective.

The applications which are used in the airports and imply particular concerns in the terms of this research are facial recognition and behaviour analysis. It should be noted that according to the information of the author, these functions are not used in Norway.[173] In contrast to that, study of different materials allows one to presume that they may be used in the airports in the UK, USA and Russia. However, no official documents can confirm this. Thus, their analysis is based on the assumption that these functions *can be used* rather than established facts. They will be discussed in more detail below.

6.4.2.2 Effectiveness

As discussed in the General Part, airports can be considered as public places with enhanced potential terrorist threats. According to aviation security experts, a well-designed integrated CCTV system and control room should increase effectiveness and achieve a high level of security, reduce staffing, turnover and costs by allowing the operator to concentrate on what humans do best, provide unparalleled accountability, and improve response times.[174] They are supposed to combine the best capabilities of humans and the best of technologies: while the latter can work 24/7, human elements are still essential in order to decide on the right course of action. The following assumptions can be helpful in evaluation.

---

[171] Millward. *Airports to get 'virtual tripwire' CCTV*. The Telegraph, 12 April 2008, http://www.telegraph.co.uk/news/uknews/1556555/Airports-to-get-virtual-tripwire-CCTV.html-

[172] Houlis (2010).

[173] Interview with Avinor representative, 22 Oct 2015.

[174] Houlis (2010).

First, it is a deterrent. A clearly visible CCTV, even faked, signals the presence of a capable guardian and alerts the public that they are being monitored, which may deter criminal activity[175] and enhances a feeling of security in passengers. The potential (rational) offenders know they are being watched, assess the risks of being caught and may choose not to offend or to offend elsewhere. As a result, even if the terrorist act occurs, the consequences may be less destructive than they might have been if the initial plans were realized. This was the case in Moscow in 2003 at a rock festival at the Tushino airfield, when terrorists did not go through the security checks and blew themselves up in front of a police cordon, killing 14 people. According to experts, if the terrorists entered the airfield and blew themselves up there, where thousands of people were present, the number of victims would have been much greater.

However, the mere presence of security equipment is not enough; it must also work. Despite evidence that visible CCTV presence deters crime to some extent, airport crimes continue to occur, and security professionals argue that using CCTV systems is not optimally effective.[176] Thus, other applications are relevant too, especially those which exploit the additional advanced CCTV features that are discussed below.

Secondly, efficient deployment is necessary. A good quality CCTV system can aid police investigations.[177] CCTV cameras allow operators to determine whether police assistance is required. This ensures that police resources are summoned only when necessary.[178]

Thirdly, it is detection and prevention. Obviously, CCTV can help to catch suspects or manage incidents after the event by reviewing the video recording. In some cases, detection may lead to punishment and the removal of the offenders' ability to offend. Images of the criminal can help to identify him, arrest him and can be used as evidence. In addition, due to active real time monitoring, CCTV can help to detect crimes and possible threats in progress and with immediate reaction, i.e. at earlier stages, preventing them from turning into incidents.

For instance, in 2002, security officers at Denver airport discovered a suspicious man, but the (then) imperfect system of CCTV failed to trace him and he entered the departure zone. In connection with this incident, eight hundred flights were cancelled; losses suffered by the airport reached two million dollars.[179]

Fourthly, intelligent CCTV in airports can serve as a pro-active tool that is able to identify people and keep detailed records of people's activities.[180] Video Analytics not only requires fewer people to do the job, but it also leads to a level of security

---

[175] Department of Homeland Security (2012b).

[176] Donald and Corkill. *Exploiting CCTV systems: Enhancing surveillance, situational awareness and intelligence acquisition*. In: Aviation Security International. Vol. 18 (2012) p. 14.

[177] Ratcliffe (2006) p. 35.

[178] Armitage (2002).

[179] Ampersecurity. Video surveillance in airport. http://www.ampersecurity.ru/novosti-industrii/videonablyudenie-v-aeroportu (date accessed: 24 Dec 2014).

[180] Information Commissioner's Office (2014).

that would not be possible using people alone: it is common that after 20 min monitoring a screen, people "turn off"[181] while cameras can monitor 24/7.[182]

In addition, human operators may have insufficient expertise to intercept the data. For instance, it was reported that during interviews, just a few operators could satisfactory answer "What are you actually looking at?"[183] while Video Analytics is programmed with definite tasks. In this way, intelligent solutions for CCTV lead to fewer alarms, identifying only specific, pre-qualified 'real' incidents, making it easier to validate and qualify events.[184] For instance, an airport case study showed that for a human operator it took over 40 min to spot an abandoned bag, while with the aid of Video Analytics, any disastrous consequences could have been prevented.[185] Manufacturers claim that Video Analytics systems could either have headed off the attack on Glasgow airport[186] or made it possible to react almost instantly.[187]

Moreover, Video Analytics can serve passenger convenience and provide a clearer understanding of airport activity. It can assist airport operators in tracking, managing and sharing information about passengers and their luggage, inform passengers about anticipated waiting times, optimize passenger flow and the entire airport experience.[188]

At the same time, although Video Analytics is able to reduce security staff in the airport, it cannot totally replace the operator. It is more appropriate to view it as a way to improve the efficiency of humans: they can be used in highlighting notable events, but it is ultimately the operator who will take immediate and appropriate action and to decide whether an incident is suspicious or not.[189] Namely, the operator can further investigate the event using the camera's pan, tilt and zoom controls, or playback the image to see what caused the alarm,[190] at the same time reducing false alarms.[191] For instance, CCTV footage can be immediately reviewed in order to identify the person who left the isolated bag and possibly locate that person.[192] Thus, human factor is still important.

Moreover, "smart" CCTV presents both a challenge and an opportunity. The challenge is primarily related to the development of appropriate models and algorithms for analysis. If the models and algorithms are adequate, they may contribute

---

[181] Medwecki (2013) p. 34.

[182] *Ibid.* p. 35.

[183] Medwecki (2013) p. 34.

[184] Batt (2013) pp. 18–19.

[185] Houlis (2010).

[186] The Glasgow International Airport attack occurred on 30 June 2007.

[187] Millward (2008).

[188] Batt (2013) pp. 18–19.

[189] Wan. *Video analytics: from access control to behaviour detection and beyond.* In: Aviation Security International. Vol. 17 (2011). p. 37.

[190] Karl. *Integrated CCTV: more than a surveillance tool.* In: Aviation Security International. Vol. 15 (2009). p. 35.

[191] Houlis (2010).

[192] Medwecki (2013) p. 36.

towards addressing security threats;[193] if not, they may produce false alarms, disturbing innocent people, or missing a real threat. Similar to other detection technologies, Video Analytics performance can be measured via detection and false alarm rates. In practice, it is problematic to have a detection rate of 100% and a false-alarm rate of 0%, but is can be accepted if detection rate is high, while the false alarm rates is low.[194] Concrete numbers may differ depending on application and will be discussed with reference to facial recognition and behaviour analysis below.

Overall, it can be seen that today, CCTV in general can be considered as an effective tool at airports due to its capabilities to deter, monitor, detect and prevent. More controversial effectiveness concern CCTV's advanced features such as Video Analytics. Modern CCTV depends on these features greatly. However, at the current stage, despite technological advancements, the security experts admit that the most effective way to collect and analyse the data from intelligent CCTV is still to be learned.[195] Respective problems will be discussed below.

6.4.2.3 Data Protection Issues

Effectiveness of CCTV depends on its capability to capture, process, analyse and store images and information at a quality which is suitable for its intended purpose.[196]

As discussed in Sect. 2.2.2, a principle issue for images to amount to personal data is that natural persons should be identified or identifiable. Guidance on CCTV adopted by WP provides further rules: images data that relate to identified or identifiable natural persons is personal data a) even if the images are used within the framework of a closed circuit system, even if they are not associated with a person's particulars, b) even if they do not concern individuals whose faces have been filmed, though they contain other information, c) irrespective of the media used for the processing, the technique used, the type of equipment, the features applying to image acquisition, etc.[197] Moreover, identifiability may also result from matching the data with information held by third parties, or else from the application of specific techniques and/or devices.[198]

This gives broad interpretation to images being "personal data": the images do not need to be associated with a particular person, any type of surveillance technology can be used, and identifiability can be the result of other – not CCTV as such – technologies. The latter may allow identification via, for example, databases of facial images, such as of known criminals, frequent flyers, etc. (this will be discussed

[193] Regazzoni et al. (2010) p. 16.

[194] Gagvani (2008).

[195] Meads (2015) p. 21.

[196] Home Office (2013).

[197] Article 29 Data Protection Working Party (2004).

[198] *Ibid.*

below). All this is relevant for the airports. If the persons were not identifiable, then the system would not be fit for the purpose.

Thus, it is presumed that images of people captured by CCTV are personal data. The next issue is processing.

CCTV in the airports may process data in a wide range of options. For instance, Cardiff Airport CCTV Code of Practice stipulates the following types of data processing:

(i) Organisation, adaptation or alteration of the data.
(ii) Retrieval, consultation or use of the data.
(iii) Disclosure of the data by transmission, dissemination or otherwise making available.
(iv) Alignment, combination, blocking, erasure or destruction of the data.[199]

Similarly to production of images, technical advances increase the ability to store, share, analyse, and carry on other actions with regard to images – i.e. personal data – and other data. From data protection perspective, the use of IP and wireless networks, the appearance of digital high-quality images and other developments allowed the images to be stored and searched digitally, analysed, reproduced and manipulated with increasing ease; images from any camera can be made available instantly and transferred to anyone, creating data protection risks[200] (will be discussed in Chap. 7).

If Video Analytics is used, CCTV becomes "smart" and provides more opportunities for analysis of data. The process is shown in Fig. 6.6.

Processing includes four steps: (i) segmentation – detecting activities and extracting relevant items for further analysis and qualification, (ii) classification – qualifying each item, (iii) tracking – monitoring of objects on the move, and (iv) activity recognition which combines the results of classification and tracking to infer the activity.[201] Influential factors for these processing steps are background model, camera model, appearance models, motion models and shape models. It is reported that

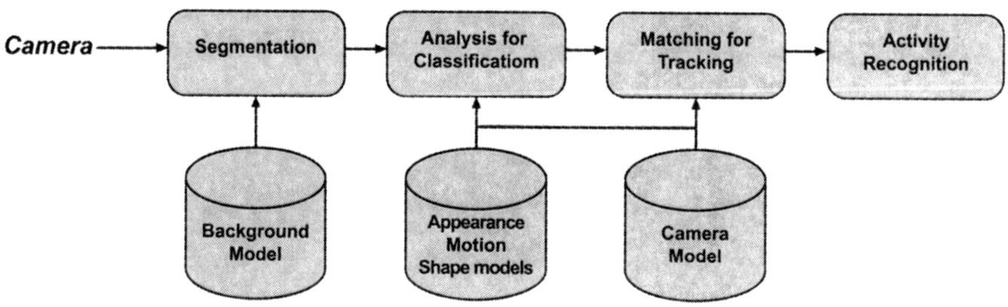

Fig. 6.6 Video Analytics Processing (Based on Figure 1 in Gagvani (2008))

[199] Cardiff Airport (2013).
[200] Gilbert (2007) p. 33.
[201] Gagvani (2008).

more than fifty independent objects can be detected, classified and tracked at the same time by one camera.[202] This figure will be applied to CCTV's facial recognition and behaviour analysis and will be discussed in more details below. In summary, CCTV, no matter how smart it is, undoubtedly deals with processing and data protection regime is applicable. Particular concerns will be analysed in Chap. 7.

6.4.2.4 Privacy Issues

As for privacy, CCTV has impact on it in two ways: through its surveillance role and through its information collecting role.

Of course, for some people the use of CCTV can make them feel safer, but for others, it is the constant feeling of being monitored, in the worst case, a paranoia that an attack may occur.[203] The latter case is extreme, but in general, interference to privacy occurs since CCTV puts people who are presumed innocent under surveillance and records their movements as they move in the airport, i.e. perform lawful activities. If seven types of privacy are taken into account, relevant types of privacy will be privacy of behaviour and action, since people have to adjust their behaviour, and privacy of location and space due to the possibilities of tracking. Moreover, CCTV as such, via ability to monitor individuals and crowds, may have impact on privacy of association.[204] Although airports probably are not the most appropriate places for protests or other gatherings of people, potentially, they can take place.

Moreover, for CCTV in general, exploiting technical capabilities by humans with harmful intentions can lead to criminal abuse, institutional abuse, abuse for personal purposes, discriminatory targeting, and voyeurism.[205] There are documented incidents in which CCTV system operators abused their powers to invade individuals' privacy.[206]

One type of privacy close to data protection is informational privacy. This type of privacy is probably the most bothersome in connection with CCTV in general: "Although the majority of the public may have become accustomed to 'being watched', those who do express concern do so mainly over matters pertaining to the processing of the information/data, i.e. what happens to recorded material".[207]

For CCTV, implications for privacy may differ if CCTV records or does not record personal data. The ECtHR stated that "the monitoring of the actions of an individual in a public place by the use of photographic equipment which does not record the visual data does not, as such, give rise to an interference with the

[202] Batt (2013) pp. 18–19.

[203] Crystal. *FAQ: UK Citizens and CCTV Privacy Rights*. Civil Rights Movement, 7 October 2015, http://www.civilrightsmovement.co.uk/faq-uk-citizens-cctv-privacy-rights.html

[204] Finn et al. (2013) p. 16.

[205] ACLU (2002).

[206] EPIC (2008).

[207] Cardiff Airport (2013).

individual's private life",[208] making public space camera surveillance that monitors public space but does not record lawful.[209] However, if materials obtained through CCTV are made public in a manner or degree beyond that normally foreseeable, interference to privacy may occur.[210] National legislation can go further and provide equal privacy protection no matter if there is any live viewing, if image is recorded or not.[211]

6.4.2.5 Other Human Rights Issues

Other human rights issues, in particular the rights to free movement and to equality and non-discrimination will be discussed in separate sections below.

6.4.3 Biometrics, in Particular Facial Recognition

6.4.3.1 General

The use of biometric data is not new, and refers back to the fourteenth century in China;[212] however, biometrics systems close to modern – i.e. automated – started to be used in the 1960s, with great increase in in 1990s and early 2000s.[213] Today, the use of biometrics is growing all around the world due to first the security concerns, and secondly due to biometrics' opportunities for identification and authentication based on unique characteristics of human bodies.[214] Making these characteristics machine-readable and subject for further use, the biometric technology "changes irrevocably the relation between body and identity."[215]

As mentioned in Chap. 2, biometric data includes "personal data resulting from specific technical processing relating to the physical, physiological or behavioural characteristics of a natural person, which allow or confirm the unique identification of that natural person, such as facial images or dactyloscopic data".[216]

---

[208] *Peck v. the United Kingdom*, No. 44647/98, 28 Jan 2003, § 59.

[209] Finn and Wright. *Unmanned aircraft systems: Surveillance, ethics and privacy in civil applications*. In: Computer Law & Security Review. Vol. 28 (2012b) p. 192.

[210] Vermeulen and Bellanova (2013) p. 304.

[211] Home Office (2013).

[212] Fingerprints were used by Chinese merchants to settle business transactions; fingerprints and footprints were used by parents to differentiate children. Liu. *Bio-privacy: legal challenges for privacy regulations of biometric identification and authentication* (2010) p. 3.

[213] *Ibid.* p. 3.

[214] *Ibid.* p. 1.

[215] Article 29 Data Protection Working Party (2012b).

[216] GDPR Art.4(14).

Physical/physiological characteristics are presented by fingerprint, finger vein, finger geometry, finger wrinkles, palm print, odour, pores, bite marks, hand geometry, dental geometry, ear, facial geometry, facial thermogram, DNA, iris, skin patterns, smile, lips, etc. Behaviour characteristics include dynamic grip recognition, handwriting, tapping, eye movement tracking, keystroke dynamics, gait, mouse dynamics, etc.[217] Despite these many types, the most common used today are facial, iris or finger recognition.

Two basic types of biometrics use can be distinguished: identification of individuals and identity verification. Identification (one-to-many) means an act of identifying a person, i.e. to establish that a passenger is a particular person. It is typically the process of comparing biometric data of an individual (acquired at the time of the identification) to a number of biometric templates stored in a database.[218] For instance, it can be used where authorities aim to identify criminals or potential offenders among the passengers through comparison against a watch list. This means that biometric data are stored on a database.

Identity verification (one-to-one) means that the identity of the person is compared to a claimed identity. It is typically the process of comparing the biometric data of an individual acquired at the time of the verification to a single biometric template stored in a device.[219] For instance, this can be relevant for checking passenger documents containing biometrics against live input data.

A type of the biometric technology which will be discussed here in more detail is facial recognition used by intelligent camera surveillance systems. The main idea is to collect face images of persons by CCTV and compare them to existing biometric data stored in the form of high-resolution images, allowing to verify or identity a person.

In general, facial recognition implies the use of the unique configuration of a person's facial features for identification and verification. A number of technologies are used, including 2D, 3D, infra-red facial scans.[220] The most systems work with two properties of the face: geometry – the configuration and placement of features, and the texture – colours, tones and condition.

In aviation security, facial recognition was initially developed in order to identify dangerous persons by comparing the live biometric input data – passenger's face image captured on CCTV – with the databases of known criminals or suspects.[221] Such identification is still one of the main purposes of facial recognition technology at the airports.

In addition, the use of facial recognition for passenger verification grows rapidly. One of the reasons for this is the fact that the ICAO selected face image as the primary biometric identifier for travel documents,[222] with iris and fingerprints being

---

[217] Liu (2010) p. 45.

[218] Article 29 Data Protection Working Party (2012b).

[219] *Ibid.*

[220] Liu (2010) p. 50.

[221] Meads (2015) p. 19.

[222] Travel documents include identification documents such as passports and supporting documents such as visas.

optional.[223] Accordingly, many states issue biometric passports, e.g. Russia and the EU Member States,[224] allowing comparison of input data with faces in travel documents at border control. Biometric passports have mainly been used in automated border control and/or Trusted Traveller Programs (TTP).[225]

There are several other steps where biometric solution, including facial recognition can be helpful at the airports. They can be used all through the airport to check individuals entering a number of specific locations: at check-in, bag drop, prior to security screening when checking boarding pass, border control, and at boarding. In addition, biometrics can be used to ensure access control for airline/airport employees and other contractors.[226]

The concept of on-the-fly biometrics suggests that the capture of biometric details (primarily in the form of fingerprints, face or iris recognition, but also vein or gait recognition) is executed automatically while the passenger is walking through the airport. The idea is that each one of the biometrics solutions should complement the others, so that the same biometric data and token can be used throughout the system.[227] This should facilitate identity verification at several points in the infrastructure, enabling person-tracking capabilities.[228] The expectation is that within the next 10 years, the majority of airports will trial single biometric travel tokens[229] and 54% airlines plan to evaluate the technology.[230] But today, although it has great potential, this technology is still being developed, with a number of challenges regarding performance (accuracy, speed) and operational requirements. Thus, further research and development is needed.[231]

Nevertheless, according to a report, in 2016, plans of 34% of all airports include interactive way-finding using biometrics; 24% – identity management using biometrics; and these numbers are growing.[232] Today, some segments of the scheme above are partly realized in some states.

In the USA, some airlines test passengers' fingerprint screening for check-in registration and to identify passengers before they board planes (e.g. Alaska Airlines).[233] In Amsterdam's Schiphol, members of the Privium programme, via iris

[223] ICAO, Doc 9303, Machine Readable Travel Document, Seventh Edition, 2015.

[224] See, e.g. Commission Decision of 28 June 2006 laying down the technical specifications on the standards for security features and biometrics in passports and travel documents issued by Member States.

[225] See more details in Finn. *The identity challenge: verifying travel document integrity, quashing breeder document fraud.* In: Aviation Security International. Vol. 21 (2015).

[226] This use is beyond the scope of this book, which, as noted in Chap. 1, is focusing on measures relating to passengers.

[227] Airports Council International (2005).

[228] COPRA (2013).

[229] SITA. The Airport IT Trends Survey 2016.

[230] SITA. The Airline IT Trends Survey 2016.

[231] COPRA (2013).

[232] SITA. The Airport IT Trends Survey 2016.

[233] Martin. *Alaska Airlines to consider using fingerprints to screen passengers.* Los Angeles Times, 24 October 2014, http://www.latimes.com/business/la-fi-alaska-airlines-fingerprint-scanners-20141024-story.html

recognition and for a fee, can get such benefits as fast border passage, priority parking and business-class check-in.[234] Today, 50,000 people use that service.[235] In the USA, users of CLEAR system, via fingerprints or iris and also for a fee, get the privilege to bypass long security waits through a separate CLEAR line.[236] However, these programmes, although reducing waiting times, do not provide privileges in terms of the scope of security screening.

The latter purpose is realized within the concept of trusted traveller programs (TTP) such as American PreCheck, which, in exchange for biometrics and other personal data, allows a person to be included in a "trusted" category with a minimum amount of screening (see Sect. 6.8.2).

At immigration/passport control, the main purposes are to identify/verify a person and to check the documents for authenticity. Biometric data can be captured and analysed. At the security screening stage, some states such as Russia check both travel documents and boarding passes; others, such as the EU states, check only boarding passes.

The point is that immigration control usually is carried out only for international flights, and the respective organs are concerned more about arriving passengers rather than those departing, since they see their role as border protection rather than aviation security.[237] At check-in, airlines also usually check travel documents in order to avoid being fined for transporting improperly documented passengers, but this check is not as comprehensive as that of immigration authorities. For instance, airlines are not connected to international databases of lost and stolen passports and may therefore allow a person with forged documents to board a plane.[238]

Thus, security experts suggest that security controls should refocus from screening to identification and authentication of passengers via biometry and automated identity control,[239] since it provides "a good starting point" for risk assessment.[240] Indeed, checking documents at security checkpoints may eliminate possible security gaps.

At the boarding gate, the technology can help to verify whether or not the person boarding the aircraft is the one who is registered,[241] to ensure that passengers did not exchange boarding passes.

---

[234] Schiphol. Privium: your benefits. http://www.schiphol.nl/Travellers/AtSchiphol/Privium/PriviumBenefits.htm

[235] Bradley (2015).

[236] Clear. https://www.clearme.com

[237] Baum. *Malaysian Airlines Flight MH370: A "deliberate act" but how, by whom, and why?* In: Aviation Security International. Vol. 20 (2014) p. 21.

[238] *Ibid.*

[239] Naudin (2014) p. 44.

[240] Finn. *Enhanced documentation control: An aviation security perspective.* In: Aviation Security International. Vol. 18 (2012) p. 38.

[241] COPRA (2013).

If the stages of possible biometrics collection are combined, using CCTV with facial recognition along with the object tracking function, it may be possible in real time or in review mode to watch a video recording and track a person and, in some cases, catch up with the subject through real time tracking.[242]

There are also other developing technologies, which may go beyond the scheme above. For instance, one of them suggests scanning fingerprint traces left on in-checked baggage and comparing the data against a database of criminals.[243] However, one can only imagine how many resources would be needed to realize this plan, especially in huge, busy airports.

Another developing issue is that in addition to the identification and verification applications of biometrics, which are relatively old technologies, in particular, fingerprints, iris scan, and facial recognition, there are also some emerging technologies which either collect new physiological characteristics or process traditional biometrics in new ways.[244]

A new trend is the use of the so-called soft biometrics, defined by the use of very common traits not suitable to clearly distinguish or identify an individual but that allow other applications or enhance the performance of other identification systems.

For instance, one more application is biometric categorization/segregation of an individual by a biometric system. It is the process of establishing whether the individual's biometric data belongs to a group with some predefined characteristic in order to take a specific action. Therefore, for this application, it is important to automatically assign the person to a certain category, rather than identify or verify him/her. This technology can be used in advertising, for example, enabling targeting of potential customers by showing different advertisements depending on the individual's age or gender.[245]

As a result, there is a change in the focus on development in biometric systems from identifying tools to soft recognition purposes, in other words, from identification to detection of behaviour or specific needs of people; an enhanced man-machine interaction allows more than identification, or categorization of an individual.[246]

In aviation security, soft biometrics could be helpful in automated profiling based on definite characteristics and already used via CCTV's behaviour analysis (see below). It could be used, for instance, to verify that luggage is picked up by the same person who sent it. A positive feature of soft biometrics in privacy/data protection terms is that it usually does not imply processing of personal data, being "soft" on privacy and data protection too.

---

[242] Medwecki (2013) p. 36.

[243] Pocs. *Legally compatible design of future biometric systems for crime prevention*. In: Innovation: The European Journal of Social Science Research. Vol. 26 (2013). p. 36.

[244] Article 29 Data Protection Working Party (2012b).

[245] *Ibid.*

[246] Article 29 Data Protection Working Party (2012b).

6.4.3.2 Effectiveness

Biometrics technologies in general provide very reliable means to perform identity verification or identification of individuals.[247] At the same time, not much assessment is done on the use of any types of biometrics in airports, where high reliability and low failures are particularly required.[248] According to some experts, large-scale testing of face recognition systems proved a very high accuracy of the technology enhancing its effectiveness for aviation security.[249] In 2014, it was reported that in the USA, CCTV's facial recognition system for the first time caught and convicted a thief.[250] Other researchers note that facial recognition is subject to high false rates,[251] and is less reliable in comparison with iris, fingerprints, or finger vein technology.[252]

The problem is that for successful work, the system requires the highest possible level of accuracy. Despite the technological advances, facial recognition technology still suffers in accuracy, however. Many image characteristics and image environments work against the visual classifications: noise, poor lighting or low visibility, low image quality, inappropriate camera positions and orientation, people constantly on move, high crowd density, etc.[253] In addition, the effects of age and facial expression, wearing glasses, hats, cultural head- or facial dress, etc. can alter the way a face looks on camera.[254]

It is a challenge to develop facial recognition techniques that can tolerate all these factors. Some technological advances enhanced the mitigation of potential difficulties. For instance, systems were developed to recognize that some factors of persons' faces can change, to tolerate poor quality, highly compressed videos and images, etc.[255] However, still, it is a problem to recognize moving images and compare them to a control image. Even in perfect conditions (stand still, right angle, lighting and background) facial recognition performance is around 90%. This means that even under perfect conditions, both false positive and false negatives may take place. There were cases when a failure of face recognition technology allowed women mistakenly holding a daughter's or husband's passport to pass through border control.[256]

---

[247] COPRA (2013).

[248] Bradley (2015).

[249] Vasilyev (2010).

[250] Woollacott. *Life after privacy: the next generation of public surveillance technology is already here*. NewStatesman, 7 July 2014, http://www.newstatesman.com/politics/2014/07/life-after-privacy-introducing-next-generation-public-surveillance-technology

[251] Badii et al. (2012) p. 367.

[252] Meads (2015) p. 20.

[253] Badii et al. (2012) p. 368.

[254] Liu (2010) p. 50.

[255] Watson. *Facial recognition – the death knell of anonymity?* BBC, 26 August 2014, http://www.bbc.com/news/uk-england-28307929

[256] Gold. *Border control biometrics and surveillance*. In: Biometric Technology Today. Vol. 2012 (2012).

This is particularly relevant to matching biometrics with a database: the more biometric identification used and the larger the database, the greater the likelihood of false matches.[257] Facial recognition becomes less reliable the larger the database gets.[258] Of course, with reference to accuracy, the most disturbing point is false positives, resulting in detention and questioning of innocent individuals. Apart from serious privacy intrusiveness, this may result in more serious consequences, such as identity theft based on stolen biometric sources. It is a problem to simply provide individual with a new identification just because it is compromised.[259] The context can make a situation worse, for example a person mistakenly identified as a criminal in a foreign country lacking human rights protections may face criminal charges.[260]

In 2002, it was reported that facial recognition technologies in an airport produced one false positive for every 700 passengers, in particular, triggering a false positive when evaluating persons with Middle Eastern appearance.[261] Even though the current rates, with technological developments, have become better, taking into account hundreds of thousands of passengers the airports process daily, one can imagine how many irrelevant individuals are subject to being bothered. For instance, in Heathrow, about 201,000 passengers arrive and depart per day,[262] hypothetically, if all of them are covered by CCTV with facial recognition, with one false positive for 700 passengers, up to 287 persons could be stopped mistakenly every day.

An option that may contribute to reducing failure rates is the use of multimodal systems using different biometrics in a simultaneous way or multiple readings/units of the same biometrics that can be adjusted to optimize the trade of security / convenience of the biometric systems. This can reduce the false acceptance rate, improve the results of a recognition system or can facilitate the collection of data of a larger population by balancing the non-universality of one source of biometric data by combining it with another.[263] But, for aviation security, the combination of CCTV's facial recognition with other biometrics types – and the use of these other types as such – can be difficult to integrate with existing airport/airline processes and systems.

Other factors that slow down the use of biometrics in aviation – in addition to the accuracy and integration problems – include the lack of industry standards governing how biometrics should be used and privacy concerns.[264] Security experts note that currently, the use of facial recognition rather than more accurate iris recogni-

---

[257] Gilbert (2007) p. 23.

[258] Watson (2014).

[259] Article 29 Data Protection Working Party (2012b).

[260] Porter. *Surveillance technologies in society*. Speech of October 2014, published on 18 November 2014, https://www.gov.uk/government/speeches/iriss-respect-and-surveille-event

[261] Badii et al. (2012) p. 367.

[262] About Heathrow Airport. http://www.heathrowairport.com/about-us/company-news-and-information/company-information/facts-and-figures

[263] Article 29 Data Protection Working Party (2012b).

[264] Bradley (2015).

tion, fingerprints or the most accurate, finger vein technology is a trade-off between accuracy and the convenience of adapting existing systems and procedures.[265] Clearly, many factors play a role, including the availability of cameras, passports with biometrics, current security procedures, etc. For the purposes of this research, however, the privacy concerns deserve special attention. The point is that they are able to prevent biometrics from being used.

6.4.3.3 Human Rights Issues

As discussed in Chap. 2, biometrics may constitute personal data, but of a specific type, since sources of data are presented by material items. Via biometrics, sensitive personal data can be collected. In the case of CCTV's facial recognition, facial images of persons constitute personal data – clearly, they allow identification of people, and thus the right to data protection is applicable.

Accordingly, the scheme of processing is quite similar to a general Video Analytics scheme above. Processing includes the following steps: segmentation – detecting facial images; analysis and matching – comparing with biometric template(s) stored in a database (one-to-many or one-to-one); and recognition – identification or verification of a person.

Overall, with CCTV, biometrics can be used for automated tracking, tracing, targeting and profiling purposes[266] (profiling will be discussed below), implying risks to the right to privacy. Relevant types of privacy include privacy of person (bodily), of location and space, of association and of data and image. Discrimination and freedom of movement issues may also apply (will be discussed below).

What is important, the concerns may become more serious due to a number of factors. First, facial images can be collected covertly and remotely, from a distance or in motion, repeatedly and on a mass scale, without the need of cooperation or action required from the individual. Camera surveillance at airports may take images of passengers without their knowledge, carrying out facial recognition in real time. Secondly, sensitive information can be revealed from biometrics data, which, as discussed, requires particular protection.

Thirdly, this technology can be used in combination with the use of personal data from other sources. As it will be shown in Sect. 6.5, there is already a huge amount of passenger personal data collected for security purposes. It is argued that collection of faces is the next step, thus, one may wonder if one day, passengers will be required to have their faces scanned, stored in a database by Interpol, and subsequently shared with law enforcement agencies worldwide in order to fly.[267] It looks like the world is moving in this direction. Huge databases of human faces are already in place, both criminals and the public at large.

---

[265] Meads (2015) p. 20.

[266] Watson (2014).

[267] Porter (2014).

Indeed, Interpol has huge biometric databases. It is reported that it produced a 'best practice guide' for 190 Interpol member states for the quality, format and transmission of images to be used in facial recognition, and encourages the sharing of facial images with the Interpol database.[268] It is unknown to me whether images are limited to criminals/suspects or not. However, there are many other systems that collect faces of ordinary people.

Collecting faces and fingerprints of foreign people at the stage of visa application is a practice common to all the selected states (USA, UK, Norway and Russia). The main idea is to check at the border whether the person holding the biometric visa is the person who applied for it. In the USA, the relevant program is US-VISIT, where the biometric data is shared with DHS agency components, other federal agencies, "federal, state, and local law enforcement agencies", the "federal intelligence community" as well with some foreign governments, including Canada, Australia, and the United Kingdom.[269]

In the EU, a similar role is carried out by Visa Information System, which concerns individuals entering Schengen area. A proposed EU Entry/Exit System is supposed to extend biometric ID checks to all non-EU nationals entering or leaving the EU.[270] Moreover, there is a proposal to expand the system to include EU nationals, due to the risk of threat from terrorists in the Schengen area, both from refugees and EU nationals who join terror groups.[271]

In the USA, there is already a national biometric database within the Next Generation Identification program aiming to protect public safety and national security maintained by FBI. It includes fingerprint, iris, facial and voice recognition data containing private data on millions of Americans,[272] and the database was estimated to reach 52 million in 2015, combining images from both criminal and non-criminal sources.[273] In its turn, the UK's Police National Database equipped with automated facial recognition technology contains 18 million photos, including of people never charged, and others cleared of an offence.[274]

Apparently, all these databases can potentially be shared and combined. As the result, comparison of the live input data with photos from all these databases can be done too.

---

[268] *Ibid.* Another question is under what legal basis Interpol has the authority to collect private data on individuals without their consent given its status as a non-governmental organization.

[269] EPIC. *EPIC Opposes DHS Biometric Collection.* http://epic.org/2013/06/epic-opposes-dhs-biometric-col.html

[270] EDRi-gram 13.21, 4 Nov 2015 https://edri.org/edri-gram/13-21/

[271] Tung. From fingerprints to facial scans: Why the French want biometrics on all EU travelers. ZDNet. 13 October 2015. http://www.zdnet.com/article/from-fingerprints-to-facial-scans-why-the-french-want-biometrics-on-all-eu-travellers/

[272] Porter (2014) and EPIC, *Biometric Identifiers.* http://epic.org/privacy/biometrics/

[273] Watson (2014).

[274] Hopkins and Morris. *'Innocent people' on police photos database.* BBC. 3 February 2015. http://www.bbc.com/news/uk-31105678

At the same time, it is reported that some jurisdictions such as the USA and the UK develop databases of images of persons taken by CCTV as they pass through immigration controls.[275] Thus, there is the possibility that the input data ends up in some database and maybe used for some purposes unknown to the individual whose image was collected. The point is that in contrast to other biometric identification such as fingerprints and iris scans requiring the subject's cooperation, facial recognition does not require consent: images can be recorded anywhere, including covertly and remotely without the person knowing it. This leads to increased privacy/data protection concerns.

Finally, CCTV's face recognition can potentially use faces from social networks such as Facebook. As Working Party noted, equipping camera surveillance systems with facial recognition systems based on social network databases will put an end to anonymity and untraced movement of individuals.[276] As discussed, air passengers are already identified and their movements are traced, but with enhanced application of the technology, this will be to a much serious extent.

There are also other factors of risk which refer to more general problems such as false positives, risks of unauthorized access, function creep, etc.[277]

The positive side of facial recognition is that just as software algorithms can be used to identify facial features, algorithms can be developed to use facial recognition technology only on the most suspicious persons, blocking or blurring facial features of other people ("anonymization" or "video masking"). Alternatively, with reference to image sharpness, in order to mitigate the privacy concerns, it is possible to have blurred images as input.[278] Other methods try to completely remove persons from video and replace them with geometric shapes, decreasing considerably the risks of privacy infringement.

Of course, if the system can be programmed only to look for individuals wanted on a warrant and to block the facial features of all other persons,[279] it would benefit privacy. The problem is that in reality, "[d]ata can be either useful or perfectly anonymous but never both".[280] In case if the image can be de-anonymized later, e.g. for the purposes of a crime investigation, they can still be considered as personal data.[281] Thus, such "anonymization" does not work. The researchers propose some solutions, e.g. contextual risk assessment, considering data handling techniques, private versus public release, quantity, motive, and trust; and balance benefits of access to information against risk of privacy harm.[282] For instance,

Novel Video Analytics architecture and processes for preserving privacy in a durable and customisable way which uses User-Intimate Requirements Hierarchy

---

[275] Medwecki (2013) p. 36.

[276] Article 29 Data Protection Working Party (2012b).

[277] Pocs (2013) p. 37.

[278] Badii et al. (2012) p. 369.

[279] Zoufal (2008) p. 93.

[280] Ohm. *Broken promises of privacy: Responding to the surprising failure of anonymization*. In: UCLA Law Review. Vol. 57 (2010). p. 1704.

[281] Vermeulen and Bellanova (2013) p. 303, Badii et al. (2012) p. 367, for more details, see Ohm (2010).

[282] For more details, see Ohm (2010).

Resolution Framework (UI REF) for engaging discourse with the relevant actors, topic maps for representing semantic collateral and complex personal privacy rules. The application of the privacy filters built based on the user requirements relies on the automatic detection of contexts for applying specific privacy-preserving rules and on image descriptors for evaluating the level of precision that the computer vision algorithms could reasonably achieve in their specific context.[283]

Accordingly, technologies may both invade privacy and protect it, but these issues, as well as the appropriateness of relying on biometrics to resolve problems or make decisions about individuals, are the issues that will need to be considered as biometric technologies evolve,[284] i.e. subject to future research.

In the meanwhile, in order to establish whether a particular biometric system is acceptable, it is suggested that one evaluate the concrete context, the purpose of processing, the amount of false positive and false negative rates, the population size, and other factors.[285] As Airports Council International (ACI) suggests, "regulators should adopt a policy which balances the goals of protecting personal data when used and processed for biometric systems with the potential benefits that such systems can deliver in terms of increasing security for the passenger and facilitating more efficient passenger flows".[286] For instance, in the UK, with regard to the facial recognition technology used by police's databases, it is suggested that the development of an appropriate regulatory regime should be initiated by proposal followed by public consultation.[287] This order can be appropriate to use of any technology.

Applicability of the rights to free movement and to equality and non-discrimination will be discussed below.

6.4.4 *Profiling*

6.4.4.1 General

Profiling, similar to biometrics, is not new and refers back to the Middle Ages, when the inquisitors tried to "profile" heretics.[288] It is believed that the first "profile" for criminal purposes was created by British detectives in 1880s with regard to serial killer Jack the Ripper.[289] Today, the term "profiling" is used in relation to different issues and in different spheres. In addition to offender profiling, there is DNA profiling, molecular profiling, profiling in computer programming, etc.

[283] Badii et al. (2012) p. 370.

[284] OECD Privacy Framework 2013.

[285] Article 29 Data Protection Working Party (2012b).

[286] Airports Council International (2005).

[287] Commissioner for the use and retention of biometric material. *Annual Report 2014* (2014).

[288] Wikipedia, Offender profiling. https://en.wikipedia.org/wiki/Offender_profiling

[289] *Ibid.*

In terms of aviation security, in general, profiling is the practice of categorizing people according to particular characteristics. Such characteristics may include passenger's actual or perceived race, ethnicity, religion, nationality, age, gender, behaviour characteristics, separately or in combination, or together with other factors.[290]

The core idea is to create profiles or dossiers – a set of definite characteristics, associations, activities, etc. to predict behaviour and more, and then search for individuals with a close fit to that set of characteristics.[291] For instance, the profile of the terrorist can be deduced by cross-indexing information from various databases, the population roll, the use of credit cards, movements detected through use of mobile phones, brand name discount cards, use of medications, etc.[292]

Three types of profiling can be noted: manual profiling performed by a human being, automated profiling performed by computers, and semi-manual, combining the first two. For any type of profiling, particular characteristics may serve as indicators of potentially dangerous behaviour, assisting in determining whether to stop, search, or question a person. In aviation security, all types of profiling are used extensively and may refer to all the selected measures, as shown in Table 6.1.

With regard to body scanners, profiling can be relevant if passengers are selected to be scanned by security agents (manually) or automatically via a computer prescreening system. For CCTV, behaviour analysis can be conducted manually by a CCTV operator or automatically by algorithms of Video Analytics; however, as it will be discussed below, the human element cannot be excluded. Thus, semi-manual profiling can take place too. For Passenger Name Record (PNR) systems, all forms of profiling can be relevant;[293] however, the automated variant is the most widespread.

Profiling, especially the automated one, can also be defined as any form of processing of personal data intended to analyse or predict the personality or certain personal aspects relating to a natural person.[294] This technique uses statistical methods and programmatic algorithms that cross-index randomly selected information

Table 6.1 Profiling in relation to selected aviation security measures

	Manual profiling	Semi-manual	Automated profiling
Body scanners	X		X
Behaviour analysis via CCTV	X	X	X
PNR	X	X	X

[290] Quinlan (2015) p. 36.

[291] Clarke (1993) p. 405.

[292] Poullet (2009) p. 214.

[293] E.g. the UK Border Force conducted manual analysis of PNR data. See Vine. *'Exporting the border'? An inspection of e-Borders.* March 2013. http://icinspector.independent.gov.uk/wp-content/uploads/2013/10/An-Inspection-of-eborders.pdf, p. 17.

[294] Article 29 Data Protection Working Party *Advice paper on essential elements of a definition and a provision on profiling within the EU General Data Protection Regulation* (2013) §1. It is based on definition of Art. 1(e) of Council of Europe Recommendation CM/Rec(2010)13 of the Committee of Ministers to member states on the protection of individuals with regard to automatic processing of personal data in the context of profiling of 23 November 2010.

from large databases and provide risk assessment of individuals or predict their future behaviour.[295]

The work with information – its collection and analysis – plays the key role in modern profiling. It is clear that "99.9999% (if not more) of passengers and freight pose no threat to aviation".[296] Therefore, the information that is available can be used to assess the risk of the people, objects or situations that can pose threats. According to Interpol Secretary General Ronald K. Noble, "Whenever attacks have been successful, it was because we collectively failed to collect, analyse and share the information we already had.[297] And whenever attacks have been prevented, it was because we succeeded in exploiting the information available to us concerning a specific passenger, group or threat".[298] Aviation security measures which deal with personal data "increase the breadth and depth of intelligence gathering and sharing worldwide",[299] thus, are of major importance.

The first system of computer-assisted passenger profiling and other appropriate passenger profiling programs refer to the USA, where the airlines, assisted by the state aviation and transport security organs and the intelligence community, were recommended to use such programs already in 1996.[300]

Today, use of profiling in aviation security is a global strategy. For instance, ACI, together with the ICAO and IATA, develop the concept for a Next Generation Security Process (NGSP), which: "(i) looks at the security of the end-to-end passenger journey; (ii) leverages intelligence and advanced passenger information to assess individual passengers; (iii) filters passengers according to their risk profiles; and (iv) focuses additional security screening measures on the passengers that present the higher risk".[301] This means comparing passenger details with existing watch lists and the use of profiling.

Another example is the US CBP's Analytical Framework for Intelligence (AFI), a program that incorporates personal information from DHS, other federal agencies, and commercially available databases.[302] Detailed list of anticipated personal data stored by AFI includes multiple data fields, allowing further inclusion of data types via formulations "including, but not limited to".[303] As EPIC reports, the CBP uses

---

[295] Poullet (2009) p. 214.

[296] IATA Press release, 5 Mar 2013. http://www.iata.org/pressroom/pr/Pages/2013-03-05-02.aspx

[297] A good example of such "failure" is 22 July 2011 in Norway.

[298] Interpol. Media release. *Global aviation security requires the industry to step in where governments fail, Interpol Chief tells IATA meeting*. 7 June 2011. http://www.interpol.int/News-and-media/News/2011/PR048

[299] Forest. *The Modern Terrorist Threat to Aviation Security*. In: Perspectives on Terrorism. Vol. 1 (2007) p. 12.

[300] Pub. L. 104–264, title III, §307, Oct. 9, 1996, 110 Stat. 3253.

[301] ACI position brief. November 2010. http://www.aci.aero/Media/aci/file/Position%20Briefs/2010/position%20brief_SECURITY%202010.pdf

[302] Department of Homeland Security. *Privacy Impact Assessment for Analytical Framework for Intelligence (AFI)* (2012a) p 1.

[303] *Ibid.* p 27.

"law enforcement or intelligence information regarding an individual" and "risk-based rules developed by analysts to assess and identify high-risk cargo, conveyances, or travellers that should be subject to further scrutiny or examination".[304]

As the result, today, with huge personal data databases and computational power, profiling is a key part of security governance. Advanced algorithmic analytics, which are unknown to the public due to security needs, enable authorities to make sense of unprecedented amounts of personal data and derive patterns in a data-driven fashion.[305]

6.4.4.2 Effectiveness

The proponents state that profiling can be extremely powerful and useful for security purposes, offering a huge potential for preventing terrorist acts,[306] because certain identifiable characteristics and behaviours are likely to be associated with terrorist activity.[307] It is possible to discover previously unnoticed relationships within the data and discover facts about a person that they might not be consciously aware of themselves.

According to many security experts, air passenger profiling represents a necessary policy decision.[308] If properly used, the data can help not only to catch terrorists whose immediate target is aircraft, but also those who travel by air to meet to plan and train for terrorist actions.

The problem is that profiling and data mining techniques are not very accurate. For instance, if there are 1.8 million airline passengers each day, a data mining program to identify terrorists with a false positive rate of only 1% would indicate eighteen thousand people as false positives.[309] Accordingly, while criminals may be missed, many innocent people can be mistakenly disturbed: searched, detained, appear on inaccurate watch lists that are used to decide whether a person can fly (see more details in Sect. 6.6 below).

Thus, similar to other technologies, profiling, although deemed to be effective, may potentially disturb many innocent people.

---

[304] EPIC Alert, Volume 22.15, 14 Aug 2015. EPIC Pursues Lawsuit over Secret Government Profiling Program.

[305] Leese. *The new profiling: Algorithms, black boxes, and the failure of anti-discriminatory safeguards in the European Union*. In: Security Dialogue. Vol. 45 (2014) p. 494.

[306] Yehoshua. *Terrorist profiling: analysing our adversaries personalities*. In: Aviation Security International. Vol. 17 (2011) p. 23.

[307] Solove. *Data mining and the security-liberty debate*. In: The University of Chicago Law Review (2008a) p. 345.

[308] Ravich. *Is Airline Passenger Profiling Necessary*. In: U. Miami L. Rev. Vol. 62 (2007) p. 56.

[309] Solove (2008a) p. 353.

6.4.4.3 Human Rights Issues

Profiling uses data mining and profiles for classification of people into different categories for "social sorting" – in order to implement a differential treatment of each category, to indicate potentially dangerous persons considered as "deserving further attention" or "special treatment".[310] Simplified, the processing scheme can be illustrated like this (Fig. 6.7).

It can be seen that personal data is first collected. Then, the data is analysed and classified, with the ultimate aim to match potentially dangerous persons and select them for questioning, further screening, etc. During these processes, it is necessary to find meaning in enormous amounts of data, to decide which data is worth analysing,[311] and then interpret the data, see connections and make predictions.[312] The ability to store the data is important too, since the data can then be re-used in a new circle, realizing the main idea of Big Data: massive merging of data generating more data, more assumptions and more knowledge about persons.[313]

Back in 1993, Clarke noted that profiling is little-understood and ill-documented, but increasingly used.[314] After more than 20 years, this statement is still relevant, and apparently, the concerns became more serious due to phenomenon of Big Data. The most important challenge of the latter is profiling: with Big Data, the "I've nothing to hide" argument makes no sense at all, since you never know what guesses are made on the basis of the data about you.[315]

Any type of profiling incurs enhanced risks in privacy, discrimination risks, and risks to the freedom of movement (see below). Profiling of personal data and behaviour profiling via CCTV, since it uses images of persons, in addition, raise data protection concerns. Further, behaviour analysis will be discussed in more detail. The use of PNR for profiling will be discussed in Sect. 6.5.3.

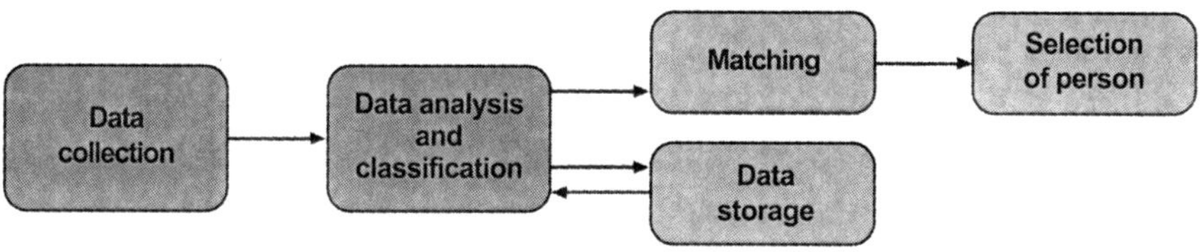

Fig. 6.7 Profiling processing

---

[310] Fuster [et al.] *Profiling in the European Union: A high-risk practice.* In: INEX Policy Brief. Vol. 10 (2010) p. 2 citing Taipale. *The privacy implications of government data mining programs.* In: Testimony before the US Senate Committee on the Judiciary. Vol. 10 (2007) p. 6 and Lyon *Surveillance as social sorting: Privacy, risk, and digital discrimination* (2003b) p. 20.

[311] Raffel (2007) p. 2.

[312] Schneier (2008) p. 7.

[313] See Chap. 2.

[314] Clarke (1993) p. 403.

[315] EDRi-gram 13.21, 4 Nov 2015 https://edri.org/edri-gram/13-21/

6.4.5 *Behaviour Analysis, in Particular via CCTV*

6.4.5.1 General

In general, behavioural analysis is a technique intended to help the professionally trained security staff to spot passengers acting unusually and target them for a further search. This includes coordinated and active monitoring of the reaction of passengers to potential detection, searching, questioning techniques, and potential avoidance or anxiety behaviour.[316] This technique may take different forms, in aviation security, three of them can be noted: monitoring by security agents, by CCTV human operator, and by CCTV's algorithms.

Behaviour analysis is often seen as a more appropriate alternative to racial and similar profiling.[317] However, as discussed, behavioural analysis is a type of profiling; these terms are often used synonymously.[318] In addition, although this technique is seen as separate from racial, ethnical, religious and other types of profiling, all these profiling types are often used in combination, making it difficult to separate one from another.

An example of such overlap are practices of the Israeli El Al, which is recognized for dealing with exceptional security concerns and has been using these techniques for years.[319] The Israeli aviation security system is often considered the best in the world, and it is profiling that makes it such.[320] The core idea is that all passengers are interviewed individually by trained security agents who all are elite and highly trained. Questions can be asked during the entire time a person is at the airport. Not only actual verbal responses, but body language and facial expressions are assessed, and other criteria play roles too. The exact criteria are not publicly available, but in addition to behavioural patterns, considerations include the ethnicity, religion, nationality, travel information, and intelligence.[321]

This way, around 2– 5 percent of all passengers are chosen for additional screening.[322] It is noteworthy that the latter can involve not only hand searches, but taking apart luggage item by item, hours of questioning, checking all of a passenger's digital photographs in a lap top, or even confiscation of all the luggage and keeping it for several days.[323]

---

[316] Donald and Corkill (2012) pp. 12, 15.

[317] Quinlan (2015) p. 36.

[318] E.g. Order of the Ministry of Transport of 25 July 2007 N 104 On approval of rules for pre- and post-flight inspections, Art.51.

[319] Karber (2001) p. 802.

[320] Palmer. *What's So Great About Israel Security?* Slate. 3 January 2011. http://www.slate.com/articles/news_and_politics/explainer/2011/01/whats_so_great_about_israeli_security.html

[321] Guttman. *Israel's Airport Security, Object of Envy, Is Hard to Emulate Here*. Forward. 6 January 2010. http://forward.com/news/122781/israel-s-airport-security-object-of-envy-is-hard/

[322] Palmer (2011).

[323] *Ibid.*

Other states adopt similar models, e.g. the UK began implementing Behaviour Analysis Screening System in 2010; in the USA, TSA's behaviour detection program Screening Passengers by Observation Techniques (SPOT) is implemented in selected US airports from 2003. In Russia, "During the pre-flight inspection in order to identify indicators of a potential threat, it is recommended to pay attention to such behaviours of passengers, as increased nervousness, concern, fussiness. In case of such signs it is necessary to conduct a psychological survey (profiling) of the passenger to determine the extent of its potential danger".[324]

A similar technique is the psychologically informed conversational style Controlled Cognitive Engagement (CCE), which allows aviation security staff to control the narrative, impose cognitive load and be unpredictable during security screening. In practice, security agents initiate informal conversation with passengers, asking such questions as whom the person is visiting. Instead of accurate answers, agents are looking for changes in the passenger's behaviour, such as increasingly shorter responses, or evasive or erratic responses.[325] According to its developers, the technique is more effective (66% success rate for detecting risk) if compared with current behavioural analysis methods which look for signs of suspicious body language such as lack of eye contact and fidgeting (less than 5% success rate).[326]

Behavioural analysis technique can be also carried out automatically rather than by human beings, specifically, one of the functions of the CCTV's Video Analytics is detection of suspicious (unwanted, abnormal) behaviour.

Similar to facial recognition, CCTV, via software and algorithms, are trained to indicate people acting suspiciously. If integrated with crime prediction capabilities based on the belief that criminals can be recognized by suspicious actions, CCTV can identify crimes before they occur. Although this is reminiscent of the plot in the movie "Minority report", experts claim that criminals do use certain patterns before committing a crime.[327] In aviation security, if conducted properly, behavioural analysis may be the only way to identify a potential body cavity bomber – undetectable by current technical means – prior to possible attack.[328]

---

[324] Order of the Ministry of Transport of 25.07.2007 N 104 On approval of rules for pre- and post flight inspections. Art. 51.

[325] Rettner. *New Airplane Passenger Screening Method Outshines Old.* Live Science. 6 November 2014. http://www.livescience.com/48661-airport-screening-conversation.html

[326] For more details, see Ormerod. *Finding a Needle in a Haystack: Toward a Psychologically Informed Method for Aviation Security Screening* (2014). In: Journal of Experimental Psychology: General, Vol. 144, No. 1 (2015) pp. 76 – 84.

[327] Crystal (2015).

[328] Bunker (2014) p. 20.

6.4.5.2 Effectiveness

With reference to the Israeli profiling programme, it has a declared 100% success rate, since no aircraft has been hijacked at an Israeli airport and no Israeli aircraft has been hijacked in over 40 years.[329] Profiling helped, for instance, to prevent the attempted bombing of an El Al flight from London to Tel Aviv in 1986. At the same time, Israeli behaviour analysis techniques are criticized for disproportionally targeting young Arab men and being simply a form of racial, religious, and national profiling.[330] Thus, it can be questioned whether the success is the result of behaviour analysis or the use of racial, religious, and national profiling – apparently, a combination.

As for SPOT and similar programmes, their effectiveness is subject to criticism. In contrast to the Israeli method, behaviour technique is not applied to all passengers and the officers are usually not as qualified as the Israelis. With reference to SPOT, for example, in the period from May 2004 to August 2008, less than 1% of all passengers identified by behaviour detection officers for secondary screening were arrested for criminal activities.[331] This means that about 99% of all passengers selected for additional screening were innocent, constituting very high false positive. According to another report, within this period, "at least 17 known terrorists have flown on 24 different occasions, passing through security at eight SPOT airports".[332] According to GAO, TSA failed to demonstrate that behavioural indicators can be used to reliably and effectively identify high risk passengers.[333]

Security experts argue that behaviour analysis makes CCTV a tool for a greater situational awareness and surveillance, as behaviour analysis is becoming a key element in target selection.[334]

According to some developers, CCTV behaviour analysis detects aggressive behaviour with 90% accuracy, "distinguishing e.g. between hugging and throwing a punch.[335] But according to other sources, it is still imperfect and may lead to high false alarms. Similar to CCTV's facial recognition, image characteristics and environment may reduce the technology performance. Another factor is the problem of adopting proper rules and algorithms for "abnormal behaviour" discussed below. A failure in the person tracker may impact the behaviour analysis.[336] All this can easily

---

[329] Quinlan (2015) p. 36.

[330] *Ibid.*

[331] Perry. *The Screening of Passengers by Observation Techniques Programme: analysing the issues*. In: Aviation Security International. Vol. 17 (2011). p. 13.

[332] Richardson. *Congressional report: TSA suffers from "bureaucratic morass and mismanagement"*. Secuirty Director News, 22 November 2011 (2011a), http://www.securitydirectornews.com/public-sector/congressional-report-tsa-suffers-bureaucratic-morass-and-mismanagement

[333] GAO *TSA's Managed Inclusion Process Expands Passenger Expedited Screening, But TSA Has Not Tested Its Security Effectiveness* (2015b).

[334] Donald and Corkill (2012) pp. 12, 15.

[335] Woollacott (2014).

[336] Badii et al. (2012) p. 367.

flag innocent people as suspicious, resulting in following, detaining, questioning and searching innocent persons.

6.4.5.3 Human Rights Issues

Similar to the general scheme of Video Analytics processing indicated above, processing in behaviour analysis via CCTV includes four steps: segmentation, classification, tracking and recognition – detecting people acting suspiciously.

The most problematic issues in this process are how to pre-define and identify what constitutes suspicious or abnormal behaviour, and who decides what exactly it indicates. For instance, in the TSA's guidelines, behavioural "rules" are established by airport security.[337] According to the EU research projects ADABTS and INDECT, the abnormal behaviour indicators are provided by the end-users (the police, public authority, as well as CCTV operators and security managers in airports and others) or based on "objective scientific analysis".[338] The point is that much of the computer's ability to detect threats relies on sensitive information from intelligence security analysts,[339] i.e. refers to restricted information. However, some general notes can be made.

In some situations it can be judged quite objectively, for instance, in case of violence, using of knifes or guns, unattended luggage,[340] sudden movements, repeated transit at given intervals,[341] or if a person runs against the general flow of passengers. In the course of in-flight camera surveillance, this can be running in the cabin, standing near the cockpit for long periods of time, etc.[342] Research project ADABTS made an effort to identify "objective data on abnormal behaviour" based on concepts from clinical psychology, where distinct and visible behaviour are all whole-body behaviours, including movement about a space, excessive body gestures or gait.[343]

In other situations, it can be more complicated. For instance, the following types of stress indicators can be noted: physiological, emotional, behavioural, and mental.[344] In order to recognize a person's emotional state, algorithms allow analysis of facial expressions, how fast they are talking, the tone and pitch of their voice,

337 Transportation Security Administration (2011) p. 124.

338 Vermeulen and Bellanova (2013) p. 299.

339 Daily news. *In-flight surveillance could foil terrorists in the sky.* New Scientist. 29 May 2008. https://www.newscientist.com/article/dn14013-in-flight-surveillance-could-foil-terrorists-in-the-sky/

340 Vermeulen and Bellanova (2013) p. 299.

341 Article 29 Data Protection Working Party (2004).

342 Daily news (2008).

343 Vermeulen and Bellanova (2013) p. 299.

344 Griffiths. *Stress response: a physiological analysis.* In: Aviation Security International. Vol. 17 (2011) p. 40.

their posture, a person nervously touching their face, or sweating excessively,[345] other signs of stress, rapid eye movements, blinking, mumbling and perspiration.[346] Such indicators can be less obvious and confusing. Thus, the experts warn that blanket generalizations should be avoided and it cannot be assumed that certain indicators equals stress, or that stress equals suspicious behaviour, since all individuals are different.[347]

In addition to general CCTV's privacy/data protection concerns, use of behaviour analysis means more serious concerns regarding privacy of behaviour and action, since behaviour is not simply monitored, but analysed, classified and detected using algorithms.

Moreover, the algorithms can be used to detect stress levels, and as a result, a person can be selected and asked additional questions regarding his/her emotional state, e.g. what exactly was on his/her mind. Thus, impact on privacy of thoughts and feelings may occur. Discrimination and freedom of movement risks appear as well (see below).

6.4.6 In-Flight Camera Surveillance

As discussed in the General Part, in-flight surveillance systems include cockpit door surveillance, cabin surveillance systems and cargo surveillance. They can be used independently and can be integrated in one common system monitoring all areas in the aircraft. The purpose is to alert the flight crew within the cockpit and/or ground services of any emergency activity including, but not limited to, terrorist activity.[348]

Similar to cameras at airports, modern cabin surveillance systems use IP-based cameras. A company that won a contract to supply cabin surveillance equipment for the Airbus A350 family of aircraft[349] states that the systems include up to 40 high definition cameras, with zoom functionality and motion detection. The video streams can be recorded and – in the event of passenger misbehaviour or theft – it can be used as evidence; connectivity during flight and on ground enables airline security to access videos during operation of the aircraft.[350] This means that access to videos can be provided to security services on the ground and, probably, other organs, both in real time and recorded time.

---

[345] Vlahos. *Surveillance society: New high-tech cameras are watching you*. In: Popular Mechanics (2008).

[346] Vermeulen and Bellanova (2013) p. 300.

[347] Griffiths (2011) p. 40.

[348] Surveillance system for aircraft interior. http://www.google.com/patents/US6864805

[349] The Aircraft Interiors International. News. http://www.aircraftinteriorsinternational.com/news.php?NewsID=11709

[350] KID-System. Video based Security System. http://www.kid-systeme.de/products-videosurveillance.html

The A350 cabin surveillance is based on network technology that consists of cameras and microphones, network switches and a video processing unit.[351] It can be seen that in contrast to cameras at airports, sound recording function can be used.

Clearly, there can be differences in where exactly cameras are located. If the systems are limited to cockpit door surveillance, cameras will only be situated near the cockpit door allowing the immediate area to be monitored. Thus, surveillance of people is limited: in order to be filmed, passengers must stand or walk in the area close to the door.

In case of JetBlue, which installed them in 2002, the cameras were positioned to cover the entire cabin, with the exception of the toilets. An announcement at the beginning of a flight about cameras on board was introduced. Video was not recorded, although it was possible to do so, and there was no sound. Some cameras were visible, others hidden.[352]

The European Union's Security of Aircraft in the Future European Environment (SAFEE)[353] project "uses a camera in every passenger's seat, with six wide-angle cameras to survey the aisles. Software then analyses the footage to detect developing terrorist activity or "air-rage" incidents, by tracking passengers' facial expressions".[354] The idea to place cameras at every passenger seat may sound as an advantage in security terms, but entails increased risks in privacy terms. The author did not find any evidence on presence of such cameras on board any aircraft, but the possibility for this still cannot be excluded.

Similarly to cameras at airports, cameras on board aircraft produce images, making the right to data protection applicable, and have impact on privacy in multiple ways, in particular if behaviour analysis and facial recognition functions are used, as in the case of tracking facial expressions to identify the level of stress and thus suspicious behaviour.

Additional impact on privacy and data protection may arise if cameras use the microphones. Sound functions cannot be excluded, since respective equipment exists. In general, audio surveillance allows one not only to see, but also to hear what is going on, presenting advantages in terms of security. In contrast to airports, it can be easier to mitigate problems of noise on board aircraft, for example, by locating microphones close to passengers.

In case of audio surveillance, there can be possible processing of an additional category of biometric data – human voices (see above). Thus, additionally applicable types of privacy are first physical privacy and secondly privacy of communication in case of surveillance of private conversations between persons.

If private conversations are recorded, it can constitute both eavesdropping and wiretapping, which in general is not allowed or allowed in some states only on a

---

[351] The Aircraft Interiors International. News. http://www.aircraftinteriorsinternational.com/news.php?NewsID=11709

[352] Blank (2002).

[353] SAFEE. http://ec.europa.eu/research/transport/projects/items/safee_en.htm

[354] Daily news (2008).

limited basis (police wires).[355] Thus, CCTV experts advise that if you have cameras with microphones, the best is to avoid audio surveillance in general if your state does not allow it, i.e. either by turning it off, or buying another one without audio.[356] I did not find any laws authorizing audio surveillance on board aircraft, thus, this advice can be relevant, and it may be the case that microphones are turned off. In uncertainty, the discussions risk being speculative.

If video or audio surveillance is covert, or is located in places like bathrooms, this means more serious privacy and data protection concerns. In general, it is prohibited, but non-location of cameras in aircraft toilets and announcement about cameras mentioned above, may mitigate these risks.

6.4.7 Concluding Remarks

CCTV technology is constantly developing and improving, shifting from analogue to digital and IP networked systems with Video Analytics. This enables integrating CCTV with other technologies such as biometrics and behaviour analysis and with the use of huge computerized databases.

On the one hand, this provides increased security opportunities. In general, CCTV are considered effective tools for airports, not only in terms of security, but for other purposes, including passenger convenience by for example informing about expected waiting times. However, the accuracy of facial recognition and behaviour analysis techniques is low, with high failure rates, which may result in disturbing innocent passengers. Although algorithms may enhance privacy by for example depersonalization or limiting technology to targeted persons, this is subject to future research.

On the other hand, privacy concerns and other human rights concern, in particular, the right to equality and non-discrimination prevent deployment of very promising technologies, such as other types of biometrics and all forms of profiling, including based on race, religion, etc.

At the same time, available technologies potentially offer unique surveillance opportunities. According to privacy advocates, new technologies contribute towards creating a system of total surveillance of citizens by means of human identification technologies in the physical and behavioural traits.[357] Thus, not only CCTV as such, but all its features and possible additional features should be considered when establishing privacy/data protection applicability.

The impact on privacy/data protection differs depending on the CCTV use, with less impact if used for detection of dangerous items and persons, area observation

---

355 CCTV Camera Pros. Audio Surveillance Equipment Laws. http://www.cctvcamerapros.com/Audio-Surveillance-Equipment-Laws-s/244.htm

356 *Ibid.*

357 Borogan and Soldatov. *Magic of persecution*. Bolshoy Gorod. 24 November 2011. http://bg.ru/society/magiya_presledovaniya-9636/

Table 6.2 Privacy/data protection application for CCTV[a]

	Basic CCTV	Sound option	Facial recognition	Behaviour analysis
Privacy of the person		X	X	
Privacy of behaviour and action	X			X
Privacy of communication		X		
Privacy of data and image	X	X	X	X
Privacy of thought and feelings				X
Privacy of location and space	X		X	
Privacy of association	X		X	
Data protection	X	X	X	X

[a]Based on seven types of privacy discussed in Sect. 2.2.1

and monitoring, and with serious impact if used for tracking persons, is integrated and combined with other technologies such as biometrics, behaviour analysis, if sound option is used. This can be seen in Table 6.2.

Modern CCTV technologies provide an increasing potential for the collection, analysis, sharing and other types of exploitation of images and data, including biometrics, creating or exploiting huge databases against which CCTV images can be compared. In addition, the seriousness of privacy/data protection concerns depends on the use of CCTV in combination with other various data sources, in particular computerized databases, storage or non-storage of data, informed or covert form of surveillance, etc.

All this makes CCTV increasingly invasive, with the risks of function creep and enhanced data mining, sharing with foreign states with inadequate privacy guarantees, etc. All these features are especially dangerous for privacy/data protection. The concrete concerns will be analysed below. In addition, the rights to free movement and to equality and non-discrimination may suffer too (see below).

6.5 Passenger Name Records Systems

6.5.1 Introduction

In aviation security, advanced collection and analysis of passenger personal data is important since the vast majority of passengers pose no threat to civil aviation, and information is critical to assess the risk. A special role in these processes is played by Passenger Name Record (PNR).

According to ICAO PNR Guidelines, PNR is the common name given to records created by aircraft operators or their authorized agents for each journey booked by or on behalf of any passenger.[358]

PNR are contained in operators' CRS, Departure Control Systems (DCS), databases of frequent flyer programs (FFP), customer management systems, customer feedback systems and so on. These data are used by aircraft operators on commercial and operational purposes while providing air transportation services. Properly managed, these data can be used to segment passengers, observe their behavioural patterns, make targeted offers, improve customer service, etc. According to IATA, airlines treat passenger data as an asset and use IT systems in order to mine the information effectively.[359]

Usage of PNR for non-transportation purposes is not something new. Customs and law enforcement authorities around the world have been using PNR manually for about 60 years.[360] However, technological developments made it possible to use PNR data for law enforcement purposes on a more advanced level. In the 1980s, the first computerized ticketing systems were developed, allowing the collection and analysis of patterns of travelling.[361]

As discussed in Chap. 4, in the late 1990s, the first computer airline passenger profiling system was introduced in the USA (CAPPS). This system used approximately 39 pieces of pre-boarding passenger data collected from the reservation systems of airlines, making it possible to automatically single out particular passengers and put them through stricter security controls. CAPPS profiles were confidential, and this gave rise to concerns about privacy as guaranteed by the Constitution.[362] The system was blamed for being over-inclusive, "flagging up to half of all passengers yet missing vital targets", that the system simply did not work, etc.[363]

CAPPS was later redesigned and expanded to CAPPS II in 2003 and to Secure Flight in 2004. In addition, after 9/11, the USA demanded that all international airlines had to provide the US government full electronic access to detailed airline passenger data on all travellers registered in the airline's computer system.

Most of EU data protection concerns were beyond the competence of European airlines. They found themselves in a difficult situation: to fly from EU to the US, they would need to comply with either EU or US law, but they could not comply with both. They could either refuse to transmit the data, thereby becoming subject

[358] ICAO PNR Guidelines (2010) §2.1.1.

[359] Airlines International, Airline Data – Know Your Passengers. 1 December 2010. http://airlines.iata.org/analysis/airline-data-know-your-passengers

[360] European Commission. Passenger Name Record (PNR). http://ec.europa.eu/dgs/home-affairs/what-we-do/policies/police-cooperation/passenger-name-record/index_en.htm

[361] See Lyon (2003a) pp. 13–20.

[362] See more details: Ravich. *Airline passenger profiling systems after 9/11: Personal privacy versus National Security*. In: Journal of the Transportation Research Forum, Vol. 44, No. 2 (2010) p. 131 citing Rhee. *Rational and Constitutional Approaches to Airline Safety in the Face of Terrorist Threats*. In: DePaul L. Rev. Vol. 49 (2000).

[363] Ravich (2010) p. 131.

to US authorities' sanctions, or they could deliver the data in violation of the EU law. As a reaction, EU Member States issued national decisions, from prohibitions to acceptance, with or without conditions.[364] At that time, most EU airlines chose to provide PNR to the US authorities, being unable simply to stop flying across the Atlantic.[365] Later, this problem was regulated by EU-US PNR Agreement.

Other states started establishing PNR regimes too. The countries that currently use/are about to use PNR for law enforcement purposes include the USA, Canada, Australia, New Zealand, South Korea, Russia, Mexico, Argentina, United Arab Emirates, Brazil, Japan, and Saudi Arabia.[366]

In the EU, the UK has a PNR system from 2004 and others, in particularly, Denmark, Belgium, France, Sweden, and Spain use it in limited ways or are testing its use; many other EU Member States receive funding from the EU.[367]

Many other states have either enacted relevant legislation and/or are currently testing potential uses of PNR data; others are considering setting up PNR systems. According to Dutch Liberal Member of European Parliament Sophie in't Veld, the countries also planning to implement PNR regimes include India, Malaysia, Qatar, and it is only a matter of time before China does the same.[368] This means that an increasing number of countries are asking for PNR data from the EU, leading to problems similar to the EU-US above. For instance, in 2013, a problem arose with reference to the Russian PNR system,[369] in 2015 – Mexico.[370] Moreover, in 2016, a PNR system was adopted on the EU level too.[371]

Of course, individual PNR systems, despite ICAO's endeavours on harmonization, are different. The USA requires all airlines flying to the USA from foreign countries to provide PNR prior to departure.

---

[364] European Commission. Roadmap. Legislative proposal and Communication on the transfer of passenger data to third countries. 10/2013.

[365] See: Ntouvas. *Air Passenger Data Transfer to the USA: the Decision of the ECJ and latest developments*. In: International Journal of Law and Information Technology. Vol. 16 (2008).

[366] Council of the European Union, Information by the Commission on the PNR legislation adopted by Mexico and the Republic of Argentina requesting the transfer of PNR data from the EU, note of 5 March 2015.

[367] In 2013, a total of €50 million, made available by the Commission, was distributed among 14 EU countries which presented projects for developing their national PNR schemes. European Parliamentary Research Service *The proposed EU passenger name records (PNR) directive. Revived in the new security context* (2015).

[368] See Nielsen and Rettman. *Russia blames EU for airline data fiasco*. EUobserver, 11 June 2013, http://euobserver.com/justice/120450

[369] For more details, see Enerstvedt (2014).

[370] On 14 July 2015, negotiations for an EU-Mexico PNR data transfer deal were formally launched (the Council gave a mandate to the Commission on 23 June 2015). http://www.europarl.europa.eu/news/en/news-room/content/20150123BKG12902/html/EU-Passenger-Name-Record-(PNR)-proposal-an-overview.

[371] Directive (EU) 2016/681 of the European Parliament and of the Council of 27 April 2016 on the use of passenger name record (PNR) data for the prevention, detection, investigation and prosecution of terrorist offences and serious crime.

The EU PNR regime requires personal data from air passengers of all extra-EU flights – voluntary – of intra-EU flights.[372] However, according to a report, considering the current security situation in Europe, all EU Member States agreed to include all intra-EU flights.[373]

The Directive obliges airlines to transfer PNR data to a single designated Passenger Information Unit (PIU) in the relevant Member State. Both the EU and EEA states are applicable. This means that the plan is to have a decentralized system of data collection.

While the EU PNR Directive and the EU-US scheme are limited to air transport, the Russian PNR system covers all modes of transport: domestic and international air transport, including flights into, out of, and over Russia, long-distance rail transport, international transport by sea, inland waterway and road transport.

What is important is that despite international standards, the existing PNR systems differ greatly in terms of IT development, data extraction and transmission. As IATA notes, the cost of non-standard PNR programs "has risen to unacceptable levels in recent years."[374]

As long as the enhanced surveillance and increased collection of personal data for security purposes, including PNR, reflect the worldwide tendencies, the problem of PNR is very wide and involves the proposed EU PNR system, the relations regarding PNR transfer between the EU and individual countries like the USA and Russia, as well as all the countries, both those requiring PNR and those whose airlines fly to/from/over these territories and have to provide PNR.

There are at present no PNR regimes which would be acknowledged by all the parties as satisfactory in all data protection terms. All existing systems are subject to criticism. I chose three PNR systems as examples.

First, there is the much discussed and criticized EU-US PNR Agreement as an example of a bilateral solution. Secondly, the proposed EU system will be discussed as a regime originating from the EU – i.e., which is supposed to be based on the EU standards rather than repeat weaknesses of the EU-US regime – but which, however, is blamed for weaknesses too. The EU PNR regime is one of the most controversial issues of EU legislation currently on the agenda.

Finally, there is the Russian system. All airlines operating domestic or international flights or passing Russia have to hand over passenger data to Russian security authorities. Many foreign airlines carry out flights into and out of Russia;[375] in addition, around 53,000 European flights transit over Russia to Asia each year. To my knowledge, the system functions; relevant airlines are connected to the system and provide PNR to Russia.[376] This regime will be discussed as an example of a country

[372] Article 2.

[373] Statewatch. EU-PNR: European Parliament has "egg on its face". 4 April 2016. http://statewatch.org/news/2016/apr/eu-pnr-ep-egg.htm

[374] IATA. Aviation security and facilitation. http://www.iata.org/whatwedo/security/Pages/index.aspx

[375] Currently, foreign air carries do not have access to the Russian domestic aviation market.

[376] Confirmed during telephone conversation with the Operator's employee 20 Nov 2015.

that requests PNR but whose democratic standards may be questionable – i.e., an example of many countries globally.

Apparently, the number of states with PNR regimes will continue to grow. The Russian system not only depicted a tendency, but raised the concern that it may be followed by other states outside Europe. All these states can be encouraged to act unilaterally and may cause the EU (or any state with data protection standards) to be faced with similar problems.

6.5.2 *Effectiveness*

Today, analysis of PNR data is considered necessary from a law enforcement perspective. It can be valuable for security purposes, especially, for threat and risk assessment and management. However, similar to other techniques discussed in this work, there are different views on how effective the use of PNR can be.

All the above-mentioned instruments on PNR, including the ICAO PNR Guidelines, the EU PNR Strategy, the EU PNR Directive, bilateral PNR agreements and Russia law acknowledge and recognize the necessity and importance of the use of PNR data in the fight against terrorism and serious crime. Taking into account the growing importance of PNR data transmission for aviation security purposes, the ICAO urges states to use PNR as an aid to aviation security.[377] The United Nations Security Council[378] encourages its members to provide PNR too.[379]

While the representatives of travel industry are mostly opposed to PNR systems and raise concerns on costs and obligations on the travel agents,[380] the aviation industry in general supports the introduction of PNR systems: ICAO and IATA issue recommendations aimed at harmonization, on the EU level; airlines consider the EU PNR framework necessary to avoid discrepancies between national provisions.[381] However, the EU airlines are against including intra-EU flights in the provisions for PNR data collection because of heavy costs for airlines and other factors.[382]

---

[377] ICAO, 37th Assembly (2010) Resolutions.

[378] Under the UN Charter, the Security Council has primary responsibility for the maintenance of international peace and security. Its five permanent members are the USA, the UK, Russia, France and China. http://www.un.org/en/sc/

[379] United Nations Security Council. Statement by the President of the Security Council. http://www.un.org/en/sc/ctc/docs/2014/PRST_EN.pdf, 2014.

[380] European Parliamentary Research Service (2015).

[381] Association of European Airlines. *Ensuring secure aviation while minimising the hassle for passengers*, 2014, http://www.aea.be/component/attachments/attachments.html?id=93&task=view. IATA and AEA. Joint letter to the European Parliament, Sept 2014, https://netzpolitik.org/wp-upload/2014-09_AEA_IATA_Joint_letter_on_EU_PNR.pdf

[382] Lufthansa Group. *Airline Perspective on Handling Data Protection Considerations*, 13 Oct 2014, http://www.icao.int/EURNAT/Other%20Meetings%20Seminars%20and%20Workshops/2014%20FAL%20Seminar%20Paris/Day1%20PPT27%20E.pdf

From an aviation security point of view, as discussed in the section on Profiling, the most effective use of PNR is via targeted passenger profiling, using as much personal data as possible, including sensitive, and along with intelligence. Security experts note that if properly used, PNR are extremely valuable for combatting terrorism and crime, with a potential to reveal "clean skin" terrorists.[383] In the case of the EU PNR system, the most effective would be use of PNR for intra-EU flights too; otherwise, as reported from the UK, it will lead to losing profiling ability for drugs targets and inadequately documented passengers.[384]

Of course, the strongest arguments for PNR systems lie in demonstrating how effective the system can be in numbers, statistics and similar.

It was reported that PNR data was "instrumental" in capturing collaborators of the 7 July 2005 London bombers and the 2008 Mumbai terror attackers, and "led to the capture of dozens of murderers, pedophiles and rapists" and "95% of all drug captures in Belgium and 85% in Sweden are caught using PNR data".[385] Profiling of passengers via PNR prevented 97 inadequately documented passengers from flying to the UK in 2011/12.[386]

According to the US DHS, in 2008–2009, PNR helped to identify people with potential ties to terrorism in over 3000 cases; in 2010, a quarter of people denied entry to the USA for having ties to terrorism were identified through PNR analysis.[387] PNR analysis played a critical role in the investigations of Najibullah Zazi, who pled guilty to plotting to bomb New York subways; David Headley, who pled guilty for his role in the 2008 Mumbai terrorist attacks and was suspected of planning attacks in Europe; and Faisal Shahzad, who pled guilty in the 2010 bombing attempt against Times Square in New York.[388]

The EU PNR Directive's Impact Assessment states that the use of PNR together with API[389] in certain cases helped to verify the identity of an individual and provides some documentation and legal arguments to justify the PNR system. However, the Commission admitted that relevant detailed statistics are not available, and points to information from individual countries, some figures related to drug crimes

---

[383] Wolff (2012) p. 4.

[384] Vine (2013).

[385] BBC. *MEPs back deal to give air passenger data to US*. 12 April 2012. http://www.bbc.co.uk/news/world-europe-17764365

[386] Vine (2013).

[387] DHS. Preventing Terrorism and Enhancing Security. http://www.dhs.gov/preventing-terrorism-and-enhancing-security. FRA. *Opinion of the European Union Agency for Fundamental Rights on the Proposal for a Directive on the use of Passenger Name Record (PNR) data for the prevention, detection, investigation and prosecution of terrorist offences and serious crime (COM(2011) 32 final)* (2011b).

[388] Omar. *Risk Based Airport Security Approach*. In: International Journal of Computer Science: Theory, Technology and Applications (IJCS). Vol. 3 (2014). p. 21.

[389] Council Directive 2004/82/EC of 29 April 2004 on the obligation of carriers to communicate passenger data.

in Belgium, Sweden and the UK mentioned above and experience of the UK PNR system.[390]

Thus, opponents, mainly privacy advocates and data protection authorities, criticize the Commission for "anecdotal evidence"[391] and argue that the EU Directive does not give substantial evidence and statistics. There are examples only of cases in which PNR data were successfully used in the course of investigations rather than cases where the use of data was misleading and led to the investigation of innocent people; no statistics showing the ratio between the number of innocent travellers whose PNR data was collected to the number of law enforcement outcomes resulting from that PNR data.[392] Although some terrorist attacks could have been avoided, if a PNR scheme had already been in place, no evidence or facts were given.[393]

Similarly, opponents state that in the EU-US Agreement, the necessity for mass collection and storage of PNR data is not fully demonstrated, there is lack of detailed justification and factual evidence for the stated purposes and for each of the methods of processing (re-actively, real time and pro-actively).[394] Reports of the US Government Accountability Office did not confirm the efficiency of PNR.[395] As for the Russian PNR system, no facts and statistics were found.

As for failure rates, the UK for the period from April 2009 to February 2010 flagged 48,682 people, leading to the arrest of 2000 people and to the denied entry into the UK of 1000 people. In the US, for the period from August 2008 to January 2010, 216 PNR data shared by the EU with the US led to one criminal judicial proceeding.[396] The problem is that the consequences of using PNR for counterterrorism purposes can be serious, including arrest, deportation, loss of a job, more enhanced search at screening, investigation or surveillance, or being added to a watch list.[397] However, the risk of failures means that many innocent people can be flagged wrongly and thus may potentially experience all these consequences.

---

[390] European Parliamentary Research Service (2015).

[391] *Ibid.*

[392] Article 29 Data Protection Working Party *Opinion 10/2011 on the proposal for a Directive of the European Parliament and of the Council on the use of passenger name record data for the prevention, detection, investigation and prosecution of terrorist offences and serious crime* (2011), European Data Protection Supervisor. *Opinion on the Proposal for a Directive of the European Parliament and of the Council on the use of Passenger Name Record data for the prevention, detection, investigation and prosecution of terrorist offences and serious crime* (2011b).

[393] Article 29 Data Protection Working Party *Letter on EU PNR to Chairman of the LIBE Committee of the European Parliament* (2015).

[394] European Parliament. Draft recommendation on the draft Council decision on the conclusion of the Agreement between the United States of America and the European Union on the use and transfer of Passenger Name Records (PNR) to the United States Department of Homeland Security, 30 January 2012.

[395] GAO *Secure Flight Development and Testing Under Way, but Risks Should Be Managed as System Is Further Developed* (2005); GAO *Efforts to Strengthen International Passenger Prescreening are Under Way, but Planning and Implementation Issues Remain* (2007).

[396] FRA (2011b).

[397] Dempsey and Flint (2004) p. 1471.

In summary, PNR apparently has value for security purposes, in particular if used for profiling and along with intelligence, but there are false positives. As discussed, the latter is a general feature of any type of profiling, where a 100% success rate is not realistic. The question is whether profiling and the failure rates are acceptable or not, and here, the view may differ: what is acceptable for security proponents can sound unacceptable for privacy advocates.

6.5.3 Data Protection and Privacy Issues

PNR as such are not aviation security equipment or specific technology. They are personal data which airlines collect on passengers for own needs. But, due to security needs, the state required these data from airlines, and PNR systems were developed, working as complicated systems of collection, storage, analysis and share vast amounts of air passenger personal data for purposes other than transportation. A general model of a PNR system is shown in Fig. 6.8 (simplified).

Accordingly, airlines receive PNR and other data from a passenger, usually at the stage of booking a flight. Then, the airlines forward PNR data to security and other organs. In general, there exist two methods of PNR transfer. The so-called "push" method of transfer implies that the data are selected and transferred by airlines to the authorities upon request of the latter. The "pull" method means that the authorities have direct and immediate access to airlines' databases. Obviously, it is the "pull" method which entails more data protection risks since it may allow access to any possible data obtained by airline companies mentioned above. Currently, the proposed EU, Russian and EU-US systems provide "push" method, but in the EU-US case, "pull" method is also possible (see Chap. 7). In the UK, within e-Borders program, analysis of PNR data done by "pull" method, but it was planned to cease after implementing the EU PNR directive.[398] After the organs receive PNR, they use them for various purposes, analyse, share with other entities, etc.

It is quite clear that PNR data amount to personal data, since PNR makes passengers identifiable and processing of personal data take place. Accordingly, data protection rights and the informational dimension of privacy are applicable.

But additionally, with reference to PNR systems, there are at least three factors which determine the essence and the role of PNR and at the same contribute to

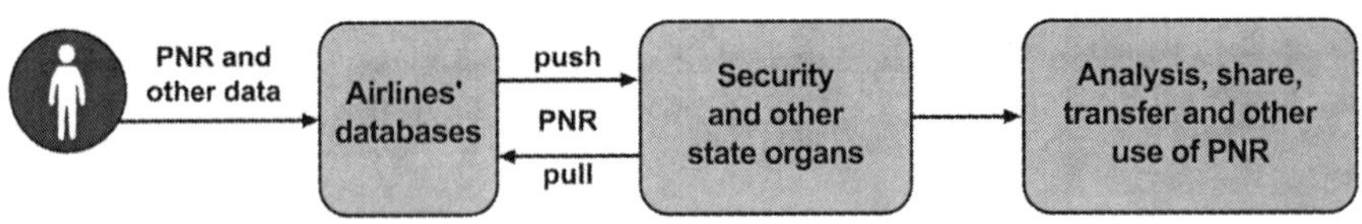

Fig. 6.8 Standard PNR system

[398] Vine (2013).

further explain particular privacy and data protection concerns. They are (i) technological advances, (ii) huge amounts of data processed by airlines in general and broad scope of possible PNR data types in particular, and (iii) wide scope of possible use of PNR data for non-transportation purposes. These factors will be discussed below.

Electronic processing of data and other technological advances similar to those discussed above, in particular IP and wireless technologies, various algorithms for automatic analysis, etc. allow instant access to, collection, analysis, integration of PNR databases with various other databases, the sharing and transfer of PNR within one entity, between different organs and across state boarders, as well as combination of PNR use with other security measures. All these factors greatly increase PNR's potential for enhancing security and other uses.

Although different systems provide varying facilities, and the number and nature of fields vary from airline to airline and even among individual PNRs from the same airline, all PNRs contain at least passenger name(s), itinerary, and contact information.[399] The Annex to ICAO PNR Guidelines provides a list of possible PNR data elements (see Annex 1 to this work). They can be categorized in the following groups: (i) Machine Readable Travel Document (MRTD) details (names, date of birth, etc.), (ii) contact details, (iii) passenger details; (iv) payment details; (v) other information (name of person making the booking, travel agent information); and (vi) data related to aircraft flight.

PNR can be captured up to 360 days in advance of flight and are created every time a traveller makes a reservation. Technically, they are not deleted from CRS and can be viewed even if a person never bought a ticket or cancelled the reservation. The basic record may contain multiple passengers within the same record. But each entry, even for one passenger, contains data on other people as well: the passenger, the travel arranger or requester, the travel agent or airline employee, a person paying for the ticket, etc. The PNR system contains all passenger data of the whole airline company, thus, the system is not restricted to a specific flight.

Most airlines and travel agencies also use the CRS as their primary customer database and accounting system and store all customer data in CRS profiles. Thus PNR also contain data on individuals who never travel by air at all, since lots of travel services, such as car rental and hotel reservations, are made through CRS.[400] The system contains all information about the services provided to a person in the past (e.g. previous travel arrangements, such as flights and other bookings, and related matters, such as upgrades, details of missing luggage, etc.[401]).

Passenger details include OSI – Other service related information, SSI – Special Services Information, SSR – Special Service Requests, and General remarks.

---

[399] IATA. Passenger Services Conference Resolutions Manual (PSCRM). 01June 2007–31 May 2008 27th Edition.

[400] See Hasbrouck. *What's in a Passenger Name Record (PNR)?* 2009. http://hasbrouck.org/articles/PNR.html

[401] British Airways. Privacy Policy. http://www.britishairways.com/en-no/information/legal/privacy-policy

Through OSI/SSI/SSR, PNR may include requests for special medical service (e.g. a person requesting provision of wheelchair, or sought clearance to fly with a medical condition or because she is more than 28 weeks pregnant) or special dietary meals. Request of a special meal may imply that a person has particular religious beliefs or a particular medical condition. This means that OSI/SSI/SSR may contain details of travellers' physical and medical conditions, indications of travellers' religious practices, that is, data of a sensitive nature.

General remarks may contain data on internal conversations and contacts between airline company's employees and agents, including various comments and abbreviations.[402] Obviously, general remarks may include whatever possible and sensitive data as well.

Moreover, airlines may process such information as data about your use of the airlines' websites, call centres and mobile applications, including information about the pages viewed, information about interactions of a person with the airline and its staff, e.g. on social media or on board one of our flights.[403] Passengers' photographs and other images, including biometric templates, can also be processed.[404] Some airlines, as discussed above, collect and use biometrical identification of passengers for facilitating registration, etc., and biometry is a special category of data requiring additional protection.

All this means that airlines actually collect and process or are able to collect and process huge amounts of personal data, including different categories of data. Not all of them fall within PNR, but in any case the range of PNR is very wide and may constitute up to hundreds elements of data. Nevertheless, although some data may go beyond any official PNR lists, another question is how much of this data ultimately may become subject to analysis by state organs (for instance, if all information is kept altogether, full access to airline's CRS means that the security organ has access to everything including data other than PNR).

It should be noted that in relation to completeness or accuracy, two types of information can be distinguished. The first group includes MRTD details (also known as API – Advance Passenger Information) which derives from travel documents information. These data are official and validated, spellings and dates are transcribed accurately, offering objective and permanently valid information. Such information is typically used within border control to check against watch lists, that is, to identify already known persons, or to deal with irregular migration issues.

The second group includes the information that the passenger submits to the CRS himself or herself. In addition, as mentioned above, PNR can be captured much in advance. Hence, PNR data are dynamic and are subject to change; these data cannot be guaranteed in completeness or accuracy, and may not be fully updated on the date of departure.

---

[402] ICAO PNR Guidelines (2010) §2.1.6.

[403] British Airways. Privacy Policy. http://www.britishairways.com/en-no/information/legal/privacy-policy

[404] *Ibid.*

Nevertheless, PNR are unique in their nature and provide a comprehensive and extremely detailed record of every entry and include data on the basis of which aspects of the passenger's history, conduct and behaviour can be deduced. As the result, PNR may be very useful for intelligence in identifying both known criminals and potentially dangerous persons who are not yet known from databases. This means huge potential for security purposes. PNR can thus be used in profiling discussed above, offering information on the background of the individuals and their possible relationship to other persons being investigated.

Today's enhanced usage, share, transfer and other actions upon passenger data on purposes other than air transportation is reflected at the airlines' general conditions of carriage (here, a practical example from the USA will be given), which usually warn the air passengers about the following:

> Upon booking a ticket for transportation, purchasing other services, or participating in any UA [United Airlines] program or service such as MileagePlus or the United Club, you hereby authorize UA and its affiliates and authorized agents to (i) collect, process, retain and use, and (ii) transfer to third parties, including, but not limited to, subcontractors, agents, affiliates, marketing partners, other carriers, and government agencies, for their use, processing and retention, any and all personal data you provide when UA believes in good faith that it is in the interests of aviation security or that disclosure is otherwise necessary or advisable or as UA deems necessary to carry out any and all business purposes related to the program or services being requested and/or in the promotion of other information, goods, and services that may be of interest to you, including, but not limited to, the following purposes: [...]facilitating travel, including obtaining immigration, security, and customs clearance; complying with applicable laws, regulations, government requests, law enforcement requests, and/or valid court orders; providing data to third parties or governmental or law enforcement agencies to comply with, or assist in the development of, security, safety, or health measures for passengers, baggage or cargo, or to provide for the prevention or detection of imminent criminal acts or the apprehension or prosecution of offenders; protecting the legal rights of UA and/or its affiliates.[405]

The "authorization" above allows the air company to collect and transfer passenger personal data for a very wide range of purposes and to a very wide range of agencies.

As a result, purposes of PNR use become far wider than those they were initially collected for. The EU Commission distinguishes three types of PNR data use:

- Re-active (historical data): use in investigations, prosecutions, unravelling of networks after a crime has been committed. In order to allow law enforcement authorities to sufficiently go back in time, a commensurate period of retention of the data by law enforcement authorities is necessary in such cases.
- Real time (present data): use in order to prevent a crime, survey or arrest persons before a crime has been committed or because a crime has been or is being committed. In such cases PNR are necessary for running against predetermined fact-based risk indicators in order to identify the previously "unknown" suspects and for running against various databases of persons and objects sought.

[405] United Air Lines, Inc. Contract of Carriage (revised 31 December 2015), https://www.united.com/web/format/pdf/Contract_of_Carriage.pdf

- Pro-active (patterns): use for trend analysis and creation of fact-based travel and general behaviour patterns, which can then be used in real time use. In order to establish travel and behaviour patterns, trend analysts need to be allowed to use the data over a sufficiently long period of time. A commensurate period of retention of the data by law enforcement authorities is necessary in such cases.[406]

Real time and pro-active use imply the use of PNR for profiling discussed above. Profiling which includes PNR analysis is often referred to as targeted passenger profiling or computer-assisted passenger profiling. It supports automated decisions on the level of screening each passenger receives, including those who should be subjected to additional security measures. This means leveraging all the passenger information provided by airlines to governments before the start of the trip, including to immigration, customs and security authorities. The overview of data must then be analysed by intelligence agencies. Detailed results of the passenger's vetting will be made known to screeners at the checkpoint, determining whether additional screening is needed, or whether the person is able to fly at all.

As it will be discussed in Chap. 7, the EU-US, the EU proposed and the Russian PNR systems all use or potentially may use profiling. Globally, this concerns other existing or planned PNR systems too. It was reported that in the UK, for example, within e-Borders program, PNR was used to identify individuals who fitted profiles which intelligence suggested were indicative of higher risk.[407] In 2013, the French parliament approved a bill of law regarding national defence programs, including creation of a PNR system which uses profiling techniques on a large scale in order to indicate "normal" and "suspect" persons according to criteria such as their declared destinations and means of payment.[408]

Information about the use of profiling can also be found on the airlines' websites. British Airways informs its passengers that their PNR data "is used for enforcement purposes, including use in threat analysis to identify and interdict potential terrorists, and other threats to national and public security; and to focus government resources on high risk concerns, thereby facilitating and safeguarding bona-fide travellers."[409] Thus, in this research, it is presumed that PNR are used for profiling. Particular privacy/data protection concerns will be discussed in Chap. 7.

[406] §2.1 of Communication from the Commission On the global approach to transfers of Passenger Name Record (PNR) data to third countries. Brussels, 21 September 2010, COM(2010) 492 final.

[407] Vine (2013).

[408] Fessard. *France jumps EU law and follows UK with mass surveillance of air travelers*. 26 December 2013. http://www.mediapart.fr/en/journal/international/261213/france-jumps-eu-law-and-follows-uk-mass-surveillance-air-travellers

[409] British Airways. Government access to British Airways booking records. Answer Id 1621. Updated 15/07/2015. http://www.britishairways.com/travel/gsanswer/public/no_no (date accessed: 20 September 2015).

6.5.4 Other Human Rights Issues

Other applicable human rights include the rights to free movement and to equality and non-discrimination. They will be discussed in separate sections below.

6.5.5 Concluding Remarks

Today, enormous amounts of passenger personal data in civil aviation are circulated, including PNR. There is a trend of growing collection and use of PNR for security purposes nationally, regionally and globally. This is already a reality worldwide and common practice, where PNR are transferred by the airlines to or are open for access by the state authorities which collect, storage, share and transfer passenger personal data via multiple PNR systems. Accordingly, data protection rights and the informational dimension of privacy are applicable.

Technological developments, wide scope of data collected by airlines and wide scope of possible PNR use define the dangers in privacy/data protection terms associated with PNR, including expanded data fields, use of sensitive data, unlimited share, including across borders, and usage, including for profiling and data mining, etc.

Apparently, the most effective usage of PNR is for passenger profiling, which, while welcomed by security experts, presents serious privacy/data protection risks, as well as the risks of discrimination and limitation of freedom of movement (see below).

6.6 Technology and Freedom of Movement

As discussed in Sect. 3.5, freedom of movement is not completely free and is subject to various limitations. Some of these are quite obvious, such as the limitation to freely enter a state where you are not a citizen. Aviation security as such, making people choose whether to be subject to security measures or not flying, is also a limitation, but obviously, it is justified: aviation security should exist to protect peoples' lives. Thus, if a measure makes people choose whether to be scanned, hand searched, filmed by CCTV, etc. – or not fly, these requirements fall within lawful general limitations that are not normally considered as having a negative effect on this right. This basically refers to scans, hand search, or basic CCTV. The problem is that a number of other aviation security measures or additional capabilities impose further limitations.

Specifically, the existence of different watch lists, blacklists, no fly lists and so on may restrict the mobility of those concerned. The police and security organs use different methods to develop such lists that include individuals related to suspects or who are themselves suspected of terrorist activities. Then, at the checkpoints, sorting processes are carried out to identify these persons. Airports in this regards act as such data filters to monitor customers to detect and block the blacklisted people.[410]

For instance, an American program Secure Flight is designed to collect passenger data from airlines and compare them against federal government's watch lists in order to decide passenger treatment. There are two types of lists: No Fly List (individuals prohibited from flying and thereby subject to being prevented from boarding an aircraft) and Selectee List (individuals posing a threat, thereby being subject to enhanced screening).[411]

Several problems appear. The first one is the existence of false negatives and false positives. The latter can have a serious impact on the freedom of movement, causing a travel ban. The audit by the Department of Justice's Office of Inspector General (USA) examined the FBI's consolidated anti-terrorist list, which, as of December 2008, contained over 1.1 million names. According to the report, some people were listed there multiple times under different spellings; 24,000 individuals were wrongly listed.[412]

Secondly, there are concerns about how no-fly lists and blacklists are generated, how to find out that you are on the list, and how to remove yourself from the list.

The criteria for inclusion on the lists are unclear. The above-mentioned report identified more than 50,000 records with no explanation of why they were on the list, making it impossible to remove them. It is the government policy not to confirm or deny someone's placement in a list. According to former chief of the US Terrorist Screening Database, FBI agents can nominate candidates to it, and inclusion in this database depends upon "whether there is reasonable suspicion to believe that a person is a known or suspected terrorist", with periodic audits and other quality control measures.[413] Despite these assurances, according to the American Civil Rights Union (ACLU), "It is not at all clear what separates a reasonable-suspicion-based-on-a-reasonable-suspicion from a simple hunch",[414] famously the database included Nelson Mandela. Moreover, a number of lawsuits allege that innocent people can be put on the list as coercion or punishment by the FBI, e.g. it is alleged the FBI attempted to leverage no-fly selectees into becoming informants.[415]

---

[410] Amicelli (2012b) p. 274.

[411] TSA. Secure Flight Program. http://www.tsa.gov/stakeholders/secure-flight-program (date accessed: 7 August 2015).

[412] U.S. Department of Justice, Office of the Inspector General Audit Division. The Federal Bureau of Investigation's Terrorist Watchlist Nomination Process. http://www.usdoj.gov/oig/reports/FBI/a0925/final.pdf, 2009.

[413] Ackerman. *No-fly list used by FBI to coerce Muslims into informing, lawsuit claims*. The Guardian. 23 April 2014. http://www.theguardian.com/world/2014/apr/23/no-fly-list-fbi-coerce-muslims/

[414] *Ibid.*

[415] *Ibid.*

With reference to redress, several lawsuits attempted to get people off the no-fly list, and the processes turned out to be long, bureaucratic, and secretive.[416] There were some successful results: e.g., in 2014, Rahinah Ibrahim, who was unable to fly from 2004, won after demonstrating that the FBI added her name by mistake.[417] However, the problem remains: according to the Justice Department Inspector General's report, since non-investigative subjects may be retained on the watch list for an extended period of time, this subset of watch list practices will continue to grow throughout the years.[418]

Further, surveillance technologies may result in what Lyon calls social sorting – categorizing persons and populations[419] – and may have both positive and negative impact on freedom of movement. Connections between social status and movement opportunities can be seen for example when systems give mobility privileges to definite individuals via biometrics, as in the case of members of Privium or CLEAR programmes.

A negative impact on the freedom of movement can be seen in the wide use of airport CCTV equipped with Video Analytic which employs facial recognition and can automatically interpret human behaviour. As Article 29 Working Party notes, persons

> have the right to exercise their freedom of movement without undergoing excessive psychological conditioning as regards their movement and conduct as well as without being the subject of detailed monitoring such as to allow tracking their movement and/or triggering "alarms" based on software that automatically "interprets" an individual's supposedly suspicious conduct without any human intervention...[420]

As mentioned above, both facial recognition and behaviour analysis technologies suffer false positives. Accordingly, wide use of Video Analytics may interfere with the freedom of movement, if innocent persons are mistakenly detained.

The consequences of using PNR for counterterrorism purposes, as discussed, can be serious, including arrest, deportation, being added to a watch list, etc. Thus, freedom of movement can be limited. Some other aspects of PNR use, for instance, if applied for intra-EU flights, may present limitations on the freedom of movement as well, but on a more general basis. As mentioned in Chap. 3, EU Member States may restrict the freedom of movement of EU citizens on grounds of public policy or public security – but such restrictions must comply with the principle of proportionality and must be based exclusively on the citizen's personal conduct, representing

---

[416] Choudhury. *Ninth Circuit Gives ACLU's No Fly List Clients Their Day in Court*. ACLU. 30 July 2012. http://www.aclu.org/blog/national-security-technology-and-liberty/ninth-circuit-gives-aclus-no-fly-list-clients-their

[417] Ackerman (2014).

[418] *Ibid.*

[419] Lyon. *Surveillance, security and social sorting emerging research priorities*. In: International criminal justice review. Vol. 17 (2007) p. 162.

[420] Article 29 Data Protection Working Party (2004).

a genuine, present and sufficiently serious threat.[421] A risk of the breach of this right is highlighted in case if the EU PNR scheme applies to intra-EU flights.[422]

At the same time, "one is still not allowed to deny boarding if there is no evidence of an immediate danger, or if the identified traveller is not already flagged by law enforcement databases", thus, it is proposed that the airport checkpoint should be able to "clear" a passenger whose data or behaviour raised suspicion, but no grounds to deny flying actually exist.[423] Accordingly, sensors and security equipment should assess the type of suspicion: forged document, wrong identity, hidden item or simply "bad intent". For the latter case, a more severe referral procedure should be put in place in order to confirm or resolve potential threats.[424] However, these procedures are subject to future rules and procedures.

Hence, it can be concluded that the right to free movement is applicable to many profiling practices, and can be limited. According to the regulators, the restrictions are reasonable, since they are aimed at protecting transportation and national security, as stated, for example, by the TSA.[425] To evaluate whether or not there is a violation, as with other rights in this research other issues such as legality, necessity and proportionality apply. They will be discussed below.

6.7 Technology and Discrimination

As discussed in Sect. 3.6, discrimination means unfounded, unequal treatment. Human rights instruments prohibit any discrimination and provide that the enjoyment of other rights and freedoms shall be secured without discrimination on any ground. Relevant grounds, as discussed, refer to personal characteristics usually related to sensitive data, such as race, ethnicity, religion, nationality, sexual preference, etc. Aviation security measures can be discriminatory as such and can also be used in discriminatory manner; discrimination can take place directly, if sensitive data is used for categorizing passengers, or indirectly, if neutral data is used.

With reference to discriminatory measures as such, relevant measures include first of all body scanners which can discriminate against passengers by the process of scanning. As discussed above, particular groups of passengers can be marked by technology as "abnormal". As the result, the privacy violation on people whose bodies are tagged as being abnormal is greater than on those with "normal" bodies: for the specific groups, the right to privacy is allocated unequally.[426] This way, the discrimination issues arise both within the right to non-discrimination as such and one

---

[421] Art.27 of Directive 2004/38/EC of 29 April 2004.

[422] European Parliamentary Research Service (2015).

[423] Couessin. *Passenger name records and advanced passenger information: data privacy concerns identified and resolved.* In: Aviation Security International. Vol. 21 (2015) p. 39.

[424] *Ibid.*

[425] The Identity Project. Secure flight. http://papersplease.org/secureflight.html

[426] Tirosh and Birnhack (2013) p. 1299.

of the general principles that all human rights, including the rights to privacy, apply for individuals regardless of physical characteristics.

Further, computer passenger profiling and Trusted Traveller Programmes (TTP, see Sect. 6.8.2) are discriminatory as such too. Classification, division and different treatment are at the "heart" of profiling, hence, discrimination is its integral part.[427] The core idea of both profiling and TTP is classification of people into categories in order to implement a differential treatment, to indicate potentially dangerous persons (in the case of passenger profiling) and low risk persons (TTP). Here, definite sensitive factors may mean higher or lower risks for the security professionals.

These practices can have a real impact upon the individuals being profiled, e.g. they can be banned from travelling. Moreover, the focus on technological extrapolations from data may lead to principles of action that negatively discriminate against particular groups or individuals; only matching with a specific group is sufficient to attract categorical suspicion and possible prejudice.[428] According to researchers, such "surveillance as social sorting" is related to "discriminatory" technologies since they rely on decision support systems that are precisely designed to differentiate and discriminate.[429]

With reference to TTP, it is their nature to provide advantages only to a limited group of passengers who are usually more mobile, are frequent flyers, upper classes of people. The poorer and less mobile groups are supposed to be considered more risky "only because they do not leave data trails in the international aviation system. Moreover, in risk governance, leaving no paper trail at all equals being a potential offender".[430] All these techniques are hence discriminatory, creating special treatment.

As for the second group, all the selected technologies – body scanning, CCTV practices and PNR systems – can be used in a discriminatory manner. As noted above, all of them may involve categorization and different treatment. Specifically, a passenger can be chosen for secondary screening via scanner by human officer or prescreening passenger program;[431] for some scanner models, the standard of scanning (Standard 1 or 2) can be chosen manually or automatically depending on a passenger or a group of passengers; particular persons are monitored and targeted via CCTV more than others; behavioural analysis done by humans or by algorithms is aimed at selecting persons too. The common idea is that some persons deserve special attention, and the latter may be drawn on the basis of definite criteria.

As a rule, it is prohibited to exercise aviation security measures in a discriminatory manner using only sensitive criteria. For instance, the UK's scanners' Code of

---

[427] Schermer. *The limits of privacy in automated profiling and data mining*. In: Computer Law & Security Review. Vol. 27 (2011). p. 47.

[428] Amicelle *Surveillance and discrimination* In: Surveillance, fighting crime and violence (2012) p. 279.

[429] *Ibid.* p. 275 citing Lyon (2007) and other authors.

[430] Leese (2013) p. 487.

[431] Applicable if not 100% passengers are scanned. For instance, while in Russia all passengers must go through the scans, in the UK scanners are applied only to selected passengers.

Practice states that persons must not be selected on the basis of personal characteristics that may constitute discrimination such as disability, sex, gender reassignment, age, race, religion or belief, pregnancy and maternity and sexual orientation. Airport operators are required to follow all the requirements relating to selection that are contained in the public and restricted parts of the security scanner direction, and detailed protocols (which are not published) contain the further information on selection criteria for those to be scanned.

Similarly, CCTV activities where the processing of the data would result in discrimination against certain people exclusively on account of these factors should be avoided.[432] According to Article 6(4) of the EU PNR Directive, the assessment of the passengers carried out against pre-determined criteria must be carried out in a non-discriminatory manner on the basis of assessment criteria which in no circumstances can be based on a person's race or ethnic origin, political opinions, religion or philosophical beliefs, trade union membership, health, sexual life or sexual orientation.

In some cases, profiling can be prohibited, but with broad exceptions.[433] For instance, the TSA's documents permit profiling based on nationality or country of origin and passengers from at least 14 countries, mostly Middle East or Africa, are usually subject to enhanced security screening.[434]

It is argued that in the UK, Schedule 7 of the UK's Terrorist Act 2000, which gives broad powers to stop and search including at airport, is used to discriminatorily screen passengers.[435] According to reports, 32 percent of British Muslims were subjected to discrimination at airports,[436] and persons from ethnic minorities are 42 times more likely to be stopped by police in comparison with white people.[437] Interestingly, as a reaction, airport personnel unnecessarily stop and search white airline passengers in order not to be accused of discrimination and to achieve more favourable statistics.[438]

But even if using exclusively sensitive criteria is prohibited, in many cases it simply does not work. The prohibitive rules are very difficult to enforce in practice, and not all security practices are as neutral as metal detectors. For instance, accord-

---

[432] European Committee on Legal Co-operation (2003) and Article 29 Data Protection Working Party (2004).

[433] E.g. U.S. Department of Justice. Guidance for federal law enforcement agencies regarding the use of race, ethnicity, gender, national origin, religion, sexual orientation, or gender identity. December 2014.

[434] Quinlan (2015) p. 36.

[435] *Ibid.* p. 37.

[436] Open Society Institute. *Ethnic Profiling in the European Union: Pervasive, Ineffective, and Discriminatory* (2009). https://www.opensocietyfoundations.org/sites/default/files/profiling_20090526.pdf, p. 7.

[437] White. *Airport security checks: More offensive to some than to others*. The Guardian. 24 May 2011. http://www.theguardian.com/politics/blog/2011/may/24/airport-security-checks-terrorism-act

[438] Littlejohn. *How to avoid airport security: Wear a burka*. Daily Mail, 10 May 2012. http://www.dailymail.co.uk/debate/article-2142619/How-avoid-airport-security-Wear-burka.html#ixzz2idYLMjYs

ing to researchers, the application of secondary screening in the USA which is formally based on randomness is focusing on Muslims, only giving the impression of randomness, leading to "randomness of Islamophobia at US Airports."[439] Studies found that CCTV operators were over-scrutinizing some persons/groups of persons, in particular, minorities, youths, the poor and black males.[440]

The problem is that while watching passengers, security officers need to make a decisions and they ask themselves: "Is there anything obvious about the person that might indicate the need for special screening?"[441] But security officers, as all people, have stereotypes, and criteria for selection are – consciously or unconsciously – race, religion, age, gender and so on.[442] It is not a secret that when certain nationalities or religions are mentioned, they automatically create certain images: "...a Nigerian man; Thai girl; Jewish man; Colombian man; Albanian man; Buddhist man; Japanese woman. Unless one has had exposure to these ethnic and religious groups one might be forgiven in thinking of (in order) a fraudster; a prostitute; or masseuse; a Rabbi or money-orientated individual; a drug dealer; a human trafficker; a monk; and, a geisha".[443]

Once a person draws attention, the officer begins to interact with this person. From this point, it is very easy for the latter to accuse the screener of being a racist.[444] Even if a person was chosen on neutral grounds, if a person is male, young and Muslim, screening can easily be seen as discriminatory, and it would be difficult if not impossible to prove the opposite. Thus, even without a prior intent to judge people on the basis of sensitive characteristics, there is the risk of discriminating against particular groups or individuals.[445]

There are some technological solutions proposed to mitigate discrimination issues. For instance, Israeli researchers are developing the technology of check-in kiosks to replace the human selectors.

> When a traveler steps up to the machine, it senses his body temperature, blood pressure, and heart rate, just as in a polygraph exam. At some point during the interaction, the kiosk presents a statement that would elicit a reaction from a would-be terrorist. It might instruct him to see an agent, or just remind the passenger that flight security is everyone's responsibility. If the flyer's vital signs shift, he would be subject to secondary screening.[446]

But this is subject to future technologies.

Similar problems refer to automated profiling. Today, equipped with intelligent technologies, one might think that there should be no risk of discrimination if it is a

[439] Bazian (2014).

[440] Carli (2008) p. 12.

[441] Cohen. *Intercultural differences: the challenge for the screeners*. In: Aviation Security International. Vol. 18 (2012) p. 14.

[442] Perry. *The Screening of Passengers by Observation Techniques Programme: analysing the issues*. In: Aviation Security International. Vol. 17 (2011). p. 13 and Wagenaar (2012) p. 10.

[443] Cohen (2012) p. 14.

[444] *Ibid.*

[445] Schermer (2011) p. 47.

[446] Palmer (2011).

machine that selects persons for further investigation, and not an operator. Ideally, technology could be used to be programmed on the basis of fact rather than prejudice.

However, even in automated profiling, the core idea is still classification, division and different treatment. Algorithms are developed by humans, thus, they "learn" to discriminate on the basis of data used to train the algorithm.[447] Thus, the above-mentioned human stereotypes appear in software code, so that discrimination takes place within "automated, algorithmic, invisible systems of (attempted) social control".[448] In this case, it can be seen that discrimination is to some extent "documented", since its "traces" can be found in the code.

Typical examples are choosing a traveller for body scanning by a pre-screening program which indicated a passenger as high risk selection or determining which scanner standard should be used – based on certain criteria.

Particular risks arise in conjunction with "smart" CCTV. The researchers compare the latter with "predictive data-mining", which aims to predict events based on patterns or "classifiers" determined using known information.[449] Via behaviour analysis, CCTV evaluates and makes inferences about a person's actions and intentions, drawing on stereotypes and profiling methods[450] described above. If classification of a situation of a person as abnormal depends on a wide variety of factors, there exists a risk that the use of certain indicators can amount to discrimination by singling out individuals or social groups for adverse treatment.[451] Even such apparently neutral criteria as specific movements can in practice discriminate and disproportionally affect the right to privacy of a specific group of persons, e.g. if CCTV alerts an operator frequently on the basis of suspicious movements that are in fact linked to practicing a specific faith.[452]

Indeed, the technology only assists the operator to focus his attention on specific events, and does not make any decision without verification by an operator.[453] Moreover, human verification is just an instrument, and not the definitive solution; it cannot ensure ex-post that discrimination is not taking place.[454] Thus, this can hardly help to *prevent* discrimination.

In relation to the usage of PNR, the EU-US PNR system allows in some cases use of sensitive data, thus, discrimination is possible. If sensitive data is not processed (Russian system) or prohibited from being processed (the EU Directive),[455] then the risk of discrimination is reduced, although not removed because such data

---

[447] Schermer (2011) p. 47.

[448] Graham. *CCTV: The stealthy emergence of a fifth utility?* In: Planning Theory &Practice. Vol. 3 (2002). pp. 239–240.

[449] Vermeulen and Bellanova (2013) p. 305.

[450] Gilbert (2007) p. 33.

[451] Vermeulen and Bellanova (2013) p. 306.

[452] *Ibid.*

[453] *Ibid.* p. 310.

[454] *Ibid.* p. 311.

[455] See Chap. 7.

in some cases may be processed, as in the case of EU.[456] Thus, there are concerns related to the use of PNR data in an indiscriminate way; EDPS suggests the use of PNR data only on cases when there is a serious threat established by concrete indicators on a case-by-case basis.[457]

With reference to any technique, even if sensitive criteria or sensitive data are not used, indirect discrimination may still apply, since profiling is able to use neutral data for analysis. For instance, a person may be selected for additional searches because of his/her name. The EU Fundamental Rights Agency (FRA) suggests that for detection of indirect discrimination – in the case of PNR – "it would be useful to create suitable aggregate statistics based on PNR data to detect discriminatory patterns and trends in the application of the PNR system; these statistics must, however, be created anonymously and in a non-identifiable manner in order to comply with EU data protection principles".[458] Another common suggestion is to develop a means to "tell" the algorithm that its predictions should be independent of any sensitive criteria.[459] However, this is a matter of future research.

At the same time, from the perspective of effectiveness of profiling and data mining for aviation security, limiting the use of particular data – sensitive data – due to privacy/data protection concerns will reduce the effectiveness of these techniques. It will first reduce the accuracy and secondly increase false positives and false negatives.[460]

Today, some aviation security experts, taking into account both effectiveness and the impossibility of excluding discrimination, suggest one should admit that sensitive characteristics constitute an essential element of aviation security measure, and that discriminatory screening measures should be authorized, since "mathematically, we all know that this is the most effective method".[461]

Of course, the latter would be illegal in terms of privacy/data protection, but in some sense it would just confirm the situation which exists de facto. The point is that the use of profiling as part of different measures, use of PNR for profiling leads to a situation whereby privacy and the data protection framework "lose their applicability, leading to a diminishing role of the tools of the anti-discrimination framework".[462]

At the same time, it is also believed that the use of PNR for profiling without intelligence is not so effective. For instance, according to a study, about 171 persons who were convicted from 1997 to 2011 of terrorism-related crimes in the USA, among the convicted there was considerable diversity in age (but mostly young), gender (mostly male, but 8 female), race, ethnicity, education, and employment. Without intelligence and its knowledge of terrorist groups, such diversity makes it

---

[456] FRA (2011b).

[457] European Data Protection Supervisor (2011b).

[458] FRA (2011b).

[459] Schermer (2011) p. 51.

[460] *Ibid.* p. 49.

[461] Naudin (2014) p. 44.

[462] Leese (2014) p. 494.

difficult to develop a profile with which PNR data might be usefully compared.[463] But this only means that sensitive criteria may be useless without intelligence; this does not imply that they should not be used.

Other experts suggest that risk assessment cannot be based on sensitive characteristics, and should rather be based on travel data, intelligence gathering, voluntary contributed information and behaviour detection technique.[464] For instance, non-discriminative "suspicious" facts include flying one-way with no luggage, buying expensive tickets at the last minute and paying by cash, etc.

A compromise might be to make it permissible to use race or other sensitive criteria in combination with other information, if it is one of characteristics to describe a particular suspect,[465] that is, in concrete situations rather than generally, when the information regarding ethnic origin, gender, age of possible terrorists exists, as in the case of Paris incidents of November 2015. However, the permissible cases may still turn to broader application, stigmatizing particular groups.

Another compromise is making profiling techniques more transparent. With reference to CCTV, for instance, it is proposed that the algorithms used for identifying suspicious behaviour could be open-source and open to public review to avoid prejudices creeping into the system.[466] But in order to achieve this, further research and improvement of the technology are needed in order to characterize and distinguish between suspicious and acceptable behaviour more accurately, in order to target only objectively selected individuals.

Nevertheless, before any substantial changes, it can be concluded that currently all the considered practices in the airports may involve discrimination issues, based on either human involvement (selection for screening via body scanner, monitoring via CCTV or behaviour analysis) or the use of technologies. But in contrast to other limitations of human rights, the use of sensitive criteria a priori violates applicable norms on equality, since as a rule, their use is prohibited. This can hardly be substantiated by security needs and thus constitutes a strong argument for disproportionality of the regimes. This will be taken into account in the proportionality test in the next chapters.

---

[463] Flynn. *Intelligence: a pre-requisite for risk-based security screening*. In: Aviation Security International. Vol. 19 (2012). p. 19.

[464] See Fatokun. *Airport security: current practice and perspectives for 2013*. In: Journal of Air Transport Studies. Vol. 4 (2013). p. 36 referring to IATA's Checkpoint of the future: Executive Summary (2013).

[465] ACLU. Airport security: increased safety need not come at the expense of civil liberties. https://www.aclu.org/airport-securityincreased-safety-need-not-come-expense-civil-liberties

[466] Gilbert (2007) p. 42.

6.8 Modern Trends in Aviation Security

6.8.1 Introduction

Historically, as discussed above, the dominant activity of aviation policing has been item-by-item screening – baggage, individuals, identity documents, authorizations and passes, data.[467] All air passengers and all baggage must pass strict security checks in order to make sure that illegal and potentially harmful objects are not brought on board aircraft. In general, this strategy considers the aviation security as "a series of concentric rings around the aircraft – like the layers of an onion".[468]

So far, aviation security regulators have been adding such layers of security focusing on responses to the latest threat. In other words, most of the measures are being implemented and enhanced in response to situations that have already occurred. Therefore a reactive "layered approach" to security was adopted: When there has been a new type of attack or threat, the regulators have responded by applying additional measures.[469]

The reactive "layered" approach has been subject to criticism for a long time. The problem is that the whole concept of constantly adding layers and item-by-item screening, requiring more time (before 9/11 the average checkpoint processed 350 passengers per hour; today the number is below 150)[470] and costs, does not necessarily improve security. The reactive approach can hardly help to discover terrorists' future plans. Nor can it help to identify the intentions of people coming to the airports.

Thus, the security experts agree that the reactive direction of the current legislation is not the long-term answer to the industry's security challenges, and what is really needed is to change the system. Namely, a shift is required from reactive to intelligence-led, pro-active, outcome-based approach, from a one-size-fits-all (suggesting that each passenger presents an equal probability of being a terrorist or someone who could present a danger. Thus all passengers are treated same and are subjects to same security measures) to a risk-based approach.[471] It is declared that with all these concepts, traditional detection methods and prohibiting specific items will become less relevant.[472] Accordingly, a new set of layers for aviation security is suggested (Fig. 6.9).

All these ideas imply a number of strategies and tactics. Some of them were discussed in Chap. 4, some of them in this chapter above. In this section, the following layers will be discussed, summarizing modern features of aviation security:

---

[467] Salter (2008) p. 256.

[468] Preto. *Aviation security – priorities of the Commission* (2009) p. 2.

[469] ACI Policy and Recommended Practices Handbook. Seventh edition. November 2009.

[470] IATA. Remarks of Tony Tyler at the IATA Ops Conference in Vienna, 15 April 2013. http://www.iata.org/pressroom/speeches/Pages/2013-04-15-01.aspx

[471] See, e.g. Recommendations of the 37th ICAO General Assembly (October 2010).

[472] Brooks (2015) p. 31.

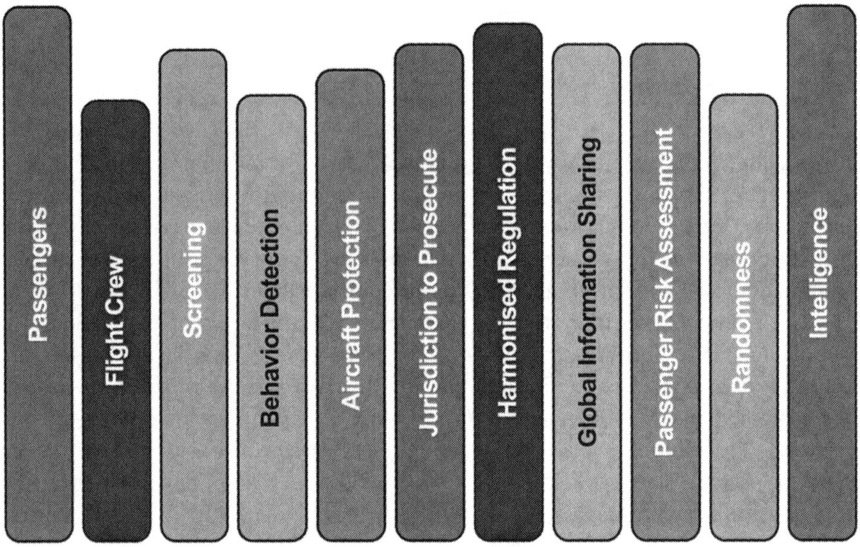

Fig. 6.9 New security layers (Brooks (2015) p. 31. www.asi-mag.com)

1. Intelligence-led, pro-active and risk-based approach
2. Global information sharing
3. Randomness and unpredictability in security
4. Combination of aviation security measures

The division into these groups is mainly for the convenience. These items are very interrelated, overlapping and inseparable from one another. All these issues have additional impact on the security-privacy dilemma thus are essential for this research.

6.8.2 Intelligence-Led, Pro-active and Risk-Based Approach

The ICAO High Level Conference on Aviation Security held in September 2012 recommended the concept of "outcome" which takes precedence over implemented procedures, and a list of modules to be taken into account included Passenger Data, Known Traveller, Identity Management, Behaviour Analysis, etc.[473]

The 9/11 Commission indicated four fundamental failures in US aviation security: a failure of imagination, a failure of policy, a failure of capabilities, and a failure of management.[474] The first point here is critical: without the imagination to

[473] Couessin (2015) p. 37.

[474] 9/11 Commission. The 9/11 Commission Report (2004) p. 339.

conceive different forms of attacks, it is impossible to create policies, develop capabilities and train/direct the management.[475]

New methodologies with pro-active intelligence approaches engaged, involve securing against terrorist acts by way of police methods such as surveillance of potential terrorist groups. The aim here is to take action against potential terrorists before they are able to attack.[476] An intelligence-led system is supposed to enable recognition of individuals who pose a possible threat from the moment they appear at the airport. Local intelligence can be very valuable, due to its relative ease of collection and immediate applicability to the individual airport.[477] According to Poole, it is necessary that the defensive measures are "dynamic and flexible, rather than static and predictable".[478] Within the pro-active approach, the "old" measures may be exploited more effectively. A good example is CCTV's Video Analytics.

Noting the adaptive nature of terrorists who always try to attack unexpectedly, it is proposed not only to examine the potential capabilities of terrorists with regard to aviation, but to study terrorist ideologies, strategies, and motivations, to educate law enforcement and intelligence agencies in all countries about how terrorists might try to "game the system" and exploit new vulnerabilities in aviation security.[479]

While "old" security methods rely heavily on detection of illegal objects, it is believed that detection of intent is more important than detection of means. This is especially relevant at present, with the lower limit of criminalization; theoretically, almost anyone may appear as "a wolf in sheep's clothing", including insider threats (see Sect. 4.5.4 Threats).

The intent is the key factor: "We can take all the steps in the world to try to secure our aircraft yet still fall victim to the actions of one individual who does not appear on our radar screen as a threat".[480] It is proclaimed that instead of looking for bad objects, security should look for bad people; instead of "detection" – "identification". Today, checking the passenger identification documents is usually done by government agents rather than security staff; according to IATA, this constitutes a weakness. The "document check" should be an integral part of a risk-based programme.[481]

According to security specialists, only pro-active approaches, i.e. only intelligence, is capable of discovering the intent, specifically, the intent not only of Islamic fundamentalists who can be boarding flights as passengers, but "the intent of extremists of all colours and creeds who may be shipping goods, flying our aircraft, working airsides at our airports and accessing the public areas of terminal buildings".[482]

---

[475] Price (2013) p. 36.

[476] Bjørnskau et al. (2007) p. iii.

[477] Raffel (2007) p. 6.

[478] Poole (2008) p. 23.

[479] Forest (2007) p. 12.

[480] Baum (2015a).

[481] IATA Recommended Practice (RP) 1701a.

[482] Baum. *Christian Fundamentalism and Justiciar Knights: proof of the foolhardness of ethic profiling*. In: Aviation Security International. Vol. 17 (2011) p. 1.

The core role here derives from a risk-based approach. For instance, one of IATA's SmartS' primary goals includes strengthening security by focusing resources based on risk levels.[483] Risk-based concept here should be distinguished from the concept of threat and risk assessment discussed above: here, it implies a shift from general risk assessment to personalized risk assessment.

One concept is that passengers could be divided into categories with different risk levels assigned: so-called trusted travellers (low risk), moderate risk, and high risk. Within this approach, there are two basic types of different treatments in screening. The first is profiling discussed above – a process aimed at identifying individuals who may pose more of a risk than others. The second is trusted traveller programmes which are aimed at identifying individuals who likely pose less risk than others.[484] They will be discussed below.

The question which arises is how the passenger risks are actually allocated. Obviously, the idea of placing known terrorists and criminals in "high risk" groups sounds reasonable. In addition, intelligence can contribute; different watch lists and databases of law enforcement and other organs. The inclusion of a person in the high risk group can also be the result of different techniques such as profiling discussed above, meaning that apart from objective criteria, many "soft" aspects can play a role, with the risks of flagging as risky for an innocent person. Moreover, inclusion in watch lists can also be the result of a mistake.

Similarly, one may wonder who actually can be "trusted". Are they the people whom everybody knows? However, there have been many cases in which even celebrities were causes of in-flight disturbances, not by virtue of being members of terrorist groups, but because they caused a threat to the flight.[485] Different Trusted Traveller Programmes (TTP) are aimed at allocating low risk to their participants.

Any such system is dependent on the quality of vetting done at the stage of enrolment.[486] In general, trusted traveller is a voluntary option for passengers who are usually frequent flyers. As a rule, they should apply for "trusted" low risk status, and in exchange for less distressing travel, passengers must provide personal data in an enhanced amount, including biometrics, and pass a stringent trustworthiness check. The security organ then evaluates applications and creates approved lists of known travellers. The latter obtain low risk status and, when identified via biometrics as low risk, can be offered a differentiated experience, particularly at security and the border crossing points.[487]

---

[483] IATA Press release, 5 Mar 2013. http://www.iata.org/pressroom/pr/Pages/2013-03-05-02.aspx

[484] Jackson [et al.]. *Assessing the security benefits of a trusted traveler program in the presence of attempted attacker exploitation and compromise*. In: Journal of Transportation Security. Vol. 5 (2012) p. 33.

[485] Baum. *Pilots, flight attendants and Welsh rugby stars: when those we "trust" are out of control*. In: Aviation Security International. Vol. 18 (2012a) p. 1.

[486] Bradley (2015).

[487] *Ibid.*

An example of TTP in practice is the US PreCheck program, which has been in operation in the USA since 2011.[488] US citizens and lawful permanent residents can get low risk status via PreCheck application, or by being members of other similar programmes such as Global Entry.[489] US citizens of frequent-flyer programmes can also be invited by a participating airline.[490] Some travellers may get the status due to their membership in a specific group, e.g. active duty military members in the USA. PreCheck passengers do not have to remove shoes, belts, liquids, laptops and jackets while screened and skip body scans, although they can be subject to random scans. It is reported that since 2011, more than 2 million passengers have been involved. PreCheck is currently available in 18 US airports.[491]

There are also regional and global plans: e.g. in the EU, a Registered Traveller Programme is proposed, allowing third-country nationals who have been pre-vetted and granted access to the Registered Traveller Programme to benefit from facilitation of border checks at the EU external border. Both travel document data and fingerprints are required.[492] The IATA Risk Based Screening is meant to be implemented globally.[493]

Voluntary provision of passenger data means that TTP provides a kind of self-governance regime. In contrast to automatic and invisible collection of passenger data by PNR systems and checks against remote databases, "TTP bring the disclosure of data back into the sphere of visibility and individual awareness," and passengers "pay" for low-risk status with their personal data.[494] Thus, although applicability of data protection and informational and physical (due to the use of biometrics) privacy are clear, the concerns can be less serious than in the case of profiling and other techniques. However, in addition to passengers' cooperation, TTP may use additional, different methods to assess whether a passenger is low risk.

The point is that TTP is argued to be effective and necessary tool in aviation security. From a security perspective, even if some terrorists use security gaps and become trusted travellers – and even if these people commit an attack, the security benefits of the TTP may still be positive.[495] In addition, it is argued that TTP as such does not address the risk posed by determined terrorists, and abandons the potential

---

[488] DHS, Aviation Security. http://www.dhs.gov/aviation-security

[489] Global Entry is a U.S. Customs and Border Protection program that allows expedited clearance for pre-approved, low-risk travellers upon arrival in the USA. http://www.cbp.gov/travel/trusted-traveler-programs/global-entry

[490] TSA PreCheck FAQ. http://www.tsa.gov/tsa-precheck/frequently-asked-questions

[491] DHS, Aviation Security. http://www.dhs.gov/aviation-security

[492] Articles 3, 7–8 of Proposal for a Regulation of the European Parliament and of the Council establishing a Registered Traveller Programme, COM(2013) 97 final – 2013/0059 (COD), Brussels, 28 Feb 2013.

[493] See also IATA's project SmartS which is being developed in cooperation with governments.

[494] Leese (2013) p. 486.

[495] Jackson et al. (2012) p. 33.

for "clean skin" terrorists. It is believed that only combined with profiling, such terrorists would likely be revealed.[496]

Thus, within PreCheck, the TSA uses, first, automated pre-screening system for risk assessments of all passengers. The aim is to identify low risk passengers eligible for expedited screening for a specific flight prior to the passengers' arrival at the airport.[497] Secondly, it is real time threat assessments through so-called Managed Inclusion. These assessments use procedures that randomly select passengers for expedited screening, behaviour detection officers who observe passengers to identify high-risk behaviours, canine teams, explosives trace detection devices in order to ensure that "low risk" passengers have no explosive material.[498]

This means the possible use of any other aviation security methods and technologies discussed above – from PNR used for profiling within pre-screening to behaviour analysis within Managed Inclusion – as a result, with corresponding human rights concerns and effectiveness issues. In addition, issues of equality may arise, since the core idea of the TTP is non-equal treatment (see above).

6.8.3 Global Information Sharing

Another key idea of modern aviation security is that enhanced data collection should be combined with better communication,[499] i.e. proper sharing of the information. It is important that different governmental agencies and relevant entities should have access to the data collected and analysed for security purposes. The data shared falls within the official strategy of cooperation and coordination of efforts at national and international levels for the key aims of the fight against terrorism: effective detection, investigation and neutralization of terrorist plots.[500]

Taking into account the restricted character of aviation security and the need to safeguard security (see Chap. 5), a negative side of increased sharing can be that valuable information can end up in the wrong hands. Hence, some of the most dangerous terrorists never appear on watch lists, to avoid having information about them transferred to airport employees in other countries.[501]

In the meantime, global data sharing is deemed to be unsatisfactory, in particular concerning data in different watch lists. For instance, with reference to the Christmas

---

[496] Wolff (2012) p. 4.

[497] GAO (2015b).

[498] *Ibid.*

[499] Wolff. *A passenger screening checkpoint of the 21st century.* Conference Transport Security Expo, London, September 2010.

[500] Aviation Security against Terrorist Threats – Conclusions of the conference of 31 October 2012, Nicosia, Cyprus. http://www.statewatch.org/news/2013/jan/eu-council-nicosia-aviation-security-16252-12.pdf

[501] Soghoian. *Insecure flight: Broken boarding passes and ineffective terrorist watch lists* In: Protecting Airline Passengers in the Age of Terrorism (2009) p. 15.

bomber in 2009, the information about the terrorist was located in databases but was not disseminated to the appropriate agencies.[502]

Another example is Interpol's Stolen and Lost Travel Documents (SLTD) database established in 2002. It is a global repository accessible by officers at airports, borders and other field points. The database holds over 40 million records including passports, identity cards, visas and UN *laissez-passer*(s). Despite increased use – e.g. in 2013, it was accessed more than 800 million times worldwide, about four out of every ten passports of international passengers are still not screened against this database.[503]

The share problems concerns intelligence too: despite efforts to share such information with airports, there is a communications problem between intelligence organs and airports due to the lack of standardization, consistency, or a designated programme.[504] It is suggested that innovation and partnerships with local and federal officials, proper foundation and development of a suitable process can contribute to appropriate use of information;[505] intelligence should cooperate with counterparts internationally.[506] But today, it is shown that, despite the current availability and possibility of using information/intelligence, actual usage is far from what is desirable, and there is a need for much better communication between the stakeholders.

While probably this is bad for security, it can be seen as positive from a data protection perspective where collection and sharing of data should be limited to absolute necessary. However, problems with communication mean problems with proper sharing rather than *less* sharing as such.

6.8.4 Randomness and Unpredictability

In modern aviation security, an essential element is the element of randomness and unpredictability in security screening and the use of technology.[507] Randomness is thought to enhance security and implies that consequences of measures can be allocated "at random."

The Aviation Security Panel of ICAO in 2009 suggested that unpredictability should be built into the aviation security regime, but in such a way that it does not diminish the level of security or result in delays for passengers – i.e. ensuring a

---

502 Holbrook. *Airport Security: Privacy vs. Safety*. In: Risk Management. Vol. 57 (2010).

503 Noble. *Passenger screening: empowering the industry to plug government security gaps*. In: Aviation Security International. Vol. 20 (2014) p. 25.

504 Raffel (2007) p. 6.

505 *Ibid.* p. 2.

506 ACI position brief. November 2010. http://www.aci.aero/Media/aci/file/Position%20Briefs/2010/position%20brief_SECURITY%202010.pdf

507 Annex 17, §4.2.6.

balance between certainty and unpredictability.[508] Increasing unpredictability is one of the goals of IATA's SmartS;[509] ECAC Aviation Security Handbook contains detailed description of the use of unpredictability for aviation security in Part II – Security Measures.[510] TSA declares that it "incorporates unpredictable security measures throughout the airport and no passenger is guaranteed expedited screening".[511] However, further details refer to restricted information.

But, for this research, the most interesting issue in this regard will be that, according to the "randomness and unpredictability" concept, a questionable measure as well as respective privacy/data protection concerns will be applicable to certain citizens at random. From a psychological point of view, one may ask: Are you ready to give up some of your privacy if this is done only randomly? Will it benefit privacy?

The law cannot operate with "random" criteria, however. Thus, irrespective of whether the security measure is used randomly or not, if it raises privacy/data protection concerns, the law must be applied in general, without exceptions.

Another question is how the randomness and unpredictability are applied in reality. As mentioned in section on discrimination above, in the USA, for instance, the security screenings are formally done with random criteria, but in fact they focus on particular categories of people such as Muslims and nationals of the Middle East or Africa, only giving the impression of randomness.

6.8.5 Combination of Security Measures

The text above shows that a growing trend in aviation security is interconnection, integration and combination of different aviation security measures and technologies which would improve and enhance each other. The increase in (inter)connectivity, interactivity and interacting capabilities through technology and also the increased dependency on such capabilities alters the shape of aviation security.[512]

Today, from the aviation security point of view, the technological developments provide ample possibilities for providing better security. This includes the use of various information-gathering technologies, which can complement each other with additional information. Fig. 6.10 below shows these interconnections as well as use of other data sources.

Of course, many of the indicated items overlap, as in the case of profiling and other methods. In general, however, Fig. 6.10 presents various aviation security measures and technologies, in particular those that collect personal data, as the whole mechanism. Accordingly, when all these technologies and data sources are

---

[508] Abeyratne (2010) p. 20.

[509] IATA. Press release, 5 Mar 2013. http://www.iata.org/pressroom/pr/Pages/2013-03-05-02.aspx

[510] European Civil Aviation Conference. ECAC Aviation Security Handbook (restricted), 2014.

[511] TSA. PreCheck. http://www.tsa.gov/tsa-precheck

[512] COPRA (2013).

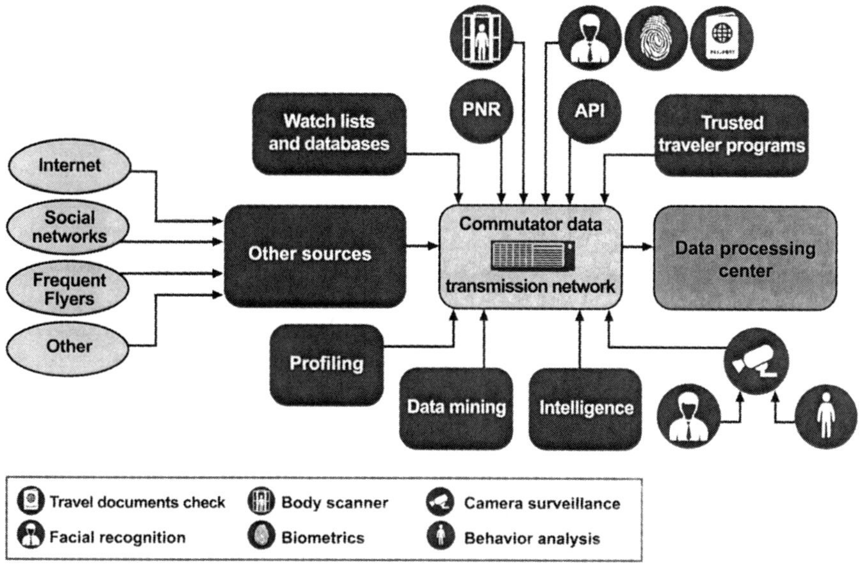

Fig. 6.10 Combination of aviation security technologies and sources

interconnected, the ability to analyse personal data increases significantly both quantitatively and qualitatively.

A typical example is the combined use of intelligent CCTV cameras with other tools like an automated recognition of persons via biometric identifiers, behaviour analysis, along with intelligence gathering and data mining, which can complement CCTV footage with additional information.[513] Specifically, intelligence gathering from various multiple sources can be combined with automated recognition of persons via biometric identifiers, behaviour analysis, and be applied at many points all through the airport: access control, passenger registration, security screening, immigration control, boarding gates, etc.

For instance, the photos captured from passports can be automatically loaded into CCTV's Visual Analytics.[514] Capturing face image of a passenger via CCTV and comparing this data against databases permits investigators to ascertain biographical information and to identify whether a person is dangerous. As mentioned above, CCTV can potentially use databases of faces from social networks, create own databases, etc. The researchers note that the emergence of new image analysis algorithms in CCTV and new data integration capabilities that combine traditional surveillance with advanced profiling and data mining techniques are among major technical trends for contemporary surveillance.[515]

---

[513] Wright et al. (2010) p. 346.

[514] Finn (2012) p. 40.

[515] Wright et al. (2010) p. 346.

An example of how body scanners can be integrated with other systems is NATO's STANDEX (Stand-off Detection of Explosives) program involving explosive stand-off detection equipment, video tracking system and data processing software – trials were executed in a metropolitan subway station.[516] This concept integrates into a system technologically advanced prototype detection instruments, based on MMW scans and laser spectroscopy for the real time and continuous inspection of people and objects for finding suicide bombers.

The key idea of modern aviation security is that passenger information and intelligence data must be integrated with screening.[517] In order to have reliable pre-screening, at least three systems should be combined: (i) matching passengers against watch lists, (ii) use of PNR for profiling and (iii) behaviour analysis; each of them could independently assess the passenger, and categorization as high risk by any of them could be decisive.[518] Moreover, this should be augmented by random selection discussed above and with many other parameters, such as destination country, geographic context and many other data to be processed. It is emphasized that this is not an alternative to security equipment, but it complements and determines the depth of the control procedure.[519] In addition to profiling, combination also implies wider use of TTP. Both are based on combination of intelligence and other collected personal data, including biometrics, combined by behaviour analysis and other methods.

Integrations contribute to the creation of comprehensive profiles and huge databases that combine multiple sources,[520] and exploiting the phenomenon of Big Data, they will generate more data. The scheme is as follows: first, all data is used within multiple methods, helping to determine which people must be selected for additional screening, detained, etc. All data may count, such as data from watch and no-fly lists, national and international databases, including image databases, airlines' API and PNR, biometric data, video footage with facial recognition and behaviour analysis, etc. While analysing all the information, new predictions, assumptions and more knowledge about people – thus, more data will be generated and used further.

The researchers note that in the future, authorities can exploit various sources of information that individuals put on the Internet, e.g. participation in online forums, and in social networks such as Facebook, Google, Twitter, or LinkedIn. This information is public if accessed without restriction. By using these open data sources, authorities will benefit from a very comprehensive pattern which can be further exploited.[521] It is suggested that the use of all these data sources along with other

[516] Jezierska-Switala. *Big City Trials: from laboratory explosives detection to the field applications (a STANDEX programme).* Body Search 2015 Conference Program, http://network.icom.museum/fileadmin/user_upload/minisites/icms/pdfs/Body_Search_SPEAKER_PROGRAMME.pdf

[517] Neeman (2013) p. 15.

[518] Flynn. *A personal view*. In: Aviation Security International. Vol. 17 (2011) p. 48.

[519] Couessin (2015) p. 37.

[520] Woodward (2002) p. 6.

[521] Couessin (2015) p. 39.

Big Data tools will contribute to the accuracy and usefulness of the risk-based approach, enabling authorities to acquire "valid and relevant information" on every passenger in real time.[522] This means more opportunities for profiling at any stage of the journey; for example, a person may instantly release a post on the Internet while waiting for a flight. At the same time, the experts warn that for effective use, Big Data should not be used for differentiating passengers alone, it should be a complementary layer combined with other methods such as random checks, observation and behaviour analysis.[523] This again proves the benefits of combination approach.

Another point is that along with security advantages, Big Data approach implies enhanced privacy/data protection/human rights risks. They include unlimited and uncontrolled multiple databases with personal data on all individuals, including sensitive data, data sharing and data retention, including data transfer across borders,[524] discrimination and freedom of movement issues, etc. Within privacy/data protection/human rights framework, Big Data approach cannot be unlimited. However, as discussed in the General Part, Big Data is not yet adequately addressed by current data protection laws and present the biggest challenge for modernization of current regimes, thus, the challenges are still to be addressed.

Overall, although interconnected and combined capabilities of different security measures are capable to provide advantages in the terms of security, it is a question of how far the combination of technologies can go. The problem is that the forms and possibilities of interference into privacy and other human rights by every technological feature of every individual aviation security measure are combined too. Thus, the impact grows in geometric progression, increasing the risks.

6.8.6 Concluding Remarks

According to the pro-active and risk-based approach, the security provision should be focusing on intelligence, increased usage of personal data, including biometrics, profiling, behavioural analysis and other techniques. Combination of different security measures, with use of personal data from different sources, randomness in screening are the global trends.

The aviation security measure capturing these trends is CCTV equipped with Visual Analytics: it is intelligent, pro-active and risk-based, collects and uses personal data, including biometrics, carries out behaviour analysis, may be connected to other measures and use personal data from other sources, including PNR, may export data to other databases.

---

522 Laviv. *Big Data: aiding aviation and border security*. In: Aviation Security International. Vol. 20 (2014). p. 11.

523 Couessin (2015) p. 39.

524 These points will be discussed in more detail in Chap. 7.

PNR, if used for profiling, is also pro-active and risk-based, can be shared with various entities, can be used in combination with TTP, CCTV and other measures.

Body scanners' role, since they are aimed to deter prohibited items, belong to old-style reactive approach. However, scans can be conducted randomly and if used only in high risk security lines, the use of scans can combine with other processes/methods, such as TTP, behaviour analysis, selection of a person by human operator or algorithm of CCTV.

All these trends are supposed to enhance security, but apparently, human rights risks will constitute additional costs that will have to be paid for the advantages gained. Thus, the main conclusion is that aviation security today deploys or plans to deploy many of techniques and methods that are in tension with passenger rights. Moreover, the intrusive features will apparently be further developed and used, and new, more advanced methods may appear, accordingly, with growing or additional concerns.

6.9 Concluding Remarks

While the previous section concerned findings on a broader perspective – common modern technological trends in aviation security, this section will summarize findings regarding the selected technologies' impact on privacy, data protection, and other human rights.

It is clear that the use of body scanners and CCTV at airports is booming, and PNR exchange is becoming worldwide practice. Not only the USA, the EU and Russia, but many other countries are using or planning to use body scans, CCTV, and impose PNR regimes.

Although this research selected these three aviation security measures for analysis, the discussion above shows that other techniques may be part of them or be connected directly or indirectly, thus be applicable too, namely, biometrics, profiling, including behaviour analysis, and TTP.

There are of course other technologies which can cause privacy/data protection concerns, such as computer topography of in-checked baggage, which enables one to see every detail inside the bag. In some states, if any "anomality" is recovered, the owner or police must be present at the moment of opening the bag; in others, a bag can be opened without this. If this technology is used for hand baggage, the owner is nearby, but this does not eliminate privacy concerns and may lead to disturbances.

All the measures have one common objective, the main one – to protect the public. Although there are different opinions on the effectiveness of scans, CCTV and PNR systems, and some weakness such as detection failures may serve contra arguments, with different application, all of them constitute "layers" of aviation security which are deemed to be necessary by security experts. Driven by modern trends to provide pro-active and risk-based security, and in order to enrich each other, various technologies are also combined and interconnected.

On the one hand, the process of technological developments which are becoming more sophisticated and "smart" is quite unambiguous and can be noted for all discussed technologies, leading to human right concerns.

On the other hand, as a reaction to raised concerns, regulators and manufacturers endeavour to provide guarantees and safeguards. The problems is that the provided safeguards may be inadequate or insufficient, subject to further development and research, as in the case of, for example, facial recognition or behaviour analysis, or they may only "mask" the concerns rather than mitigate them, as in the case of questionable "image-free" scanners or blurring of CCTV images which can be de-anonymized later. Moreover, the safeguards, no matter if they work or not, may have negative effects on security benefits, such as ATR and reduced visibility of scanners' images.

At the same time, human rights concerns may prevent the usage of effective security measures or features. Examples include backscatter and X-ray scanners, wider use of biometrics, unlimited profiling with use of sensitive data, etc. At the same time, the reasons of non-use can also include difficulties in incorporating the measure into airport and airlines' existing practices, financial and other issues.

The considered aviation security measures – body scans, CCTV and PNR – are a good illustration of close connection and overlapping between different types of privacy and between privacy and data protection rights, as shown in Table 6.3.

It can be seen that to a different extent, with respect to all the selected measures, interference with privacy is accompanied and overlapped by data protection concerns. Accordingly, they are a good illustration of a situation when norms on privacy can be applied with respect to data protection, and data protection mechanisms can be relevant for privacy too. Therefore, in order to achieve a maximum protection, each case should be analysed from both sides in order to see which protection is afforded by each of the regimes, which one offers the best protection, and if possible, apply both.

Moreover, it is a typical situation that one and the same measure leads to multiple human rights concerns, proving that all human rights are interdependent, indivisible and interrelated. Table 6.4 below helps to illustrate how the human rights discussed relate to each other with regard to the selected measures:

Table 6.3 Aviation security measures' impact on privacy and data protection

	Body scanners	CCTV (with all possible features)	PNR
Privacy of the person	X	X	
Privacy of behaviour and action	X	X	
Privacy of communication		X	
Privacy of data and image	X	X	X
Privacy of thought and feelings		X	
Privacy of location and space		X	
Privacy of association		X	
Data protection	X	X	X

Table 6.4 Aviation security measures' impact on human rights

	Scanners	CCTV basic[a]	Smart CCTV[b]	PNR
Right to life	Probably positive	Positive	Positive, but subject to future development	Positive, in particular if used for profiling
Right to health	Probably no	No		
Privacy	Negative			
Data protection	Negative			
Freedom of movement	No	No	Negative	Negative
Non-discrimination	Negative			
Freedom of thought, conscience and religion	Indirect negative			
Rights of the child	Any negative impacts become more serious for a child			

[a]I.e. without additional Video Analytics features such as facial recognition, behaviour analysis and use of external databases
[b]I.e. with the indicated additional features

It is assumed that all the indicated measures are supposed to assist security and thus contribute to the right to life, although another question is how effective they are. While basic CCTV is deemed to be effective and the positive effect on the right to life is quite definite, "smart" CCTV is more questionable: although it has great potential, CCTV is subject to further improvements. The use of body scans is indicated as probably effective due to failures in detection. PNR is considered effective, in particular if used for profiling.

Currently used MMW scanners may have a negative impact on health, but apparently, it is extremely low and difficult to prove. However, a person may opt out of a hand search, and if really worried about effects of MMW scanners on health, should be advised that flying as such and standard daily devices such as mobile phones present more health risks.

The right to freedom of movement may suffer as well, in particular with regard to PNR and "smart" CCTV, if innocent persons are mistakenly stopped, searched and detained, mainly as the result of mistaken inclusion on various watch lists, inaccuracy of facial recognition and behaviour analysis techniques, etc.

All the measures raise concerns with respect to the right to non-discrimination. This can be done directly (if sensitive criteria are used) or indirectly (if neutral criteria are used). Body scanners may discriminate against "abnormal" persons and may be used discriminatively if a person is selected by security officers to go through the device or automatically chosen due to specific characteristics. In the case of CCTV, some people are over-scrutinized by either human security staff or by automatic tools. The use of PNR may also involve the use of sensitive data and profiling which is discriminative by nature.

With regard to the freedom of thought, conscience and religion, all the measures can have a negative impact indirectly, mainly via religion as a criterion for discrimination. Additionally, if "smart" CCTV detects movements which are quite neutral but are linked to practising a specific faith but which are deemed suspicious, this can discriminate and disproportionally affect the right to privacy of a specific group of persons.

Aviation security measures are applied to children in the same scope as to adults. However, due to the stages in their physical and mental development, children are vulnerable. Children have a limited ability to understand what happens in the course of security, how their data is used; they are very dependent with regard to enforcement of their rights, etc. Thus, the general human rights concerns raised by aviation security measures are becoming more severe for children, increasing aggregated harm, if any. Thus, any interference requires in-depth substantiation in addition to general substantiation.

Further, the use of profiling, TTP and other systems within the pro-active and risk-based approach constitute a problem from the perspective of privacy/data protection/human rights. As discussed, they imply that more and more personal data is collected, stored and shared. In contrast to profiling, TTP implies some individual's control over his/her data, at least on the stage of submitting. However, in both cases, persons have no idea on which criteria they are being assessed for; a risk-based passenger differentiation can be based on such factors as religion, race, gender, causing discrimination issues. At the same time, there are as a rule no statistics on their effectiveness for combatting terrorism and crime.

A positive sign for privacy– but probably negative for security – is that global fulfilment of the pro-active and risk-based approach is still subject to future undertakings, e.g. not all the states operate PNR systems. In the meantime, despite the endeavours of security professionals coming up with new ideas, the security organs fail to use the new ideas effectively. The majority of security activities in the airports are based on the old, traditional reactive approach, with item-by-item screening. Surprisingly, this concerns even such "pioneers" in aviation security as the USA. For instance, according to a report by two committees in the US House of Representatives, the TSA "has grown into an enormous, inflexible and distracted bureaucracy, more concerned with human resource management and consolidating power, and acting reactively instead of proactively".[525] Moreover, since many elements of pro-active and risk-based approach are already in place in many states, the concerns apply.

Similarly, global information sharing presents value for security, but additional risks for passenger's rights, and at the same time faces challenges in implementation globally.

Finally, the combination of various measures may contribute not only to the effectiveness of aviation security regimes, but also to the aggregated effect, making concerns more serious, and promoting them as issues that must be taken into account.

---

[525] Richardson (2011a).

Chapter 7
Analysis of Privacy and Data Protection Principles

7.1 Introduction

Principles of privacy and data protection provide – or at least are aimed at providing – concrete mechanisms for the evaluation of aviation security measures' impact on privacy and data protection. In this chapter, legal principles of privacy and data protection from Chap. 5 will be considered in respect of the selected aviation security measures. The privacy/data protection concerns of body scanners, camera surveillance at airports and PNR regimes will be analysed from a more narrow point of view, presenting, as examples, concrete relevant details of the regulations in the discussed jurisdictions. As noted above, since in Norway the selected measures are used to a lesser extent in comparison with other states (no scanners, no PNR yet), Norway is discussed less in comparison with the EU, the UK, the USA and Russia.

The main purpose is to analyse whether the regulations of the selected technologies on different levels correspond to the principles. In addition, the discussion may also contribute to discussing whether the technologies' regulations represent some special challenges for the principles' application. In particular, how well the regulations of the analysed technologies, including applicable technology-neutral and technology-specific regulation on privacy/data protection are "suited" for the principles and vice versa.

It should be noted that the general principle of non-discrimination, which overlays with the right to equality and non-discrimination, was discussed in Sect. 6.7, with the conclusion that all the selected measures may be discriminatory and/or be used in a discriminatory manner.

O.M. Enerstvedt, *Aviation Security, Privacy, Data Protection and Other Human Rights: Technologies and Legal Principles*, Law, Governance and Technology Series 37, DOI 10.1007/978-3-319-58139-2_7

7.2 Legality

7.2.1 Introduction

As discussed in Chap. 5, the legality principle is very broad and the requirements of the principle that are important for this research fall within several groups. First, people should be informed – but not jeopardizing security – about the use of scans/CCTV/PNR as well as their consequences. Secondly, the criteria "in accordance with law" implies that aviation security measures which result in limitations on human rights must have a legal basis in law (precise, specific and clearly defined: accessible and foreseeable by persons – these requirements overlay with transparency) and the law must contain judicial protection.

Thirdly, aviation security measures must serve a legitimate aim. Fourthly, data processing should be lawful. The latter generally means that other data protection principles should be respected, thus, this condition can be evaluated only after discussing all applicable principles of data protection. Any concerns may lead to non-compliance with lawful processing.

7.2.2 Information Openness and Transparency

7.2.2.1 Introduction

Ideally, passengers should be provided with appropriate, comprehensive and clear information about aviation security measures – along with information regarding the protection of their rights – before travelling and before purchasing the tickets. However, as discussed, there is much amount of restricted information in aviation security.

Moreover, "too much" transparency can be dangerous. "There is no questioning the good that transparency creates in a wide range of contexts, government especially. But we should also recognize that the collateral consequence of that good need not itself be good".[1] An example of the latter are possible negative consequences in the case of "too much" transparency about aviation security. The proliferation of technology and information may enable public use new technologies: it is now easier to be aware of how new technology may be misused, there is an increase in available weapons, and the technologies may be misused to instigate threats.[2] Consequently, the technologies which are designed to protect civil aviation may, if used by criminals, serve as threats against it.

[1] Lessig. *Against Transparency*. The New Republic. 9 October 2009. https://newrepublic.com/article/70097/against-transparency.

[2] COPRA (2013).

The problem is that in some cases it can be difficult to determine if the lack of information is justified by the security needs or constitute non-compliance with the principle of transparency. Below, this will be considered with respect to body scanners, CCTV and PNR.

On the international level, the ICAO provides some rules on passenger notification, with a typical form of such notification, only in respect to PNR (see below); on the other measures, the EU or national rules apply.

7.2.2.2 Scans

As for body scanners, some stories in the press still show "naked" pictures produced by old generation scanners as illustrations to the text about new scanners. As discussed in Chap. 6, while this may be true, it can be the case that quality images are inside the machines and are never seen by humans – however, the potential opportunity for the latter probably exists. In addition, scanners may deal with sensitive data. In any case, when going through a scan, travellers may wonder about the devices' nature. Thus, legal information about the scanners is critical.

According to the EU Impact Assessment, appropriate, comprehensive and clear information on all aspects of scanner usage should be provided to the public at airports, before travelling.[3] Such actors as Article 29 Working Party, EDPS and FRA suggest that information should be given on the data that is being processed, image produced, data storage, what the scanner does, functioning, purposes, benefits, responsible persons for the data processing, prohibited items, health risks, possibility for refusal, including alternatives and consequences of not using the scanner, info point for more detailed information; information should be provided at airlines, travel agent, check-in, screening point, general media, on airline/airport and government authority websites.[4]

Some states make efforts to provide some details about scanners. The UK Code of Practice stipulates that information should be provided ideally before ticket purchase, and in any event prior to entering the passenger screening area, in a number of languages. In the USA, it is required that DHS should provide notice to public regarding its collection, use, dissemination, and maintenance of personal data. Technologies or systems using personal data must be described in PIA, and no system can exist in a secret.

In the PIA regarding scanners, it is stated that this document exists to ensure TSA's continued transparency on its use of AIT; information on AIT technologies is available on TSA's website; TSA publishes PIA on AIT updates reflecting operational or technology changes; in 2013, TSA published a Notice of Proposed Rule Making on the use of AIT in screening operations which received more than 5500 comments from the public; and that TSA plans to publish its Final Rule.[5]

---

[3] European Commission (2011), §136.

[4] See, e.g. FRA (2010b) p. 12.

[5] Department of Homeland Security (2015) p. 3.

In Russia, general rules require that air passengers should be informed about the transport requirements. In places selling tickets, airports, check-in areas, before the control zones, checkpoints, in the places where customs declarations are filled out, there should be displays and posters with special information on aviation security.[6] However, in practice, these displays are dedicated mainly to the lists of prohibited items and other security rules. I did not find any more detailed rules containing privacy/data protection safeguards for scanners. However, this information is available on the websites of the manufacturers or sellers of scanners and is quite easily obtainable on request from, for instance, the airports.

It can be seen that the regulators try to satisfy the transparency principle and much progress has been made, in particular in the EU, USA and UK. However, not all the items recommended by Article 29 Working Party, EDPS and FRA above are covered, meaning that more details could be provided.

In addition, as discussed in Chap. 6, the nature of scanners and their capabilities are quite unclear. The critical point is whether the information provided is appropriate, comprehensive and reliable, in particular the information concerning how the scans actually work. As noted above, the governments and producers' statements and assurances regarding image-free scanners and data storage option seem to be difficult to rely on. One may wonder if the information available is aimed at concealing privacy concerns.

Finally, if a person is chosen for scanning due to particular criteria, there is a lack of information about the applicable algorithms (see Sect. 7.2.2.5). All hese factors can be additional arguments pointing to the fact that the requirements of the transparency principle are not properly satisfied.

7.2.2.3 CCTV

While it is clear that airports greatly rely on camera surveillance, travellers do not know exactly how this works,[7] in particular, the nature of surveillance, its purposes, dangers and capabilities.[8] Persons who are watched cannot see the watcher, do not know who is watching, what they are watching for, or how the data being recorded, stored or used.[9] As discussed, the technical advances made the term "CCTV" misleading, and the continued use of this term indicates a lack of awareness. If you refer to the websites of the respective organizations such as airports or aviation security organs, whereas typically there is comprehensive information about screening rules or prohibited items, information on CCTV is scarce or not present.

The Council of Europe and Article 29 Working Party endeavoured to provide further details on the transparency principle for CCTV. The Council of Europe

[6] Order of the Ministry of Transport of 25 July 2007 N 104 On approval of rules for pre- and post-flight inspections, §47.

[7] Wagenaar and Boersma (2012) p. 8.

[8] Gilbert (2007) p. 33.

[9] EPIC (2008).

suggests that the operator should make clearly discernible, in an appropriate manner, that camera surveillance is taking place, its purpose and the identity of the controller (but in some cases the purpose and the identity of the controller are clear from the circumstances) or by informing the data subject beforehand of the above. Other information pertaining to the specific circumstances (e.g. technical specifications of the chosen system) should be provided to the data subject where this is necessary to guarantee fair processing of personal data and does not jeopardize the purpose of the surveillance.[10]

The most usual means by which to give information to data subjects are signs. Article 29 Working Party lists the following requirements to signs: they should (i) be clearly visible and readable, (ii) specify the places monitored (not necessary to specify the precise location of the equipment); (iii) include the details of the operator, the purposes of the camera surveillance; (iv) only specific, well-grounded limitations to the information requirements may be allowed.[11]

The UK Surveillance CCTV Code of Practice contains two principles regarding transparency: Principle 3 – There must be as much transparency in the use of a surveillance camera system as possible, including a published contact point for access to information and complaints, and Principle 5 – Clear rules, policies and procedures must be in place before a surveillance camera system is used, and these must be communicated to all who need to comply with them.[12] In addition to Article 29 Working Party's suggestions, it states that good practice includes considering the publication of information on the procedures and safeguards in place, impact assessments, performance statistics and other management information and any reviews or audits undertaken.[13]

In Norway, the general Personal Data Act contains the requirement to give clear notice as well, by way of signs or other means, to make people clearly aware that the place is under surveillance, whether the monitoring includes audio recording and who is the controller.[14]

The US DHS' PIA requires notice signs, with an example of the wording: "This Area Under Video Surveillance," and published PIA providing that federal properties are under surveillance by CCTV cameras and why the cameras are necessary.[15]

In Russia, according to federal law, when security services are assisted by camera surveillance, the public should be informed of this by posting relevant information in places where it is visible daytime and night-time, and prior to entry into the protected area. This information should contain information about the conditions of the regimes inside the building and entrance requirements.[16] However, the law does

[10] European Committee on Legal Co-operation (2003).

[11] Article 29 Data Protection Working Party (2004).

[12] Home Office (2013).

[13] *Ibid.*

[14] §40 of Personal Data Act 2000.

[15] Department of Homeland Security (2012b).

[16] §3 of Article 12 of Federal law of 11 March 1992 N 2487-I On private detective and security activity in the Russian Federation.

not provide any additional special requirements for a more specific content or format of these signs.

One example of good practice is the UK Cardiff Airport CCTV Code of Practice. The information to public is provided in two ways: (i) publishing this Code of practice and its availability in various places, and (ii) clearly visible and legible signs located at designated points so that people are aware that they are entering a zone covered by CCTV. The signs contain an image of a camera and the following information: (a) operator (Cardiff Airport Limited), (b) the purpose of the CCTV system, (c) contact details.[17]

At the same time, despite the universal requirement to post clear signs about CCTV at airports, and despite good practices in some jurisdictions, problems arise. Technology-specific laws with provisions on transparency are lacking. Not all the airports have a CCTV Code of Practices or established procedures ensuring transparency. General provisions of national data protection laws can hardly ensure that the airports' practices are satisfactory. Consequently, although the signs are usually required and are in place, the information is not provided elsewhere; for instance, not all security organs or airports' websites contain comprehensive information on CCTV.

Moreover, information on signs may be insufficient, e.g. they may lack the details about the operator, purposes of CCTV, etc. Strictly speaking, information about any other specific measures that deal with personal data and are used in connection with CCTV are also required. This is the case if CCTV uses biometrics, or leads to creation and/or implies use of external databases. However, it can be problematic to inform the public about all smart functions of CCTV via signs.[18] Indeed, it could be impractical to provide comprehensive information via posted signs. Thus, other possibilities should be explored. In addition, if CCTV is used for behaviour analysis, there is lack of information about the respective algorithms (see Sect. 7.2.2.5). Thus, currently, there is a lack of transparency with reference to CCTV at airports.

7.2.2.4 PNR

The ICAO PNR Guidelines stipulate that air carriers or their agents must properly notify passengers, for example, at the time of booking of flight or ticket purchase, that the carrier may be required to provide any or all of its available data PNR to the authorities of the state of departure, arrival or transit, and that this information may be shared with other authorities. This notice must include the specific purpose of data collection and guidance for passengers on how they can be access their data and to demand redress. ICAO also suggests a typical form of such notification.

[17] Cardiff Airport (2013).

[18] Thon. *Kameraovervåking – fra VHS-kassetter til intelligent videoanalyse*. 9 July 2015. https://www.personvernbloggen.no/2015/07/09/kameraovervaking-fra-vhs-kassetter-til-intelligent-videoanalyse/

The EU Strategy provides that every individual shall be informed at least about the purpose of processing of personal data, the persons who will be processing that data, under what rules or laws, the types of third parties to whom data is disclosed and how and from whom redress can be sought.

In order to enhance data protection guarantees, FRA suggests that states, in cooperation with DPAs, should provide information to passengers on the collection, storage and processing of PNR, on the applicable data protection principles and their rights, especially on the redress mechanisms available to passengers under the laws of the state; this can be provided on the websites, via leaflets and posters displayed in travel agencies, at airport counters and at border crossing points, via airlines at the time of booking and check-in.[19]

According to the EU Directive Recital 25, Member States must ensure that passengers are provided with "accurate information that is easily accessible and easy to understand about the collection of PNR data, their transfer to the PIU and their rights as data subjects". It is noteworthy that in one of the previous versions of Directive (2011), there was Article 11(5) providing more details on such information.[20] Therefore, the final version provides fewer details and the regulator relies on national regulators, two factors that can be seen as a weakness.

The EU-US Agreement Article 10 states that DHS must provide information regarding its use and processing of PNR via publications in the Federal Register; publications on its website; notices that may be incorporated by the carriers into contracts of carriage; statutorily required reporting to Congress; and other appropriate measures. DHS shall publish and provide to the EU for possible publication its procedures and modalities regarding access, correction or rectification, and redress procedures. The Parties shall work with the aviation industry to encourage greater visibility to passengers at the time of booking on the purpose of the collection, processing and use of PNR by DHS, and on how to request access, correction and redress.

The rules about the Russian PNR system refer to general data protection law[21] rather than provide specific details about data protection issues and data subjects' rights. The general law provides that data subject has the right to be informed about processing of his/her personal data, including information about the legal basis, purposes of processing, the controller, terms of processing and storage period, etc. (Article 14(7). Accordingly, the controller must, upon request of the data subject, inform him/her of processing of personal data (Article 18(1)). However, the Russian PNR system does not provide any specific rules about the air passenger notification. Clearly, the general rules obliging the controller to provide data "upon request of the data subject"

---

[19] FRA. *Twelve operational fundamental rights considerations for law enforcement when processing Passenger Name Record (PNR) data* (2014).

[20] Stating that information must be provided by the states and by airlines, or their agents at the time of booking a flight and at the time of purchase of a ticket in a clear and precise manner about the transmission of data, the purposes of their processing, the period of data retention, their possible use, the possibility of exchanging and sharing such data and their data protection rights, in particular the right to complain to the competent national DPA.

[21] Federal law of 27 July 2006 N 152-FZ On personal data.

cannot ensure proper notification of every individual involved. This constitutes a weakness in comparison with the ICAO and the EU's recommendations and the EU-US scheme. The legislation should oblige the authorities to ensure that the passengers are informed about the data processing at the earlier stages mentioned above.

In addition, in all the systems, exhaustive lists of databases against which PNR data may be run are absent,[22] as well as the list of authorities that might receive PNR. This constitutes an unfavourable factor for transparency too. Moreover, in the EU Directive, the existence of "General remarks"[23] in the list of PNR data, specifically, the wording "such as" makes it very broad, without limitation.[24]

Finally, although the discussed undertaking seems to be reasonable at least in some jurisdictions and may allow passengers to know about use, storage and transfer of their data, there is no information on how the system will assess the passengers with regard to profiling techniques.[25] In other words, passengers will not know about the assessment criteria and whether or not they have been chosen by the system for further screening.[26] This falls within a broader term "algorithmic transparency", which apart from the use of PNR for profiling concerns any profiling technique (see Chap. 6), will be discussed in section below with reference to aviation security technologies in general.

But overall, the information on PNR provided to passengers is insufficient and lacks very important items which, as discussed in Chap. 6, may have the most serious impact on privacy/data protection.

7.2.2.5 Algorithmic Transparency

As discussed in Chap. 6, profiling can be a part of all the selected measures: scans, if a person is chosen to be scanned on certain criteria; CCTV's behaviour analysis, and PNR. One of the key problems concerning profiling is the lack of transparency, in particular, on algorithms used to process information and make automated decisions on individuals. With this regard, and also due to the use of algorithmic decisions in an increasingly wider range of applications (the rise of a "black box" society[27]), there is a call for "algorithmic transparency" of any secret government profiling systems, implying that data processes that impact individuals must be made public.[28]

[22] FRA (2011b).

[23] According to the EU PNR Directive Annex 1, data field 12 is "General remarks (including all available information on unaccompanied minors under 18 years, such as name and gender of the minor, age, language(s) spoken, name and contact details of guardian on departure and relationship to the minor, name and contact details of guardian on arrival and relationship to the minor, departure and arrival agent)" (see Annex 3 below).

[24] FRA (2011b).

[25] Article 29 Data Protection Working Party (2011).

[26] FRA (2011b).

[27] Pasquale (2015). The Black Box Society: The Secret Algorithms That Control Money and Information. Harvard University Press, 1 edition edition.

[28] EPIC Alert Volume 22.23, 11 Dec 2015 https://www.epic.org/alert/epic_alert_22.23.html

Another proposal is the concept "transparency by design". According to it, the legal duty to be transparent should be integrated into the design of algorithms, e.g. by providing the persons with the option to verify certain data about themselves and check whether automated decision processes are indeed correct.[29] However, this overlays with the principle of data subject participation discussed below – and, as it will be noted, several problems appear.

There are some positive developments. The GDPR requires the controller to provide the data subject with information about the existence of profiling, and meaningful information about the logic involved, as well as the significance and the envisaged consequences of such processing for the data subject.[30] This concerns both cases where personal data are collected from the data subject and cases where personal data have not been obtained from the data subject.

However, one may wonder what the requirement of "meaningful information about the logic involved" means, and what is needed to explain an algorithm's decision.[31] It is noted that "correlations stand for a probability that things will turn out the same in the future. What they do not reveal is why this should be the case".[32] As the result, the nature of the requirement of the increased transparency in algorithmic decision-making is not very clear. Burrell notes three barriers to transparency[33]: intentional concealment by decision makers; gaps in technical literacy; and a "mismatch between the mathematical optimization in high-dimensionality characteristic of machine learning and the demands of human-scale reasoning and styles of interpretation".

While the GDPR Articles 13 and 14 are meant to deal with the first barrier (the controllers must inform data subjects), Article 12 – with the second one (requiring to provide information "in a concise, transparent, intelligible and easily accessible form, using clear and plain language"), there are challenges associated with the third barrier – particularly relevant to algorithmic selection and design.[34] It is proposed that any adequate explanation would at least provide an account of how input features relate to predictions.[35] In the aviation security, this would require to answer such questions as: Is the model more or less likely to require secondary screening if the individual is a minority? Which characteristics play the largest role in prediction?, etc.

In any case, the assessment of air passengers raises the question of predictability. It would be preferable to have predefined risk indicators and to check the relevance of individual results of the assessments on a case-by-case basis and by a person.[36]

---

[29] Schermer (2011) p. 51.

[30] Articles 13(2(f)) and 14(2(g)).

[31] Goodman and Flaxman (2016). EU regulations on algorithmic decision-making and a "right to explanation". p. 6

[32] Hildebrandt (2008). Defining profiling: a new type of knowledge? Springer. p. 1745.

[33] Burrell (2016). How the machine "thinks": Understanding opacity in machine learning algorithms. Big Data & Society, 3(1).

[34] Goodman and Flaxman (2016) p. 6.

[35] Goodman and Flaxman (2016) p. 6.

[36] Consultative Committee of the Convention for the protection of individuals with regard to automatic processing of personal data (T-PD). Opinion on the Data protection implications of the processing of Passenger Name Records. 19 August 2016, p. 8.

However, the challenge is that predictive algorithms use dynamic criteria which are able of self-learning and constantly evolve.

In summary, in general, although there are some attempts to solve the algorithmic transparency issues, and there are beliefs that if properly applied, algorithms can offer increased transparency over their human counterparts, the challenges are still subject to further research.[37]

In the meantime, with reference to any of the selected techniques, there are no specific legal rules describing the algorithms used. Even if we apply the GDPR provisions to these technologies, they address only two of three barriers to transparency; moreover, it is not clear whether (and how) the GDPR requirements be fulfilled and enforced. This allows to conclude that today, there is a problem of algorithmic transparency with respect of all the selected measures.

7.2.2.6 Conclusion

Overall, today, the situation concerning information about aviation security is changing, since many states are trying to adopt rules and procedures according to transparency principle. Positive developments are provision of information with supporting pictures near the screening points, including body scanners' sample images, clear and visible signs about CCTV at the airports, information at airlines, airports, security organs and other entities' websites and in press, etc.

However, as discovered, the information provided may be incomplete, inadequate, or controversial. As the result, despite increasing passenger awareness of the need for tight security controls, people know little about the details of aviation security measures, including information about the effects on and protection of their rights. It is reported that the passengers quite simply do not know what to expect from airport security.[38] The majority of people know that CCTV at airports is in place but have no idea of how exactly they work and which capabilities are used. Not many people have a clear idea about the nature of body scanners or where their personal details released while booking a flight can appear later. Thus, passengers are subject to aviation security technologies whose use and operation they do not fully know or understand.

To some extent, the unawareness can be justified by the need for security. As discussed above, precise details and equipment specifications on aviation security cannot be made publicly available without jeopardizing security because this would enable criminals to ascertain the capabilities and limitations of the devices and make it easier for them to plan their acts.

At the same time, the nature of aviation security implies a power imbalance. As discussed in Chap. 4, it is the states and security organs who decide how aviation security measures will be used, the safeguards to employ and how much information on these will be shared with the public. As a result, it may be the case that the

---

[37] See more detail in Goodman and Flaxman (2016) pp. 6–7.

[38] Ghee (2015).

lack of information or uncertainty is caused by privacy/data protection/other human rights concerns rather than by security needs[39] and is instead intended to hide a respective problem or violation.

In other words, in order to avoid human rights problems, the easiest way may be simply to conceal information on what really happens: the less you know, the less you worry. This concerns multiple issues: from controversial questions about scanned images and saving options to use of PNR for profiling, CCTV's enhanced capabilities, combined databases, integrated technologies, Big Data and other issues discussed in Chap. 6.

Thus, while the lack of transparency can partly be justified due to security needs, in the case of concealing violations, it cannot be justified.[40] The question that emerges is how to differentiate the one from the other, and here again, a secrecy issue can be a problem. This leads to the rhetorical question as to how one controls the controllers. In this situation, full transparency can hardly be achieved.

Nevertheless, taking into account the justifiable limitations due to security needs, individuals must be provided with as open and adequate information as possible. Ideally, complete and reliable information regarding the functioning and technical specifications of the devices/capabilities of the technologies, reviewed and examined by competent authorities, including independent reviewers, should be provided and made publicly available in an appropriate form. It might also be recommended that procedural requirements be improved regarding details about how the information can be provided to the passengers in reality, including responsible persons/entities. Thus, there is much room for improvement in aviation security with regard to transparency.

7.2.3 In Accordance with the Law

With regard to the legal basis in law internationally, the ICAO, which sets out the basis for aviation security, currently does not require member states to use body scanners, CCTV or PNR. At the same time, some general international guidelines may be given, but this concerns mainly PNR (see below). As discussed in Chap. 4, some common aviation security requirements are provided at the level of the EU (obligatory for EU/EEA Member States, i.e. the UK and Norway). For the other selected states, the usage or non-usage of the selected measures is based on the decision of a particular state. Below, "in accordance with the law" will be considered measure by measure. After that, more general judicial/administrative protection with reference to all the selected measures will be analysed.

---

[39] Courtney (2012) p. 60.

[40] Brodeur. *Introduction*. In: Canadian Journal of Criminology and Criminal Justice. Vol. 48 (2006). p. 330.

7.2.3.1 Scans

As for scans, in the EU, as mentioned above, the Commission Regulation No 1141/2011 permits the use of scanners which do not use ionising radiation, i.e. MMW scans. EU/EAA Member States and airports do not have an obligation to deploy security scanners, but if they decide to use them, they must comply with the operational conditions and performance standards set at European level. It is up to the state to use them individually or in combination, as a primary or secondary means.

Nevertheless, the EU indirectly presses the Member States to use the devices. The rule is: if no scanners are used, other means should provide an adequate level of security. These other means may be more expensive and demanding, difficult to adapt for airport operations, etc., thus, scans can be preferable. This was the main reason why scanners are planned to be used in Norway.[41] Moreover – and this concerns states outside Europe – there is pressure from the USA, which, as discussed in Chap. 4, has global influence in aviation security, and endeavours to assist in providing adequate security in foreign international airports.

In the UK, a further document on scans is the Code of practice which contains privacy safeguards.[42] In contrast to previous practice which did not provide an alternative method of screening,[43] passenger may opt out of being scanned; an alternative method includes at least an enhanced hand search in private.[44]

In the USA, regulation on scans includes TSA's security rules and DHS' Privacy Impact Assessment (PIA) on body scanners. Pursuant case law,[45] the scans are installed and operated lawfully. The courts considered them as administrative searchers and not in violation of the Fourth Amendment, the Video Voyeurism Prevention Act, and the Privacy Act.[46] The Court decided that the TSA should provide notice and invite comments on the use of the technology for primary screening, however, it was not required to stop using the devices: "[D]ue to the obvious need for the TSA to continue its airport security operations without interruption, we remand the rule to the TSA but do not vacate it…"[47]

Nevertheless, backscatter scanners were removed from US airports in 2013 and replaced by MMW. According to EPIC, the TSA removed backscatter scans only following the EPIC lawsuit.[48] In turn, the TSA explained that the decision came

---

[41] Interview with Avinor representative, 22 Oct 2015.

[42] UK Department for Transport (2015a).

[43] UK Department for Transport (2013).

[44] UK Department for Transport (2015a).

[45] *EPIC v. DHS*, 653 F.3d 1 (D.C. Cir. 2011), 15 July 2011.

[46] For more details about the case, see Gusella. *Violating Privacy in Private: How Epic v. DHS Creates an Impossible Burden on Plaintiffs Trying to Demonstrate a Privacy Act Violation*. In: Boston College Law Review. Vol. 53 (2012).

[47] *EPIC v. DHS*, at 8.

[48] EPIC news, 13 Jan 2014. EPIC Settles FOIA Case, Obtains Body Scanner Radiation Fact Sheets. See also: EPIC. FOIA Documents on Body Scanner Radiation (Jan 2014) http://epic.org/redirect/011314-epic-radiation-foia.html

after the supplier of the backscatter machines did not meet a deadline to install a technological fix that would show only the generic image.[49] Apparently, actual removal was caused by a number of factors, including availability of less questionable and more privacy-friendly alternatives in the form of MMW.

In Russia, the legal basis for scanning is contained in the very general Order of the Ministry of Transport of 25 July 2007 N 104 "On approval of rules for pre- and post-flight inspections" which describes very broadly technologies and special equipment used for inspection at airports. Apart from general data protection law, I did not find any more detailed regulation or soft law containing respective privacy/data protection safeguards in respect of scans.

In summary, all the considered jurisdictions have some legal basis for body scanners. In addition, the EU, UK and USA provide special privacy safeguards for scans. However, as discussed in the previous section, the laws do not cover all the details about the scans (at least those items recommended by Article 29 Working Party, EDPS and FRA). This means that there is lack of legal transparency. Consequently, the requirements of accessibility, foreseeability and specifically, precisely and clearly defined law are not always satisfied.

7.2.3.2 CCTV

In general, a common EU legislative framework on the use of CCTV as such and on its use in the EU airports is non-existent. Technology-specific soft law instruments include Article 29 Working Party Opinion on the processing of personal data by means of video surveillance (2004) and Council of Europe's guiding principles for video surveillance (2003).[50] Since members of Council of Europe include non-EU Member States (e.g. Russia), the latter instrument has the potential to play an important role in the development of national principles for CCTV outside Europe. However, they are recommendations only, and it is up to the states to enforce them nationally. Thus, national law applies. In addition, these recommendations are already outdated since they do not address some new CCTV features. In some EU countries, national DPA issue also opinions, guidelines or codes of conduct on CCTV,[51] which contribute to lawful use of CCTV, e.g. the UK and Norway discussed below, but they face challenges with regard to new features too.

The DPD provides procedures of licensing or notification schemes[52] which may apply to CCTV systems, but they are described very broadly. In some EU/EEA countries, CCTV at airports is subject to licensing – prior approval from DPA or other authorized authority (e.g. the UK, Denmark, France and Spain[53]), or prior notification (Norway). The latter procedure is also in effect in Russia.

---

[49] Tirosh and Birnhack (2013) p. 1264.

[50] European Committee on Legal Co-operation (2003).

[51] Article 29 Data Protection Working Party (2004).

[52] DPD Articles 18–20.

[53] Agustina and Clavell (2011) p. 171.

In the UK, CCTV systems can be subject to a license by Security Industry Authority (SIA) pursuant Private Security Industry Act 2001, which requires that members of the private security industry must be licensed. It is a criminal offence to carry out licensable activities without an SIA license.[54] Also, conditions may be attached such as where the camera is placed, how it is monitored, etc.

The UK DPA 1998 is designed to cover the processing of CCTV image data, but it is not specifically designed to deal with sophisticated data-matching processes carried out on a large scale.[55] In 2011, it was reported that although there were restrictions on the use of public area CCTV cameras in the UK, the most were not operated in compliance with requirements of the DPA.[56]

In order to rectify the situation, a Surveillance Camera Code of Practice (2013) and a CCTV Code of Practice (2014)[57] were issued. CCTV Code 2014 sets out "good practice advice for those involved in operating CCTV and other surveillance camera devices". It is claimed to reflect the advances in surveillance technologies and their implementation, and explain the key data protection issues for CCTV operators.[58]

The Surveillance Camera Code came into effect after concerns over the potential for the abuse or misuse of surveillance by the state in public places, with appointment of a Surveillance Camera Commissioner to promote the code and review its operation and impact. He is appointed by the Home Secretary, but he is independent from government. According to the Surveillance Camera Commissioner, the code contains 12 guiding principles, which, if followed, will mean cameras are only ever used proportionately, transparently and effectively.[59]

However, compliance with the Code is voluntary and no enforcement or penalties are adopted for breach of its directions.[60] It is argued that the regulator's powers to enforce the rules are limited, and the surveillance camera commissioner must be given the powers and the resources to take action to ensure accountability.[61] Apparently, however, one commissioner cannot oversee the actions of every CCTV operator, including those at airports.

In Norway, the controller is required to notify the DPA prior to physically installing cameras[62]; hence, the CCTV systems at airports must be registered at the

[54] Home Office (2013).

[55] Colvin (2007).

[56] Privacy International (2011) p. 822.

[57] Full name: "In the picture: A data protection code of practice for surveillance cameras and personal information".

[58] Reeve. *Information Commissioner Releases New CCTV Code of Practice*. 15 October 2014. http://www.securitynewsdesk.com/information-commissioners-office-releases-new-cctv-code/

[59] Porter (2014).

[60] The SecurityLion. *New Surveillance Code of Practice "will not slow UK CCTV market growth"*. 21 August 2014. https://thesecuritylion.wordpress.com/2013/08/21/new-surveillance-code-of-practice-will-not-slow-uk-cctv-market-growth/

[61] BBC. Surveillance camera code of practice comes into force. 12 August 2013. http://www.bbc.com/news/uk-23636462.

[62] Norwegian Personal Data Act Section 37.

DPA. The latter also issues recommendations regarding use of CCTV in practice, containing different tips on how to operate CCTV lawfully.[63] However, many cameras were still installed without notification and DPA reported about 1150 complains on CCTV received in 2014.[64]

According to the US DHS, it has legal authority permitting and defining the collection of CCTV under 40 U.S.C. § 1315 to protect the buildings, grounds, and property owned, occupied or secured by the federal government, and the persons on the property.[65]

The DHS conducted a Privacy Impact Assessment (PIA).[66] However, this PIA concerns CCTV at DHS' facilities rather than CCTV in airports. I did not find any specific PIA for the latter. Nevertheless, some privacy provisions are contained in TSA's guidelines called "Recommended Security Guidelines for Airport Planning, Design and Construction" issued in 2011, which covers airport CCTV.

As the result of the fragmental protection of general US law and absence of specific law protecting privacy while operating CCTV, American privacy advocates claim that the DHS and TSA's practices are not satisfactory, and suggest a better regulation. For instance, EPIC suggested "EPIC Framework for Protecting Privacy and Civil Liberties If CCTV Systems Are Contemplated".[67]

In Russia, there is no separate law dedicated to regulation of CCTV. Regulation relating to certain legal aspects of CCTV is contained fragmentarily in codified acts (Civil Code, Labour Code, the Administrative Code[68]) and special legislation, such as the Personal Data Law,[69] law On Operational-Search Activities,[70] law On Private Detective and Security Activity in the Russian Federation,[71] etc. One of the Personal Data Act requirements is the notification of DPA (Article 22(1)) before any processing operations, i.e. before installation of CCTV system. DPA includes the notification information into the register of operators, which is publicly available. Information obtained from CCTV in the absence of such notice will not have legal effect.[72]

Consequently, camera surveillance in Russia is not covered by specific law, and there is no any CCTV Code of Practice; fragmental regulation and general Data

---

[63] Datatilsynet. *Kameraovervåking – hva er lov?* September 2015. http://www.datatilsynet.no/Global/04_veiledere/Kameraovervåkingsveileder.pdf

[64] Thon (2015).

[65] Department of Homeland Security (2012b). In addition, according to the DHS, the Physical Security Criteria for Federal Facilities, An Interagency Security Committee Standard, Report issued on April 12, 2010 and successor documents require CCTV monitors for the majority of federal buildings from low security requirements to very high.

[66] Department of Homeland Security (2012b).

[67] EPIC (2008).

[68] Civil Code of 30 November 1994 N 51-FZ, Labour Code of 30 December 2001 N 197-FZ, Administrative Code of 30 December 2001 N 195-FZ.

[69] Federal law of 27 July 2006 N 152-FZ On personal data.

[70] Federal law of 12 August 1995 N 144-FZ On operational-search activities.

[71] Federal law of 11 March 1992 N 2487-I On private detective and security activity in the Russian Federation.

[72] Art. 55(2) of the Civil Procedure Code, Art. 50(2) of the Constitution of the Russian Federation.

Protection Law provisions allow various legal interpretations depending on different interests. For instance, some privacy advocates argue that much of CCTV use is illegal.[73] CCTV producers and sellers publish "legal instructions" on their websites on how to install and use CCTV lawfully, focusing on notification to DPA, prohibition of usage of covert cameras as well as some other selected requirements.

In summary, the advantages of licensing or notification procedures are as follows: provision of ex ante control, transparency, contact between CCTV operators and DPA or other authorized authority, learning and sensory mechanisms.[74] These procedures could be especially valuable in cases of "smart" CCTV, in the event that every additional feature is declared in advance and registered.

The first problem is that currently, the prior authorization requirement is the exception rather than the rule. If the law does not require an operator to apply before setting up the camera, but simply notifies the authority about it, less control and transparency can be ensured.

Secondly, the procedures may lead to increased bureaucracy.[75] In order to avoid resource- and time-consuming procedures, many actors, especially smaller ones, may choose not to register cameras at all. This leads to a high percentage of unregistered cameras. For instance, FRA's report reveals that, in practice, CCTV cameras are often not registered and/or monitored in some EU Member States.[76] According to Norwegian Datatilsynet (DPA), there is a significant number of unregistered cameras in Norway.[77] The picture is similar in Russia.

The most critical question concerns regulation of CCTV's additional capabilities such as facial recognition and behaviour analysis discussed in Chap. 6, which may be in use in the UK, USA and Russia. The problem is that, as noted in Chap. 1, in aviation security, there is a tendency to deploy new technologies as soon as they are available. But, as ACLU warns, as long as there is no clear consensus about the limits of CCTV, it is "in danger of evolving into a surveillance monster".[78]

Ideally, every new CCTV feature should be evaluated and regulated before its introduction. For instance, the general rule is that if the EU Member States want to use a new type of technology, they must comply with all relevant EU regulation on privacy/data protection. But the concrete procedures are described very broadly/vaguely, like prior checking mechanisms DPD Art.20; recommendations of Article 29 Working Party are outdated; national laws differ greatly and are insufficient, etc.

Another example is that of Russia. With reference to the use of biometrics by CCTV, Personal Data Law Art.11(1) requires that, to process biometric data, the data controller must obtain the data subject's consent in writing. Article 11(2) provides exemptions, including cases provided for by the RF laws on defence, anti-

---

[73] See, e.g. Puzanov (2012) p. 10.

[74] Agustina and Clavell (2011) p. 171.

[75] Agustina and Clavell (2011) p. 171.

[76] FRA (2010a) p. 7.

[77] Aftenposten, 21 May 2014. http://www.aftenposten.no/nyheter/iriks/Minst-150-borettslag-og-sameier-har-kameraovervaking-7572162.html

[78] ACLU (2002).

terrorism, transport security, anti-corruption, etc. But none of them are applicable. The laws On Combating Terrorism and On Transport Security, which, theoretically, could regulate these issues in relation to transport security (see Chap. 4), do not contain any relevant provisions. Thus, for CCTV processing biometric data to be lawful under Personal Data Law the CCTV operator must obtain data subject's consent in writing, which is not done, hence, processing of biometric is unlawful.

Consequently, CCTV generally has a basis in law. Existing regulations, to different extents, satisfy the "accordance with the law" criteria; however, various deficiencies regarding accessibility, foreseeability, specific, precise and clearly defined law can be noted. The most critical point, however, is that current laws fail to provide a legal basis and sufficient safeguards to cover some of the more advanced technologies that can be used in "smart" CCTV and/or in combination with CCTV.

7.2.3.3 PNR

As discussed in Chap. 6, many states currently have established or are planning to establish PNR systems. Article 13 of the Chicago Convention stipulates that the laws and regulations of a state as to the admission to or departure from its territory of passengers shall be complied with by or on behalf of such passengers upon entrance into or departure from, or while within the territory of that state. Therefore, the state itself determines what information it requires from persons entering, departing or staying in this state.[79]

With the intent to harmonize PNR usage worldwide, the ICAO issued PNR Guidelines (2010) establishing uniform measures for PNR data transfer and the subsequent handling by the states; IATA – Recommended Practice PNRGOV.[80] The former suggest that states should not be overly restrictive concerning the transfer of PNR data by air carriers to foreign authorities, and states should ensure the protection of PNR.[81]

But the ICAO documents, as discussed in Chap. 4, are not binding on the states, thus, it is up to the latter to establish concrete requirements and guarantees. In reality, different states establish different and sometimes conflicting PNR demands, and full harmonization is not achieved. The problems include various data exchange requirements (e.g. formats and methods of transfer), requests for data elements beyond existing international standards; absence of common objectives and clear agreement on process among states.[82] As a result, air carriers may face legal, technical and financial problems. For the purposes of this research, however, the most relevant problems are privacy/data protection issues.

Apparently, different states provide, if any, different data protection guarantees and have different opportunities to enforce them in reality. The point is that in many jurisdictions, including the EU, the UK, Norway, and Russia, cross-border transfer

[79] ICAO PNR Guidelines (2010) §1.2.

[80] IATA Recommended Practice 1701a, 2011 (PNRGOV).

[81] §§2.12.1-3.

[82] PNRGOV.

of personal data to the countries lacking an adequate level of protection is prohibited.[83] Thus, in general, in the case of PNR, if an airline transfers passengers' personal data to a country lacking an adequate level of protection, it may violate national data protection legislation and may incur respective liability. But at the same time, if airlines do not provide PNR, they risk being grounded according to the laws of the state requesting PNR.

For instance, airlines flying to/from/over Russia may incur liability for non-compliance with the transport security requirements,[84] namely, administrative and criminal liability, depending on the consequences of the violation. If the carrier simply did not transfer the PNR data, the penalty is a fine or grounding of the aircraft.[85] If there are serious consequences of violation, e.g. large-scale damage, grave injury to human health, death of persons, then the carrier may incur criminal liability, including imprisonment up to 7 years.[86]

How to deal with this problem? Apparently, the most logical way is to enter into an agreement. This is currently reflected in international documents. The ICAO Guidelines on PNR stipulate that air carrier must comply with the laws of the state of departure and the state destination. If the laws of the state of departure do not allow an air carrier to comply with the requirements of the state of destination, both countries should settle the conflict of laws. Prior to the settlement, states are advised to apply no fines or other sanctions against air carriers taking into account the specific circumstances of the case.[87]

To avoid frustration and create a legal basis for the transfer, the EU followed the practice of concluding bilateral agreements with the states in question. Accordingly, it was sought to solve the problem of inadequacy by establishing, ensuring and guaranteeing an adequate level of data protection in the agreements.

The EU-US case was regulated by a series of bilateral EU-US PNR agreements.[88] The first PNR agreement between the EU and the USA was concluded in 2004.[89] It was then ruled invalid by the European Court of Justice,[90] and in 2006, an "Interim

[83] See Art. 25 of DPD and respective articles of national laws of the UK, Norway and Russia.

[84] Article 12(3) of the Law on Transport Security.

[85] Shadrina. *Will not go far: From July next year it will not be possible to buy a ticket for a single mode of transport without a passport.* In: Rossiyskaya Gazeta. 26 Sep 2012. http://www.rg.ru/2012/09/26/transp.html

[86] Article 263.1 of the RF Criminal Code.

[87] §§2.4.3-5.

[88] For an overview of EU-US PNR agreements 2004–2007, see Mironenko. *Air passenger data protection: Data transfer from the European Union to the United States* (2010).

[89] Agreement between the European Community and the USA on the Processing and Transfer of PNR Data by Air Carriers to the United States Department of Homeland Security and Bureau of Customs and Border Protection of 28 May 2004.

[90] ECJ Judgment of 30 May 2006 on joint cases C-317/04 *European Parliament v. Council of the European Union* and C-318/04 *European Parliament v. Commission* (OJ C 228 of 11 September 2004), §§ 61, 70.

Agreement"[91] was signed, followed by 2007 agreement.[92] Current agreement entered into force on 1 July 2012.[93] The EU concluded bilateral PNR agreement also with Canada in 2006[94] and with Australia (the first one of 2008[95] and a new one of 2011).[96]

The arrangements, having an international dimension, provoked widespread privacy and security concerns. The problem arose that EU PNR agreements were concluded on a case-by-case basis, and despite the fact that all the agreements addressed the same issues, the provisions were not identical, leading to different rules for air carriers and for data protection. Clearly, compromises were made due to political and commercial needs: flights must go on.

In 2010, in order to provide for an appropriate legal basis for PNR transfers to a third country, to harmonize the PNR transfer and establish common requirements, the European Commission published a strategy on the global approach to transfers of PNR to non-EU countries (the EU Strategy).[97] For the medium term, the EU Strategy declared that the EU should explore the possibility of replacing bilateral agreements by a multilateral agreement between all countries that use PNR data. For the longer term, if many more countries become involved with PNR, the Strategy declared the EU's aim to set these standards on an international level.[98] However, this strategy is a recommendation and a non-obligatory instrument. Another problem is that reaching agreement on PNR deals which would satisfy the EU data protection standards can take years especially for countries whose democratic standards and privacy laws may be questionable.[99]

[91] Agreement between the European Union and the United States of America on the processing and transfer of passenger name record (PNR) data by air carriers to the United States Department of Homeland Security, 2006 O.J. (L 298) 29. This agreement was valid until 31 July 2007.

[92] Agreement between the European Union and the United States of America on the processing and transfer of Passenger Name Record (PNR) data by air carriers to the United States Department of Homeland Security (DHS) of 29 June 2007. 4 August 2007. (L 204/18).

[93] Agreement between the United States of America and the European Union on the use and transfer of Passenger Name Records to the United States Department of Homeland Security Council of the EU (17434/11), adopted by Council 26 April 2012.

[94] Agreement between the European Community and the Government of Canada on the processing of Advance Passenger Information and Passenger Name Record data. 21 March 2006. (L 82/15).

[95] Agreement between the European Union and Australia on the processing and transfer of European Union-sourced passenger name record (PNR) data by air carriers to the Australian Customs Service, 8 August 2008. (L 213/51).

[96] Agreement between the European Union and Australia on the processing and transfer of Passenger Name Record (PNR) data by air carriers to the Australian Customs and Border Protection Service, 29 September 2011. (L 186/4, 14 July 2012).

[97] Communication from the Commission On the global approach to transfers of Passenger Name Record (PNR) data to third countries. Brussels, 21 September 2010, COM(2010) 492 final.

[98] Page 10 of Communication from the Commission On the global approach to transfers of Passenger Name Record (PNR) data to third countries. Brussels, 21 September 2010, COM(2010) 492 final.

[99] Statewatch news, 8 March 2015. http://www.statewatch.org/news/2015/mar/eu-mexico-agentina-pnr.htm

As for the existing agreements, on the one hand, formally, the agreements constitute legal basis for PNR transfer between the EU and USA/Canada/Australia, and state that they ensure an adequate level of data protection, e.g. Article 19 of EU-US Agreement 2012 states that DHS is "deemed to provide, within the meaning of relevant EU data protection law, an adequate level of protection for PNR processing and use".

On the other hand, the adequacy decisions of the EU do not necessarily mean that the regimes are adequate, as it was demonstrated in the *Schrems* case where the ECJ found "Safe Harbour" agreement on transferring personal data to the USA invalid (see Chap. 5). The data transferred to US companies within the "Safe Harbour" concerned personal data from the EU citizens. The types of data were critical to everyday commerce, including payment data.[100] In many points, it is the same as or equivalent to some of the data types covered by the PNR agreement, e.g. names, contact and payment details (see Annex 2 to this work). Thus, in case if no substantial changes to the US regime are made, it can be argued that the EU-US PNR regime is inadequate, due to similar reasons.

The inadequacy of PNR agreements also has some concrete argumentation. Data privacy advocates argue that these agreements, especially the EU-US one,[101] contain a number of serious deficiencies and disturbing points. Specifically, it is claimed that they fail to ensure an adequate level of data protection or proof that they are necessary and proportionate[102]; they raised concerns about purpose limitation, profiling, data retention periods, etc. (see analysis of data protection principles below). Thus, the legality of the EU-US PNR agreement is highly questionable.

The EU PNR proposal was first presented in the form of a Framework decision,[103] then in 2011 in the form of a Directive.[104] Although it is not mentioned, the EU-USA PNR scheme was used as a model. The status and prospects of the Directive were controversial. Specifically, in 2013, the EP Civil Liberties Committee (LIBE) rejected the proposal because it considered it to be disproportionate and privacy invasive. At the same time, before the legislation was agreed, the European Commission was funding national PNR implementations on an unsystematic basis.[105] In August 2014, the European Council called for finalization of work on the EU PNR proposal.[106] In July 2015, despite criticism, the EP approved a proposed

[100] Taylor. *It's Clear the U.S. and EU Economies Need a Safe Harbor 2.0.* ITI. 4 November 2015. https://www.itic.org/news-events/techwonk-blog/its-clear-the-us-and-eu-economies-need-a-safe-harbor-20.

[101] Agreements for the transmission of PNR between the EU and Australia and the EU and Canada are often argued to provide a higher level of protection than the EU-US deal.

[102] E.g. Article 29 Working Party on data protection: Letter to the Civil Liberties Committee of the European Parliament, Brussels, 6 January 2012. Ref. Ares(2012)15841 – 06/01/2012.

[103] Proposal for a Council Framework Decision on the use of Passenger Name Record (PNR) for law enforcement purposes 5618/2/09, 29 June 2009.

[104] Proposal for a Directive on use of PNR data for purposes of combating terrorism (Brussels, 2 Feb 2011, COM (2011) 32 final).

[105] European Parliamentary Research Service (2015).

[106] §18 of the Summit Conclusions, Special meeting of the European Council (30 August 2014).

text for the PNR Directive,[107] and the Directive was adopted on 27 April 2016.[108] It is a compromise version reached by the Commission, Council and the EP and is mandatory for EU/EEA states.

Article 14 states that Member States shall lay down rules on penalties, including financial penalties, against air carriers which do not transmit data or do not do so in the required format. Accordingly, airlines refusing to provide the requested data may face fines.

What is important, the EU PNR Strategy mentioned above does not discuss the EU scheme at all, concentrating on international agreements. And while the EU develops PNR system, the EU does not seem to be considering the implications of potential reciprocity requirements. As the Article 29 Working Party states, "A European PNR model could result in similar requirements being raised reciprocally by non-democratic countries or countries that do not provide an adequate level of protection of fundamental rights and freedoms, including personal data and privacy. It is clear that there could be serious consequences for individuals should such countries receive EU PNR data."[109] These cases are already in place. In addition, the disputable European PNR system may further weaken the EU's position on any negotiations on PNR, already weakened by accepting the EU-US terms.

An example of a country that requires PNR and from the EU view lacks an adequate level of protection is Russia. In 2007, the Ministry of Transport was required to create a unified state information system of transport security (USISTS), with automated centralized databases of personal data on passengers (ACDPDP) being its integrated part.[110] The corresponding provisions were incorporated into the Russian Air Code[111] and other regulations. However, only in 2012, the concrete provisions on ACDPDP were stipulated by an order of the Ministry of Transport,[112] and the system entered into force 1 December 2013.

It is noteworthy that the details on the PNR regime are contained in an order of executive body rather than law. As noted in Sect. 4.7.3, this may have both advantages and disadvantages. The latter include the risks of compromising the legality of the regimes and the concept of separation of powers, the risk of bureaucratic abuse of power, etc.

---

[107] Statewatch news, 16 July 2015. http://www.statewatch.org/news/2015/jul/eu-meps-agree-pnr.html

[108] Directive (EU) 2016/681 of the European Parliament and of the Council of 27 April 2016 on the use of passenger name record (PNR) data for the prevention, detection, investigation and prosecution of terrorist offences and serious crime.

[109] Article 29 Data Protection Working Party, Opinion 10/2011 on the proposal for a Directive of the European Parliament and of the Council on the use of passenger name record data for the prevention, detection, investigation and prosecution of terrorist offences and serious crime, 5 April 2011.

[110] Federal law of 9 February 2007 N 16-FZ On transport security, Article 11.

[111] Air Code of 19 March 1997 N 60-FZ, Article 85(1).

[112] Order of the Ministry of Transport of 19 July 2012 N 243 On approval of the formation and maintenance of automated centralized databases of personal data on passengers and personnel (crew) of vehicles, as well as providing the data contained therein.

Nevertheless, while the PNR regime formally has its legal basis in Russia, PNR transfer to Russia from other states, such as EU Member States, would be illegal since, as discussed, it is prohibited to transfer personal data to countries lacking an adequate level of data protection, and there is no bilateral agreement between Russia and the EU.

With respect to any PNR system, their legality can be questioned in the light of the ECJ's ruling of 2014 finding invalid the EU data retention directive (see Chap. 5). In the USA, similarly, with regard to data retention by telephone operators entailing further usage by law enforcement authorities and intelligence services, in 2015, the US Second Circuit Court of Appeals ruled that the NSA's telephone record collection program exceeds legal authority.[113]

In the ECJ case, one of the indicated problems was that the directive applied even to "persons for whom there is no evidence capable of suggesting that their conduct might have a link, even an indirect or remote one, with serious crime."[114] PNR regimes also apply to such persons.

In defence of the PNR system, the Commission points out differences between the two systems that should be taken into account in the legal assessment of any PNR system. They concern the category of persons whose data are processed, the frequency of the data collection and the nature of the collected data: (1) the collection of PNR from a person who takes an international flight and is therefore already registered for border control purposes; it thus reveals less detail, (2) PNR only concerns a more narrow set of persons (cf. §59 of the judgment), (3) the processing of PNR data does not risk interfering with any obligation of professional secrecy (§58 of the judgment), (4) there are substantive and procedural safeguards relating to the access and subsequent use of the data retained (§61 of the judgment), etc.[115]

Of course, data retention of all persons' electronic communications is more serious than collection of PNR, but in the light of data retention case, it is suggested that the regulators need to justify why the PNR regimes – massive, non-targeted and indiscriminate collection of data of individuals[116] is really needed.[117] And here, many other considerations, such as proportionality, data protections principles, should be taken into account.

The legality of any PNR regime can also be questioned in the light of the status of a new EU-Canada PNR agreement (pending). In September 2016, Paolo Mengozzi, advocate general of the ECJ, indicated that the EU cannot ratify it because a number of its provisions do not respect the EU Charter of Fundamental

[113] EPIC Alert, Volume 22.09, 13 May 2015 https://epic.org/alert/epic_alert_22.09.html

[114] *Digital Rights Ireland Ltd v. Minister for Communications, etc.* Joined Cases C-293/12 and C-594/12, 8 April 2014, §58.

[115] The EU Commission letter of 11 March 2015. Response to the resolution of the European Parliament of 11 February 2015 concerning the consequences of the judgment of the Court of Justice on the Data Retention Directive and its possible impact on the proposed EU Directive on Passenger Name Records (PNR).

[116] *Digital Rights Ireland Ltd v. Minister for Communications, etc.* Joined Cases C-293/12 and C-594/12, 8 April 2014, § 17.

[117] European Data Protection Supervisor *Counter-terrorism, De-Radicalisation and Foreign Fighters. Joint debate during the extraordinary meeting of the LIBE Committee* (2015a).

Rights.[118] Taking into account that the EU-Canada PNR agreement is considered the least privacy-intrusive of current PNR schemes, the conclusions in the opinion of advocate general can apply to any of them.

Further, there is another problem highlighted in connection with the *Schrems* case and common for any PNR regime. Today, there are large Global Distribution Systems (GDS) which book and sell tickets for multiple airlines. There are a number of major GDS platforms in the world. While Amadeus is based in the EU, such GDS as Sabre, Galileo and Worldspan are based in the USA. Many airlines and travel agencies are subscribed to the US GDS, even if they do not actually fly to the USA. Accordingly, if they collect information in the EU and enter it into a PNR system in the USA, there is a violation of the European data protection law.[119]

There are some radical undertakings aimed at prohibiting the storage of personal data abroad at all, for instance, a Russian so-called data localization law.[120] It requires that the data controller, while collecting personal data, ensures the recording, systematization, accumulation, storage, rectification (update, change) and extraction of Russian citizens' personal data using databases located in Russia. It is not yet clear how it will be possible to enforce this law. However, in November 2016, a court found that LinkedIn had failed to keep information about Russian citizens inside Russia, and Roskomnadzor blocked access to LinkedIn throughout Russia.[121] Theoretically, any airline using the above-mentioned GDS and transporting Russian citizens can be the next.

Finally, it is a question whether use of PNR for profiling is in accordance with law. This issue will be discussed in the purpose limitation section below.

7.2.3.4 Judicial Protection

Another requirement of "in accordance with law" is that the law must contain judicial protection against arbitrariness, including the right to complain and to compensation for damages. For instance, the EU Strategy on PNR contains a general requirement that every individual has the right to effective administrative and judicial redress. However, the concrete procedures of redress are supposed to be established on national levels.

---

[118] Opinion of advocate general Mengozzi, 8 Sept 2016. http://curia.europa.eu/juris/document/document.jsf?text=&docid=183140&pageIndex=0&doclang=EN&mode=lst&dir=&occ=first&part=1&cid=353089

[119] EDRi-gram 13.21, 4 Nov 2015 https://edri.org/edri-gram/13-21/

[120] Federal law of 21 July 2014 N 242-FZ On amendments to certain laws of the Russian Federation in order to clarify the procedure for personal data processing in information and telecommunications networks.

[121] The Telegraph. LinkedIn 'blocked in Russia' under controversial data law. 16 Nov 2016. http://www.telegraph.co.uk/news/2016/11/10/linkedin-blocked-in-russia-under-controversial-data-law/

As mentioned in Chap. 5, the following entities for complaining with reference to privacy/data protection issues usually exist: (i) airport, airline, aviation security provider and similar; (ii) national DPA or other relevant organs; and (iii) courts.

As discussed in Chap. 6, passengers may become subject to additional screenings, denial of boarding, arrest, deportation, appearing on a watch list, etc. as the result of using aviation security measures. Hence, redress would imply that they can complain and receive information about the processing of these data and the possibility to request access to and rectification of such data[122]; they have the right to redress in case of violation of data protection rights, including when no access is provided or inaccurate PNR data is not rectified or erased; the right to compensation in case of damage as a result of any other violation of individuals' rights.[123]

Of course, a passenger may complain already at the airport – to airport security department or aviation security provider, law enforcement agency who actually performs aviation security tasks or airline/operator who transfers data. But this means that these entities should have internal rules and mechanisms how complains from individuals should be addressed.

The UK Surveillance Camera Code provides some rules on this level regarding complaints. It requires that the CCTV operator have an effective procedure for handling concerns and complaints about the use of CCTV. Information about the procedures should be available to the public. Where a complaint is made and the complainant is not satisfied with the response, there should be an internal review mechanism in place using a person not involved in handling the initial complaint. Complaints must be handled in a timely fashion and complainants must be given an indication of how long a complaint at the outset will take to process. At the same time, the Surveillance Camera Commissioner has no statutory role in relation to the complaints, although operators should be prepared to share with him information about the nature of complaints on an ad hoc and anonymized basis to assist in review of the operation of the code of practice.[124] This generally means that no control under CCTV operators' complaining procedures exists.

Some UK airports have developed procedures on how to provide access to person's images (for a fee, see below), which can be considered a kind of redress mechanism. However, a problem appears that the period of time for redress for an individual is very short: in most cases, video is retained for no more than 6 months.

With regard to PNR, in most cases, the airlines provide general information about PNR transfer on their websites and advise the person in question to contact respective authorities such Border Control in the US.[125] For instance, passenger complaints with regard to EU-US PNR Agreement should be addressed to CBP or

[122] The person's rights to access, rectification and deletion will be discussed below.

[123] FRA (2014).

[124] Home Office (2013).

[125] British Airways. Government access to British Airways booking records. Answer Id 1621. Updated 15 July 2015. http://www.britishairways.com/travel/gsanswer/public/no_no (date accessed: 20 September 2015).

DHS Traveller Redress Inquiry Program.[126] However, although the agencies can receive complains, reports from privacy advocates show that most US federal agencies still ignore human rights complains.[127] Thus, in order to guarantee to EU citizens that complaints will be resolved, additional measures are required.

A positive development can be seen in the EU PNR Directive, stipulating that PIU should appoint a data protection officer; the Member States "shall ensure that a data subject has the right to contact the data protection officer, as a single point of contact, on all issues relating to the processing of that data subject's PNR data".[128]

But in general, although passengers have a right to complain to airlines, airports, security organs or other entities, the results depend greatly on the internal mechanisms established by these entities, which may not be sufficiently developed. In these cases, a person may further (or instead) contact specific authorities – DPAs and courts.

One of the tasks of DPAs is to handle complaints. The EU Strategy on PNR requires that independent DPA with effective powers of intervention and enforcement must exist to exercise supervision over those public authorities who use PNR data. They should be able to impose fines.

But as discussed in Chap. 2, statuses of DPAs differ greatly. They depend on the degree of development of privacy and data protection laws and their enforcement, DPAs' ability to impose fines, development of other civil institutions, as well as other factors. Accordingly, possibilities of administrative and judicial protection in the selected states vary, and different problems arise.

For instance, Russian regulators do not provide any additional redress guarantees with respect to scans, CCTV or PNR and rely on the general data protection law. The latter states that if the data subject believes that the data controller infringes his/her rights and liberties, he/she is entitled to contest the controller's actions or failure to act with the DPA or in court. The data subject has the right to protect his/her rights and legal interests, including the right to require compensation for losses and/or compensation for moral damage, in court (Article 17). But the Russian DPA, as discussed in Chap. 2, cannot be considered as a fully independent body either and in addition faces other difficulties: insufficient resources and personnel, it cannot initiate administrative proceedings, does not receive help from other organs, and the amount of fines is low. Thus, difficulties may appear.

National DPAs in the EU suffer different problems too: lack of independence, lack of competent staff with the technical competence or industry knowledge, lack of financial resources, limited powers.[129] Moreover, specific procedures on complains – and this may concern any aviation security measure – often lack detailed legal and administrative regulation; processes are time-consuming, expensive and not very effective. Some improvements are expected from the revised EU data pro-

---

[126] EU-US PNR Agreement Article 14.

[127] The Identity Project. *Most federal agencies still ignore human rights complains*. 6 November 2015. http://papersplease.org/wp/2015/11/06/most-federal-agencies-still-ignore-human-rights-complaints/

[128] Article 5(3).

[129] FRA (2010a) p. 8.

tection law. For instance, a new Directive on law enforcement ("Police Directive") provides that a person should have the right to a judicial remedy against both a controller or processor and decisions of supervisory organs.[130] Data subjects have the right to receive compensation from relevant entity for the damage suffered. The compensations must be determined by Member States. There are also penalties for non-compliance with the Directive.[131] These provisions may contribute to some harmonization of national rules.

Passenger complaints with regard to EU-US PNR Agreement, in the event that a complaint cannot be resolved by CBP or through the DHS program, the complaint may be directed to the CPO of the DHS.[132] In addition, the latter carries out independent review and oversight over compliance with the privacy safeguards.[133] But, as discussed in Chap. 2, in contrast to UK and Norwegian DPAs, the USA CPO DHS is not independent body.

Article 5(3-4) of the EU-US PNR Agreement provides notifications of affected individuals in the case of a privacy incident, and in the cases of "significant privacy incidents" involving PNR – notifications of relevant European authorities. The EDPS suggested that the recipients of the notification be clarified, to notify a competent US authority; to define what constitutes a "significant privacy incident"; to specify the content of the notification to individuals and to authorities.[134] Thus, further improvements can be undertaken.

FRA, with reference to the EU PNR regime, taking into account its experience with national DPAs, suggests to ensure guarantees of full independence of the national supervisory authorities, their ability to take action on their own initiative to protect proactively and effectively the interests of data subjects and ensure that they have sufficient resources to do so in practice.[135] However, I doubt that this can be done in practice via only wording revision, since much depends on national circumstances.

Problems arise as well on the level of applications to the court. Article 5(5) of the EU-US PNR Agreement states that administrative, civil, and criminal enforcement measures are available for privacy incidents under US law. Article 13 provides redress for individuals regardless of nationality, country of origin, or place of residence. But, as the EDPS noted, Article 21 states that the agreement "shall not create or confer, under US law, any right or benefit on any person", hence, even if a right to redress is granted in the US under the agreement, such right may not be equiva-

[130] Art. 51–52.

[131] Art. 54–55.

[132] DHS, CBP. U.S. Customs and Border Protection Passenger Name Record (PNR) Privacy Policy, 21 June 2013. https://www.cbp.gov/sites/default/files/documents/pnr_privacy.pdf

[133] EU-US PNR Agreement Article 14.

[134] European Data Protection Supervisor *Opinion on the Proposal for a Council Decision on the conclusion of the Agreement between the United States of America and the European Union on the use and transfer of Passenger Name Records to the United States* Department of Homeland Security (2011a).

[135] FRA (2011b).

lent to the right to redress in the EU.[136] This means that persons may be afforded weaker protection in comparison with the EU one.

Thus, in the USA, there may appear problems with reference to redress: for instance, as mentioned in Chap. 6, the judicial processes that attempted to get people off the no-fly list turned out to be lengthy, bureaucratic, and secretive. In 2013, reviewing the Agreement, the European Commission found the lack of redress mechanisms for passengers denied boarding, and advised to provide more transparency on the redress mechanisms which would allow passengers who are not US citizens or legal residents to challenge DHS decisions.[137] However, the DHS Privacy Report 2014 revealed that these recommendations are still not addressed.[138]

In Russia, formally, according to the principle of equality of individuals before the law, the right to administrative and judicial redress under the Russian law may apply for individuals regardless of race, origin, nationality, etc. However, it is unknown whether effective enforcement measures will be available for privacy incidents involving PNR, CCTV or scanners as long as there are problems with human rights enforcement in general. It is hence questionable if redress mechanisms correspond to the standards of the EU law. Accordingly, the problem of failure to provide the right to effective judicial redress may appear.

Similar problems arise in the EU too. For instance, in the UK, general weaknesses with data protection include the difficulty of claiming compensation for moral damage arising from misuse of personal data.[139] Regulator receives many complaints, showing that public awareness is high; but decisions have indicated a timidity and are often perceived as a "soft touch".[140] Moreover, one of the key concerns regarding profiling – which can be part of the selected aviation security measures – is the limits of redress from automated decision making in general.[141]

Consequently, it is a challenge in the meanwhile to ensure redress and compensations in the course of aviation security. Thus, the respective mechanisms should be further improved. The key factor is that the opportunity for administrative and judicial protection in the course of aviation security depends greatly on the administrative and judicial system in the particular state in general. Therefore, this requires a more complex approach, along with repair of existing more general gaps and problems.

---

[136] European Data Protection Supervisor (2011a).

[137] European Commission. *Report from the Commission to the European Parliament and the Council on the joint review of the implementation of the Agreement between the European Union and the United States of America on the processing and transfer of passenger name records to the United States Department of Homeland Security*. Brussels, 27 November 2013 COM(2013) 844 final.

[138] Department of Homeland Security Privacy Office. 2014 *Annual Report to Congress*. 30 September 2014. http://www.dhs.gov/sites/default/files/publications/dhs-privacy-office-2014-annual-report-FINAL.pdf

[139] European Commission, Press Release IP/10/811, 24 June 2010. http://europa.eu/rapid/press-release_IP-10-811_en.htm?locale=en

[140] Privacy International (2011).

[141] Schermer (2011) p. 51.

7.2.4 Legitimate Aim

The "legitimate aim" is one of the grounds for limitation of the right to privacy. Clearly, the aim of all the selected measures to safeguard civil aviation against acts of unlawful interference sounds reasonable; objective is legitimate and responds to "pressing social need", since it serves national security and safety.

As noted in Chap. 5, these objectives easily fall within the "legitimate aim" criteria, but security aims may be very wide in scope and may "mask" some other additional reasons and motivation. For instance, the rationale for collecting all passengers' images, i.e. innocent people, taken by CCTV with facial recognition in a database can go beyond the formal aim of aviation security; there are risks of unlawful access of the PNR database by the governments for purposes other than fighting terrorism; etc. In addition, even if the measures serve national security, the technologies are not predetermined, neutral, and fixed and this does not lead to their automatic legitimization.[142]

In this situation, when the requirement of legitimate aim can be formally satisfied for literally any aviation security measure in order to set the limits and substantiate the limitation, other criteria discussed below in this chapter should be considered along with the need to assess any excessiveness. In particular, they are the principles of purpose limitation and proportionality.

7.2.5 Conclusion

It can be seen that, apart from the legitimate aim criteria, which are formally satisfied by all the measures due to their broadness, the considered elements of the legality principle, to different extent, suffer.

As for transparency, the analysis of applicable regulations from the selected jurisdictions shows that positive developments take place. However, even taking into account the restricted information and security needs, passengers are not provided with sufficient information about the security measures, in particular, about the most intrusive features of the measures, combination of the technologies, use of profiling and different databases, etc. Transparency problems have direct impact on the requirements of the next criteria – "in accordance with the law", namely, accessible law and foreseeable interference, precise, specific and clearly defined law.

In summary, and this concerns all the discussed jurisdictions, the selected measures have some basis in law. However, a number of critical issues can be noted.

The discussion above reveals that on the EU and national levels as well there is a lack of detailed laws on the technologies. The most problematic issues in all jurisdictions concern CCTV, where the lack of regulation can be noted regarding general regulation and in particular – regarding CCTV's additional capabilities. The latter, as a rule, is not sufficiently regulated, thus, presents special concern with reference to legality.

[142] Tirosh and Birnhack (2013) p. 1305.

The situation with PNR is controversial as well, in particular in the light of the ECJ's Data Retention Directive decision (this concerns all the systems). In addition, taking into account *Schrems* decision, the legality of EU-US PNR scheme is particularly questionable due to inadequacy of the US general privacy regime. Moreover, any system storing PNR data in the USA (e.g. in USA-based GDS), can be considered violating the EU law. As for the Russian PNR system, formally, it would be illegal to transfer PNR from the EU to Russia. Another critical issue is whether it is lawfully to use PNR for profiling purposes.

But even if it is presumed that all the measures have basis in law, to be in accordance with law, presence of legal regulation is not enough. It should have a quality of law.

There is lack of technology-specific laws describing impact on privacy/data protection and providing particular safeguards. While the EU, UK and USA attempt to provide such law, at least via soft laws, Russian regulators rely mainly on general data protection law, the application of which can be problematic.[143] Overall, despite attempts in some states, the requirement of explicit, detailed provisions which are sufficiently clear, sufficiently foreseeable and meet the required degree of certainty with respect to their application is not thoroughly satisfied. In addition, the question arises as to whether or not the existing rules have the quality of law, specifically, whether the rules respect privacy and data protection principles.

If the technologies have impact on persons' rights to privacy and data protection, but sufficient safeguards are not provided, then the quality of law is questionable and the legality principle suffers, making a good argument that the introduction of body scanners/CCTV/PNR cannot be regarded as being in accordance with the law, because the regime under which this has been done fails to contain sufficient safeguards to protect fundamental rights. This will be analysed below with reference to other principles, which, in order to satisfy in accordance with the law criteria, should be respected all together.

Further, mechanisms of administrative and judicial protection are not sufficiently developed at all levels (aviation security providers, DPA and courts). The rights of passengers to redress may also suffer due to general national problems concerning DPA, courts and other elements of civil society.

In summary, there are many arguments that the regulations of contemporary aviation security measures fail to satisfy the legality principle. This will be further assessed taking into account additional principles too.

7.3 Proportionality

7.3.1 Introduction

As discussed in Chap. 5, the proportionality principle is the key principle for this research in a number of dimensions. This section will assess the proportionality of selected technologies in privacy/data protection terms.

---

[143] See sections on data protection principles below.

The following are some examples of how the regulator endeavours to prescribe the proportionality requirements. ICAO PNR Guidelines stipulate that there should be a reasonable balance between the need to protect PNR data and right of the state to require the disclosure of passenger data. Therefore, states should not be overly restrictive concerning the transfer of PNR data by air carriers to foreign authorities, and states should ensure the protection of PNR.[144] In its resolution on PNR, the EP stated that any new legislative instrument must be preceded by a proportionality test.[145] The UK Surveillance Code stipulates that "Decisions over the most appropriate technology should always take into account its potential to meet the stated purpose without unnecessary interference with the right to privacy and family life."[146]

However, these instruments lack any further explanations of concrete techniques, thus, the considerations elaborated in Chap. 5 will be applied, considering three criteria: suitability (effectiveness), necessity and excessiveness. Effectiveness was thoroughly discussed in Chap. 6, thus, the respective findings will be used.

The final question of the proportionality principle is whether the limitations on the rights of individuals are proportionate to the benefits of deploying body scans, CCTV in the airports, and PNR systems, i.e. to the threat which these measures are supposed to prevent. As noted in Chap. 5, many factors play role: privacy and data protection concerns (discussed in this chapter), other human rights concerns (Chap. 6), threats and risks in aviation (Chap. 4), security benefits of the measures (Chap. 6). Thus, the final question will be addressed in Sect. 7.10.

7.3.2 *Suitability (Effectiveness)*

The effectiveness issues were discussed in Chap. 6. In summary, there are different opinions on the effectiveness of scans, CCTV and PNR systems.

On the one hand, regulators and security experts believe that these measures are effective in protecting the travelling public, in particular if using a risk-based approach, including profiling techniques, if combining and interconnecting with other technologies and databases, intelligence gathering and so on. For instance, the most effective usage of PNR is for passenger profiling, "smart" CCTV presents huge opportunities for security purposes.

On the other hand, privacy advocates argue that the strongest arguments for effectiveness could be showing how effective they are in realizing their goals via appropriate numbers and statistics. An accurate use of statistics could remove doubts and contribute to a proper assessment of the effectiveness of security practices.[147]

---

[144] §§2.12.1-3.

[145] European Parliament resolution of 5 May 2010 on the launch of negotiations for Passenger Name Record (PNR) agreements with the United States, Australia and Canada.

[146] Home Office (2013).

[147] Vermeulen and Bellanova (2013) p. 311.

In the meantime, there are some statistics on how effective the use of the measures can be. There were cases when for instance PNR helped to catch dangerous criminals. While it is impossible to ensure 100% correct detection and 0% of failures, some failure rates can be acceptable, some of them can lead to disturbing many innocent people.

For instance, MMW fails to properly detect non-metallic weapons and explosives, confusing them with other objects; there are failures in profiling. While basic CCTV at airports can be considered as effective, thus, relevant and suitable, there are some questions regarding CCTV's Video Analytics. Although with great potential, today, there is underdevelopment of its particular features, such as facial recognition and algorithms of behaviour analysis. As discussed in Chap. 6, the most effective way to collect and analyse the data from intelligent CCTV is still to be developed. Thus, currently, some features can be excessive. This will be further discussed.

7.3.3 Necessity

7.3.3.1 Scans

As discussed in Chap. 5, the measure can be considered necessary if it has clear advantages compared with other available means and if other less intrusive methods are not available. This implies two questions: if the measures with equivalent effectiveness exist, and if yes, if they are less invasive than scans.

In 2008–2009, during the EU public consultation on scanners, while vast majority of responses regarding the proportionality of the scanners considered use of scanners as proportionate, some opponents believed that scanners were not proportional since "the current screening procedures are effective security measures" and since alternative existed.[148] Nevertheless, it was concluded that scanners offer a valuable alternative to comprehensive hand searches and metal detectors as a screening method.[149]

As discussed in Sect. 6.2, in aviation security, passenger screening can be conducted in several ways, for instance, in the EU, the allowed methods are as follows: (a) hand search; (b) walk-through metal detection equipment; (c) hand-held metal detection equipment; (d) explosive detection dogs; (e) explosive trace detection (ETD) equipment; (f) security scanners which do not use ionising radiation.[150] The question which arises is which of these methods or combination of methods may allow for effective detection similar to scans.

---

[148] Summary of the Replies to the Public Consultation on Body Scanners, 27 November 2008–19 February 2009. Annex IV to European Commission. *Impact Assessment on the possible use of security scanners at EU airports* of 23 March 2011.

[149] European Commission (2011), §135.

[150] Annex to Commission Regulation (EC) No 272/2009 of 2 April 2009 supplementing the common basic standards on civil aviation security.

Metal detectors are unable to detect non-metallic IEDs and weapons, as demonstrated by such attempted attacks as the "shoe bomber", episode with liquid explosives and the "underwear bomber" (see Chap. 4). As such, using metal detectors cannot be considered adequate alternative to scanners. Explosive trace detection cannot serve alternative either, since it is limited to those explosives which have traces.

The use of animals, mostly dogs, for illicit material detection has been common practice for decades, as their senses are highly developed and sensitive. It is scientifically proven that dogs are capable of detecting explosive and dangerous substances. For instance, Russian airline Aeroflot reported breakthrough results of scientific research involving its canine service, namely, the unique specially-trained Sulimov dogs, which is an integral part of aviation security.[151] Bio Explorers utilize mice, performing their training automatically, generating consistent results, and alarm indication is performed without the need for human-animal interface.[152] But, it is obvious that use of dogs or mice is not a matter of one minute, and it takes time and resources to establish such services universally.

Some other suggestions were made on what can be seen as alternative, e.g. behaviour profiling used in Israel discussed in Chap. 6, which did reduced terrorist threats in Israel, while there are no scanners there.[153] But, methods used in Israel, a country that has only one major international airport, can hardly be appropriately used in other, especially bigger, states. In addition, profiling as such cannot be considered as a method for detecting prohibited articles and its value more generally as a risk indicator.[154]

Of course, there may be future developments, e.g. some early-stage wide-band metal detection technology holding the promise of being able to improve detection capability and possibly provide the ability to detect even non-metallic objects.[155] Those under consideration include laser-based stand-off detection such as Surface Enhanced Raman Scattering, Quantum Cascade Lasers, Pulsed Laser Explosive Residue Detection, and Standoff Laser-Induced Breakdown Spectroscopy.[156] Promising technologies of screening people include quadrupole resonance.[157] But they are subject to future development, and it is not known whether they will be less intrusive than current MMW.

---

[151] Aeroflot News. *New scientific achievements of Aeroflot's canine service*. 15 May 2015. http://www.aeroflot.ru/cms/new/51033

[152] Amsterdam. *Mice in Portals: Incorporating Animals in Technology*. Body Search 2015 Conference Program, http://network.icom.museum/fileadmin/user_upload/minisites/icms/pdfs/Body_Search_SPEAKER_PROGRAMME.pdf

[153] For more details, see Taylor. *Touched by an Agent: Why the United States Should Look to the Rest of the World for a New Airport Security Scheme and Stop Using Full-Body Scanners*. In: Hous. J. Int'l L. Vol. 35 (2013) and Palmer (2011).

[154] Summary of 1st Body Scanner Task Force meeting, 12 December 2008. Annex IV to European Commission. *Impact Assessment on the possible use of security scanners at EU airports* of 23 March 2011.

[155] Neeman (2013) p. 13.

[156] *Ibid.* p. 15.

[157] Benda. *Quadrupole Resonance: Advantages & Possibilities*. Body Search 2015 Conference Program, http://network.icom.museum/fileadmin/user_upload/minisites/icms/pdfs/Body_Search_SPEAKER_PROGRAMME.pdf

Thus, at the current stage, it is commonly believed that comparable results to detect non-metallic items can be achieved by hand searches. Keeping in mind that MMW are not perfect at detecting non-metallic weapons and explosives, and cannot detect objects inside the body, there cannot be raised higher requirements to hand searches. Some security experts argue that metal detectors, combined with hand searches, are actually better at detecting weapons and explosives than the scanners.[158] But in any case, any other practical, adequate and universal alternatives to scans are not yet available.

As discussed in Chap. 6, hand searches are currently the commonly accepted alternative measure for passengers wishing to opt out from scanners. For instance, in Russia, passengers refusing to go through scan are subject to mandatory hand search.[159] At the same time, some regulators specify that not all types of hand search provide the same effectiveness. In the UK, it is stressed that only enhanced hand searches, i.e. full physical hand searches or strip-searches can be considered effective.[160] According to the UK Code of Practice, such enhanced hand search must take place in a private room or an area away from the main search comb and it may involve the loosening or removal of clothing.[161] Probably, only rules like that may establish adequate effectiveness. In practice, however, not all hand searches are carried out like this and can be performed more superficially, thereby reducing the effectiveness.

The problem is that even if carried out effectively, hand searches can be considered an adequate and affordable method under the proviso that only very few passengers opt out. If full hand searches are performed on all passengers instead of scans, it will require too many additional resources and much more time. Compared with scanners, which take 20 seconds to complete, hand searches take about 2 minutes. Moreover, the quality of hand searches may vary because of the pressure to screen large numbers of persons on a continuous basis, in particular at larger airports.[162] Another problem is privacy concerns prohibiting direct touching of private body areas as part of the hand search, while it is relatively easy to hide threats in sensitive body parts,[163] which can also reduce effectiveness.

Thus, although with probably similar effectiveness, hand searches are definitely less affordable and less suitable for airport security than scans. Nevertheless, the next question is what is less intrusive: scanners or hand searches. Evaluation of what is less intrusive and more effective – hand searchers or scanners – may lead to a choice of the optimal measure, "striking a better balance between security effectiveness and privacy concerns as well as passenger experience".[164]

---

158 Mick (2013).

159 Sheremetyevo. Pre-screening of passengers. http://svo.aero/flight-safety/passengers/

160 UK Department for Transport (2013).

161 UK Department for Transport (2015a).

162 European Commission (2011), §34.

163 Neeman (2013) p. 13.

164 *Ibid.*

The point is that hand searches can be intrusive. For instance, pat-downs carried out by the TSA employees are often considered by the public as intrusive and humiliating, as the officers are instructed to search intensively in every part of a human body, literarily. The EU Impact Assessment also notes that hand searches are considered by some passengers as intrusive and provoke concerns related to violation of dignity and privacy, and there were many passenger complaints related to medical and religious grounds in regard to screening by hand searches.[165]

According to the EP, a full hand search "causes more irritation and is more likely to be rejected than a scanner",[166] i.e. scans are considered as less invasive. The EU Impact Assessment states that trial results show that security scanners are perceived as a less intrusive method than full hand search, and can therefore reduce the possible discomfort for passengers.[167] In the USA, as mentioned above, according to the TSA, 99 per cent of the US passengers choose scans over hand searches. Of course, a part of the 1 per cent may choose the hand search on grounds other than privacy concerns, e.g. due to health concerns, but this may only prove that the percentage of passengers rejecting scans as more intrusive than hand search is even less. These facts seem to prove that body scanners are seen as less intrusive than hand searches. Another question is why a person has to choose at all. As Tirosh and Birnhack note, the choice between these methods is actually not a real free choice.[168] In the current situation, however, if neither method is chosen, the person will not be permitted to fly.

As mentioned in Chap. 6, scans have an impact on data protection, physical privacy, privacy of behaviour and action, and informational, as well as other human rights. But if it is presumed that the quality images are kept inside the machines only, it may be considered more privacy-friendly to be searched by a machine that cannot speak rather than a human, where in addition, a problem of touching intimate parts applies.

Keeping all this in mind, it can be presumed that scans are necessary, since there are currently no alternatives to scans which were of the same effectiveness and clearly less intrusive. Taking into account that with present regimes, only a very small number of passengers opt to have hand searches, considering it less intrusive, it would be unreasonable to use it for all travellers. This basically reflects the current approach: as mentioned above, the EU actually pressures Member States to use scanners, since providing alternatives is more demanding.

7.3.3.2 CCTV

Article 29 Working Party suggests that the deployment of CCTV should be limited to cases where alternative means and/or security measures prove clearly insufficient or inapplicable in view of the purposes of the processing.[169] One more Working

---

165 European Commission (2011).

166 European Parliament resolution of 6 July 2011 on aviation security, with a special focus on security scanners (2010/2154(INI)).

167 European Commission (2011).

168 Tirosh and Birnhack (2013) p. 1284.

169 Vermeulen and Bellanova (2013) p. 303.

Party's condition is that the use of CCTV has to be proportionate "to the prevention of concrete risks and specific offences – e.g., in premises that are exposed to such risks, or in connection with public events that are likely reasonably to result in such offences".[170] The Council of Europe also notes that the alternative measures should not involve disproportionate cost.[171]

Indeed, in general, CCTV is only one measure within a vast range of alternatives in preventing crime.[172] Other crime prevention strategies, requiring no image acquisition, include, for instance, more police officers, or more private security patrols, reinforced door or security grills,[173] installation of automatic gates and clearance devices, joint alarm systems, better and stronger lighting of streets at night etc.[174]

But these are general considerations. While it is possible to use the listed alternatives for some places, which are specific, or may incur less risk of terrorism and crime, the airports are in many parameters unique places, exposed to a high risk of terrorism (see Chap. 4). Moreover, the purposes of CCTV at airports include operational, security, safety and law enforcement purposes (Chap. 6).

The advantages of CCTV discussed above prove that this technology, for the use in the airports for the established purposes, is suitable. According to aviation security experts, effective use of CCTV technology can help achieve a level of security that is just not possible otherwise,[175] and call for further improvements for CCTV cameras in airports.[176] This leads to the conclusion that it would be very difficult, if not impossible, to find an alternative to CCTV in the airports; thus, they are necessary there.

7.3.3.3 PNR

It is in generally argued that PNR systems do not assess whether they are necessary, i.e. whether comparable results could be achieved with alternative, less privacy-intrusive measures.[177] Apart from PNR, there are many other systems that use personal data for security purposes and for the identification of suspects. In the EU, there are such measures as API Directive,[178] biometrics in passports, SIS,[179]

---

[170] *Ibid.*

[171] European Committee on Legal Co-operation (2003).

[172] Agustina and Clavell (2011) p. 169.

[173] Ratcliffe (2006) p. 21.

[174] Vermeulen and Bellanova (2013) p. 303.

[175] Medwecki (2013) p. 36.

[176] Center for Strategy and Evaluation Services. Aviation Security and Detection Systems – Case Study. January 2011.

[177] Article 29 Data Protection Working Party (2011) and European Data Protection Supervisor (2011b).

[178] Council Directive 2004/82/EC of 29 April 2004 on the obligation of carriers to communicate passenger data.

[179] Schengen Information System (SIS) is EU border-related database system established pursuant to Schengen Agreement of 14 June 1985. It allows national border control and judicial authorities

VIS[180] and national border protection schemes. In the USA, there are the Electronic System for Travel Authorization (ESTA),[181] the Visitor and Immigrant Status Indication Technology (US-VISIT) etc., and almost every year a new system is established that uses personal data and databases.

Hence, it is recommended that before introduction of a new system, there should be evaluation of the existing systems, their methods of cooperation and of possible security gaps. In order to deal with gaps, the existing mechanisms should be improved instead of introducing a whole new system.[182] Specifically, any substantiation of PNR should clarify why the existing instruments such as API and less intrusive alternatives are not sufficient to achieve the purpose pursued.[183]

With regard to API in the EU, for example, airlines are required to communicate API to law enforcement authorities from 2004. A similar US system is Secure Flight which conducts passenger watch list matching.[184] As discussed in Chap. 6, PNR data fields include API data, thus, a logical question is what would be the additional value of transfer additional categories of data – implying more intrusiveness.

The point is that due to a limited amount of data, API collection in general can be useful only to match the names of already known terrorists or suspects, while PNR allows identification of potentially dangerous persons who are not yet known. This entails the use of profiling techniques, which, if used along with intelligence, can be very effective (see Chap. 6). Thus, as mentioned above, security experts insist that the use of PNR is of a unique value, making PNR system necessary.

However, in the EU, a comprehensive evaluation of PNR in comparison with API, SIS or VIS has not been conducted, and less intrusive alternative tools have not been proposed.[185] I did not find any relevant studies proving the need of PNR systems in the USA or in Russia. Thus, although probably necessary, PNR systems lack formal proof of necessity.

---

to obtain information on persons or objects. In its new SIS II generation, the system will allow to check identities on the basis of biometric information.

[180] Visa Information System (VIS) is designed to hold biometric data to identify persons who have lodged a visa application for an EU Member State. See: Council Decision 2004/512/EC of 8 June 2004 establishing the Visa Information System (VIS); Regulation (EC) No 767/2008 of the European Parliament and of the Council of 9 July 2008 concerning the Visa Information System (VIS) and the exchange of data between Member States on short-stay visas (VIS Regulation).

[181] The Electronic System for Travel Authorization (ESTA) is an automated system that assists in determining eligibility to travel to the United States under the Visa Waiver Program (VWP) and whether such travel poses any law enforcement or security risk. Upon completion of an ESTA application, a VWP traveller is notified of his or her eligibility to travel to the United States under the VWP. See: CBP. ESTA. Frequently Asked Questions. https://esta.cbp.dhs.gov/esta/application.html?execution=e1s1

[182] Article 29 Data Protection Working Party (2011).

[183] Article 29 Data Protection Working Party (2015).

[184] TSA. Secure Flight Program. http://www.tsa.gov/stakeholders/secure-flight-program (date accessed: 7 August 2015).

[185] European Parliamentary Research Service (2015).

7.3.4 Non-excessiveness

Non-excessiveness, as discussed in Chap. 5, may refer to both a form of a particular measure and the manner of its executing. In general, the measure must not go further than is necessary to realize the goals; if a measure is nevertheless used, its scope must be limited to the absolute minimum. Many of the excessiveness issues overlap with such data protection principles as purpose limitation and minimality. These particular issues will be discussed in more detail in the respective sections below.

Moreover, excessiveness may refer to an additional feature of a measure as such, e.g. CCTV's facial recognition. In order to evaluate possible excessiveness, the feature's relations to all applicable principles should be considered, taking into account impact on other human rights too.

7.3.5 Conclusion

In order to evaluate the proportionality of the technology applications, several factors should be taken into consideration, including suitability (effectiveness), necessity and non-excessiveness.

As discussed in Chap. 6, all the measures are effective to some extent: definitely basic CCTV; "smart" CCTV is subject to further improvements; the use of body scans is probably effective due to failures in detection; PNR are considered effective, in particular if used for profiling.

All the measures are in general considered necessary, since apparently no equally effective and less intrusive practical methods can be substituted for them. However, additional proper substantiation, including statistics, can be helpful.

As for excessiveness, it will be assessed after discussing all the measures' concerns in relation to all relevant legal principles.

But the most important dimension of the proportionality principle for this research is whether the limitations on the rights of individuals are proportionate to the benefits of deploying body scans, CCTV in the airports, and PNR systems. This will be discussed at the end of this chapter, i.e. after all the legal principles are considered.

7.4 Purpose Limitation

7.4.1 Introduction

As discussed in Chap. 5, one of the fundamental principles of data protection is the purpose limitation principle which limits both purposes and use of data. Many other data protection principles depend and are connected with it. In this section, the following issues will be discussed: clear and defined purposes, domestic share and transborder share.

7.4.2 Scans

Images produced by body scanners, as noted by the EU's Impact Assessment, should only be used for aviation security purposes.[186] Specifically, as prescribed in the US PIA, they are used to identify prohibited items secreted on the body.[187] Indeed, the images are used to identify if there are any "abnormal" items on the body or under the clothes of a passenger.

If taking into account that images cannot be saved, transmitted or otherwise further used, as stated by all the states and aviation security organs, which means complete prohibition of any retention or transmission of the images made by the scanners in any format, then this principle is satisfied. The US PIA states that "Because there are no images to share, they cannot be used in any other context inside DHS or outside of the Department".[188] But, since some opinions maintain the opposite (see Chap. 6) the risks cannot be excluded (see also Data retention section).

7.4.3 CCTV

With reference to CCTV in the airports, the systems are usually installed to serve a wide range of purposes, including legitimate security purposes (see Chap. 6). While it is quite obvious that the specified, explicit and legitimate criteria are respected, the possibility of using data for new purposes in new contexts (function creep, which can be very relevant to any smart technology) is more questionable. Data sharing with third parties is a serious concern as well: while it is clear that CCTV images have potential for containing material which can become evidence for law enforcement purposes, it is not guaranteed that sharing will not exceed the established purposes.

For instance, the purpose of the UK Surveillance code is to ensure that individuals and wider communities have confidence that surveillance cameras are deployed to protect and support them, rather than spy on them.[189] The purpose limitation principle is the first principle in this Code, and additional requirements include that CCTV operators should always consider the requirements of the end user of the images, in particular if the purposes are prevention, detection and investigation of crime and the end user is the police and the criminal justice system. Any proposed extension to the purposes for which a CCTV system was established and images and information are collected should be subject to consultation before any decision is taken.[190] However, these wordings can be interpreted very broadly and it is not

[186] European Commission (2011), §136.

[187] Department of Homeland Security (2015) p.4.

[188] *Ibid*, p. 5.

[189] BBC. Surveillance camera code of practice comes into force. 12 August 2013. http://www.bbc.com/news/uk-23636462

[190] Home Office (2013).

clear what the "consultation" implies, who is consulted and who makes the decision, and how far such decisions can go. Consequently, the risks of function creep remain present.

Some rules can also be found in the DHS' PIA: the purpose of the CCTV system is to detect and respond to potentially unlawful activities in real time, to deter criminal activity, and to provide investigatory leads. The video feeds may also be used to support law enforcement investigations to the extent that they contain information relevant to a criminal (or potential criminal) activity[191]; DHS will use images to identify an individual or link an individual to a specific event or investigation.[192] The critical elements are the wide description of purposes and the possibility of using data for new purposes in new contexts.

Further, with reference to sharing the data, this can be done in several ways: by granting access to real time viewing and by sharing the already recorded images. Indeed, privacy risks for the first instance can be higher than for the second one.

Thus, with reference to real time viewing, the basic rules usually restrict monitoring to the CCTV operator and specific authorized persons.[193] However, in the case of Cardiff Airport, for instance, some other entities can view images in real time as well, since the CCTV system is linked to the police who maintain secondary monitoring facilities at their CCTV Control Room at Cardiff Airport but do not record images themselves, although they may obtain recordings from the airport.[194] Moreover, under exceptional circumstances, upon request, the police may assume direct control of the CCTV system (but in this case, the control room will be staffed and the equipment operated by only authorized and trained personnel); in very exceptional circumstances, e.g. a major/large scale incident, the police can request to take total control over the CCTV system, including the staffing.[195]

With reference to access to recorded data, the basic rule is that the recorded images should be viewed in a restricted area.[196] The UK Surveillance Code provides that access to images is restricted and there must be clearly defined rules on who can gain access and for what purpose; the disclosure of images and information should only take place when it is necessary for such a purpose or for law enforcement purposes.[197]

---

[191] "For example, if a suspicious package is placed outside a federal building that uses the system, the cameras will provide an image of this activity and allow DHS or local law enforcement to take appropriate responsive action. Additionally, if the package is determined to be an explosive device, the recordings could be used to further investigate this criminal activity, assist in identifying the suspects, and/or provide evidence that may be used in court". Department of Homeland Security (2012b).

[192] Department of Homeland Security (2012b).

[193] Information Commissioner's Office (2014). See also Data security section below.

[194] Cardiff Airport (2013).

[195] *Ibid.*

[196] Information Commissioner's Office (2014). See also Data security section below.

[197] Home Office (2013).

This Code also refers to exemptions provided by the UK legislation, which include cases where non-disclosure would be likely to prejudice the prevention and detection of crime, and for national security purposes. The operator should have policies and guidelines to deal with any requests for images disclosure, e.g. the images of other individuals can be obscured to prevent unwarranted identification, and it is the CCTV operator who decides about disclosure.[198]

The Cardiff Airport CCTV Code of Practice, based on this Code, contains further details: every movement and usage (processing) of CCTV images must be recorded; access to and use of recorded material will be strictly for the defined purposes and objectives; recorded material will not be copied, sold, released or used for commercial purposes or for the provision of entertainment.

Subject to compliance with the rules of this Code, material can be shown to witnesses, including police officers, for the purpose of obtaining identification evidence; members of the police service or other agency having a statutory authority to investigate and/or prosecute offences may release details of recorded material to the media only in an effort to identify alleged offenders and/or potential witnesses. The Code also allows the use of recorded material for the training and education of those involved in the operation and management of the CCTV system and for those involved in the investigation, prevention and detection of crime.

In the USA, airports, local law enforcement officers and the TSA have access to the footage.[199] The DHS PIA states that, consistent with the original collection, disclosure to federal, state and local law enforcement agencies must be for antiterrorism and law enforcement purposes only.[200] DHS has formal and informal agreements with these agencies in order to share information if cross-jurisdictional law enforcement is needed. These agreements specify the usage, retention period, access, dissemination and security of the system. DHS Component Privacy Offices must periodically audit these agreements to ensure accuracy and compliance.[201] Individuals who were victims of a crime, criminal defendants or members of the public through a FOIA request may request copies of the video.

In response to the privacy risk that there may occur unauthorized or inappropriate sharing of CCTV images outside of DHS, the DHS states that in most cases, external partners do not have real time access to the video feeds and may receive the video after it has been recorded which prevents inappropriate viewing of the public or specific locations by non-DHS users. External partners with access to DHS video feeds must follow specific memoranda of understanding or rules of behaviour. Requests for video must be approved by a supervisor and must be requested before the Digital Video Recorder overwrites the video.[202]

---

[198] *Ibid.*

[199] Elliott (2009).

[200] Department of Homeland Security (2012b).

[201] *Ibid.*

[202] *Ibid.*

But the wording "in most cases" does not exclude the access to real time video, and these exceptions are not defined. The Code of Cardiff looks better in this sense, referring at least to "exceptional circumstances".

In Russia, it is presumed that airports, aviation security and law enforcement agencies have access to video as well, both to real time and recorded. For instance, it is reported that in Pulkovo airport, CCTV images can be transferred over a network to any airport service, dispatching service, the police, aviation security service, and so on.[203] This is typical for CCTV systems in general: for instance, images from "smart" CCTV installed at Moscow metro station Okhotniy Ryad are transferred to the Interior Ministry, Federal Security Service and the Ministry of Emergency Situations. The computers provide biometric identification of each person captured by the camera using search databases of the Ministry of Interior.[204]

In summary, all the considered jurisdictions share images with police and other law enforcement organs for the defined purposes. An example of good practice is Cardiff Airport. Another question is whether the police/other organs may misuse the data after getting it – this will be difficult to control. If no technology-specific rules regarding transborder transfer are provided, general regulation on the latter applies. And here, again, the control during further use of images by foreign organs can hardly be enforced.

But, the most problematic issue is the existence of databases containing all CCTV images, integration of CCTV with intelligence gathering, and as a result – interconnection with different other databases as well as combined usage with other technologies (see Chap. 6). All this makes possible circulation of personal data, both coming from CCTV and from other sources, which can be very difficult to control.

7.4.4 PNR

PNR systems, which extensively use passenger personal data initially collected for commercial purposes – other than security purposes, at first sight, contravene the principle of purpose limitation. But here, a lawful exception applies: by the authority of law, data can be used for other purposes.

ICAO PNR Guidelines provides a list of purposes for PNR analysis in Sect. 2.2.2:

(a) improve aviation security;
(b) enhance national and border security;
(c) prevent and combat terrorist acts and related crimes and other serious crimes that are transnational in nature, including organized crime, and to enforce warrants and prevent flight from custody for such crimes;

---

[203] Sio.su. Building a system of video surveillance and access control at the airport Pulkovo. http://sio.su/down_asec_77_def.aspx

[204] Borogan and Soldatov (2011).

(d) protect the vital interests of passengers and the general public, including health;
(e) improve border control processing at airports;
(f) facilitate and safeguard legitimate passenger traffic.

It can be seen that the purposes are wide and apart from national, border and aviation security, include also f), which relates to aviation operations rather than security – however, this can contribute to security too.

According to the EU Strategy on PNR, the purposes for PNR data should include only law enforcement and security purposes to fight terrorism and serious transnational crime. Similarly, in the EU PNR Directive, the purposes of PNR collection are prevention, detection, investigation and prosecution of terrorist offences and serious crime.[205]

It looks like the purposes are narrower than those in ICAO's instrument, for instance, border control purposes are not mentioned. However, the key notions are terrorist offences and serious crime, which potentially may relate to any kind of national security, including aviation security, border security and so on.

According to the Directive Article 3(8), "terrorist offences" means the offences under national law referred to in Articles 1 to 4 of the Council Framework Decisions 2002/475/JHA on combating terrorism,[206] while "serious crime" encompasses the categories of offences listed in Annex II to the Directive; they are punishable by a custodial sentence or a detention order for a maximum period of at least 3 years under the national law of a Member State (Article 3(9)). The latter includes 26 types of various crimes. In general, the new wording regarding serious crime presents an improvement in comparison with the previous version of the Directive which referred to another instrument.

However, the definition of "terrorist offences" is not specified in the text but refers to another instrument, which may potentially be changed, thus creating extra risks for further extending of the aims. Specification in the text could provide more legal certainty.[207]

Moreover, the text leaves different possibilities to extend the aims of PNR processing; for example, via a wording that PNR processing is allowed for the purposes of "prevention, detection, investigation and prosecution of terrorist offences or serious crime" (Art. 1(2)) which extends the activities of national authorities during which PNR may be gathered and used. Article 29 Working Party notes that the list of crimes is long, broad and sometimes very vague, for instance, "computer-related crimes" (line 9) may cover not necessarily serious crime, and suggests reducing the list.[208]

In addition, Directive's Article 7 (5) allows the use of PNR for "other offences, or indications thereof", if they are detected in the course of enforcement action further to PNR processing. This means that PNR can be used with respect to crimes which have no relation to terrorism or serious crimes, including the minor ones.

---

[205] Art. 1(2).

[206] OJ L 164, 22 June 2002, p. 3.

[207] European Data Protection Supervisor (2010).

[208] Article 29 Data Protection Working Party (2015).

PNR data can also be used in order to identify persons who require further examination by the competent authorities and by Europol, in view of the fact that such persons may be involved in a terrorist offence or serious crime by comparing PNR data against databases relevant for the purpose of prevention, detection, investigation and prosecution of terrorist offences and serious crime, including databases on persons or objects sought or under alert, in accordance with Union, international and national rules applicable to such databases (Art. 6(3)(a)).

A question arises as to exactly which databases these are. In comparison with the previous version, the Directive describes the databases more precisely. Nevertheless, the description is still quite broad. It would be more specifically to indicate that the databases must be "set up for the purpose of" rather than "relevant for the purpose of", since the latter can be interpreted very broadly. It is also possible to add the databases to the list of competent authorities; the implementing legislation must provide the restrictions on the use of these databases.[209]

But the most disturbing point for privacy advocates is the use of PNR for profiling. As discussed in Chap. 6, profiling techniques may be very effective for security, but they produce false negatives and false positives, and traditionally raise serious concerns with fundamental rights.

Within the EU Directive, PNR data will be used to identify persons who were previously "unknown" but whom analysis of the data suggests may be involved in such crime – PNR data will be processed against pre-determined criteria, and PNR data will be used to develop profiles (Art. 6(2)(c)). This implies that PIU will conduct a "risk assessment" of the traveller, which could lead to questioning or even refusal of entry.

Art. 6(4) stresses that such assessment must be carried out in a non-discriminatory manner on the basis of assessment criteria established by PIU. The criteria must be targeted, proportionate and specific; it must be regularly reviewed in cooperation with the competent authorities and must in no circumstances be based on sensitive criteria.

Nevertheless, as EDPS notes, the use of such proactive measure – to take measures before a crime has been committed on the basis of risk indicators – in a law enforcement context is strictly defined and limited, but here, the notions of risk indicators and "risk assessment" are not sufficiently developed, and the latter may be confused with the notion of "profiling". The link between the original facts, and the patterns deduced from these facts is questionable. The main concern is the fact that decisions are taken using criteria established using the data of passengers in general, i.e. patterns deriving from the data of other individuals, and these decisions can greatly affect data subjects, while "It is extremely difficult for individuals to defend themselves against such decisions".[210] Overall, privacy advocates argue that PNR data use should be limited and under no circumstances may be used for profiling.[211]

---

[209] Article 29 Data Protection Working Party (2011).

[210] European Data Protection Supervisor (2010).

[211] European Parliament resolution of 5 May 2010 on the launch of negotiations for Passenger Name Record (PNR) agreements with the United States, Australia and Canada; EP Committee on

At the same time, the GDPR allows profiling via lawful exceptions,[212] which may include purposes of national security. According to the ACLU, "There are some security measures that are extremely intrusive and should only be used when there is good cause to suspect that an individual is a security risk".[213] If this logic is used, then, many truly effective methods, such as data profiling, may be used, despite their intrusiveness, according to proportionality principle. However, for evaluation of proportionality, as discussed, all factors should be taken into account. This question will further be addressed in the Conclusion section.

In the EU-US Agreement Article 4, PNR data are used to prevent, detect, investigate and prosecute terrorism and serious transnational crimes. Serious crimes are defined as crimes punishable by 3 years of imprisonment or more under US law. But the definition of transnational serious crime is very broad and covers all crimes where more than one jurisdiction is involved.[214]

Additionally, PNR may be used "on a case-by-case basis where necessary in view of a serious threat and for the protection of vital interests of any individual or if ordered by a court" as well as "to identify persons who would be subject to closer questioning or examination". This means that PNR can be used for other cases as well, e.g. minor immigration or customs offences, and may be used for profiling of passengers. As a result, data privacy advocates argue that the purpose limitation is too broad.[215]

In Russia, §63 of the Order stipulates that processing of passenger data in ACDPDP is carried out in accordance with Article 5(2) of Personal Data Law which provides that the processing of personal data should be limited to the achievement of specific, pre-defined, and legitimate purposes. The processing of personal data that is incompatible with the purpose of collection of personal data is not allowed. The purpose of PNR processing is "to implement measures to ensure transport security".[216]

From an EU perspective, it can be argued that the purposes are not indicated clearly or precisely. For example, no specification is made that the security purposes are restricted to combating terrorism and serious transnational crime. In practice, "measures to ensure transport security" can include a very broad category of activities, including profiling.

Moreover, different statements made by officials in the press may raise questions as well. For instance, according to the Deputy Head of the Ministry of Transport and

---

Civil Liberties, Justice and Home Affairs. Report on the proposal for a directive of the European Parliament and of the Council on the use of Passenger Name Record data for the prevention, detection, investigation and prosecution of terrorist offences and serious crime (COM(2011)0032 – C7-0039/2011 – 2011/0023(COD)). 29 April 2013.

[212] Art. 21 and 22.

[213] ACLU. Airport security: increased safety need not come at the expense of civil liberties. https://www.aclu.org/airport-securityincreased-safety-need-not-come-expense-civil-liberties

[214] Article 29 Data Protection Working Party *Letter to the Civil Liberties Committee of the European Parliament* (2012a).

[215] European Data Protection Supervisor (2011a).

[216] Article 11(1) of the Federal law of 9 February 2007 N 16-FZ On transport security.

Federal Service for the Oversight of Transport (Rostransnadzor),[217] although the main purpose of the database is transport security, probably, in the future, information from the database can be used for such non-security related cases as assistance to a passenger who has lost a ticket, or to recover damages from the carrier on request of the court.[218] Clearly, these purposes may ensure passengers' consumer rights, but go beyond the indicated purposes.

Another suggestion was made that the database should not be used for purposes unrelated to aviation security, that it must not allow law enforcement agencies to ban untargeted people, for instance those who avoid child support, etc. from the flight.[219] The rules of the Personal Data Law mentioned above prohibit the use of personal data incompatible with the purpose of collection, but it is unclear if the security organs follow without any exceptions, similar to the US case.

With regard to sharing of personal data, according to the ICAO, the state must take steps to ensure that every public authority having access to PNR must provide the appropriate level of data management and data protection.[220] As discussed, several types of share can be applicable for PNR:

(i) domestic share within the organization,
(ii) domestic share with other organizations/persons,
(iii) for the EU system – share between EU states,
(iv) transfer to other countries.

With reference to (i), the analysed systems provide technical and organizational security requirements which are aimed at preventing unauthorized access, disclosure or misuse of data (will be discussed in Sect. 7.8). In addition, the EU Directive prescribes that PIU should appoint a data protection officer for the internal supervision of data processing activities and implementing the related safeguards.[221]

In my opinion, additional requirements could be given based on the suggestions of FRA. It suggests that the PIU staff should be given the level of access to the data (e.g. analysts working only on drug-related crimes should not have access to the terrorism-related alerts); analysis of PNR should not be outsourced to private service providers; PIU staff must offer guarantees in terms of competence, integrity and respect of fundamental rights and prior to their assignment they should be trained.[222] Of course, if applied to any PNR system, these undertakings can improve the level of data management and data protection, but this implies additional costs and efforts.

With reference to (ii) the EU Strategy states that collected PNR data should be disclosed only to other government authorities with powers to combat terrorism and serious transnational crime and that afford the same protections as those afforded by

---

[217] Federal organ that will oversee the transfer of data to the database by transport companies.

[218] Shadrina (2012).

[219] Smirnov. *All the world has long been collecting the data this way*. Aviaport. 9 April 2007. http://www.aviaport.ru/digest/2007/04/09/118983.html

[220] ICAO PNR Guidelines (2010) §2.12.1.

[221] Art. 5.

[222] FRA (2014).

the recipient agency under the agreement in accordance with an undertaking to the latter. PNR data should never be disclosed in bulk but only on a case-by-case basis. FRA further suggests that the list of such authorities should be publicly available.[223]

According to the EU Directive Article 7, each Member State "shall adopt a list of the competent authorities entitled to request or receive PNR data or the result of the processing of PNR data from the PIU in order to examine that information further or to take appropriate action for the purpose of preventing, detecting, investigating and prosecuting terrorist offences and serious crime." Additionally, Europol can receive PNR data or the result of their processing within its competence. The authorities must be competent for the prevention, detection, investigation or prosecution of terrorist offences or serious crime. Each Member State must notify the list of the authorities to the Commission, with subsequent publishing the notification in the Official Journal of the European Union. If this works and the lists comprise exhaustive lists of bodies, it can be seen as a safeguard.

It is stressed that the PNR data and the result of the processing received by the PIU may be further processed by the competent authorities of the Member States only for the specific purpose of preventing, detecting, investigating or prosecuting terrorist offences or serious crime. However, this is "without prejudice to national law enforcement or judicial powers where other offences or indications thereof, are detected in the course of enforcement action further to such processing" (Art. 7(5)). The latter provision can potentially broaden the scope of data usage.

The EU-US Agreement provides provisions on domestic sharing in Article 16. In particular, the DHS may share PNR and analytical information obtained from PNR only pursuant to a careful assessment of such safeguards as consistence with the purposes stated in Article 4; receiving authorities must afford to PNR equivalent or comparable safeguards; and PNR share must be "under examination or investigation and pursuant to written understandings and US law on the exchange of information between domestic government authorities".

However, the agreement is not specific on how compliance with the safeguards can practically be ensured, particularly with respect to retention periods. The reference to the US law is not specific and it is unclear how this law may assist in enforcement. As Article 29 Working Party notes, the agreement does not provide that transfers shall be made on a case-by-case basis only.[224] The suggestion to include the case-by case requirement can be helpful as well as the EDPS's suggestion that the list of authorities that might receive PNR must be specified.[225] However, these will not solve the problem.

The problem, as noted in Sect. 2.3.3.3, is that the US Privacy Act applies only to personal data that is maintained in databases by federal agencies. If PNR data is accessed, but remains in airlines' databases, i.e. private sector databases, the Privacy Act will not apply. But even if the data is pulled by law enforcement and intelligence agencies, many key provisions of the Act will still not apply. For instance, Part 1507

[223] *Ibid.*

[224] Article 29 Data Protection Working Party (2012a).

[225] European Data Protection Supervisor (2011a).

of CFR implements provisions of the Privacy Act of 1974 that permit TSA to exempt any system of records within the agency from certain requirements of the Act.[226] In addition, information about a person is owned by the person who collects it, not by the data subject, with a few exceptions; enormous databases filled with personal data are owned by governments, marketing firms and so on.[227] All this may have effect on the enforcement of PNR agreement in general and the US domestic sharing in particular. It would be helpful if an exhaustive list of authorized agencies was provided and they were obliged to provide data protection safeguards.

In Russia, according to Article 11(4) of the Federal law On Transport Security, information resources of the USISTS are restricted information. The Order in Section 13 provides that federal executive bodies authorized by the Government of the Russian Federation to carry out functions in the field of transport security, the Russian Interior Ministry, and the FSB ("consumers of information") use the data contained in the ACDPDP.

But what actually are the "federal executive bodies authorized by the Government of the Russian Federation to carry out functions in the field of transport security"? Logically, it should be found in the Government's resolutions.

The system of aviation security organs was discussed and illustrated in Chap. 4, and multiple organs were presented. The development and implementation of the state policy in aviation security is fulfilled by the Federal Air Transport Agency.[228] Nevertheless, actual aviation security activities – providing measures to protect civil aviation against acts of unlawful interference – are performed by this agency in cooperation with the FSB, Ministry of the Interior, Ministry of Defence, Ministry of Foreign Affairs, and Federal Customs Service.[229]

In addition, according to the Program of Civil Aviation Security of the Russian Federation, the Ministry of Transport, the Ministry of Interior, the Federal Air Transport Agency, Federal Service for the Transport Oversight (Rostransnadzor) "in cooperation with interested federal executive bodies within their authority and in accordance with the legislation of the Russian Federation take measures for the protection of civil aviation against acts of unlawful interference".[230] It can be seen that the list of authorities can hardly be exhaustive.

Moreover, for other transport modes, additional organs may be relevant. Taking into account that the database is common to all transport modes and that all the organs authorized to carry out security functions on other transport modes (rail, sea, etc.) also will have access to the data, the scope of organs having access to the data is quite broad.

---

[226] Title 49 of the Code of Federal Regulations (CFR), Chapter XII, Subchapter A, Part 1507 (Privacy Act Exemptions). Section 1507.1 Scope.

[227] Schneier (2008) p. 78.

[228] Section 7, Resolution of Government of 30 July 1994 N 897 About Federal system of protection of civil aviation from acts of unlawful interference.

[229] *Ibid.* Section 8.

[230] Art. 19, Order of the Ministry of Transport of 18 April 2008 N 62 On approval of Program of civil aviation security of the Russian Federation.

According to the information of the Operator,[231] the organs authorized to use the data contained in the ACDPDP are limited to the Interior Ministry, FSB, and security department of the Ministry of Transport, i.e. not even the whole ministry, but a special department. A representative of the Ministry of Transport, in response to the author's request, noted that the list of organs authorized to access data from the ACDPDP is contained in the Order.[232] But this sends us to the start of the discussion.

It can be seen that the information is quite controversial. Apparently, the same problems as those indicated with reference to the EU-US scheme above may be relevant. It would be helpful if the regulator provided an exhaustive list of authorized agencies in respect of particularly the PNR system and obliged them to provide safeguards.

With reference to sharing between the EU states – this concerns the EU PNR system only – according to the Directive Art.9, with regard to persons identified by a PIU, the states must ensure that all relevant and necessary PNR data or the result of the processing is transmitted by that PIU to the PIU of the other member states. The receiving PIU shall transmit the data to their relevant competent authorities (Art. 9(1)).

The PIU may also request the PIU of another state to provide it with PNR data or the result of their processing. There should be a duly reasoned request for such data based on any one or a combination of data elements, as deemed necessary by the requesting PIU for a specific case of prevention, detection, investigation or prosecution of terrorist offences or serious crime. PIU must provide the requested data as soon as practicable (Art. 9(2)).

In cases of emergency only, the competent authorities may directly request the PIU of another state to provide them with PNR data (Art. 9(3)). Finally, exceptionally, if access to PNR data is necessary to respond to a specific and actual threat related to potential terrorist offences or serious crime, the PIU at any time has the right to request another PIU to obtain PNR data of flights landing at other points in time than usual (Art. 9(4).

Of course, it is a positive development that data sharing and its extent depend on the seriousness of each particular case. However, the wording leads to confusion: Art. 9(2) is applicable in a "specific case"; Art. 9(3) concerns "cases of emergency", and Art. 9(4) mentions "specific and actual threat related to terrorist offences or serious crimes". The difference between them is not clear and different threats allowing data share is not sufficiently defined. It is suggested that the conditions allowing transfers of data in each case be specified.[233] Meanwhile, it is quite obvious that in the security field, the authorities may tend to use "exceptional" cases on a broad basis.

With reference to transfers to other countries, the EU Strategy stipulates restrictions on use and further dissemination of PNR data to third country.[234] Such onward transfers shall be subject to appropriate safeguards. In particular, the receiving third country should transfer this information to a competent authority of another third country only

---

[231] Telephone conversation with the Operator's employee 4 July 2013.

[232] Letter of 5 August 2013 N 07-05-01/1277-is signed by Druzhinin A.A., Head of Department of Legal Support and Legislative Activities, Ministry of Transport of RF.

[233] European Data Protection Supervisor (2011b).

[234] Third country here means a country other than the contracting states.

if the latter undertakes to treat the data with the same level of protection as set out in the agreement and the transfer is strictly limited to the purposes of the original transfer of the data. PNR data should be disclosed only on a case-by-case basis.

The EU Directive's Article 11 states that member states will be permitted to transfer PNR data as well as the results of their processing to third countries on a case-by-case basis if:

- the conditions of Article 13 of Council Framework Decision 2008/977/JHA are fulfilled
- the transfer is necessary for the purposes specified in Article 1(2)
- the third country agrees to transfer the data to another third country only where it is strictly necessary for the purposes specified in Article 1(2) and only with the express authorization of the member state
- same conditions as those laid down in Article 9(2) are fulfilled.

However, in exceptional circumstances, transfers of PNR data without prior consent are allowed to respond to a specific and actual threat related to terrorist offences or serious crime of a Member State or a third country and prior consent cannot be obtained in good time. The responsible authority must be informed without delay and the transfer shall be duly recorded and subject to an ex post verification.

Additionally, it is stipulated that Member States transfer PNR data to competent authorities of third countries only upon ascertaining that the intended use of PNR by the recipients is consistent with all the terms and safeguards (Art.11(3)). The data protection officer of the state that has transferred the data must be informed too (Art.11(4)).

The EU-US Agreement provides rules on third countries transfer in Article 17(1). They refer to the terms of the agreement, but the latter does not specify how compliance with these terms can be ensured. Moreover, the agreement does not provide that transfers shall be made on a case-by-case basis only.[235]

In my opinion, the novelties of the EU Directive are quite promising and are a step forward in comparison with the EU-US scheme, at least as preparatory requirements before any transborder transfer. But questions remain: which authorities of third countries can obtain access to PNR, how PNR will be further used abroad, and how the latter can be controlled. In addition, third countries' authorities can forward PNR to other countries. The point is that purpose limitation also applies to all these transfers too.

Even if the third countries "promise" to follow all the rules and instructions, how the EU authorities may *enforce* the rules in foreign states – for example, prevent from further inconsistent use of PNR, determine which leverage they can use, etc. Even though some rules are provided in the wordings, e.g. penalties for the third country uses of data for something else, obligation to ensure that the onward transfer is proportionate, obligation to keep records of the transfer, as well as other obligations,[236] this can be difficult to ensure and control in practice. Apparently,

---

[235] Article 29 Data Protection Working Party (2012a).

[236] Amberhawk Training Limited. *A review of some important aspects of the EU-USA PNR agreement*, December 2011. http://amberhawk.typepad.com/files/eu-usa-pnr-deal-amberhawk-analysis.pdf

when the PNR data is already possessed by a third country's law enforcement authorities, they may use it according to their views and needs, including further transfer to other organs/other countries.

FRA suggests that data transfers to a third country should be possible either in the framework of existing EU agreements ensuring an adequate level of data protection or according to bilateral or multilateral agreements or other written understandings that ensure an adequate level of data protection.[237] The EDPS recommends that data transfers to third countries should be subject to prior judicial authorization; the DHS should not transfer the data to third countries unless they guarantee an equivalent level of protection.[238]

However, I consider that such requirements may not be very practical: the number of countries with "adequate protection" is low,[239] and the processes of concluding alternative agreements may take years.

The Russian Order does not contain any terms and provisions on transfers to other countries. The Personal Data Law Article 12 contains general rules: cross-border transfer of personal data to foreign countries that are parties to Convention 108, as well as to other foreign countries providing adequate data protection is carried out in accordance with this federal law, and may be prohibited or limited in order to protect the foundations of the constitutional system of the Russian Federation, morality, health, rights and lawful interests of citizens, national defence and state security. The list of foreign countries that are not parties to Convention 108 and provide adequate data protection is adopted by the authorized body (Roskomnadzor).

The Ministry of Transport did not provide any further details on the possibilities of further transfer of PNR referring to restricted information.[240] According to the information of the Operator, the PNR data will not be transferred from Russia to other countries.[241] No transfer means no problems similar to those indicated for the EU and the EU-US schemes. However, lack of concrete provisions does not constitute grounds for concluding that there definitely will be no transfer; additional legal guarantees are needed.

In summary, none of the considered PNR systems fully follows the purpose limitation principle, allows or may allow the use of profiling and unlimited or insufficiently limited data share on all levels. All the systems are subject to improvements in terms of data protection; however, not all improvement suggestions are very practical.

---

[237] FRA (2014).

[238] European Data Protection Supervisor (2011a).

[239] The European Commission has recognized Andorra, Argentina, Canada (commercial organisations), Faeroe Islands, Guernsey, Israel, Isle of Man, Jersey, New Zealand, Switzerland, Uruguay and the US Department of Commerce's Safe Harbour Privacy Principles as providing adequate protection. The decision on the latter, as noted above, was ruled invalid by the ECJ. Thus, the decisions on these states do not actually mean that the protection in these countries is adequate in practice.

[240] Letter to the author dated 5 August 2013 N 07-05-01/1277-is.

[241] Telephone conversation with the Operator's employee 4 July 2013.

7.4.5 Conclusion

For scans, provided the machines do not save personal data, the purpose limitation principle is respected, however, if they are saved, the risks remain.

For CCTV, the critical elements are the wide description of purposes and the possibility of the data being used for new purposes in new contexts. For instance, all the considered jurisdictions share images with police and other law enforcement organs for the defined purposes. An example of good practice is Cardiff Airport. Another question is whether the police/other organs will misuse the data after getting it – this will be difficult to control. No technology-specific rules regarding transborder transfer mean the general regulation will be applied. Here again, the control under further use of images by foreign organs can hardly be enforced.

Moreover, the most problematic issue is the existence of databases containing all CCTV images, integration of CCTV with intelligence gathering, and consequently – interconnection with different other databases as well as combined usage with other technologies. All this enables circulation of personal data, both coming from CCTV and from other sources, which can be very difficult to control.

In the EU PNR Directive, there are many improvements in comparison with the previous versions, in particular, regarding data share. However, some problems still remain, and there is lack of precise purpose and use limitation. The US and Russian systems share a similar concern: the purposes of transfer are established, but they are broad. All the systems allow or may allow the use of profiling as well as unlimited or insufficiently limited data share, in particular the US and Russian schemes, which present more serious weaknesses in comparison with the EU Directive.

Thus, there are problems with purpose limitation with respect to CCTV and PNR, while for scanners, probably no problems can be indicated provided no images are stored.

Moreover, if taking into account the enhanced global data sharing and combination of different technologies and data sources discussed in Sect. 6.8, the risks to purpose limitation may grow in geometric progression. They include unlimited and uncontrolled multiple databases with personal data on all individuals, including sensitive data, unlimited and uncontrolled data share and data retention, including data transfer across borders, etc.

7.5 Minimality

7.5.1 Introduction

As discussed in Chap. 5, the minimality principle requires that the amount of personal data collected should be limited at the initial stage. This is especially relevant for scans, which take images of all or some passengers, CCTV, where thousands of cameras record millions of peoples' images daily, and PNR systems, where the airlines collect huge amounts of passenger personal data which are then processed for security purposes, i.e. other than transportation needs. Thus, it is critical to indicate which of the data processed by scans, CCTV and PNR regimes is strictly needed and which data may be excessive.

In general, a number of undertakings types contributing to minimality can be noted: limitation of data collection wherever possible at the initial stage, rendering data anonymous and deletion of data immediately after it is used.

Limitation of data collection at the initial stage requires indicating which data is actually needed instead of collecting everything. Limitation can refer to several dimensions. It can be limiting of data subjects, i.e. evaluating whether it is necessary to scan, monitor and track all the passengers and collect PNR of all passengers instead of a more targeted approach, in particularly, in the light of the ECJ's ruling on the EU data retention directive, where the collection of electronic communication data of everybody instead of a more targeted approach was not considered to be proportionate, even for security needs.[242]

In addition, specific undertakings can contribute, such as limiting fields of collected personal data or amount of sources. An important contribution can be made via technological developments, serving as PETs. To limit data collection, the design of software may allow or prohibit certain operations that involve the collection of personal data. Examples of this include the use of "push" or "pull" method of PNR transfer; biometric identifiers can be used to store personal data in devices under control of the data subjects (i.e. smart cards) rather than in external data bases, etc.

Secondly, collected data should be rendered anonymous wherever possible. In general, data are said to be anonymous where the data subject is not identifiable. Accordingly, anonymization is a process by which identifying information is manipulated to make it difficult to identify data subjects.[243]

DPD Article 6 provides that personal data must be "kept in a form which permits identification of data subjects for no longer than is necessary for the purposes for which the data were collected or for which they are further processed". This implies that when the data are no longer necessary, they should be deleted or anonymized. Article 29 Working Party underlines the DPD encourages to have personal data anonymized "by default".[244]

The DPD states in Recital 26 that "the principles of protection shall not apply to data rendered anonymous," but it does not provide any further guidance, mentioning only that "codes of conduct may be a useful instrument for providing guidance as to the ways in which data may be rendered anonymous" (Recital 26). The Council of Europe, in contrast, stipulates in its recommendations that techniques and procedures ensuring the anonymity of data subjects should be developed by the controllers. For instance, personal data collected and processed for statistical purposes

[242] *Digital Rights Ireland Ltd v. Minister for Communications, etc.* Joined Cases C-293/12 and C-594/12, 8 April 2014. See also the Legality section above.

[243] Ohm (2010) p. 1707.

[244] Article 29 Data Protection Working Party *Opinion 05/2014 on Anonymisation Techniques* (2014b) p. 7.

should be made anonymous as soon as they are no longer necessary in an identifiable form, immediately after the end of data collection, checking or matching operations.[245]

If complete anonymity may not be appropriate because of legal constraints, and if it is permitted by law, pseudonyms may be used.[246] Pseudonymity is a compromise between anonymity and identified persons. Data is deemed to be pseudonymous if it cannot, without the use of additional information, be associated with a particular individual; it does not contain direct identifiers to a person, but can be related if very specific additional data are associated with it.[247] Thus, pseudonymity can be ensured as long as "additional information is kept separately and is subject to technical and organisational measures to ensure that the personal data are not attributed to an identified or identifiable natural person".[248] This means that the protection can be broken as the result of particular procedures, e.g. authorization of access to an otherwise closed index.[249]

Although the use of anonymous or pseudonymous data can serve as a privacy-enhancing mechanism, their use in general and with reference to aviation security measures in particular give rise to certain difficulties.

First, as discussed in Sect. 2.2.1 and in Chap. 6, in the case of aviation security, complete anonymity is impossible, since air passengers cannot travel and go through security controls without disclosing their identity. Thus, complete anonymity or pseudonymity can hardly be ensured.

Secondly, in reality and particularly in security contexts, "data can be either useful or perfectly anonymous but never both".[250] It is argued that anonymization will not work, if, for instance, CCTV image are anonymized, but can be deanonymized later, e.g. for the purposes of a crime investigation, thus they can still be considered as personal data.[251]

Thirdly, achieving full anonymization is very difficult from a technical point of view. For instance, as discussed in Chap. 6, CCTV's facial recognition can be designed in a way that the faces of individuals are not recognizable, with the exception of targeted persons such as suspects, but today, certain technological challenges remain. Therefore, true and irreversible anonymization in aviation security is hardly achievable.

---

245 Recommendation No. R (97) 18 concerning the protection of personal data collected and processed for statistical purposes, Art 8.

246 See more about pseudonymised data in Article 29 Data Protection Working Party Opinion 4/2007 on the concept of personal data, 01248/07/EN, WP 136, 20 June 2007.

247 Clarke. *Identified, Anonymous and Pseudonymous Transactions: The Spectrum of Choice*. 30 April 1999. http://www.rogerclarke.com/DV/UIPP99.html#Ps and GDPR Art.4(5).

248 GDPR Art.4(5).

249 Clarke (1999).

250 Ohm (2010) p. 1704.

251 Vermeulen and Bellanova (2013) p. 303, Badii et al. (2012) p. 367, for more details, see Ohm (2010).

Fourthly, the phenomenon of Big Data makes it possible to re-identify individuals in datasets previously believed to have been "anonymized".[252] The bigger and the more comprehensive data collection is, the longer data are retained, the more data are shared – the more the chances are that individuals can be identified.[253]

However, this "all or nothing" approach is subject to criticism. The point is that it is often difficult to determine whether data is anonymized or is still personal data, due the risk-based nature of anonymization and its dependence on various factors that are difficult to quantify, such as challenges in articulating the harms on privacy and in foreseeing the available technology that could be used for re-identification.[254]

Esayas suggests a more practical solution of establishing different layers of risks of harm that takes account the sensitivity of the data and the context of its usage. Among the four layers, the most relevant to aviation security would be the following: (i) the anonymization is applied to sensitive data and would be available with limited access, (ii) the anonymization is applied to non-sensitive data and would be available with limited access. The latter could be subject to less strict requirements of anonymization.[255]

Moreover, anonymization, despite the indicated limitations, can still play a role in ensuring the secure use or sharing of data within or among different organizations,[256] and can assist with compliance with a number of data protection principles, such as the data security and purpose limitation.[257] In this dimension, no matter if anonymization is "perfect" or not, it can contribute.

Thirdly, data should be deleted immediately after it is used. This overlaps with the principle of non-retention of data beyond a certain period of time and will be discussed below.

7.5.2 Scans

As discussed in Chap. 6, persons going through scans may in any case be identifiable, and personal data processing takes place. Moreover, the images may include sensitive material. However, with reference to the minimality principle, the situation differs if the quality images (i) do not exist, (ii) are produced, but not stored, and (iii) produced and stored, as may be the case. As discussed, the most probable cases are (ii) and (iii), but it may be the case that the quality images remain inside the machines only and are never seen by human beings. Many concerns can be eliminated by this safeguard, together with the ATR function, presentation of images as non-images or

252 European Data Protection Supervisor *Opinion 7/2015 Meeting the challenges of big data* (2015b) p. 7.

253 European Data Protection Supervisor (2015b) p. 15.

254 Esayas (2015) p. 3.

255 *Ibid.* p. 7.

256 European Data Protection Supervisor (2015b) p. 15.

257 Esayas (2015) p. 1.

generic figures, blurring faces, and not storing the images. But since, as discussed above, the storage function is questionable, the risks may remain. Thus, it will be discussed whether minimality would be served by not scanning all passengers.

The manufacturers state that body scanners are best used for targeted individuals who have been identified by other means. Thus, scanners would be used only in conjunction with other systems on a small percentage of passengers rather than on all of them.[258] With such an approach there may be no need to scan all passengers. No mass scanning of the general public would mean fewer privacy and data protection concerns.

In the light of the EU data retention directive case, scanning all passengers on all flights cannot be considered to be proportionate, especially taking into account that currently used MMW technology detects such sensitive items as prosthetics, breast implants, diapers, etc. and fails to differentiate between them and weapons, thus wrongly singling out innocent people.

Aviation security experts agree that the routine use of MMW is an invasion of civil liberties,[259] and propose to use them sparingly. This may also contribute to security: a much higher probability of detecting weapons, IEDs and their components would be achieved by implementing the risk-based screening approach.[260] As discussed in Chap. 6, according to this approach, passengers are divided into three categories of risk (high, normal, and low). Accordingly, use of body scanners is limited to the high risk group. Thus, they will be used only if another method such as passenger prescreening system, behaviour analysis, etc. gives cause for additional searches. However, although probably enhancing security and reducing aggregated privacy/data protection risks, this approach still raises human rights concerns, in particular discrimination risks (see Chap. 6).

7.5.3 CCTV

Taking into account the number of cameras installed in the airports and the fact that they cover the majority of airport locations, it would be impractical to limit them to particular persons rather than all people who are at the airport. The question that arises is whether it can be possible to limit the use of "smart" CCTV to targeted individuals.

As for CCTV's behaviour analysis, it would be impossible to limit its use because, as discussed in Chap. 6, it aims to detect potentially dangerous persons. Logically, in order to detect such persons, any and all persons must be analysed. There appear problems of false positives, but as noted above, further development of algorithms may improve the rates.

---

[258] Huey (2012).

[259] Baum (2015c) p. 1.

[260] Association of Independent Aviation Security Professionals (2011).

While detecting criminals and terrorists with the support of CCTV's facial biometrics may be reasonably acceptable, it can be considered excessive if applied to all persons no matter whether they are engaged in any suspicious activity or not, in particular if databases are created that include all passengers' images. The latter cannot be relevant to specific targets, thus their collection without requirement of relevance to any particular facts and without limitations as to subject matter or individuals covered cannot be deemed necessary (see the EU data retention directive case discussed above).

Collecting images of all people can be particularly dangerous in terms of data privacy, since data on innocent persons appear in a database and are later correlated with other data, thus becoming material for inclusion in other databases of which the aims may go beyond aviation security. This includes the risks of the legitimacy of various databases and data quality, security risks, unlimited share, function creep, misuse, etc.

Therefore, creation of image databases on all passengers going through the airports can be considered excessive, unless additional justification of the purposes, necessity and effectiveness are provided. In any case, the optimal decision would be to use facial recognition technology only on the most suspicious persons, in order to minimize the negative effect on other people.

Further, with reference to CCTV in general, according to Council of Europe, CCTV should be non-excessive with regard to the CCTV purposes sought, and those responsible for CCTV should assess and choose which technology should be used according to location of the cameras and respective requirements (image definition, zoom capacity, camera miniaturization) without using excessive measures.[261] CCTV should be operated in a way that does not make the persons recorded recognizable if the purpose of the processing does not require their possible identification.[262]

As Article 29 Working Party suggests, in order to limit personal data processing, a number of filming arrangements can help, including the visual angle (e.g. not showing private places nearby), the type of equipment (fixed or mobile), actual installation (e.g. location of cameras, use of fixed view and/or movable cameras), possibility of zooming in images in real time or stored images, possibility to blur and delete individual images, image-freezing functions, etc.[263] More attention should be paid to a number of specific CCTV features (non-exhaustive list): interconnection of CCTV as managed by different data controllers, use of biometric data, use of voice identification, indexing systems, use of facial recognition systems that are based on the targeting of suspected offenders, or else on the basis of predefined abnormal behaviour, possibility to automatically trace routes and trails and/or reconstruct or foresee a person's behaviour, taking of automated decisions based either on a person's profile or on intelligent analysis and intervention systems unrelated to standard alerts.[264]

[261] European Committee on Legal Co-operation (2003).

[262] *Ibid.*

[263] Article 29 Data Protection Working Party (2004).

[264] *Ibid.*

Any of these functions can be highly privacy-intrusive. Article 29 Working Party suggests that they should be subject to evaluation on a case-by-case basis within the framework of the prior checking of processing operations (DPD Article 20). Obviously, such analysis is needed, since what is appropriate for public places with terrorism risks may be excessive for many other locations.

A similar option is conducting Privacy Impact Assessment before implementing any CCTV system. Moreover, there should be enhanced safeguards for enhanced surveillance: any additional analysis capability added by "smart" cameras or other technology will require corresponding privacy and security safeguards.[265]

EPIC suggested that the public, the local community, and privacy and security experts should be consulted prior to any decision to introduce CCTV or implement any significant change to an existing system.[266]

Internal prior checking (and also reviewing) mechanisms can be found in the TSA's provisions regarding the planned use of Video Analytics: each of the analytics functions selected by the operator should be specified, including how the function will be assessed and validated during acceptance testing.[267] Another idea is control and oversight at the CCTV operator level, namely, human intervention before making any decisions that may affect an individual: for instance, Article 29 Working Party proposes that human intervention is appropriate in connection with the behaviour analysis, since failures can occur[268]; the UK Surveillance CCTV Code suggests this option for any use of biometric recognition systems.[269]

As discussed in Chap. 6, the angle of the view, quality images with high resolution, zoom capacity and many other technical capabilities are the fundamental requirements of effective camera surveillance systems at airports. Since the airports are objects of increased security risks (see Chap. 4), aviation security experts and manufacturers suggest to use the most advanced CCTV systems, i.e. the basic idea is that all the functions which may further enhance security, should be used.

Nevertheless, even taking into account high risks, a number of points can be considered excessive. Those which will be analysed in this section include: (i) airport locations for CCTV, (ii) visible or covert CCTV, (iii) audio/sound, and (iv) additional features, in particular, facial recognition, behaviour analysis, and the use of various databases. The main question would be what is absolutely needed for the CCTV purposes in the airports and what goes beyond what is necessary to achieve the purposes.

Airport infrastructure includes many various zones, and it may be impossible to draw the line between the airport and non-airport areas (see Chap. 4). At the same time, security should be ensured at all these places. Clearly, not all the locations where the airport's cameras can reach are equal. They can include public and private zones (e.g. residential nearby), restricted areas, some of them, although public, imply more privacy, e.g. toilets.

---

[265] EPIC (2008).

[266] *Ibid.*

[267] Transportation Security Administration (2011) p. 125.

[268] Article 29 Data Protection Working Party (2004).

[269] Home Office (2013).

Cardiff Airport CCTV Code of Practice stipulates that although the areas covered by CCTV comprise the public and private areas, both internal and external, "cameras will not be used to look in to private residential properties. 'Privacy zones' and camera end-stops are programmed into the system in order to prevent the interior of any private residential property within range of the system being surveyed by the cameras."[270]

Airport toilets are internal private areas. Deploying CCTV there would be excessive, since it leads to an inappropriate invasion of the individual's private sphere. The same purpose can be served by installing the cameras outside the entrance to the room. In different CCTV PIAs and Codes, they deserve special attention in terms of not placing CCTV cameras there at all,[271] implementing safeguards to only view recorded images taken in these areas after an incident has occurred,[272] or allowing cameras in such places only to address a particularly serious problem that cannot be addressed by less intrusive means.[273]

But in the Cardiff's Code, it is not stated explicitly that cameras will not monitor such locations; at the same time, the UK Surveillance CCTV Code of Practice allows cameras to be placed there, as mentioned above. Thus, it is not clear whether cameras are placed in the airport's toilets or not. Obviously, allowing CCTV in such areas would be excessive.

Covert cameras are also subject to strict limitations. A general rule is that no person can use covert cameras (and thus assume law enforcement powers) without involving the police. Only law enforcement authorities are the competent authorities vested with executive powers to investigate offences and prosecute them. In Russia, it is prohibited to use covert cameras and can lead to criminal punishment. In addition, in the event a (non-covert) camera should record an activity that might lead to criminal charges, the relevant extract of the camera's footage shall only be disclosed to law enforcement authorities. In practice, the airports do not use covert cameras: e.g. this is explicitly stated in §2.4.3 of Cardiff Airport CCTV Code of Practice.[274]

Furthermore, another disturbing point of CCTV would be the use of audio recording in addition to images. The regulators' views differ here. For instance, the DHS PIA expressly prohibits the use of audio, while the UK Surveillance Code, although with limitations ("a strong justification of necessity to establish its proportionality"), includes such an option.[275] Nevertheless, in airports, too much noise can negatively impact the work of the systems, making the sounds a problem rather than an advantage, and as a rule, the CCTV audio audio-recording function is disabled. No sound or voice recording means enhanced privacy protection. The sound option of CCTV on board aircraft was discussed in Chap. 6.

---

[270] Cardiff Airport (2013).

[271] Department of Homeland Security (2012b).

[272] Information Commissioner's Office (2014).

[273] Home Office (2013).

[274] Cardiff Airport (2013).

[275] Home Office (2013).

The next issue concerns CCTV's face recognition and behaviour analysis, databases of CCTV images, integration of CCTV with intelligence gathering, interconnectivity with different other databases as well as combined usage with other technologies which also may provide personal data (see Chap. 6). All this makes possible circulation of personal data, both coming from CCTV and from other sources, which can be very difficult to control.

CCTV's biometrics may contain data revealing a data subject's racial or ethnic origins or data concerning health. Sensitive data is subject to enhanced protection, but as long as CCTV's facial recognition is not thoroughly regulated, increased risks appear. Within the minimality principle, it would be preferable not to collect sensitive data at all.

The UK Surveillance CCTV Code of Practice contains principle 12: "Any information used to support a surveillance camera system which compares against a reference database for matching purposes should be accurate and kept up to date".[276] But nothing is said about the content of such databases and any limitations or safeguards. Other legal instruments or materials contain no rules either.

As mentioned above, the issue of which video should be stored and at what quality level are operational rather than technical requirements. They include assessing the frequency and consequences of potential threats. Obviously, they differ from the circumstances and location of the CCTV. In the USA, for instance, there are no approved standards or TSA regulations that govern how an airport operator is to define and apply these requirements.[277]

Thus, the operators of CCTV are responsible for establishing and limiting CCTV parameters. Accordingly, it would be preferable for every CCTV operator to establish a policy limiting CCTV systems and minimizing the collection of data and the use of the most privacy-invasive tools such as face recognition. All the limitations should also be controlled by an independent entity. Specifying the nature and extent of limits on camera use would help to prevent misuse and abuses.[278]

Some relevant regulation, for instance the UK Surveillance Camera Code, states that privacy interests are not significantly impacted since the use of surveillance data is just for observation related to activities in directing responses or in looking for suspicious and dangerous conduct, and retention of data is for a limited period. Use of technologies for anonymizing information, particularly CCTV data as it is taken or stored, provides an enhancement that protects both anonymity and the interests of individuals.[279] Similarly, the DHS states that it only uses the video feeds which contain information relevant to a criminal (or potential criminal) activity; all other video feed is automatically overwritten.[280]

Indeed, Video Analytics, which employs the facial recognition and behaviour analysis functions, is supposed to limit the collection and storage of data only to that

---

[276] *Ibid.*

[277] Transportation Security Administration (2011) p. 128.

[278] EPIC (2008).

[279] Home Office (2013).

[280] Department of Homeland Security (2012b).

which is deemed to be dangerous, i.e. to target specific individuals, thereby limiting the privacy risks and contributing to the data limitation principle.[281] Ideally, CCTV architecture should be designed to use the most intrusive algorithms only on the most suspicious persons.

But some difficulties emerge: first, as discussed earlier, full anonymization cannot be achieved; secondly, targeting specific individuals is subject to false positives. Thus, further improvements are needed. Thirdly, in the UK and USA there are databases of images of *all* passengers taken by CCTV. All this contradicts the assurances of the agencies that privacy is protected by anonymization and only data on criminals are collected. The only valid point might be that privacy is safeguarded by a relatively short data retention period. This issue – how long the data should be stored – will be discussed below.

In summary, during evaluation at any level, the security experts would no doubt argue that every new feature which has a potential to enhance security is necessary and it would be unwise not to use it. But as discussed in Chap. 6, all new features of "smart" CCTV have additional, more serious impact on privacy and data protection as well as other human rights. Opportunities for data processing, for instance, increase greatly, entailing more concerns for all data protection principles.

At the same time, as discussed above, there is lack of legal regulation of these techniques, and there are quite high rates of false positives, making these technologies highly intrusive and potentially leading to dramatic situations for innocent people who, for example, are mistakenly "recognized" abroad. Thus, the non-excessiveness requirement within proportionality and the principle of minimality may suffer.

7.5.4 PNR

According to many data protection stakeholders, it is disproportionate to collect all data on all passengers on all flights.[282] As discussed above, collection of PNR on all passengers can be compared with the EU data retention directive, which was ruled invalid by ECJ. Thus, it is suggested that the EU PNR Directive should be specific and targeted to address a particular issue, and the only measure compliant with data protection requirements would be the use of PNR-data on a case-by-case basis, when there is a serious threat established by concrete indicators.[283]

For instance, the system could target specific third countries or itineraries assessed objectively as indicating a greater risk. Article 29 Working Party, for instance, proposed to limit the data collection scheme with reference to specific criteria: "(i) to data pertaining to a particular time period and/or a particular geographical zone and/or to a circle of particular persons likely to be involved or (ii) to

---

[281] Vermeulen and Bellanova (2013) p. 310 and Home Office (2013).

[282] See, e.g. Article 29 Data Protection Working Party (2011) and European Data Protection Supervisor (2011b).

[283] European Data Protection Supervisor (2011b); Article 29 Data Protection Working Party (2011).

persons who could, for other reasons, contribute, by the retention of their data, to the prevention, detection or prosecution of serious offences."[284]

Some counter-terrorism experts also state that a more targeted approach would permit investments to be made in more intelligence resources instead of "passive large scale databases we are unable to fully analyse".[285] Similar to the scanners, however, a targeted approach implies other risks, such as discrimination. In addition, as discussed in Chap. 6, for profiling, the most effective use involves collecting as much data as possible, including sensitive data, and any limitations made based on privacy concerns may make the measure less effective.

Another measure with respect to the EU PNR system could be limiting the scope of data sources. The EU Directive covers 100% of the flights departing from and coming into the EU and possibly also intra-EU flight (and, as noted above, the EU Member States plan to include all intra-EU flights). Obviously, exclusion of intra-EU flights would serve to conform to the principle of minimality of collected data.

One more undertaking might be to limit the list of the categories of PNR data and prohibit the use of sensitive data. The ICAO PNR Guidelines contain the list of possible PNR data elements (see Annex 1 to this work). It contains multiple and detailed data fields. The key point is that operators decide themselves which of these data collect or not collect for their commercial purposes.

A practical tool for limiting PNR data fields can involve two layers. First, the airlines should evaluate and decide which information is absolutely necessary and which is not necessary for their commercial purposes and limit data collecting at this stage, either by not collecting or by limiting access to some data to passengers only. With reference to the latter, it can be the case if some passenger's personal data is not of value to the operator but can be helpful for passengers' own needs, records or statistics. As mentioned in Chap. 5, the airlines could collect these data, but organize the process in such a way that passengers can access and manage their personal details on the airline's website without anyone else having access to these data.

Secondly, it is possible to limit the scope of personal data once again at the stage of transferral to state authorities. Accordingly, the airlines transfer to security organs not all the information from their CRS, but only particular pieces of data, and instead of full access to the CRS ("pull" method), security organs would be recipients of these limited items via the "push" method.[286] These issues will be discussed below with reference to particular PNR systems.

As for sensitive data, ICAO provides that "PNRs should not contain any information that an aircraft operator does not need to facilitate a passenger's travel, e.g. racial or ethnic origin, political opinions, religious or political beliefs, trade-union membership, marital status or data relating to a person's sexual orientation. Contracting States should not require aircraft operators to collect such data in their PNRs".[287]

---

[284] Article 29 Data Protection Working Party (2015).

[285] European Data Protection Supervisor (2015a).

[286] See Chap. 6.

[287] ICAO PNR Guidelines (2010) §2.1.9.

At the same time, as discussed in Chap. 6, PNR may contain data on meal preferences or medical conditions, as well as additional text and general remarks, which are included to facilitate travel. Although these data may be useful for passenger risk assessment, and the ICAO's list of possible PNR includes the fields "General remarks" and "Free text/code fields in OSI, SSR, SSI,[288] remarks/history", ICAO provides that it is particularly important that airlines and states protect these data; such data should be used only if concrete indications exist which require the use of such data for the stated purposes.[289]

The EU Strategy requires that exchange of data should be limited to the minimum, with an exhaustive list of the categories of PNR data to be transferred. However, it does not provide any common exhaustive list. In respect of sensitive data, it prescribes that PNR containing sensitive data cannot be used only under exceptional circumstances – namely, an imminent threat that can result in loss of life – and safeguards must be provided, e.g. the data may be used only on a case-by-case basis, under the authorization of a high-ranking official, etc.

Both the EU-US Agreement and the EU Directive contain 19 PNR data elements (see Annexes 2 and 3). However, a closer look at the EU-US Agreement reveals that many data fields contain multiple data, for example, line 7: "All available contact information (including originator information)". The same applies to other lines. According to the opinions of EDPS, the lists of data in both instruments are disproportionate and should be reduced.[290] However, the data fields are within the ICAO's recommendation, and in general satisfy industry standards. However, a critical point is the issue of sensitive data.

The EU-US list's field 17 contains SSR/OSI/SSI. The EU-US Agreement provides that sensitive data could be used only when a person's life is at risk. This data will be accessed only case-by-case and will be permanently deleted after 30 days from receipt, unless it is used for a specific investigation. But the limits of such "specific investigation" are missing. This broadness is deemed excessive.

The EU Directive stipulates in Art. 13(4) that Member States should prohibit the processing of sensitive data, namely, data revealing a person's race or ethnic origin, political opinions, religion or philosophical beliefs, trade union membership, health, sexual life or sexual orientation. Data revealing such information, if received by the PIU, must be deleted immediately.

The assessment criteria shall in no circumstances be based on these criteria (Article 6(4)). If the PNR data transferred by air carriers include data beyond those listed in Annex I, the PIU must delete such data immediately and permanently upon receipt (Article 6(1)). The word "permanently" was added to the final version of the Directive. New wordings have also provided compliance of Articles 13(4) and 6(4) with the DPD Article 8 and Article 6 of Council Framework Decision 2008/977/JHA: the sensitive data list was adjusted, adding to the list the philosophical belief, trade union membership, health data, sexual life or sexual orientation of the subject. These can be seen as positive developments.

---

[288] See Chap. 6.

[289] ICAO PNR Guidelines (2010) §2.1.10.

[290] European Data Protection Supervisor (2011a) and European Data Protection Supervisor (2011b).

However, the field "general remarks" may still contain sensitive data. Privacy advocates believe that sensitive data should not be collected at all.[291] In order to achieve this, FRA proposes the following steps:

1. PIU should remove automatically from "General remarks" any sensitive data.
2. PIU should control if the sensitive data is removed and, if not, delete them, e.g. by running a matching and removal program with a regularly updated glossary of "sensitive terms" over the database.
3. Before any processing of PNR data (e.g. in reaction to an alert based on automated processing) or any onward forwarding to third parties, PNR data should be manually reviewed by the PIU to control if they contain any sensitive data in free text.[292]

Obviously, these steps can prove to be effective, but may require additional work for PIU, and the result will depend on how staffed, equipped and qualified every particular PIU is.

In Russia, there is a common list of data for all transport modes and there are also additional fields on every transport mode, hence, many data fields are repeated several times and the list looks much longer than the American or the EU ones (see Annex 4). In developing the rules of information exchange between a specific carrier and Operator of USISTS, some data elements may be excluded from the list or included, depending on technical possibilities. But in general, the list is based on the ICAO's list.

An essential point is that in contrast to other lists, the Russian list does not contain any sensitive data. This was also confirmed by letter from the Ministry of Transport.[293] No collection of sensitive data means no problem with their processing. This fact makes the Russian lists more proportional and reasonable in comparison with other regimes. Both the EU and EU-US lists are considered by many data protection advocates as excessive and recommended to be reduced.

As noted above, the use of "push" or "pull" method of data transfer has impact on the amount of collected data too. The EU Strategy and PNR Guidelines of ICAO suggest that to safeguard the data in the databases and to maintain the airlines' control thereof, data should be transmitted using the "push" method.[294] In the EU Directive, the EU states will not be able to access the databases of air carriers ("pull"). Instead, they must request the data, which are then sent by the air carriers concerned ("push").

Article 15(1) of EU-US Agreement states that data will be transferred using the "push" method. However, Article 15(5) requires carriers to "provide access" to PNR data in exceptional circumstances. It is argued that if the pulling of data remains technically and legally possible, there should be rigorous independent monitoring (of the log files).[295] EDPS suggested prohibiting the "pull" system,[296] which would provide better protection.

---

[291] European Data Protection Supervisor (2011a), European Data Protection Supervisor (2011b) and FRA (2014).

[292] FRA (2014).

[293] Letter to the author dated 5 August 2013 N 07-05-01/1277-is.

[294] For explanation of the methods, see Chap. 6.

[295] Article 29 Data Protection Working Party (2012a).

[296] European Data Protection Supervisor (2011a).

According to the Russian Order, the suppliers of information provide data to ACDPDP in electronic form automatically on a schedule on a time scale close to real by selecting the required data from their information systems and unloading them into the exchange file of agreed format. This means that the "push" method is used, thereby providing stronger protection.

In summary, the Russian and the proposed EU PNR systems appear to be more protective than the EU-US system with regard to the minimality principle.

7.5.5 Conclusion

It can be concluded that there appear to be problems with the minimality principle with respect to scanners; there are generally no problems for basic CCTV; there are problems with respect to "smart" CCTV and PNR.

The problems are generally connected to a global trend of enhanced collection of personal data. For instance, there is increased use of data originating from a private sector, such as PNR. Another example is the proposed PNR-like system which collects data from a centralized Integrated Tourist Record Computer System, i.e. data on all tourists staying at hotels, hostels, etc.[297]

The use of information focuses on earlier stages in order to prevent possible threats. Such proactive rather than reactive policing has resulted in mass surveillance and data collection. However, as discussed, despite the fact that the percentage of actual attacks is far lower than the number of people checked and data collected, mass surveillance is supposedly justified from the perspective of civil aviation, to respond to higher risks of terrorism and catastrophic consequences. Nevertheless, with one-size-fits-all approach, this results in treating innocent people in the same way as people who might be objectively identified as suspects. Although the targeted approach, which could contribute towards limiting data collection, is one of the key modern approaches in aviation security, in practice, it is quite difficult to ensure. In addition, from a security perspective in general, it may be much easier to collect all data just in case.

Moreover, as illustrated in Fig. 6.10 in Sect. 6.8.5, integration of different techniques may lead to collection of even more data, with the creation of huge databases that combine multiple sources: data from watch lists and no-fly lists, intelligence, national and international databases, including image databases, data from the commercial sector, including airlines' PNR, biometric data, video footage with facial recognition and behaviour analysis, information from the Internet, online forums and social networks, etc. While analysing all the information, new predictions, assumptions and more knowledge about people will be generated – thus, more data that can then be further used. This may include the use of sensitive data, which presents specific risks. Moreover, the information that is used can be of a different nature, presenting both objectively determined information (hard data) and information based on evaluation and analysis (soft data). The use of the latter can may cause different failures and abuses.

[297] EDRi-gram 13.17, 9 Sep 2015, https://edri.org/edri-gram/13-17/

Thus, although with great potential for security purposes, the enhanced collection of data and combination approach, as well as the exploitation of Big Data imply increasing human rights risks.

7.6 Non-retention of Data Beyond a Certain Period of Time

The Europol annual inspection report 2012 found that processing of personal data of a group of 96 individuals not representing any danger, and storage of data for 5 years without any check on their relevance, as well as taking into account the further processing, did not constitute a proper evaluation of the necessity of retaining these data.[298] As mentioned above, the selected measures are generally applied to all passengers, thus, almost all the stored data relates to innocent people.

7.6.1 Scans

With reference to scans, as discussed in Chap. 6, the storage function is one of the most controversial points. Probably, images are not stored, and it would be the best option to minimize the risks.

If data storage by the scanners is possible, then special legal provisions could at least regulate and limit such storage. For example FRA proposes that the possibility of storing images as evidence should be considered only in those cases where a suspicious object has been identified through body scanning and its existence subsequently confirmed through a hand search. In all other circumstances where no suspicious object is identified nor its existence confirmed, the storage of images of people by which they can be identified must be considered a severe infringement.[299]

Subsequently, concerns arise over the retention periods for data relating to passengers who have been found carrying prohibited articles or in situations where the security officer cannot determine whether or not the passenger is carrying prohibited articles. This retention period should be clearly specified and the data should in any case be deleted as soon as it is no longer necessary.

7.6.2 CCTV

In the case of CCTV, Article 29 Working Party suggested that while in some cases, it may be sufficient only to view images rather than record them, in other cases, a sufficient and reasonable retention period for keeping of CCTV camera recordings should be determined in advance before using video systems, but at any rate not

---

[298] Article 29 Data Protection Working Party (2014a) § 5.16.

[299] FRA (2010b).

exceeding 7 days.[300] After the lapse of this period, images must be deleted (with the exception of images that may be requested by the police or judicial authorities.)[301]

The UK CCTV Code of practice and the UK Surveillance Camera Code contain similar provisions, but do not stipulate any maximum or minimum periods because the retention period for different systems varies.[302]

The issue as to whether or not to record the images – and if yes, for what period of time – should not be determined simply by the storage capacity of a system.[303] Factors that must be assessed include the frequency and consequences of potential threats. Obviously, they differ from the circumstances and location of the CCTV. In the USA, for instance, there are no approved standards or TSA regulations that govern the storage criteria for an airport operator.[304]

As mentioned above, airports are considered critical objects with an increased risk of terrorism and crime. The common idea is that at such locations, retention might be extended.[305] At the same time, taking into account that airports have hundreds or more cameras, video storage can entail significant design, management, and cost challenges. Additionally, the retention period may depend on the concrete tasks of surveillance (see Chap. 6).

For instance, to detect dangerous persons, gathered data has little or no utility after the individual has left the premises. Thus, the retention period for maintaining this data should be relatively short.[306] The same concerns the task of monitoring an area against dangerous behaviour: as a rule, there is little need for extensive retention of data, but some relatively short retention may be required to look for behaviour patterns over a period of time. Where suspicious behaviours are identified and there is a desire for longer retention, some enhanced standard like reasonable suspicion should be applied to requests to extend retention.[307]

In practice, each airport has its own policies regarding archived videos and the storage period,[308] with a period of one month being very common. E.g. Cardiff Airport CCTV Code of Practice stipulates a maximum period of 30 days.[309]

In the USA, TSA reported that many airport operators store video recordings for 30 days; other airports store video for only 7 days while, in extreme cases, a year or more may be motivated by policy concerns including public safety and risk management.[310] The DHS stores recorded video for 6 months; records which are retrieved due to suspected criminal activity will be retained until investigative or enforcement

[300] Article 29 Data Protection Working Party (2004).

[301] *Ibid.*

[302] Home Office (2013).

[303] Information Commissioner's Office (2014).

[304] Transportation Security Administration (2011) p. 128.

[305] Zoufal (2008) p. 161.

[306] *Ibid.*

[307] *Ibid.*

[308] Elliott (2009).

[309] Cardiff Airport (2013).

[310] Transportation Security Administration (2011) p. 128.

action is completed. But, even after the retention period, footage can be retrieved upon request and approval of a supervisor and confirmation that the recordings are relevant to actual or suspected criminal activity.[311]

In Norway, the data are deleted after 7 days. In general, the technology/data are not shared with other agencies, but in conjunction with investigations, upon request by the Police, the data can be kept up to 30 days. Upon a court order, it can be kept longer.[312]

The Russian legislation does not define any retention period for camera surveillance. The Personal Data Law contains only the general rule that personal data cannot be stored longer than necessary to fulfil the purposes of data processing. However, in practice, Russian airports store the data one month.[313] There are some views, e.g. those of CCTV equipment providers, recommending selective archiving for 3 years.[314]

Certainly, unlimited retention implies increased privacy risks. Although in practice it is limited, a legal regulation, or at least a definition of the exact retention periods in internal airport codes, as in the case of Cardiff, would be a good practice. Another improvement would be storing only "tagged" frames, unless there is a compelling reason for storing all frames for more than a few days.[315] For instance, Video Analytics can provide privacy safeguards in data retention, if used for selective storage of data when the system detects something "dangerous", rather than retention of all data.

7.6.3 PNR

With reference to PNR, both the ICAO and the EU recommend that the period of retention of PNR should not be longer than necessary for the performance of the defined tasks. The EU Strategy notes that the period of retention should take into account the different ways in which PNR data are used and the possibilities of limiting access rights over the period of retention, for example by gradual anonymization of the data. ICAO adds that the state should, in accordance with national laws or regulations, have a system for monitoring, ensuring appropriate deletion of the PNR data.

Criticizing the Strategy, EDPS states that PNR data should be deleted if its transfer did not lead to any enforcement action, a maximum period of retention should be established, and gradual anonymization should be an obligation.[316]

---

[311] Department of Homeland Security (2012b).

[312] Enerstvedt. *SMART Work Package 2 – Smart Surveillance in Border Control, Country report Norway* (2012).

[313] Romanova. *The use of CCTV to monitor the behavior of employees* (2006). http://www.delo-press.ru/articles.php?n=6772

[314] Sio.su. Building a system of video surveillance and access control at the airport Pulkovo. http://sio.su/down_asec_77_def.aspx

[315] Transportation Security Administration (2011) p. 128.

[316] European Data Protection Supervisor (2010).

FRA proposes a maximum of 30 days for PNR retention period. After that, PNR data should be pseudonimyzed by masking out name, address and contact information, general remarks and any collected API. Pseudonimyzed data can be accessible to a limited number of personnel of the PIU specifically authorized to carry out analysis of PNR data and develop assessment criteria. Only the Head of the PIU is able to permit access to masked out data, only for specific investigations or to comply with data protection rules. Upon expiry of the data retention period, PNR data should be permanently deleted. PNR data used for specific criminal investigations or prosecutions may be kept by law enforcement and judicial authorities under the conditions allowed by national law.[317]

The EU Directive Article 12 took into account some of these recommendations. It provides that Member States must ensure that the PNR data provided by the air carriers to the PIU are retained in a PIU's database for a period of 5 years after their transmission to the PIU of the Member State on whose territory the flight is landing or departing. After 5 years, the data must be permanently deleted.

Upon expiry of 6 months (but not 30 days) after the transfer, all PNR data must be depersonalized through masking out name, address and contact information, payment details, general remarks and any API. But "to depersonalise through masking out of data elements" means "to render those data elements which could serve to identify directly the data subject invisible to a user" (Art. 3(10)). This means that this is done without deleting these data elements. Thus, that data is in fact never anonymized.

After 6 months, full data can be used only for responding, on a case-by-case basis, subject to duly reasoned requests from competent authorities or Europol. The disclosure of the full PNR data can be permitted only when approved by a judicial authority or by another competent national authority, subject to information and ex post review by the data protection officer of the PIU. Accordingly, despite the certain conditions, access to the full PNR data is always available.

Moreover, the result of the assessment of the passengers shall be kept by the PIU "as long as necessary" to inform the competent authorities and PIU of other Member States of a positive match (Art. 12(5)). This potentially allows unlimited storage.

The retention period is argued to be unreasonable.[318] Article 29 Working Party proposes a more flexible approach and refers to the Court of Justice of the EU stating that "the determination of the period of retention must be based on objective criteria in order to ensure that it is limited to what is strictly necessary".[319] Accordingly, it encourages the legislator to propose a shorter period, adding a recital specifying the criterion justifying the concrete need to which the chosen period corresponds.[320] However, the indicated periods are included in the final version of the Directive, meaning that the periods of 6 months and 5 years sound reasonable to the EP, Commission and the Council. Moreover, this constitutes an improvement in comparison with other PNR systems (see below).

---

[317] FRA (2014).

[318] European Data Protection Supervisor (2011b).

[319] *Digital Rights Ireland Ltd v. Minister for Communications, etc*. Joined Cases C-293/12 and C-594/12, 8 April 2014, §64.

[320] Article 29 Data Protection Working Party (2015).

Under the EU-US Agreement, US authorities will keep PNR data in an active database for up to 5 years. After the first 6 months, all information which could be used to identify a passenger would be "depersonalised". After the first 5 years, the data will be moved to a "dormant database" for up to 10 years, with stricter access requirements for US officials. Thereafter, data would be fully "anonymized" by deleting all information which could serve to identify the passenger. Data related to any specific case will be retained in an active PNR database until the investigation is archived.

In comparison with the EU Directive, the storage of all data for up to 15 years sounds excessive and disproportionate. Moreover, after 15 years, only anonymization of the data is provided. Taking into account the difficulty of truly anonymizing data and the lack of explanation for why the anonymized data is needed, it is proposed that data should be deleted.[321] The EDPS goes even further and suggests that the data should be anonymized (irreversibly) or deleted immediately after analysis or after a maximum of 6 months.[322] Moreover, it is unclear how the anonymization relates to data that have already been transferred onwards to third countries.[323] It is clear that in comparison with the EU Directive, the EU-US scheme is weaker in data retention terms.

In Russia, Article 5(7) of the Personal Data Law states that personal data shall be stored in a way that allows verification of the identity of the data subject no longer than it is necessary for processing purposes, if the retention period of personal data is not set by federal law or the treaty a party (or beneficiary, guarantor) to which is the data subject. Processed personal data shall be destroyed or anonymized upon achieving the set purposes or in the case if such purposes cease to be relevant, unless otherwise provided by federal law.

In the case of PNR data processing, the retention periods are not determined,[324] providing options for unlimited storage. Clearly, this contradicts international and the EU recommendations on data protection, and is weaker than the (although controversial) the EU-US and the EU schemes. But in summary, all the PNR systems can be improved in terms of data retention.

7.7 Data Quality

The data quality principle requires that personal data must be relevant, accurate, complete and kept up-to-date. Since the use of personal data in civil aviation is now increasing, and concrete characteristics of data as well as any inaccuracies (or "anomalities", as in the case of scans), may have direct effects on passengers – e.g.

---

321 Article 29 Data Protection Working Party (2012a).

322 European Data Protection Supervisor (2011a).

323 European Parliament. Draft recommendation on the draft Council decision on the conclusion of the Agreement between the United States of America and the European Union on the use and transfer of Passenger Name Records (PNR) to the United States Department of Homeland Security, 30 January 2012. 2011/0382(NLE)).

324 This was also stated in the Letter of 5 August 2013 N 07-05-01/1277-is.

the image generated by a scan will require further interpretation, a person may be subject to additional screening, a mistake in spelling of the name may lead to denied boarding, etc., the data quality principle is critical.

In the USA, a police officer ran an individual's name through his computer during a routine traffic stop, and the system incorrectly indicated a warrant for that person's arrest (the system was not properly updated), and police officer arrested that person.[325] The Supreme Court found that the officer's action was lawful,[326] but Justice O'Connor, in a concurrence joined by Justices Souter and Breyer, argued that "Surely it would not be reasonable for the police to rely, say, on a recordkeeping system, their own or some other agency's, that has no mechanism to ensure its accuracy over time and that routinely leads to false arrests, even years after the probable cause for any such arrest has ceased to exist (if it ever existed)".[327]

This Fourth Amendment case provided a constitutional basis for the principle that unless the databases to arrest or detain individuals and the method of searching them are accurate, the government should not rely on such databases.[328] This can also apply if a passenger is chosen for additional screening at the airport – the use of inaccurate or unreliable data for this would not be reasonable.[329]

With regard to protecting databases in general, swift correction of errors and regular checks of data to ensure accurateness[330] as well as ensuring data security discussed above are helpful. Some specific undertakings with regard to selected measures will be the following.

7.7.1 Scans

With reference to scans, according to the US PIA, "The Automatic Target Recognition generated image is accurate, timely, and complete and is directly relevant to the identification of threat objects. Potential threat items are resolved through a directed physical screening before the individual is cleared to enter the sterile area".[331]

The concern here is the inability of the machines to differentiate prohibited items and such personal characteristics or items as implants, as discussed above. This means that the quality of the images – for the intended purposes – is not very adequate.

---

[325] *Arizona v. Evans*, 514 U.S. 1 (1995), 1 March 1995 at 4.

[326] *Ibid.* at 16.

[327] *Ibid.* at 17.

[328] Dempsey and Flint (2004) p. 1487.

[329] *Ibid.*

[330] Gilbert (2007) p. 26.

[331] Department of Homeland Security (2015) p. 5.

7.7.2 CCTV

The nature of CCTV – recording real time video – implies that data is supposed to correspond to these requirements, at least at the moment of recording. For instance, the DHS explains that there is no editing feature or ability to change the image; users can zoom or pan the cameras to follow one individual but it is unlikely that incorrect information about a person is produced from the CCTV cameras.[332] At the same time, security experts admit that digital data can be altered very easily so that even experts will be unable to see that any change took place (this will be further discussed on the Data security section).

The most critical issues, however, arise when CCTV systems are used along with data from other sources, including image databases; accordingly, such additional data should satisfy to data quality requirements as well. The problem is that any external database is beyond the control of the CCTV operator and may contain inaccurate information, leading to false positives. If the images are being checked against a database that contains errors, then innocent individuals might become the subjects of law enforcement investigations or other measures.[333] As mentioned above, the risks increase if such error takes place abroad, where the opportunities to prove the opposite may be limited, and the particular state may lack respect for privacy and other human rights.

To deal with these risks, the UK Surveillance Code stipulates that any information used to support CCTV that compares against a reference database for matching purposes should be accurate and kept up-to-date. Any use of technologies such as facial recognition systems which may rely on the accuracy of information generated elsewhere, such as databases provided by others, should not be introduced without regular assessment to ensure that the underlying data is fit for the purpose.[334] Moreover, CCTV operators should consider any approved operational, technical and competency standards relevant to a system and its purpose and work to meet and maintain those standards, including additional standards if the system has specific advanced capability such as Video Analytics or facial recognition systems, etc.[335]

Of course, these undertakings are better than nothing, but still, they cannot guarantee that external data, especially from image databases, including data from foreign states, is accurate. Again, as mentioned above, the use of image databases, and facial recognition as such, as well as other additional CCTV features should be subject to better evaluation whether they are not excessive, really needed, how effective they are, whether the benefits overweigh the privacy risks.

---

[332] Department of Homeland Security (2012b).

[333] EPIC (2008).

[334] Home Office (2013).

[335] *Ibid.*

7.7.3 PNR

For PNR, accuracy of the information and completeness belong to principles of PNR transfer stipulated by ICAO.[336] The risks of incompleteness or inaccuracy of data arise due to the fact that the passengers submit information to the airlines themselves and that PNR can be captured much in advance, i.e. maybe not fully updated. Thus, the data controller may be not able to mitigate possible problems at the initial stage of data collection.

In other cases, the similarity of a person's name with names of criminals from watch lists may result in denial to travel. This is not the fault of the passenger, and in these cases, it is the data controller that must take measures to ensure that an innocent person is not inconvenienced due to error. However, as mentioned in Chap. 6, such "mistakes" may take years to correct, and during all these years, the person cannot fly.

Thus, although the data quality principle is formally introduced with reference to all the selected measures, it can be a challenge to ensure it in reality.

7.8 Data Security

As discussed in Chap. 5, personal data should be kept secure from any potential abuses such as loss, unauthorized access or use, disclosure or misuse, deletion, change, or exploitation.

The problem is that many features of modern technologies discussed in Chap. 6 such as digitalization of data, the use of IP and wireless connections imply that data can potentially be stored and accessed at any time anywhere in the world and by multiple users, and can be easily altered, stolen or used for purposes other than specified.

Abuses may take different forms and include criminal misuse of the data collected by individuals who have access to the information, institutional misuse by departments, voyeurism,[337] appearance of "naked" images from body scanners on the Internet, as has happened in the past, etc. Thus, security and integrity in transmission, databases, access to the digitally-stored data, etc. – all of these should be ensured.

Since information security also constitutes a part of aviation security, common security undertakings may be relevant, thus, broader and serious security requirements of the former may apply to security of personal data. For instance, according to the TSA, CCTV image transfers across the airport network or over the Internet should be monitored by both airport security and by the airport IT department.[338]

As discussed, security in general includes technical and organizational security methods. General technical solutions include encryption or encoding, secured storage media, backup systems, password protection, different identification/authenti-

[336] ICAO PNR Guidelines (2010) §2.3.2.

[337] EPIC (2008).

[338] Transportation Security Administration (2011) p. 129.

cation mechanisms and multilayer access control, software which is able to identify persons accessing the data, log systems automatically auditing when the data are accessed and by whom and when data are changed,[339] etc. Organizational methods primarily limit access to data to authorized persons.

7.8.1 *Scans*

For body scanners, the decision not to create or retain images is supposed to resolve data security issues. As the US PIA of 2015 states, "AIT data is transmitted in a proprietary format to the viewing monitor, and cannot be lost, modified, or disclosed. TSA's decision not to retain images mitigates further data storage security issues".[340]

However, in the previous version of the US PIA of 2011, more detailed security undertakings were described. They included the following undertakings: data transmission between the checkpoint and the viewer by a landline connection; the data transmitted through the landline is transmitted in a proprietary format that cannot be deciphered without the proprietary technology; the computers used to process and present the images are locked with both physical and software controls to prevent the insertion of any storage media or other communication devices.[341] These requirements may be more relevant as long as the status of creation and storage of images is unclear. In addition, organizational measures for Type B scans limit access to the remote image viewing rooms, secured restricted premises for viewing images, and prohibit operators from bringing photographic devices into the room.[342]

7.8.2 *CCTV*

For CCTV, the video feed can be secured during transmission, with only videos of stored data available.[343] Technological devices can be used to restrain hackers. PETs are also capable to protect data.

If the data is stored using a cloud computing system, it is required to ensure that this system is secure and/or to ensure that the provider can ensure the security of the information.[344] There are also reliable techniques for irreversible anonymization of images (see above).[345]

---

[339] Gilbert (2007) p. 26.

[340] Department of Homeland Security (2015) p.5.

[341] Department of Homeland Security (2011).

[342] *Ibid.*

[343] Department of Homeland Security (2012b).

[344] Information Commissioner's Office (2014).

[345] Badii et al. (2012) p. 367.

According to DHS, the log-in and use of the system is traceable to a particular user and periodically audited for misuse and discriminatory practices.[346]

Within organizational measures for CCTV, the DHS requires that only authorized personnel have access to the stored video data, there are privacy trainings for CCTV operators and other relevant persons (training includes privacy, technical aspects of the system, and disciplinary procedures for violations), and all DHS employees must agree to a general "Rules of Behaviour" before logging into any DHS system. The system is audited when an employee with access leaves the organization or when called for by a supervisor.[347] Effective review and audit mechanisms, regular reports and clearly defined policies are also organizational measures of security.[348]

Other methods include secured restricted premises for viewing CCTV recorded image; access to servers and digital storage volumes may require special physical storage and access control provisions such as biometric identification of authorized personnel.[349]

7.8.3 PNR

The EU PNR Directive Articles 13(2-3) specify that provisions of the Framework Decision 2008/977/JHA regarding confidentiality of processing and data security apply, as well as the respective norms of the DPD. Art. 13(5) requires states to ensure that the PIU maintains documentation of all processing systems and procedures under their responsibility, at least:

(a) the name and contact details of the organisation and personnel in the PIU entrusted with the processing of the PNR data and the different levels of access authorisation;
(b) the requests made by competent authorities and PIUs of other Member States;
(c) all requests for and transfers of PNR data to a third country.

The PIU must make all documentation available, on request, to the national supervisory authority. Moreover, PIU must keep records of at least collection, consultation, disclosure or erasure; the records must show, in particular, the purpose, date and time of such operations and, as far as possible, the identification of the person who consulted or disclosed the PNR data, and the identity and recipients of that data (Art. 13(6)). The PIU must communicate personal data breach to the data subject and to the national data protection authority (Art. 11(8)).

All this means that there should be log records of every access, modification or deletion of PNR data by the PIU staff, with internal (by Data Protection Officer, Art. 5) and external (by the national supervisory authority) control. Many of the recommendations given in respect of an earlier version of the Directive[350] were taken into account.

[346] Department of Homeland Security (2012b).

[347] *Ibid.*

[348] Home Office (2013).

[349] Transportation Security Administration (2011) p. 129.

[350] E.g. FRA (2014).

Nevertheless, there are more options for improvement. For instance, the PNR database can be separated from other databases.[351] In addition, in order to enhance responsibility to secure processing of the data, there should be an obligation of mutual information in case of security breach: authorities must inform PIU of any unlawful disclosure.[352]

The EU-US Agreement stipulates the technical measures and organizational arrangements in Article 5(1–2). There are no claims regarding security standards from privacy advocates.

As discussed in Chap. 5, the Russian Personal Data Law provides detailed and comprehensive security requirements. The Order follows this line. The Operator of USISTS is responsible for data security of ACDPDP.[353] Accordingly, he is obliged to use security equipment determined by the FSB and the FSTEC and produced by companies licensed by the FSB and the FSTEC. According to the information of the Operator, all necessary attestation and certificates for securing data in ACDPDP have been obtained.[354] The Ministry of Transport specifies that providing data to ACDPDP is carried out electronically via secure channels (VPN-channels of Internet or channels of protected branch networks).[355]

Consequently, it can be seen that all regulators pay much attention to security issues with respect to all the discussed measures. Formally, it can be argued that all the considered systems' provisions pertaining to data security fall within the requirements of the data security principle. Nevertheless, opportunities for improvement and even better security exist. It would perhaps be beneficial to use the best features from each of the measures for the rest of the regimes, whenever applicable.

At the same time, similar to the assumption that 100 per cent effective aviation security does not exist, 100 per cent effective data security does not exist either. Thus, data leakage and other misuses still might occur. Another danger is that, as the Edward Snowden case revealed, no security rules and undertakings to protect data can prohibit some agencies from gaining access to it. Moreover, in many countries, data are typically stolen by bribery rather than by exploiting security gaps.

Thus, the most promising positive developments would be to follow the minimality principle discussed above, i.e. to limit data collection to the very minimum during the initial stage and limit its storage to the minimum amount of time, thereby reducing the options to share data or combine data from different sources, etc. The best protection for body scans would be to ensure that images are not saved (as the states claim); for CCTV – similarly, no record of images; for PNR – limitation of data fields, data share, etc. But, as discussed above, problems exist, in particular with unlimited and uncontrolled data sharing, the possibility of sensitive data processing, etc.

---

[351] *Ibid.*

[352] European Data Protection Supervisor (2010).

[353] The Operator is Federal State Unitary Enterprise "ZashshitaInfoTrans", an enterprise subordinated to the Ministry of Transport.

[354] Telephone conversation with the Operator's employee 4 July 2013.

[355] Ministry of Transport. Information for entities of the transport infrastructure and carriers in connection with the entry into force of the Order of Ministry of Transport of Russia № 243. 20 June 2013. http://www.mintrans.ru/news/detail.php?ELEMENT_ID=20360

7.9 Data Subjects' Rights

The data subject influence principle refers to a number of rules providing various rights of individuals. The right to be informed about aviation security measures in general was discussed within the Legality section above. In this section, two aspects will be discussed: consent and the rights to access, rectification and deletion.

7.9.1 *Consent*

As discussed in Chap. 5, the interests of aviation security may fall within the exceptions and justify most instances of data processing without consent. Therefore, it is impractical to object to scans, CCTV or PNR transfer: the objection to a particular security measure will lead either to deployment of an alternative measure, as in the case of scanning, being denied the right to fly.

Nevertheless, it is often argued that the public accepts and consents to aviation security measures. For instance, the huge amount of camera surveillance devices led people to tolerate and accept that CCTV is always behind the scenes recording day-to-day activities.[356] With reference to body scanners, as mentioned above, the TSA states that independent surveys indicate general public support of the use of scanners.

The US PIA states that the DHS should, to the extent practical, seek individual consent for the collection, use, dissemination, and maintenance of personal data.[357] From the DHS/TSA perspective, passengers express their consent by selecting the screening method – they have the option to decline scanning in favour of physical screening.[358] At the same time, the US PIA notes that due to the use of Automatic Target Recognition (ATR), "the mitigation of privacy issues associated with the individual image generated by previous versions of AIT not using ATR", and due to the need to respond to potential security threats, TSA may mandate AIT screening for some passengers as warranted by security considerations.[359] This means that some of the passengers will be deprived of the choice at all.

Nonetheless, the fact that a person may choose between scanners and hand searches can hardly mean his/her consent to scanners. People's acceptance and/or positive attitude do not necessarily mean consent. Although there may be different understandings of consent, as discussed in Sect. 5.4.6, in the EU data protection law, consent is "any freely given specific and informed indication of his wishes by which the data subject signifies his agreement to personal data relating to him being processed".[360] It is questionable whether the public would still "accept" the regime if they were aware

[356] Annereau (2014).

[357] Department of Homeland Security (2015) p. 4.

[358] *Ibid.*

[359] *Ibid.*

[360] DPD Article 2(h).

of some concrete intrusive features such as CCTV's use of biometrics, behaviour analysis, databases, etc. With respect to scanners, similar claims about acceptance were made when the old "naked" scanners were used, and many people accepted not knowing exactly what images were processed. Thus, it is a question of whether people indeed understand what they actually accept and tolerate. Moreover, as discussed, in the air transport context, the consent can hardly be "freely given" at all.

The UK Surveillance Camera Code of Practice develops the term "surveillance by consent", implying that the consent can be presumed to be obtained if CCTV pursues legitimate aims and a pressing social need, is proportionate, where there is transparency regarding its use and where it is not intrusive.[361] This means that in general, privacy and data protection principles discussed in this work should be satisfied. However, as revealed in this work, current use of aviation security technologies in general and CCTV in particular hardly meets the demands of all privacy and data protection principles. Thus, this concept is subject to future improvement.

7.9.2 Access, Rectification and Deletion

In regards to access, rectification and deletion, the law requires that if the personal data are saved and kept, mechanisms should be provided to allow travellers to access and rectify the data. Weaknesses may exist in terms of realizing these rights already at general level, such as those reported from the UK.[362]

7.9.2.1 Scans

With reference to scans, as long as the governments state that they do not save the data, access and rectification are difficult: how can a state provide access to data that they claim do not exist? In the USA, it is required that DHS should provide mechanisms for appropriate access, correction, and redress regarding DHS's use of personal data. But in the specific passages regarding scanners, it is stated only the information about choosing between scanning and hand search (see Sect. 7.9.1).[363] No further details are provided.

7.9.2.2 CCTV

As for CCTV, the rights to access, rectification and deletion are more applicable. However, some CCTV features such as short retention period, other persons on the images, etc. may narrow the scope of application of these rights.

---

[361] Home Office (2013) and Porter (2014).

[362] European Commission, Press Release IP/10/811, 24 June 2010.

[363] Department of Homeland Security (2015) p. 4.

The Council of Europe stipulates that during the storage period, the right of access to the data, and, where appropriate, the right of rectification, blocking and/or erasure, should be granted to the data subject unless this would entail disproportionate effort. Any limitations due to such disproportionate efforts should be laid down exclusively by primary legislation (see Article 13(1) of the DPD) with due regard for the data subject's right to defence in respect of specific events that may have occurred in the period considered.[364]

Additionally, if personal data are recorded and kept, this should be done as far as possible in a way that allows data subjects to exercise their right of access without obtaining information about other people.[365] In practice, this means that instead of providing direct access to the recordings, the data controller may allow the data subject to view the relevant footage, simultaneously safeguarding the rights of the third parties on the images, e.g. via blurring in order not to disclose their identities.[366]

In the UK, the CCTV Code provides very detailed instructions and terms on the right to access. For instance, information must be provided within 40 calendar days of receiving a request; the CCTV operator may charge a fee of up to £10; those who request access must provide details on their identification and location; it is recommended to have internal procedures and a documented process for the requests; exemptions can apply. The operator is required to provide the data subject with a copy of all the information that constitutes their personal data (unless data subject agrees to receive their information by viewing the footage). Where information of third parties is also shown, the operator may need to obscure this information.[367]

Examples of good practice can be found on London Luton and Cardiff airports' websites, which both provide application forms for data subject access request and guidance. A fee of £10 is charged for a request. The request form should be completed in order to verify persons' identity and to locate the specific personal data requested.[368] Exemptions to the provision of information include cases involving personal data that is being processed for the prevention or detection of crime or the apprehension or prosecution of offenders.[369]

Nevertheless, these rules stipulate only rules on access to the records, not to rectification or deletion of them. There is a logical explanation for the problems associated with these options.

For instance, in the USA, according to DHS, given the nature of CCTV systems, access, review, and correction of the video cannot be provided. The limitations for this are, first, that it can be difficult to find and view a particular video and secondly that the storage of video is 6 months or less; a video image cannot be corrected since

[364] Article 29 Data Protection Working Party (2004).

[365] European Committee on Legal Co-operation (2003).

[366] Article 29 Data Protection Working Party (2004).

[367] Information Commissioner's Office (2014).

[368] E.g. the form requires ID with image, name, date of birth and current address, full details of the date, time and location of the images and, if possible, a description of what the person was wearing on that day. London Luton Airport Operations Limited. CCTV Subject Access Request Form.

[369] Cardiff Airport (2013).

it captures the events in real time. However, individuals may access their information by submitting a Freedom of Information Act/Privacy Act (FOIA/PA) request to the DHS FOIA Office. Similar to the UK case, an individual should provide a detailed description regarding the location of the CCTV, the date and approximate time, or other identifying information assisting to find the requested record.[370]

The DHS admits that this lack of direct access, review, correction and deletion represent a risk to individual privacy, however, the DHS considers it necessary, given the utility of CCTV systems and the retention rates. Moreover, while some individuals will not have a formal mechanism for access or redress, DHS has internal mechanisms to correct inaccuracies and protect against abuse through the auditing system and the "Rules of Behaviour", privacy trainings, etc.[371] However, similar to redress mechanisms, these are internal rules and procedures where individuals have no influence. Various data leaks and scandals in the past do not inspire more trust.

In Russia, I did not find any comparable detailed procedures or request forms.

In summary, indeed, the nature of CCTV makes it difficult to provide the individual's right to access and to demand that the data be rectified or erased. There are examples of good practices concerning the right to access. However, obviously, this does not concern all the UK airports; in the USA, it is not clear if the DHS procedures are applicable in the airports. Other airports and jurisdictions should follow good practices.

7.9.2.3 PNR

For PNR, the EU Strategy and ICAO PNR Guidelines suggest that individuals must be provided with access to their PNR data, and where appropriate, with the right to seek rectification and deletion of PNR data.

FRA suggests that air passengers should have the right to access their PNR data, including when these have been masked out, whether the data have been disclosed, and to whom, and be informed of any refusal or restriction of access, and the right to seek rectification of his/her PNR data, where the data are inaccurate, and be informed of whether his/her PNR data have been rectified or erased.[372]

According to the EU Directive, passengers' right to access, rectification, erasure and blocking should be in line with the Council Framework Decision 2008/977/JHA (Art.13(1)). However, this framework decision includes shortcomings in terms of data subjects' rights in comparison with the Directive 95/46/EC.[373] A higher standard of safeguards, based on the principles of Directive 95/46/EC, could strengthen the Directive.[374]

The EU-US Agreement Articles 11–12 state that any individual, regardless of nationality, country of origin, or place of residence will have the right to access their PNR data, correct or rectify the PNR, including the possibility of erasure or blocking, if the infor-

---

[370] Department of Homeland Security (2012b).

[371] Department of Homeland Security (2012b).

[372] FRA (2014).

[373] Article 29 Data Protection Working Party (2011).

[374] European Data Protection Supervisor (2011b).

mation is inaccurate. But some "reasonable legal limitations" under US law apply. As a result, Article 29 Working Party expressed doubts as to whether US law and the Agreement provide for the respective rights in line with requirements of the EU law.[375]

The Russian Personal Data Law stipulates the rights of the data subject to access, to correct, block or destroy personal data as well as respective obligations of data controller to provide these opportunities to data subject. However, these rights can be restricted due to transport security needs and other lawful grounds.[376] Personal data collected according to the Federal Law On Transport Security constitute elements of transport security information, thus, §5 of Article 14(8) of Personal Data Law restricts the data subject's right to access.[377]

In contrast to the EU-US Agreement, this is a general rule rather than exception. However, the risk of broad application of the restrictions and limitations in the US case makes the regimes quite similar.

But what about the passengers' opportunity to gain access to records, rectify and delete them in practice? One common problem is that these rights apparently may suffer due to the involvement of many different entities (the airlines, PIUs, national law enforcement authorities and other organs, PIUs of other states, other states' organs, etc.) and recipients of data in third countries as a result of transfer to these third countries.

Another common problem is that data controllers, despite the fact that they are obliged to provide data subjects with access/rectification/deletion, may have no established procedures and mechanisms thus are not capable to provide these rights and fulfil their duties. Notable examples are unsuccessful attempts to get air passenger personal records from American authorities and EU airlines, which will be discussed below.

As for the USA, as discussed above, no comprehensive federal privacy law currently applies to the collection, use, and disclosure of passenger travel information. Consumer advocates express concern that it is difficult for passengers to learn what information airlines and others in the travel sector are collecting, keeping, and sharing about them.[378]

In 2008, the DHS Privacy Office reported that requests for PNR data typically take more than a year to answer – many times longer than the legal time limits in the Privacy Act and Freedom of Information Act.[379]

In the case initiated by a US citizen privacy advocate Hasbrouck,[380] after a three-year delay in processing a PNR request and after a hearing before a court, only some restricted

[375] Article 29 Data Protection Working Party (2012a).

[376] §5 of Article 14(8) of Russian Personal Data Law.

[377] Letter of 5 August 2013 N 07-05-01/1277-is.

[378] US Senate Committee on Commerce, Science, and Transportation. Rockefeller Opens Inquiry Into Airline Passenger Fees and Consumer Data Privacy Policies. 18.08.2014. http://www.commerce.senate.gov/public/index.cfm?p=PressReleases&ContentRecord_id=835cc647-7697-47dc-89f0-4cfc24524a57&ContentType_id=77eb43da-aa94-497d-a73f-5c951ff72372&Group_id=4b968841-f3e8-49da-a529-7b18e32fd69d&MonthDisplay=8&YearDisplay=2014

[379] DHS Privacy Office. A report concerning Passenger Name Record Information derived from flights between the US and the European Union. 18 December 2008. https://www.dhs.gov/xlibrary/assets/privacy/privacy_pnr_report_20081218.pdf

[380] *Hasbrouck v. U.S. Customs and Border Protection*, Case number C 10-03793 RS, U.S. District Court, Northern District of California, filed 25 August 2010.

information was received. If US citizens alone have problems in enforcing their rights to access, it can be even more problematic for the EU citizens via the EU-US Agreement.

An example of the enforcement problem concerning a EU citizen was the request in 2007 from a member of the EP, Sophia In't Veld, to receive PNR information pertaining to her. At first she received a claim from DHS that they did not have any record of her trip. The PNR data was finally received after American lawyers filed a lawsuit on her behalf, but the data was "late, clearly incomplete, and inconsistently and inappropriately redacted".[381]

As for the EU, according to Hasbrouck, when he asked KLM Royal Dutch Airlines to see the records of one of his trips from the US to the EU and back, he was informed months later by KLM that they had outsourced the handling of his data to companies in the US; they did not know what data their contractors and agents had collected or retained, or with whom they might have shared the data; and they had no provisions in their contracts that would enable them to force their contractors to provide this information. When Hasbrouck asked the Dutch DPA to intervene, they admitted that they had no staff with the technical competence or sufficient knowledge of the industry to interpret the limited data that KLM had disclosed. Hasbrouck was informed that they could not help, and there was nothing to be done except hire a lawyer in the Netherlands to prepare and file a private lawsuit against the airline.[382]

Of course, the indicated cases date from some years back in time, but reports from privacy advocates show that most US federal agencies still ignore human rights complaints[383] and that there can still be problems with the rights to access, rectification and deletion of PNR data. I believe that the same difficulties may arise in Russia. The consequences of the EU Directive are not yet clear, but experience shows that airlines, PIUs and other organs involved evidently must establish necessary procedures and routines.

7.10 Concluding Remarks

It is clear that current aviation security technologies represent some special challenges for the principles' application. This is mainly due to their specific, particularly advanced features, but also due to the amount of restricted information and other factors related to aviation security. Although legal principles are general and can be applied to the regulation of the discussed technologies, additional explanation of how they can specifically be applied can be helpful, as shown by for instance the UK codes of practices at the level of national DPA and the level of airports.

In summary, almost all the selected principles suffer with respect to some of the selected aviation security measures or all of them. Despite safeguards, it is quite obvious that these technologies still imply privacy and data protection concerns. Deployment

---

[381] Hasbrouck. *Can you really see what records are kept about your travel?* 30 December 2008. http://hasbrouck.org/blog/archives/001595.html

[382] *Ibid.*

[383] The Identity Project (2015).

of scanners/CCTV/PNR cannot be regarded as in compliance with all the established data protection requirements and privacy standards. In the meantime, the governments fail to demonstrate that the security policies are currently justifiable within the law.

In addition, no state is guaranteed from more and more enhanced surveillance and possible abuses by law enforcement authorities in the name of security. Even within the established guarantees, who can guarantee that a particular state will keep its promises, and that abuses and violations will not occur? The recent cases of the NSA and other various state agencies' secret use of personal data-pursuant surveillance programs do not inspire optimism.

In this section, the most important issue is whether the scans, CCTV and PNR systems' security benefits can substantiate the problems indicated with reference to privacy and data protection principles. As noted above, this should be evaluated taking into account also the higher threat risks in civil aviation (see Chap. 4) and additional negative effect to other human rights (Chap. 6).

Positive features of the measures in security terms and, accordingly, in their ability to contribute to the right to life are summarized in Table 7.1.

The grades used are Positive, Probably positive, and Positive with additional conditions. It can be seen that all the measures are supposed to improve security and thus can contribute to the right to life and are seen as suitable from aviation security points. All the measures are in general considered necessary, since apparently, no equally effective and less intrusive practical methods can substitute them. However, additional proper substantiation, including statistics, can be helpful.

However, only basic CCTV is considered as more or less objectively effective tool at airports. CCTV's additional "smart" features, in particular, biometrics, behaviour analysis, and integration of databases, although with great potential, are subject to further development.

Scanners were indicated as "probably" suitable due to controversial and questionable issues and failures in detection. It may be the case that most effective features are disabled due to privacy or health concerns. In addition, the alternative to opt out is to undergo an advanced hand search, which is considered as intrusive as well, probably, more intrusive. As for PNR, they are effective and provide additional value in comparison with other existing measures such as API only if used for profiling.

Negative findings are summarized in Table 7.2.

As the table shows, different degrees of problems can be present:

- No problems
- Probably no problems. This means that the situation is uncertain (e.g. due to unclear nature of the measure, or use of external databases with unclear data protection terms, etc.), but the absence of problems is more likely than the existence of problems.
- May be problems. Same as above, but being problems is more likely than non-existence of problems.
- Problems in some states. This means that some states are more successful in guarantees than others.
- Problems

The suggested degrees are very approximate, since no mathematical evaluation can be made, for instance, "Problems" may differ in degree of severity and in relation to the particular measure in a particular state. However, all these difference present more minor issues for the whole picture.

"Problems summary" in numbers is given for illustration only – to roughly compare the selected measures – rather than for any mathematical measurement, since it is clear that different principles have different "weight" in violations.

In addition, findings on other human rights from Sect. 6.9 should be taken into account too:

It should be noted that all general principles in the tables above relate not only to privacy/data protection, but to other human rights too, since, as discussed in Chap. 3, similar schemes for evaluating their justification are used, i.e. evaluating "in accordance with law", "necessity", etc.

As for excessiveness, all concerns with respect to all individuals' rights should be taken into account. Due to the high number of indicated "problems", "smart" CCTV can be seen as excessive in relation to basic CCTV; PNR used for profiling – excessive in comparison with use of PNR for other purposes. In the latter case, however, as discussed, PNR can hardly provide additional advantages in relation to the already used API, and can thus lose its benefits.

In addition, as discussed, excessiveness overlays with the minimality principle, where a number of parameters can be used to limit data collection – and, accordingly, negative effects on informational privacy and data protection.

If one considers a measure's application to all passengers rather than to targeted passengers, the following conclusions can be drawn. This is impractical for basic CCTV, but its facial recognition technology can be limited to targeted individuals. With regard to body scanners, their use for a high-risk passenger only apparently can lead to reducing the human rights risks and enhancing security, i.e. contributing to the goals of both "sides" of aviation security and privacy dilemma.

It is less clear with PNR, since limiting the collection of personal data (i.e. not to include all passengers), or limiting purposes (prohibiting the use for profiling), or in scope (e.g. excluding intra-flights), or limiting data fields, in particular sensitive – can be problematic without decreasing the effectiveness of PNR. Thus, endeavours to ensure non-excessiveness in privacy/data protection terms may have negative impact on effectiveness. If effectiveness is decreased due to privacy concerns, security opponents get additional grounds to state that the measures should not be used.

Certainly, if an ineffective security measure is cancelled, it is often of benefit not only to human rights, but to security as well, since better alternatives might be found.[384] However, what about effective, but intrusive methods?

There definitely should be some limits on how far security measures can go. Security experts agree that some techniques, such as "routine full body scanning, embarrassingly intimate pat-downs and racial profiling" have no place in the airports of any democratic state.[385] Nevertheless, others probably can be used: as noted

[384] Solove (2008a) p. 362.

[385] Baum (2015c) p. 1.

above, privacy advocates themselves believe that some security measures that are extremely intrusive can be used in some circumstances. If this logic is used, then, many really effective methods, such as data profiling, can be used. Nevertheless, the opposition from civil liberty groups may freeze effective security solutions, including the use of PNR for profiling, the use of facial biometrics, behaviour analysis, etc.

Thus, security experts encourage these groups to follow their own logic: "We need to start selling the public a system that works and we need to get groups, such as the ACLU, to work with us by focusing on their very own message of performing "security measures that are extremely intrusive… when there is good cause to suspect that an individual is a security risk".[386]

At the same time, there are a lot of political and economic issues and policies which greatly affect security and privacy regimes. There is a huge industry behind the screening, surveillance and other security equipment, with their own lobbies in legislatures and governmental agencies. While terrorist threats seem to be highlighted, the governments need to ensure the public that they "are doing something" and the scanners, for instance, seem to be perfect means for this. Security experts only assist the states in their decisions about which devices or methods to use. Various other motivations may lay behind a particular technology or particular decisions.

In order to try to resolve the dispute as objectively as possible, I tried to evaluate the aviation security measures using both aviation security tools and principles of privacy and data protection, and then, apply the principle of proportionality to all factors, such as suitability and necessity, threats, breaches of privacy/data protection and other human rights.

Thus, taking into account Tables 7.1, 7.2 and 7.3, I tried to "pinpoint" the measures on Fig. 7.1 based on Fig. 5.1.

Mathematically, the more security benefits the measure provides, the more limitations to human rights may be permitted. The diagonal blue line is the proportionality line and indicates rough borders between "proportionate" and "disproportionate" zones: ideally, the security measures should be under or somewhere on this line.

In comparison with Fig. 5.1, the "acceptability" line of maximum risks to human rights allowed is indicated in three variants:

- red solid line – privacy advocates point of view, implying almost perfect conditions with almost no concerns,
- red dash line – regulators' point of view; some concerns are incorporated by existing regulation (but it is another question if the legal basis satisfies to the quality of law – see Sect. 7.2),
- red dot line – pro-security point of view: concerns are allowed as long as the measure enhances security.

The location of measures is very approximate and rough, but shows some general considerations and assumptions.

[386] *Ibid.*

Table 7.1 Aviation security measures and suitability and necessity

	Scanners	CCTV basic	Smart CCTV	PNR
Suitability (effectiveness)	Probably positive	Positive	Positive, but subject to future development	Positive, in particular if used for profiling
Necessity	Positive			

Table 7.2 Relation between aviation security measures and privacy and data protection principles

	Scanners	CCTV basic	Smart CCTV	PNR
Transparency	Problems			
In accordance with the law	Problems	May be problems	Problems	Problems
Legitimate aim	No problems			
Lawful processing	Problems			
Excessiveness	May be problems	No problems	Problems	May be problems
Non-discrimination	Problems			
Purpose limitation	Probably no problems	Problems		
Minimality	May be problems	No problems	Problems	Problems
Data retention	Probably no problems	Problems	Problems	Problems
Data quality	Problems	No problems	Problems	Problems
Data security	May be problems			
Data subject influence	Probably no problems	Problems in some states	Problems	May be problems
Summary	Problems: 5	Problems: 5–6	Problems: 10	Problems: 8
	May be problems: 3	May be problems: 2	May be problems: 1	May be problems: 3
	Probably no problems: 3			

Privacy advocates tend to state that CCTV benefits are disproportionately small to the privacy risks[387] and argue that all available studies fail to provide evidence that surveillance brings significant benefits, or that any of its benefits outweigh the limitations on privacy.[388] But as discussed, the context is essential. Basic CCTV systems in the airports can be considered suitable, relevant and necessary, with no visible alternatives, although it is questionable that some features are not excessive. Thus, while human rights risks exist, they are substantiated by necessity and effectiveness of CCTV: evidently, in the airports, they are a must.

387 E.g. EPIC (2008) and ACLU (2002).

388 Gilbert (2007) p. 34.

Table 7.3 Relation between aviation security measures and other human rights

	Scanners	CCTV basic	Smart CCTV	PNR
Right to health	Probably no problems	No problems		
Freedom of movement	No problems	No problems	Problems	Problems
Right to equal treatment and non-discrimination	Problems			
Freedom of thought, conscience and religion	May be problems			
Rights of the child	Problems: any negative impacts become more serious for a child			
Summary	Problems: 2	Problems: 2	Problems: 3	Problems: 3
	May be problems: 1	May be problems: 1	May be problems: 1	May be problems: 1
	Probably no problems 1			

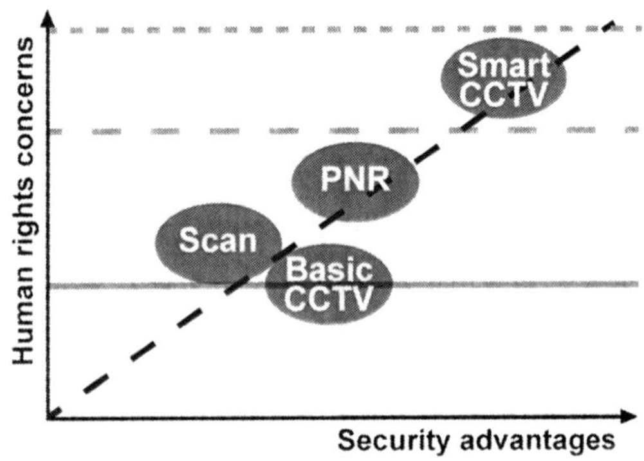

Fig. 7.1 Proportionality principle applied to scans, CCTV and PNR

Thus, basic CCTV systems can be placed within the proportionate zone. Among the considered measures, it is a practice with the least amount of aggregated concerns. Nevertheless, human rights advocates may still argue that CCTV is beyond the acceptable level of risks, thus, CCTV are placed *on* the line, which can be interpreted as being both below or above the line, but in any case very close.

Even in the event the body scanner does not save data and quality images are inside the machines and are never seen by humans, the scanners still may be privacy-intrusive and lead to discrimination, in particular for some groups of population. This constitutes a minimum impact on human rights if all declared safeguards work and do not simply hide violations. Due to these risks and "probable" effectiveness, the scans are place beyond the proportionality zone, but close to the proportionality line. It is surely above the privacy advocates' acceptability line, but does not exceed the regulators' line, since scanners are accommodated by current laws. It should be noted, however, that if all *possible* indicated risks are taken into account, the breach of human rights can be more serious, moving scanners higher.

Human rights concerns of PNR systems are particularly serious if PNR are used for profiling. In rough numbers they exceed those of scans, thus, PNR are placed higher, but still under the regulators' line of acceptability. PNR systems can be valuable for security, thus, are placed close to the proportionality line.

A "champion" in human rights problems is "smart" CCTV. If properly developed, CCTV can potentially be very effective. A solution could be avoiding the use of additional intrusive features such as facial recognition, behaviour analysis, and external databases, but from a security point of view, this will reduce effectiveness, "falling" to the level of basic CCTV. Thus, further development and research are needed, including further technical improvements to diminish risks of false positives, ensuring targeting only particular individuals. Other tools such as safeguards neutralizing the negative impact on privacy, or at least minimizing it, should be used as along with PIA for any new features in order to exclude or minimize negative impact as well, so that all applicable human rights and principles are respected, and regular assessments of CCTV use should be conducted. In the meantime, "smart" CCTV is placed above the acceptability lines of both privacy advocates and regulators, due to a clear lack of regulation. Nevertheless, due to great potential, they are placed close to the proportionality line, since high risks are supposedly "justified", to some extent, by high benefits.

It should also be noted that the DPD provides exemptions in Article 13, which stipulates that the member states may restrict the scope of the data protection principles when such a restriction constitutes a necessary measure to safeguard, among other things, national security, defence, public security, the protection of the data subject or of the rights and freedoms of others. According to some views, the indication of "a necessary measure" means that these exemptions are restricted only for specific investigations, a case-by-case request, and not to cases where the personal data processing is systematic as it is foreseen by the use of scanners, CCTV or PNR.[389] This approach is questionable: as discussed, there are a number of restrictions due to security needs with regard to "systematic" measures, concerning for example the right to access, consent, etc. However, even if we consider that all exemptions are applicable, many of the issues which have been discussed under the analysis of Article 8 ECHR may still be applicable to data protection legislation issues as well.

Thus, overall, analysis of modern aviation security measures within legal principles reveals that there are breaches of privacy, data protection and other human rights. Further improvement of technologies and their regulation, practices and further justification of measures, including every additional feature, are needed. At the same time, due to the security benefits and the seriousness of the threats they are supposed to prevent, all the measures are close to the proportionality line.

Another question is where to place the acceptability line. As indicated, the level of acceptability seen by regulators is higher than that of privacy advocates and cov-

---

[389] See, e.g. Article 29 Data Protection Working Party. Opinion 6/2002 on transmission of Passenger Manifest Information and other data from Airlines to the United States, 24 October 2002, § 2.2; EPIC. EU-US Airline Passenger Data Disclosure. http://www.epic.org/privacy/intl/passenger_data.html

ers all the measures apart from "smart" CCTV, the regulation of which is not yet sufficiently developed. Aviation security experts apparently view all the indicated measures as acceptable. While narrowing the scope of aviation security would likely be inadequate to meet modern threats to civil aviation, as mentioned above, privacy advocates should listen to their own words – if measures are intrusive, but effective, they probably can be used, and the red solid line in Fig. 7.1 can be moved up.

Part III
Conclusion

Chapter 8
Conclusion

8.1 Returning to Research Aims

The first aim was to discuss whether the use of the selected aviation security measures corresponds to legal principles of privacy and data protection. For this evaluation, it was important first to know the context – i.e., to use the results of the second aim, describing the measures, their background, emergence and contemporary usage, and secondly, to understand which exact legal principles should be used, as well as applicable regulations (examples from selected jurisdictions). In addition, since all human rights are interconnected, the impact on other human rights was assessed.

Keeping all this in mind, Chap. 7 concluded that the selected measures, in varying degrees, mainly *do not* correspond to legal principles and constitute breaches to a number of human rights. Nevertheless, due to security benefits and the seriousness of the threats they are supposed to prevent, all the selected aviation security measures are closer to being proportionate rather than disproportionate. For the majority of the measures, "conditional" proportionality is indicated, implying that they are subject to further legal and technological improvement.

This has direct relation to the third aim: analysis of options for rectification and recommendation. Apparently, it would be unwise to simply prohibit aviation security technologies that can protect the public, but are intrusive. It is more appropriate to allow improvements and protect fundamental rights at the same time, ensuring aviation security *plus* privacy. It is a matter of how to ensure this in practice, how to apply the existing laws to new technologies and of other practices that can contribute. The discussion reveals that although some potential for improvement by means of both technologies and legal means exists, not all suggestions are practical or possible.

Another question is whether current processes and tendencies, such as technological developments, globalization, harmonization of regulation, modernization of

O.M. Enerstvedt, *Aviation Security, Privacy, Data Protection and Other Human Rights: Technologies and Legal Principles*, Law, Governance and Technology Series 37, DOI 10.1007/978-3-319-58139-2_8

existing data privacy regulation, developing technology-specific regulation and mechanisms of self-regulation, growing use of Privacy Impact Assessment (PIA) and Privacy by Design (PBD), attention to passenger experience, etc. can contribute to solving the existing issues or to creating additional problematic issues for the aviation security-privacy dilemma, thus, making the situation and future prospects more complicated.

This chapter will discuss selected findings from this work relating to legal, technological and other developments in aviation security and privacy/data protection which, in my opinion, present the most controversial and problematic issues relating to aviation security-privacy dilemma. They will reflect current and future prospects.

8.2 Regulation Issues

8.2.1 Introduction

As discussed in this work, in particular, in Chaps. 2 and 4, both data protection and aviation security becomes more and more regulated at all levels (national, regional, and international). Nevertheless, the discussion in Chap. 7 shows that despite some positive features in regulation such as global harmonization, modernization of general data protection laws, appearance of technology-specific guidelines, emerging mechanisms of self-regulation, there are a number of factors having negative impact on the protection of individuals in the course of aviation security.

The point is that the appearance of new technologies implies fundamentally different requirements to the protection of privacy, involves the expansion of traditional notions of law and establishment of appropriate protection mechanisms. As the UK Surveillance Camera Commissioner states,

> I believe issues of privacy will come to the fore the more these technologies are used – and communities will challenge its use. I truly believe that we should embrace what these developments offer, but it must be done within the right regulatory framework. A framework that ensures we don't run before we can walk. We must work with manufacturers, installers and end users so we can use these new systems well with the support of communities.[1]

As Walker suggests with regard to anti-terrorism legislation, "a rational code not panic legislation; a criminal justice model not a war model; the language of rights not the language of balance; international cooperation not unilateralism…"[2] Since, as discussed in Chap. 4, anti-terrorism and aviation security legislation are related and overlap, these suggestions are relevant too.

---

[1] Porter (2014).

[2] Walker (2006) p. 1137.

The discussion below will be divided into several sections combining findings on regulation problems and suggestions for improvement, concerning both regulation and related procedures and mechanisms.

8.2.2 Legal Regulation

As discussed in this research, protection of individuals' rights in the course of aviation security via legal measures at any level can be done by general technology-neutral regulation and technology-specific regulation, including industry-developed instruments such as Codes of practices. The former is supposed to address common challenges of modern technologies and seems to take the form of primary legislation, while technology-specific regulation is supposed to increase legal certainty in respect of particular technologies and takes the form of delegated/secondary legislation.

As discussed, since many of modern aviation security issues are of a global nature, international harmonization of regulations is indicated as one of the drivers and trends in future aviation. Thus, the role of international aviation security regulation is becoming more and more important. Similarly, since globalization and technological developments made personal data processing a global phenomenon, international regulation of data protection is important too. With this regard, the presence and development of international instruments in both aviation security and privacy/data protection, including technology-specific rules for technologies used in aviation security, sound promising.

As illustrated by aviation security regimes in the selected states, despite global endeavours in international aviation security regulation, there are still non-harmonization and divergences. Negative factors include poor cooperation between governments, lack of international standards on all the technologies, lack of enforcement mechanisms, and differences in technical, economic, and threat environments of the ICAO member states leading to differing national aviation security practices.[3]

The problem is that when international agreement is lacking or problematic to enforce, unilateral or bilateral action can be taken.[4] An example is the operation of various PNR systems. As discussed, international standards exist for PNR, established by the ICAO and IATA. Some regulators take them into account, to different extents. Nevertheless, current PNR programs are non-standard, leading to various problems in terms of technological development, data extraction, transmission, etc. This constitutes an unfavourable factor not only for airlines that have to comply with different rules, but for privacy and data protection in aviation security in general. Clearly, it is much more difficult to solve privacy and data protection concerns

[3] Chaps. 4, 6 and 7.

[4] Walker (2006) p. 1150.

in complicated and differing systems, without universal standards and with internal problems.

With regard to international regulation of privacy/data protection, more endeavours have been initiated with regard to technology-neutral rather than technology-specific regulation. However, as this research reveals, more harmonized technology-neutral regulation is still to be achieved. There is the lack of a single mandatory regulatory framework and of legal framework for integration of different levels: global, regional, national.[5] There is a lack of responsibility, accountability and monitoring at the international level. Privacy experts are quite pessimistic in this regard and note that the chances of extensive harmonisation at the global level are extremely unlikely in the short-term.[6] This is due in part to political, cultural and other differences, including between the Western states too (see Chap. 2). Another reason is the lack of the centre of initiative – an international body with sufficient powers.

In this situation, I believe that technology-specific instruments will aid in addressing the security-privacy dilemma. There is need for further international cooperation and dialogue between the states concerning privacy and aviation security. For example, in order to standardize and harmonize aviation security measures, in addition to Guidelines on PNR, ICAO could adopt instruments with respect to scans, CCTV and other technologies. It can be helpful to legally establish common universal principles of data protection for definite technologies. They may contribute in efforts to establish global, standard approaches in the aviation industry, including human rights issues. The states would benefit from the use of technologies for security purposes, airlines and airports – from common requirements, passengers – from basic protection of their rights.

As noted in Chap. 4, in practice, ICAO standards are seen by the states as mandatory, while its recommendations, guidelines and the like are not. Probably, the instruments above could be adopted in the form of the standards. At the same time, as discussed, the ICAO lacks enforcement powers. In this regard, the ICAO's role could likely be revised so that it might be enabled not only to establish, but to enforce such rules. However, this can be problematic in practice due to the sovereignty issues.

In the EU, better harmonization has been achieved in aviation security and data protection in terms of both technology-neutral and -specific instruments than what has been achieved elsewhere in the international arena. The EU presses hard to ensure that its member states have the same level of data protection and aviation of security at the airports. There are endeavours to modernize general data protection laws according to new challenges, and the EU adopts some instruments on aviation security measures with common privacy/data protection guarantees (e.g. body scanners and PNR). However, due to various interests pertaining to the reforms, modernization, at least with regard to the GDPR, has not been realized to its full potential[7];

[5] Kovaleva (2007) p. 14.

[6] Bygrave (2014) p. 205.

[7] See Chap. 2.

due to the differences between member states and the gaps in privacy/data protection regulation related to particular aviation security technologies, some challenges remain.

Moreover, despite the leadership in data protection, and increasing adoption of national laws based on the EU standards, surveillance practices in the EU are increasing both in general and in aviation security in particular.[8] These tendencies are alarming: if this happens to the data protection "leader", what about other states, which, from the EU's point of view, are inadequate in terms of data protection?

On the national level, in general, the selected measures have some basis in law. But as discussed, as a rule, the main type of regulation of aviation security technologies is general law rather than specific. Due to the need for quick reaction, aviation security regimes may put some technologies or its features into practice before their specific regulation is in place, relying on general laws which can be broad and "accommodate" a wide range of activities. The same applies in terms of reliance on general data protection law rather than providing technology-specific provisions.

As the result, there is a lack of specific legal regulation of some of the activities in aviation security, including provisions pertaining to privacy/data protection. In particular, this concerns the newest, most technically advanced and comprehensive measures/features such as "smart" CCTV.

Another alarming tendency is the provision of details on the regime through orders or other instruments issued by executive bodies rather than laws. As discussed, although the former are more dynamic and operative in reacting to security needs and may have a positive effect on security, this may compromise the legality of the regimes, the concept of separation of powers in a democratic state, create the risk of bureaucratic abuse of power, etc. The restricted character of regulation which can be substantiated not only by security needs, but the intention to hide privacy concerns (see Sect. 8.2.4.2) is problematic too. All this leads to problems with such requirements as the laws being accessible, interference being foreseeable by persons, laws being precise, specific and clearly defined.

The technology-specific laws that exist are very fragmented, both concerning areas of application and concrete relevant measures. They may already be outdated since they do not address new features of aviation security technologies on a proper detailed level. Some of them are recommendations only and are not enforceable. Even if such specific legislation provides privacy guarantees and safeguards, it is a question whether these can be considered adequate – they may provide weak protection in practice.

An illustration is the bilateral EU-US PNR agreement: despite endeavours of the regulators, full compliance with the EU data protection requirements has not been achieved, and in the light of recent ECJ cases, it may be the case that the agreement is illegal. Moreover, even when negotiating a "domestic" PNR regime within the EU, the regulators fail to ensure compliance with own data protection principles. It is thus a question of whether it is possible to provide adequate safeguards at all, and of whether the strict data protection requirements are realistic.

---

[8] See Chaps. 2 and 6.

With respect to legal regulation in general, it can be concluded that general technology-neutral regulation (as primary legislation) can be better improved by appropriate modernization according to new challenges. Technology-specific laws, which should be based on and derive from primary laws, must specify precisely and clearly the technologies, their necessity, purposes, limits and other requirements. It is proposed that the regulators should first identify best practices (including prohibited practices, required practices, and recommended practices); and secondly adopt an innovative regulatory framework for these best practices that promotes experimentation with new technologies and engineering practices and encourages regulatory agreements through stakeholder representation, face-to-face negotiations, and consensus-based decision making, and supports flexible, incentive-driven mechanisms.[9]

The specific rules, apart from laws, may take other forms, such as mechanisms of voluntary self-regulation, including industry-developed codes of conduct. The government can choose to have the industry – which has both interests and incentives for having their initiatives adopted – to establish rules for appropriate use of technologies.[10] This is especially relevant for aviation security having a plethora of such influential actors as airports, airlines, equipment producers, etc., as well as their industry associations, global and regional, who can contribute to harmonized development and application of the rules.

A combination of updated technology-neutral regulation, developed technology-specific regulations, and privacy/data protection norms specified in industry-developed codes of conduct may be a promising solution. However, other problems and undertakings are essential too; they will be discussed below.

8.2.3 *Enforcement of Regulation*

In addition to formal regulation, a prerequisite is that the law should work: "A law of privacy, to be worth anything, ought to have one essential and simple characteristic: it ought to work".[11] The most important issue is not only adopting instruments – the latter concerns all levels – but in addition, following-up and ensuring that implementation occurs. Thus, the ability of national state authorities to make laws related to aviation security and privacy work is a separate question.

For instance, national regulators may adopt rules on PNR which would formally satisfy the EU data protection standards, but will they be implemented? The law-in-books versus law-in-action problem is particularly relevant for states like Russia, with relatively newly established democratic regimes and democratic values, where many legal rules are written on paper but are not fully enforced in reality, where the laws simply do not work.

---

[9] Rubinstein. *Regulating privacy by design*. In: Berkeley Tech. LJ. Vol. 26 (2011) p. 1413.

[10] Teknologirådet and Datatilsynet (2014) p. 43.

[11] Prescott. *Kaye v Robertson—a reply*. In: The Modern Law Review. Vol. 54 (1991) p. 454.

A common suggestion is to establish liability for use of privacy-invading technology. For instance, a liability similar to product liability could be imposed on those who produce and sell such technologies.[12]

However, experience with fines for violating data protection laws in general shows that this is not the most effective way to ensure good practices. There is the risk that sanctions will be ignored; DPAs may lack the powers or resources to "catch" the violators, etc. In particular, this is relevant for the states like Russia, with general problems of law enforcement and long traditions of corruption. As discussed, the amounts of fines for data protection violations may be quite low. The data controller may consider that paying fines would be cheaper than the price of resources needed to comply with the data protection requirements.

Another risk – irrespective of whether the fines are high or low – is that instead of complying with the rules or paying fines, the controller may prefer to pay a bribe. Moreover, if DPAs have lack the authority to impose fines or have insufficient control, resources and personnel, there is always the chance that the violation will never be discovered. Of course, airports, airlines and various security entities – who may be data controllers in respect of aviation security measures – are prominent actors, thus, with less of a chance to be unnoticed. Regardless, they may frequently use various types of leverage, find gaps in imperfect legislation to circumvent regulation, etc.

Of course, Russia is a specific state in terms of law enforcement. But even the Western states face the problem of ignoring the sanctions, a typical illustration being the large numbers of unregistered CCTV cameras. Probably, instead of relying on the fear of sanctions, a more promising solution would be to strengthen transparency and accountability (see below).

If going from the enforcement of laws within the state to a higher level – enforcement of undertaken obligations by the states – there appear questions as well. As noted in Chap. 4, it is extremely difficult to make the countries follow the extradition and prosecution requirements as well as other requirements imposed by international aviation security conventions and instruments.

As mentioned in Chap. 3, in respect of human rights law, monitoring and holding states accountable are procedures taking place at three levels (national, regional and international), involving various actors (the states, non-governmental organizations, national human rights institutions, international treaty bodies, international courts etc.), and using various mechanisms (reviews of policy, budgets or public expenditure, governmental monitoring mechanisms, impact assessments, democratic processes, judicial redress etc.). This overlaps with the principle of accountability discussed in Sect. 8.2.4.2, but in general, all these procedures are subject to improvement.

The problem is that the regimes like the one adopted by the USA raise doubts about proper enforcement of security and privacy laws and lack of abuses, in particular in the light of the Snowden revelations. It is a question of whether anyone can

---

[12] Hildebrandt and Tielemans. *Data protection by design and technology neutral law*. In: Computer Law & Security Review. Vol. 29 (2013) p. 520.

stop a sovereign state if it suddenly decides to enhance security measures violating its previous promises on data privacy and beyond, in particular if this is done in secret.

In some cases, the courts may aid in stopping or limiting such actions. For instance, in a number of cases in conjunction with ECHR Article 8 involving the UK, the ECtHR decisions elucidated some important principles on surveillance limits.[13] However, those limits are often highly contested and continue to cause fissures, since legal certainty via legislation is not always provided.[14] For instance, the ECtHR ruled that searches undertaken under section 44 of the UK Terrorism Act 2000 were unlawful.[15] However, although the ruling had important implications for the police and for civil liberties, there are still much wider problems with the use of stop and search powers by British police.[16] The problem is that even if ECtHR case law with reference to Article 8 ECHR sets limits and offers guidelines for the implementation of member states' legislation, it is a challenge to make the states comply with the decisions.[17]

Thus, overall, problems of law enforcement may arise as well as difficulty enforcing obligations undertaken by the states. Weak leverage in enforcing mandates on sovereign states makes the problem even more complicated.

8.2.4 Legal Principles

8.2.4.1 Introduction

It is clear that technologies develop much faster than legislation/soft law of any kind, whether general or technology-specific. Probably, a more practical solution is adopting laws outlining and enforcing general privacy and data protection principles.[18]

As discussed in this research, legal principles of privacy and data protection are important, especially taking into account the international nature of aviation security. The current principles, such as transparency, proportionality and purpose limitation, require that fundamental rights should to be protected more dynamically when faced with new technologies and challenges.[19] Thus, more appropriate laws,

---

[13] E.g. *Gillan and Quinton v. United Kingdom* (2010), *S. and Marper v. the United Kingdom* (2008), *Peck v. the United Kingdom* (2003), etc.

[14] Pearse. *Investigating Terrorism: Current Political, Legal and Psychological Issues*. John Wiley & Sons, 2015, p. 225.

[15] *Gillan and Quinton v. United Kingdom*, No. 4158/05, 12 Jan 2010.

[16] Equality and Human Rights Commission. *Stop and think A critical review of the use of stop and search powers in England and Wales*. 2010. http://www.equalityhumanrights.com/sites/default/files/documents/raceinbritain/ehrc_stop_and_search_report.pdf.

[17] Wright et al. (2010) p. 351.

[18] For more details, see Solove (2011).

[19] European Data Protection Supervisor (2015b) p. 4.

especially international laws outlining and enforcing general privacy and data protection principles are required, in particular, with a focus on how the principles can be applied to new technologies.

This way, the law would clarify the application of the right to privacy/data protection to aviation security. Accordingly, the principles could be applied on a case-by-case basis and adapted for every new technology without the need to wait for detailed legislation on a specific technology. Another point is that existing principles may be complemented by new principles such as accountability and privacy by design.[20] Below, two of the promising principles – transparency and accountability – will be discussed in more detail. Privacy by design will be discussed separately in Sect. 8.3.

8.2.4.2 Accountability

As noted above, various aviation security technologies are important to protect the flying public, and to try to halt technological deployment would be unrealistic. A more practical solution would be to adopt appropriate legal, policy and technical measures focusing on key aspects: to regulate and control first the adoption of the measure and secondly the use of the measure in a responsible manner. As Solove argues, liberty interests "are generally not achieved by eliminating particular security programs but by placing them under oversight, limiting future uses of personal data, and ensuring that they are carried out in a balanced and controlled manner".[21]

An accountability relationship is one in which one party makes demands on the other party to report on his or her activities, and has the ability to impose costs on that party.[22] It is possible to distinguish authorized or institutionalized accountability, when the requirement to report and the right to sanction are mutually understood and accepted, from other types of accountability, where the one party claims the right to hold the other party accountable, but the latter does not recognize a corresponding, mutual obligation.[23] Clearly, only the first type can be considered as "proper" and "strong" accountability. Only such accountability is able to ensure control mechanisms, e.g. an independent authority for evaluation/assessment of the measure, when the results of such assessment are obligatory for the entity operating these measures.

As Etzioni argues, if accountability is strong enough, powers that might otherwise be excessive might be acceptable. Of course, some powers may be inappropriate despite any accountability, however, "The remedy, if accountability is found deficient (or excessive), is to adjust accountability and not to deny the measure altogether".[24] But here, the question of "who will control the controllers" arises.

---

[20] *Ibid.*

[21] Solove (2008a) p. 362.

[22] Keohane. *Global governance and democratic accountability*, Duke University (2003) p. 12.

[23] *Ibid.* p. 12.

[24] Etzioni (2002) p. 40.

These conclusions work only if the system is deemed to be basically trustworthy, ignoring "both claims by public authorities that no strengthening of accountability is needed and the shrillest civil libertarian outcries that sound as if no one is to be trusted".[25]

According to the researchers, in general, the most serious problems arise with respect to accountability to persons outside the acting entity, whose lives are affected by it.[26] This is definitely applicable to the situation in aviation security, where passengers are affected by security decisions, but are outside the system. The question of accountability here is very difficult, since merely being affected cannot be sufficient grounds to create a valid claim, and the decision makers are simply not able to consult with everybody affected.[27]

At the same time, within the field of data privacy, some emerging mechanisms can be noted which address accountability in respect of relations between data subjects (in the case of aviation security, passengers) and data controllers (typically, airlines, aviation security providers, operators of the technologies).

The new GDPR puts great emphasis on the accountability.[28] Article 5 (2) stipulates that the data controller is responsible and must be able to demonstrate compliance with all principles of personal data processing ("accountability"). To be able to demonstrate such compliance, the controller must implement appropriate technical and organizational measures, including, if applicable, appropriate data protection policies. Adherence to approved codes of conduct or an approved certification mechanism may be used as an element to demonstrate compliance as well (Art.24(3)). When compared with the DPD, the GDPR also imposes new requirements and liabilities on data processors (Art.28).[29]

Mechanisms that increase the accountability are also among the proposed additional and alternative strategies to mitigate the risks of automated profiling and other techniques, both through ex ante screening of data mining applications for possible risks and ex post checking of results.[30]

8.2.4.3 Transparency

The principle of transparency requires that persons be informed about aviation security measures, the consequences to their rights, purposes of data processing, the logic used in algorithms to determine assumptions and predictions about them, etc. Currently, as discussed, despite some positive developments, there is a lack of transparency. Two reasons for this are the large amount of restricted information in

[25] *Ibid.* p. 41.

[26] Keohane (2003) p. 14.

[27] Keohane (2003) p. 15.

[28] Gellert and Gutwirth (2013) p. 529.

[29] Processor means a natural or legal person, public authority, agency or other body which processes personal data on behalf of the controller (GDPR Art. 4(8)).

[30] Schermer (2011) p. 52.

aviation security and the fact that "too much" transparency and information overload can be dangerous. Another reason is that the nature of aviation security implies a power imbalance, leading to informational imbalance between the aviation security organs and individuals. It is the organs that decide how much information is to be shared publicly.

Indeed, a general assumption is that aviation security should be secured, thus, much information about aviation security issues has restricted access. While keeping some details secret can be substantiated by security needs; in other cases, unnecessary secrecy can occur. As a result, a lot of methods, including collecting and processing personal data, take place in secret with little or no monitoring by the courts, by independent authorities, or by the public.

A problem is that information about the capabilities and operation of aviation security measures can be restricted not only for reasons of security, but in order to avoid human rights concerns. Thus, there are risks of hiding violations. This concerns multiple issues: from controversial questions about scans' images and saving options to the use of PNR for profiling, CCTV's enhanced capabilities, combined databases, integrated technologies, etc.

Some technological developments are able to make the situation more serious. As discussed, via camera surveillance at airports, biometric data can be collected in real time covertly and remotely, from a distant location or in motion. If the idea of integrating the body scans devices in airport building structures or in different objects at the checkpoints is realized, passengers will not even know at which place and moment they are inspected. Such invisible systems will be all the more capable of hiding the risks of privacy infringements.

Moreover, in civil aviation, information issues may spread beyond security: as noted in Chap. 3, many actors may be interested in silencing or in creating uncertainty about flying and related health issues: flying as such incurs more risks than body scanners.

How is it possible to have transparency if the processes of risk assessment and details about aviation security measures refer to a restricted category and are not known? How can one find out which algorithms are used to analyse passenger data or to determine "abnormal behaviour"? How is one to know in which databases one's personal data are kept, if databases are integrated and shared? How can you know about human rights risks if the risks are hidden and some data processing can be carried out invisibly? Of course, many of these questions are quite rhetoric. Nevertheless, some recommendations can be given.

First, complete and reliable information regarding the functioning and technical specifications of the technologies and their capabilities, reviewed and examined by competent authorities, including independent reviewers, should be provided and made publicly available in an appropriate form. At the same time, release of such information should be done with due caution, in order not to constitute a threat to transportation security by enabling terrorists to ascertain all the capabilities and limitations of the devices.

There should be specific requirements regarding the quality of the used equipment/technology. Legislators must specify the technology design, defining the goals that the technology must achieve.[31] Technical specifications should include limitations on personal data collection, storage or other type of processing. The requirement of a system of encryption, password or similar security, different identification/authentication mechanisms as well as other technical measures enhancing privacy and data protection should be provided and regulated by law. All technical requirements should not only be established by law, but also be accompanied by control and enforcement mechanisms.

For instance, rules on CCTV should contain clear limitations on CCTV capabilities and operation rules. A separate EU legislative measure and more detailed national provisions are also recommended at the airport level – CCTV Codes of Practices at individual airports. Examples of good practice (although, with a number of shortcomings) are the UK Codes and some airports' practices.

With reference to automated profiling, "algorithmic transparency" is essential. The knowledge about data processes that impacts individuals, the limits of profiling and data mining should be raised, including the logic used in algorithms to determine assumptions and predictions about them, and statistics on the effectiveness and failure rates, checked by a panel of experts and legal professionals.[32] As noted in Chap. 7, according to "transparency by design", the legal duty to be transparent should be integrated into the design of algorithms.

Certainly, providing knowledge about technical/operative features requires cooperation of legal professionals and aviation security professionals, technologists, computer scientists and other experts. Today, they tend to use different vocabulary. Thus, in order to provide transparency and establish other necessary safeguards, it is recommended that the ties between these different fields of science be strengthened.[33]

In summary, taking into account the justifiable limitations due to security needs, individuals must be provided with appropriate, comprehensive and clear information about the applicable security measures – as open and adequate as possible. They must be given clear information on what data is processed, including data observed or inferred about them. They must be better informed on how and for what purposes their information is used, and they should be informed regarding the protection of their rights and respective mechanisms. Ideally, they should be informed before purchasing the tickets and before travelling.

It is also recommended that procedural requirements be improved regarding details about how the information can be provided to the passengers in reality, including responsible persons/entities.

With this regard, websites of all relevant stakeholders, in particular, state organs, airports, and airlines, should contain updated and complete information about the applicable measures. Relevant brochures should be available at offices of airlines or tour operators and at the airports. At the airports, display boards near the screening points can contribute too. While it may be impractical to reveal all the details about

[31] Pocs (2013) p. 39.

[32] Schermer (2011) p. 52.

[33] *Ibid.*

technical capabilities on posted signs concerning CCTV, some alternative/supporting forms should be considered and developed.

Technology will very likely also contribute towards ensuring transparency. New, innovative ways should be developed to inform the public, including innovative and privacy-friendly engineering and organizational measures and business practices.[34] The latter can be developed, for instance, in the airports.

8.2.4.4 Conclusion

Transparency and accountability belong to the measures that are supposed to enhance resilience in surveillance societies.[35] A better transparency and accountability regarding privacy can be helpful for security as well: "By exposing security interests to sunlight and heeding liberty interests, the government could ultimately be more accountable to the people. The result might be not only better protection of liberty but also more thoughtful and effective security".[36] Thus, the regulation should be combined with transparency and control mechanisms. However, another question is how to ensure them in reality – apparently, problems similar to those discussed in respect of law enforcement in Sect 8.2.3 may arise.

Other procedures intended to contribute to safeguard privacy in the use of technology may include different undertakings falling within policy and technical measures: privacy trainings, internal privacy rules, the use of Privacy Impact Assessment (PIA) and Privacy by Design (PBD), etc. PIA and PBD will be discussed in more detail below.

8.2.5 Privacy Impact Assessment

A useful procedure for technology-specific regulation and beyond is Privacy Impact Assessment (PIA).[37] PIA may be defined as "a systematic process for evaluating the potential effects on privacy and data protection of a project, initiative, proposed system or scheme and finding ways to mitigate or avoid any adverse effects".[38] Certainly, PIA can be applied in many different spheres and concern laws, technologies, business strategies, etc.

With respect to technologies, PIA is designed to engage the public in consultation aimed at considering the impacts and concerns, possible alternatives or

[34] European Data Protection Supervisor (2015b) p. 14.

[35] For more details, see Wright et al. (2015) p. 287.

[36] Solove (2008a) p. 362.

[37] Another term in use is data protection impact assessment (DPIA). In this work, PIA and DPIA are synonymous.

[38] De Hert. *The proposed data protection Regulation replacing Directive 95/46/EC: A sound system for the protection of individuals*. In: Computer Law & Security Review. Vol. 28 (2012) p. 140.

safeguards before use of any new technology, assisting to anticipate the public's reaction and prevent costly redesign. The UK CCTV Code 2014, for instance, states that surveillance cameras should be used before deploying thorough PIA, with the aim to assist in complying with the law.[39] PIA includes analysis of less intrusive alternatives and solutions and establishes whether such measures may be limited to certain periods and/or locations.[40]

Wright notes two reasons for PIA emergence: first, public reaction against the increasingly privacy-invasive actions of governments and secondly, governments and organizations became interested in PIA too, appreciating that privacy is now a strategic variable and putting it into their risk assessment and risk-management frameworks.[41] The results of a PIA may differ: from "yes or "no" to a proposed technology/measure to freezing the whole method or its parts until further developments, implementation of additional safeguards, restrictions, etc. Accordingly, PIAs have close relation to PBD discussed below, since it is much easier to develop privacy-friendly technologies if privacy/data protection issues are considered early in their design stage.[42]

PIA processes mentioned in this research include those in respect of body scanners, PNR, and CCTV. They relate to the EU, UK and USA. Other states which actively use PIA are Australia, Canada, Hong Kong, and New Zealand.[43] There are differences in approach between the existing PIA methodologies. The one developed by the UK Information Commissioner's Office (ICO), for example, places an emphasis on consultations with relevant stakeholders. For instance, it is common to organize public consultations with relevant stakeholders, e.g. such consultations were carried out with reference to body scanners and CCTV.

In the USA, PIAs are to be posted on the websites of the government departments that undertake them. A number of PIAs can be found on the DHS' website. According to the DHS, it conducts a PIA when:

- Developing or procuring any new technologies or systems that handle or collect personal data.
- Creating a new program, system, technology, or information collection that may have privacy implications.
- Updating a system that results in new privacy risks.
- Issuing a new or updated rulemaking that entails the collection of personal data.[44]

A number of shortcomings can be noted. Certainly, PIAs are best before implementation. In reality, PIAs are often carried out post-factum, as for example in the case of scanners in the EU. In some EU member states, scanners were used before

[39] Reeve (2014).

[40] Aquilina (2010) p. 139.

[41] Wright et al. (2010) p. 352.

[42] De Hert and Papakonstantinou (2012) p. 140.

[43] Wright et al. (2010) p. 352.

[44] DHS. PIA. http://www.dhs.gov/privacy-impact-assessments.

any PIA. EU PIA was performed before common EU regulation on scanners.[45] After implementation, as experience shows, it is possible to rectify something via a more complicated procedure: public debate, followed by (i) developing technological/operational solution by the manufacturers/security organs and then (ii) appropriate regulative initiatives/adoption of regulation, or vice versa: first (ii) then (i).

Other problems include the fact that PIAs may assess the impact on individuals, rather than on society as a whole[46]; they may focus on data protection rather than privacy; PIAs can be carried out in a non-transparent way; they may deal with existing technologies rather than those expected to emerge[47]; PIAs are usually adopted in the form of soft law and are, hence, only recommendations and lack sufficient control and enforcement mechanisms etc.

In general, a regulatory requirement should make it mandatory to conduct a PIA before use (or amendment of use) of any new technology/feature of the existing technology. An important step is the GDPR provision which states: "Where a type of processing in particular using new technologies, and taking into account the nature, scope, context and purposes of the processing, is likely to result in a high risk to the rights and freedoms of natural persons, the controller shall, prior to the processing, carry out an assessment of the impact of the envisaged processing operations on the protection of personal data".[48] PIA must contain at least a description of the processing operations and their purposes; assessment of the necessity and proportionality; assessment of the risks to the rights and freedoms of data subjects; measures envisaged to address the risks.[49]

However, it cites cases when PIAs are particularly required (evaluation of persons based on automated processing, including profiling, processing of large scale of sensitive data, a systematic monitoring of a publicly accessible area on a large scale).[50] Although it is stated that the supervisory authority shall make a list of types of processing operations that are subject to PIA,[51] this can be considered as a narrowing of the potential scope of application of obligatory PIAs.

Further, PIAs should be conducted at the national, regional and international levels. Nevertheless, international PIAs are particularly important, since aviation security is an international area, and there is "a need to be aware of the international context, of the development of transnational networks of technology providers and of the international efforts"[52] in aviation security. The use of PIA should be broadened in all dimensions: it should use positive examples of environmental and other

[45] European Commission. *Impact Assessment on the possible use of security scanners at EU airports* of 23 March 2011.

[46] Raab (2012b) p. 382.

[47] Wright et al. (2010) p. 353.

[48] Art.35(1).

[49] Art.35(7).

[50] Art.35(3).

[51] Art.35(4).

[52] Wright et al. (2010) p. 353.

impact assessments[53]; cover all dimensions/types of privacy[54]; address all areas and emerging technologies.

Moreover, since the technology may have impact on other human rights, PIA should include also assessment the compatibility of a proposed measure with all applicable individual's rights.[55] In civil aviation, an effective way of using PIAs could be evaluating aviation security technology's impact on all human rights, including the right to life/security. Thus, PIA can be broadened and transformed to Human Rights Impact Assessment (HRIA).

Accordingly, during HRIA, the security benefits of the technology should be evaluated and taken into account too. Thus, impact assessment should include consideration of the concrete technology's all-sides impact, including the measure's ability to protect the right to life and security.

In addition, in order to facilitate both security effectiveness and respect for human rights, such HRIA could be integrated with improved processes for evaluations of effectiveness. As noted in Sect. 4.5.5, testing, evaluation, and performance measurement can enhance effectiveness, but this is not always sufficiently performed. Of course, many problems may appear with this regard due to differences in interests, motivations, vocabulary, etc. For instance, it is clear that legal professionals, technologists, aviation security professionals and other interested persons use different vocabularies. This requires procedures for the interdisciplinary cooperation between all these experts.[56] Thus, proper cooperation can contribute.

8.2.6 Concluding Remarks

For the aviation security-privacy dilemma, a promising solution may be a combination of all the discussed forms of regulation and other mechanisms, with clear police guidelines and new laws.

Currently, in aviation security, there is some progress in respect of specific regulation of some of the measures, e.g. detailed provisions on PNR transfer in the EU Directive and EU-US Agreement, PIAs and Code of practices in some of the states. Nevertheless, as this research reveals, in general, the basic nature of legal regulation of technology- and security-driven processes and privacy in aviation security can be described as relying heavily on general technology-neutral legislation, lacking technology-specific legislation at all levels. Moreover, the existing regulation lack precise detail not only on the protection of individual's privacy and other rights, but also on the technology itself and how it is supposed to protect – life – even when taking into account the need not to jeopardise the security.

[53] De Hert and Papakonstantinou (2012) p. 141.

[54] Wright et al. (2010) p. 353.

[55] Gellert and Gutwirth (2013) p. 529.

[56] Pocs (2013) p. 48.

Thus, the advantages of using both technology-neutral regulation and technology-specific regulation in an optimal way – the former as primary legislation, establishing common rules by making them neutral, and the latter as secondary, increasing legal certainty in respect of particular technologies – has not yet been tackled adequately in the civil aviation context.

Consequently, many aviation security practices constitute examples of Space-Age technology, but Stone-Age law.[57] The legal basis that regulates and describes technology and privacy protection is antiquated in contrast to modern technology. Accordingly, the current law fails to adequately define both the amount of desired privacy and the amount/limits of the security measure involved.

8.3 Privacy by Design

The Privacy by Design (PBD) concept suggests that the technologies and relevant organizational measures could be developed in a privacy-friendly way from the start, covering all life cycle from design to deployment and operation until discarding and replacement.[58] PBD aims at building privacy – and data protection – into the design specifications and architecture of information and communication systems and technologies.[59] The key idea is that protection should be ensured proactively instead of reactively and to prevent privacy issues instead of devising remedial solutions.[60]

This concept – called "Data protection by design and by default" is stipulated in the GDPR. According to its Recital 78 and Article 25, the controller must take appropriate technical and organizational measures at the time of the determination of the means for processing and at the time of the processing itself to ensure that the data protection requirements are met. These measures include minimizing personal data processing, pseudonymising personal data, transparency with regard to the functions and processing, improving security features, etc. Moreover, certification mechanism pursuant to GDPR Article 42, which must be voluntary and transparent, may be used in order to ensure compliance with the requirements. The aim is to ensure that data protection is included in the design phase of a new technology and that, by default, only data which are necessary are processed.

---

[57] Blank (2002).

[58] This concept was proposed in 2009 by Ontario's Information and Privacy Commissioner Anne Cavoukian. It implies that privacy cannot be assured solely by compliance with regulatory frameworks; rather, privacy assurance must become an organization's default mode of operation. It suggests that privacy is best protected by a comprehensive strategy that embraces organizational, technical, and legal measures. This concept therefore extends to a "trilogy" of encompassing applications:(1) IT systems; (2) accountable business practices; and (3) physical design and networked infrastructure.

[59] European Data Protection Supervisor (2015b) p. 14.

[60] Badii et al. (2012) p. 366.

As revealed by this research, aviation security relies heavily on technology to minimize human interference in the security process and particularly making security decisions.[61] Technologies are the result of design and other decisions made by people; instructions are specified in the computer software, etc.[62] A key word here is automation, which, as discussed, may refer to any of the selected measures. Instead of humans, the systems rely on automatic threat recognition, automatic facial recognition and behaviour analysis, automated decisions and predictive analysis, automatic categorization of a person to a high or low risk, etc. Another positive factor of automation and minimizing the human factor is reduction or elimination of the possibility of corruption, e.g. "agreement/collusion" between terrorists and the security officer at the check point. Of course, a human may still be required to take part in the processes, to control the operations, but the main work load is done by the machine that is able to work 24/7.

Although these automatic applications can be effective, they are subject to various failures, leading for example to loss or leaks of data, function creep, etc. They are not very accurate, leading to false negatives and false positives, which appear to be a problem for many aviation security technologies discussed in this work. "It is therefore essential that good engineering practice be put into place to develop contingency plans for potential failures, as well to encourage system design to minimise the chance of failure and to make systems failsafe where faults or failures are inevitable".[63] But, as discussed, although technologies are potentially able to improve performance and reduce failures and failure rates, it is difficult if not impossible to exclude them. False alarms will apparently continue to be a critical issue for the security industry.

From security prospects, the biggest problem of false alarms is false negative, meaning that a dangerous person can go undetected. From human rights prospects, the main problem is false positives. Clearly, a minor error may cause negative consequences for a passenger, from going through additional security checks to denial of boarding, arrest and so on. At the same time, as discussed, these people may have a limited opportunity to find out why they are selected and the data that is collected pertaining to them, and to have incorrect data deleted.

Another problem in respect of human rights is that automatic techniques may stigmatize persons with certain background or characteristics, since algorithms are based on definite criteria, possibly including sensitive information. Thus, they may result in discrimination, re-enforcement of stereotypes, social and cultural segregation and exclusion.[64] Also, as noted above, the algorithms of automation processing and decision making are not known to the public, as a rule. If documentation does not exist and decisions become increasingly automated, as EPIC notes, "processes

[61] Kirschenbaum. *A personal view*. In: Aviation Security International. Vol. 20 (2014) p. 44.

[62] Schartum. *Making Privacy By Design Operative*. In: International Journal of Law and Information Technology (2016).

[63] Gilbert (2007) p. 21.

[64] European Data Protection Supervisor (2015b) p. 8.

would become increasingly opaque and less accountable, posing risks of secret profiling".[65]

A positive side is that people's instructions to software determine many parameters, including those of data processing. Privacy rules can hence be programmed to be automated. In contrast to automation of security checks and other procedures (technology automation), this will result in automated application of the law – "legal automation".[66] If applied to privacy rules, the latter will imply "support of privacy without allocating time, knowledge, consciousness and motivation in each individual case to realize privacy laws".[67] Thus, the use of PBD in aviation security, especially in the context of automation systems – both technological and legal automation – can be relevant and promising. Necessary prerequisites for this would be integration into the design of legal knowledge of substantive privacy/data protection issues and knowledge of technology.[68]

If applied properly, PBD may contribute towards solving or mitigating some of the privacy/data protection problems, in particular by limiting collection and processing of personal data. For instance, the systems can be programmed to target and collect data on only specific individuals rather than all persons; storage of data can be selective where the system indicated unusual events; automatically encrypted, anonymized or pseudonymous data are generally less vulnerable than open data,[69] etc.

Automated access to personal data only, i.e. without any persons involved, can be seen as less intrusive than a situation that permits humans to access these data. As discussed, machine assessment of quality images produced by scanners or manual searches conducted by a robot can be seen as improvements in this direction. The discrimination issues that are probably impossible to exclude (see Chap 6) can be further improved technologically in order to better target specific persons on more objective grounds and reduce failure rates.

Currently, however, although there are some positive examples of how PBD works in reality,[70] it is more usual to try to fix a technology by adding privacy safeguards ex post.[71] For instance, in the case of body scanners, safeguards were developed and introduced post-factum.

A more serious risk is that some technical or organizational undertakings can hide or mask privacy/data protection threats. For instance, as discussed in Chap. 6, the language used for body scanners presents the scanning process as automatic, anonymized, universal, neutral, routine, and professional. It is claimed that no

---

[65] EPIC Alert Volume 22.23, 11 Dec 2015. https://www.epic.org/alert/epic_alert_22.23.html.

[66] See more detail on "legal automation" in Schartum (2016).

[67] *Ibid.*

[68] *Ibid.*

[69] Schartum. *Personvern og transportsikkerhet: personvernmessige spørsmål knyttet til tiltak for å sikre transportmidler mot fiendtlige anslag* (2007) pp. 169–170.

[70] See, e.g. Balboni and Macenaite. *Privacy by design and anonymisation techniques in action: Case study of Ma3tch technology*. In: Computer Law & Security Review. Vol. 29 (2013).

[71] Tirosh and Birnhack (2013) p. 1270.

images are created. It is believed that modern millimetre wave scans are much more privacy-friendly than the old "naked" devices. But the controversial issues about the existence of images and saving options constitute ground for questioning whether the technological undertakings fall within PBD or are simply hiding old concerns.

The same may refer to anonymization, masking, blurring faces and similar procedures which are supposed to be used for images of CCTV and scanners, sensitive data processed in the course of PNR usage, etc. On closer examination, these undertakings can easily be reversed, de-anonymized, making a person identifiable again. In addition, the use of Big Data allows identification even using non-personal data. However, as discussed, this is not the reason not for using anonymization at all, since even if imperfect, it can contribute to complying with a number of data protection principles.

Overall, due to the controversial issues, one may ask whether the PBD is applied in a process that "ultimately enhances organizational power to the detriment of individuals," whether PBD is used for the introduction of a privacy-invasive technology; which data protection standards it reflects, and how open, inclusive, and transparent the process is.[72]

In the situation when there is imbalance of powers in aviation security, and it is the security regulators who possess true information about the measure and decide how much of this information to make available to the public (see above), it may difficult if not impossible to get reliable information about aviation security technologies and to differ whether a particular feature is safeguarding privacy or hiding privacy threat.

As a result, although the ideas, aims and attempts to adopt laws on PBD application are considered positive, the concept of PBD seems to be easier declared on paper than implemented in reality, and it is still "a provocative concept and a challenging obligation".[73] Moreover, laws play a dual role with respect to PBD: they can encourage using it, but at the same time may incidentally hinder certain of its forms due to the basic character of the norms, vague and discretionary nature, absence of significant incentives, etc.[74] This duality leads to the difficulty that PBD will be developed to its full potential.[75]

Nevertheless, some recommendation can be given. First, in terms of globalization, the use of aviation security technologies internationally and when the laws generally fail to provide adequate instructions how to implement PBD, the use of general legal principles might be more fruitful. In Chaps. 1 and 5, I explain why and which exact legal principles were chosen for this research as well as methodological difficulties associated with the search and choice of concrete principles. Similarly, for PBD, legal privacy/data protection principles, carefully selected and checked against applicable law, can be used as the basis. Accordingly, PBD can be used for

[72] Bygrave. *Hardwiring privacy (forthcoming)* In: The Oxford Handbook on the Law and Regulation of Technologies (2016).

[73] Hildebrandt and Tielemans (2013) p. 520.

[74] Bygrave (2016).

[75] Schartum (2016).

ensuring legal principles, in particularly on the global level. With this regard, further regulation on proper interpretation and use of these principles mentioned above can be helpful.

Secondly, taking into account interconnection and application of various human rights in the course of aviation security, similarly to PIA/HRIA, PBD should be used more broadly in respect of all human rights, extending to Human Rights by Design.

Thirdly, PBD should be supported by appropriate organizational and technical procedures to ensure that the technology solutions aimed to protect privacy would actually do so.[76] Thus, when implementing PBD, it is necessary to go beyond the idea of PBD by looking towards the underlying issues involved in a particular technology. For instance, with reference to automated profiling, the creation of independency constraints can minimize the risk of discrimination.[77] In particular, the transparency and accountability undertakings discussed above may contribute greatly.

In the introduction phase, a clear description should be presented of the security measure and its objectives as well as its advantages in relation to the existing measures. The preparatory process should include an open and transparent debate. If despite these undertakings there are still privacy or other human-rights concerns, it should be analysed whether the measure can be redesigned, supplemented or substituted by another measure, or whether new measures can be developed to remove these concerns, i.e. before its implementation. Moreover, since technologies of the future are not static and are subject to quick changes and upgrades that may invade privacy, their development should be monitored in order to foresee potential abuses.[78]

If the measure or its feature is already widespread, and serious concerns are found at this stage, a proper decision would be to limit or not use the feature/technology, but this can be done only by considering the effectiveness, necessity and non-excessiveness. This is not so easy to be implemented in reality: despite concerns, body scans, CCTV and PNR are used globally, and their non-usage may reduce aviation security (see more detail below).

8.4 Improving Passenger Experience

> One of the biggest challenges facing security agencies is how to effectively screen people – ideally on the move – without hindering flow, without subjecting people to health risks, without invading their privacy and, most importantly, whilst ensuring that we can actually identify prohibited items.[79]

---

[76] Balboni and Macenaite (2013) p. 340.

[77] Schermer (2011) p. 52.

[78] Woodward (2002) p. 7.

[79] Body Search 2015, London, June 2015, www.bodysearchworld.com.

Two controversial aspects can be noted. On the one hand, as mentioned above, passenger convenience is indicated among drivers and future trends in aviation. The core idea is that time spent on screening and intrusive methods should be reduced to minimum, and the airport should be transformed into a great place to be. On the other hand, human rights concerns – that also relate to passenger experience – can cancel or otherwise prevent from using aviation security measures which are deemed to be effective for security.

8.4.1 Impact on Passengers

In general, airlines and airports have economic interest in passenger convenience, first of all economically: more satisfied air passengers mean better conditions for business. They already provide a multitude of services including hotel rooms, conference facilities, shopping malls, etc. Indeed, passenger experience is worth investing: for instance, people will then spend more time in retail than in security lines.[80]

Passenger experience is supposed to be improved by using more appropriate, less time consuming and less intrusive technologies. As discussed above, procedures within PIA/HRIA and PBD/HRBD, including increased use of automation can have positive impact on passengers' rights too.

For instance, the use of more privacy-friendly scanners instead of "naked" ones and the use of scanners instead of hand searches are often considered positive steps. Intelligent CCTV in airports can serve passenger convenience and provide clearer understanding of airport activity. It can assist airport operators in tracking, managing and sharing information about passengers and their luggage, inform passengers about expected waiting times, optimize passenger flow and the entire airport experience.[81]

Other common ideas are appropriate use of biometrics, which, as discussed in Chap. 6, can improve the passenger experience through automatic boarding, entrance to lounges and providing other benefits.

The key role in this regard belongs to trusted traveller programmes, which, as discussed in Chap. 6, allow the participating persons going through a reduced and expedited security screening which allows for a friendlier process and better passenger experience using more automation. According to TSA, other advantages of PreCheck are that a person may plan his/her travel with confidence and enjoy a more predictable experience; there is no need to remove shoes, laptops, liquids, belts, or light jackets.[82] However, the principle of unpredictability ensured by TSA means that no one can be guaranteed expeditious screening, and as discussed, the nature of TTP is discriminatory.

---

[80] Bradley (2015).

[81] Batt (2013) pp. 18–19.

[82] TSA PreCheck. http://www.tsa.gov/tsa-precheck .

But, although the technologies are able to make aviation security measures in some terms better, as discussed above, in most cases, they do not eliminate all the problems, appear to be insufficient, inadequate, future-oriented and dependent on technologies which are yet to be developed.

As the result, declarations coming from state authorities and security producers stating that privacy/data protection with regard to a particular technique will be respected, in reality, often turn out to be weak and insufficient, in the worst case – only "masking" infringements. Some state authorities or producers may state directly that this particular equipment, device, or technology "respects privacy", but it is a question of whether these statements can be trusted, as in the case of scanners. While some good practices exist, these are not enough.[83]

Moreover, improving one feature – e.g. time spent on screening – may lead to the use of security methods within the concepts of risk-based approach, on-the-fly biometric, and integrating multiple security systems. All these systems imply enhanced usage of profiling and biometrics thus raising additional human rights concerns.

As a result, it can be concluded that improving passenger experience via technologies does not necessarily lead to less privacy/data protection/human rights concerns.

One more idea to improve passenger experience is a shift from viewing the passenger as a dangerous threat to viewing the passenger as a valuable client.[84] The key role in realizing this will be that of security employees who will be able to both enforce security and treat passengers as valued clients.[85] Accordingly, appropriate training of employees can be helpful. This sounds promising, since it does not require any additional technologies and does not incur human rights concerns, but instead relies on communication and ways of treating customers. Who knows? Perhaps this "non-technological" approach – viewing the passenger as a valued client rather than a potential threat or as a person who might be carrying a bomb – is the future. But appropriate training and methods for this are yet to be developed,[86] and obviously, this will require many resources, both human and otherwise, resulting in an expensive regime, maybe much more expensive than current approaches.

8.4.2 Impact of Human Rights Concerns on Security

Taking into account the passenger's experience may not only lead to fewer, or more, human rights concerns, but may also lead to less effective measures. In some cases, a promising, necessary and effective measure in terms of technology may prove to be limited, amended, frozen, or prohibited. From the perspective of security experts, this may be a negative factor for security.

[83] Increasing Resilience in Surveillance Societies (IRISS) (2013) pp. 10–11.

[84] Kirschenbaum (2014) p. 44

[85] *Ibid.*

[86] *Ibid.*

For instance, as a consequence of a number of factors, including privacy concerns, the backscatter scanners that were more effective than MMW were removed from many world airports; mannequin figures are used for analysis instead of actual images providing less details; X-ray transmission body scans are prevented from being used even on limited basis; there are problems with enhanced use of data profiling, wider use of biometrics, etc. For instance, during 4 months, in 2006–2007, there was a test of biometrics use (fingerprints and iris) for the employees' access-control In Norway. The program was discontinued due to privacy/data protection concerns.[87]

As mentioned above, the reasons for non-use of effective practices can also be difficulties in incorporating the measure into airport and airlines' existing practices, financial concerns and other issues. But if the limitations or prohibition are mainly based on human rights concerns, in some cases this can be justified, as in the case of prohibition of routine use of scanners with high radiation risks. In other cases there might be a risk the decisions were based on less objective or speculative facts. As Baum notes:

> Sensationalist media campaigns have been "successful" in preventing the deployment of the most effective passenger screening technologies, thereby limiting our capabilities to prevent the next terrorist attack.[88]

As the result, "we continue to utilise the least effective systems and pin our future aspirations on bureaucratic, data-driven processes which imply that they "know" people. Most disturbing of all, data profiling cannot be applied worldwide…".[89]

The problem is that privacy advocates campaigns can be based on negative, erroneous information rather than on facts. For instance, in 2007 a Norwegian airport operator was preparing for a body scanner test in cooperation with Norwegian DPA. However, it so happened that the media got hold of the information and many errors appeared, wrong pictures from the backscatter, use of words such as naked scanning, etc. Thus, under the pressure from the Transport Minister, and under claim that the machines usage infringes privacy, the tests were never implemented.[90]

The states are obliged to provide aviation security as effectively as possible. If a terrorist attack occurs, no one will listen to explanations that a particular technique which might help in preventing the threat was not used due to privacy concern. Everybody will just say their security organs failed to do their job. In this situation, a proportionality test should be applied – as discussed in Chap. 7, if threats and benefits are proportionate to human rights concerns, then the measure can probably be used.

---

[87] Enerstvedt (2012).

[88] Baum. *Emerging technologies: it's not all about X-rays*. In: Aviation Security International. Vol. 19 (2013). p. 12.

[89] Baum (2015c) p. 1.

[90] Communication with Avinor representative 28 June 2011.

8.5 Concluding Remarks

The key roles in aviation security and privacy dilemma, apparently, belong to both the regulation and the technologies. In theory, if properly developed and applied, they are able to protect both security – and the right to life – along with other rights of the individual, ensuring the main thesis of this work: aviation security *plus* privacy. As discussed, the law has a potential to define both the amount of desired privacy as well as data protection requirements and the amount/limits of the security measure involved. The technology is able to ensure not only aviation security benefits, but contribute to data minimization, data security, non-discrimination, etc.

Certainly, these issues should be handled in the stage of technology development, with peer review and assessment, including both security benefits and privacy impact assessment. But as discussed, it is easier said than done. It is common that a technology is first developed, and only afterwards the regulators try to introduce legislation, consider safeguards, etc. The discussed mechanisms and procedures which can be helpful are subject to improvements.

The selected technologies – scans, CCTV, and PNR are arguably more or less proportionate if violations are compared with the level of threats and security benefits. However, this does not mean that situation should be *status quo*. In all the jurisdictions, there is a potential for improvement in terms of privacy and data protection, via both legal and non-legal means. However, there are certain difficulties, and not all proposals for improvement are practical.

Moreover, the technological developments can be both positive and negative for both aviation security and individual's rights, making the situation more complicated. The current processes as well as suggestions for the future show that privacy/data protection rights may be strengthened and weakened at the same time.

The key point is that the latter processes are developing to a much greater extent than the former, not only through development of more invasive technologies as such, but also by more general tendencies such as pro-active and risk-based approaches, which means enhanced use of profiling, increased global data share, combination of various measures, endeavours to make the security process invisible, etc. Although all this may enhance security, the negative impact on individuals may grow in geometric progression.

In addition, while the technological safeguards to protect individuals can be improved, there are certain limitations. Thus, it may be easier to develop measures that simply mask human rights infringements. The latter tendency is particular source of worry.

Thus, despite privacy and data protection regulation, despite the endeavours of aviation security stakeholders to improve passenger experience, including privacy, and despite the pressures of privacy advocates and non-governmental organizations, the respective rights of air passengers are still vulnerable.

A more global point is that surveillance is increasing worldwide. The analysed states, in particular, the USA, the UK and Russia are "pioneers" with respect of many surveillance methods, including the most enhanced aviation security technologies. The number of states deploying similar regimes is growing. At the same time, the states with "adequate data protection level" in the EU terms[91] or, broader, adequate human rights situation are the vast minority in global terms. The majority may face similar challenges and problems as those encountered by the selected states, both with regard to legislation gaps, abuses by the state, or the fact that the laws do not work. All this creates global possibilities for abuses and violations of air passengers' rights. The selected regimes can thus be considered as only examples of many regimes, including future regimes.

Globally, the cases of scans/CCTV/PNR, upon closer look, reveal a number of critical issues: aviation security versus privacy dilemma, privacy and data protection concerns, problems of internal regulation and law enforcement, enhanced and unlimited surveillance, underdevelopment of democratic values, etc.

The point is that the selected measures form a part of national security strategies, where the powers of the international community, other states or courts are limited. The majority of the problems have internal, national roots. Thus, national endeavours constitute the key factors, and a broader, more complex approach is needed.

Moreover, due to the fact that civil aviation is widely recognized as carrying an enhanced risk of terrorism and crime, it presents an environment in which the most advanced technologies are deployed. Paramount in importance – after successfully implementing these technologies – is that they may then be spread to other areas of transport, to other crowded places, to other new areas of the public realm. This means that if no proper limits on surveillance are set in the earlier stages, privacy and other human rights may ultimately be affected to an undesirable extent. The watchful eyes of the authorities may realistically cover much broader areas than civil aviation. By extension, these other areas may represent domains of personal privacy into which so-called 'justified' encroachment by public authorities would be a fundamental violation of freedom.

[91] For the list of these states, see Sect. 7.4.4. As noted, the decisions on the adequacy do not actually mean that the protection in these countries is adequate in practice.

Abbreviations

ACDPDP	Automated centralized databases of personal data on passengers (Russia)
ACI	Airports Council International
ACLU	American Civil Rights Union
AIASP	Association of Independent Aviation Security Professionals
AIT	Advanced Imaging Technology
API	Advanced Passenger Information
ATR	Automated Target Recognition
CAPPS	Computer Assisted Passenger Prescreening System
CBP	Customs and Border Protection (USA)
CCTV	Closed Circuit Television
CFREU	Charter of Fundamental Rights of the European Union
CPO	Chief Privacy Officer (USA)
CRS	Computer Reservation System
DCS	Departure Control System
DfT	Department for Transport (UK)
DHS	Department for Homeland Security (USA)
DPA	Data Protection Authority
DPA 1998	Data Protection Act 1998 (UK)
DPD	Data Protection Directive 95/46/EC (EU)
EC	European Community
ECAC	European Civil Aviation Conference
ECHR	European Convention on Human Rights
ECtHR	European Court of Human Rights
ECJ	European Court of Justice
EDRI	European Digital Rights Initiative
EDPS	European Data Protection Supervisor
EEA	European Economic Area
EOS	European Organization for Security

O.M. Enerstvedt, *Aviation Security, Privacy, Data Protection and Other Human Rights: Technologies and Legal Principles*, Law, Governance and Technology Series 37, DOI 10.1007/978-3-319-58139-2

EP	European Parliament
EPIC	Electronic Privacy Information Centre
EU	European Union
FBI	Federal Bureau of Investigation (USA)
FRA	Fundamental Rights Agency (EU)
FSB	Federal Security Service of the Russian Federation
FOIA	Freedom of Information Act
GAO	Government Accountability Office (USA)
GDPR	General Data Protection Regulation (EU)
GDS	Global Distribution Systems
HRBD	Human Rights by Design
HRIA	Human Rights Impact Assessment
IATA	International Air Transportation Association
ICAO	International Civil Aviation Organization
ICCPR	International Covenant on Civil and Political Rights
ICESCR	International Covenant on Economic, Social and Cultural Rights
ICO	Information Commissioner's Office (UK)
ICT	Information and Communication Technologies
IED	Improvised Explosive Device
IP	Internet Protocol
ISIL	The Islamic State of Iraq and the Levant, also known as IS (The Islamic State)
IT	Information Technology
MMW	Millimetre wave scanners
NSA	National Security Agency (USA)
OECD	Organization for Economic Cooperation and Development
OSI	Other service related information
PBD	Privacy by Design
PIA	Privacy Impact Assessment
PIU	Passenger Information Units
PNR	Passenger Name Records
RF	Russian Federation
SARPs	Standards and Recommended Practices (ICAO)
SmartS	Smart Security
SPOT	Screening of Passengers by Observation Techniques
SSI	Special Services Information
SSR	Special Service Requests
TFEU	Treaty on Functioning of the European Union
TSA	Transportation Security Administration (USA)
TTP	Trusted Traveller Programmes
UDHR	Universal Declaration of Human Rights
UK	United Kingdom of Great Britain and Northern Ireland
UN	United Nations
USA	United States of America
USISTS	Unified state information system of transport security (Russia)
USSR	The Union of Soviet Socialist Republics

Annexes

Annex 1: ICAO PNR Guidelines, PNR Data Elements

An operator's system(s) may include the following data elements:

Data groups or categories	Component data elements
PNR name details	Passenger name, family name, given name/initial, title, other names on PNR
Address details	Contact address, billing address, emergency contact, email address, mailing address, home address, intended address [in State requiring PNR data transfer]
Contact telephone number(s)	[Telephone details]
Any collected API data	Any collected API data, e.g. name on passport, date of birth, sex, nationality, passport number
Frequent flyer information	Frequent flyer account number and elite level status
PNR locator code	File locater number, booking reference and reservation tracking number
Number of passengers on PNR	[Number]
Passenger travel status	Standby information
All date information	PNR creation date, booking date, reservation date, departure date, arrival date, PNR first travel date, PNR last modification date, ticket issue date, "first intended" travel date, date of first arrival [in State requiring PNR data transfer], late booking date for flight
Split/divided PNR information	Multiple passengers on PNR, other passengers on PNR, other PNR reference, single passenger on booking
All ticketing field information	Date of ticket issue/purchase, selling class of travel, issue city, ticket number, one-way ticket, ticket issue city, automatic fare quote (ATFQ) fields

(continued)

O.M. Enerstvedt, *Aviation Security, Privacy, Data Protection and Other Human Rights: Technologies and Legal Principles*, Law, Governance and Technology Series 37, DOI 10.1007/978-3-319-58139-2

Data groups or categories	Component data elements
All travel itinerary for PNR	PNR flight itinerary segments/ports, itinerary history, origin city/ board point, destination city, active itinerary segments, cancelled segments, layover days, flown segments, flight information, flight departure date, board point, arrival port, open segments, alternate routing unknown (ARNK) segments, non-air segments, inbound flight connection details, on-carriage information, confirmation status
Form of payment (FOP) information	All FOP (cash, electronic, credit card number and expiry date, prepaid ticket advice (PTA), exchange), details of person/agency paying for ticket, staff rebate codes
All check-in information*	Generally available only after flight close-out: check-in security number, check-in agent I.D., check-in time, check-in status, confirmation status, boarding number, boarding indicator, check-in order
All seat information	Seats requested in advance; actual seats only after flight close-out*
All baggage information*	Generally available from DCS only after flight close-out: number of bags, bag tag number(s), weight of bag(s), all pooled baggage information, head of pool, number of bags in pool, bag carrier code, bag status, bag destination/ offload point
Travel agent information	Travel agency details, name, address, contact details, IATA code
Received-from information	Name of person making the booking
Go-show information*	Generally available only after check-in and flight close-out: go-show identifier
No-show information*	Only available after flight close-out: no-show history
General remarks	All information in general remarks section
Free text/code fields in OSI, SSR, SSI, remarks/ history	All IATA codes

*These elements are contained in the DCS and are not available prior to departure. A recommendation has been made to the World Customs Organization (WCO) to consider incorporating these elements in future API messaging. Depending on the airline system, these elements may or may not be part of a PNR

Annex 2: EU-US PNR Agreement, PNR Data Types

1. PNR record locator code
2. Date of reservation/issue of ticket
3. Date(s) of intended travel
4. Name(s)
5. Available frequent flier and benefit information (i.e., free tickets, upgrades, etc.)
6. Other names on PNR, including number of travellers on PNR
7. All available contact information (including originator information)

8. All available payment/billing information (not including other transaction details linked to a credit card or account and not connected to the travel transaction)
9. Travel itinerary for specific PNR
10. Travel agency/travel agent
11. Code share information
12. Split/divided information
13. Travel status of passenger (including confirmations and check-in status)
14. Ticketing information, including ticket number, one way tickets and Automated Ticket Fare Quote
15. All baggage information
16. Seat information, including seat number
17. General remarks including OSI, SSI and SSR information
18. Any collected APIS information
19. All historical changes to the PNR listed under points 1 to 18

Annex 3: EU PNR Directive, Passenger Name Record Data as Far as Collected by Air Carriers

(1) PNR record locator
(2) Date of reservation/issue of ticket
(3) Date(s) of intended travel
(4) Name(s)
(5) Address and contact information (telephone number, e-mail address)
(6) All forms of payment information, including billing address
(7) Complete travel itinerary for specific PNR
(8) Frequent flyer information
(9) Travel agency/travel agent
(10) Travel status of passenger, including confirmations, check-in status, no show or go show information
(11) Split/divided PNR information
(12) General remarks (including all available information on unaccompanied minors under 18 years, such as name and gender of the minor, age, language(s) spoken, name and contact details of guardian on departure and relationship to the minor, name and contact details of guardian on arrival and relationship to the minor, departure and arrival agent)
(13) Ticketing field information, including ticket number, date of ticket issuance and one-way tickets, automated ticket fare quote fields
(14) Seat number and other seat information
(15) Code share information
(16) All baggage information
(17) Number and other names of travellers on the PNR

(18) Any Advance Passenger Information (API) data collected (inter alia document type, document number, nationality, country of issuance, date of document expiration, family name, given name, gender, date of birth, airline, flight number, departure date, arrival date, departure port, arrival port, departure time, arrival time)
(19) All historical changes to the PNR listed in numbers 1–18.

Annex 4: Russian PNR System, PNR Data Types

General list of data collected for all transport modes:

(a) passenger personal data:

1. first name, last name, patronymic;
2. date and place of birth;
3. type and number of identity document which was used to purchase travel document (ticket);
4. place of departure, destination, type of the route (direct, transit);
5. date of travel.[1]

(b) data describing transactions recorded in the information system of the carrier or of the participant of transport infrastructure (the data on the operation recorded):

1. the type of recorded transactions;
2. the name of the carrier;
3. the date and time of the transaction in an automated information system of the carrier.[2]

Data for air transport:
API data fields:[3]

(a) passenger personal data:

1. names (first name, last name, patronymic or middle name, if applicable);
2. date and place of birth;
3. type and number of identity document which was used to purchase travel document (ticket);
4. type and number of identity document, which is acquired by a travel document (ticket);

---

[1]Article 11(5) of the Federal law of 9 February 2007 N 16-FZ On transport security and §8 of the Order of the Ministry of Transport of 19 July 2012 N 243 (Order).

[2]§9 of the Order.

[3]§51 of the Order.

5. place of departure, destination, type of the route (direct, transit, transfer);
6. date and time of aircraft departure.

(b) data describing transactions recorded in the information system of the carrier or of the participant of transport infrastructure (the data on the operation recorded):

1. name or code of the carrier which actually performs the flight;
2. flight number
3. arrival date and time.

PNR data fields:[4]

(a) passenger personal data:

surname,
name,
patronymic (if available, or middle name);
date of birth
place of birth (if available);
type of identity document which was used to purchase travel document (ticket);
number of identity document which was used to purchase travel document (ticket);
place of departure,
place of destination,
type of the route (direct, transit, transfer);
date and time of aircraft departure.

(b) data on the operation recorded:

name or code of the carrier which actually performs the flight:
flight number
code indicator of PNR;
unique number of the transport document;
number of passengers in the PNR entry, name, surname and patronymic of these passengers;
information about requested (actual) seats on the aircraft (if available);
reservation status code (through the route);
stopover points for flights with a stopover;
information about the dates (date of creation PNR, date of the reservation, date of last change in the PNR, date of formation of transport document, date of late booking on the flight);
time of formation of transport document;
data and code of the agency/carrier, who drew up the transport document;

[4] §52 of the Order.

information about the luggage and carry-on baggage (number of pieces, weight, bag tag numbers, weight of hand luggage);
information on check-in (control number of check-in, identifier of the agent upon check-in, number of check-in desk, check-in time, boarding number);
date and time of the aircraft arrival;
information about the passenger no show at the departure.

Selected Sources and Materials

Table of Legislation and Other Legal Texts

International

Agreement between the United States of America and the European Union on the use and transfer of Passenger Name Records to the United States Department of Homeland Security, the Council of the EU decision 17434/11 of 26 April 2012.
Convention for the Protection of Human Rights and Fundamental Freedoms of 4 November 1950 (Rome Convention).
Convention for the Suppression of Unlawful Acts against the Safety of Civil Aviation of 23 September 1971 (Sabotage Convention or Montreal Convention).
Convention for the Suppression of Unlawful Seizure of Aircraft of 16 December 1970 (Hague Convention).
Convention on the Marking of Plastic Explosives for the Purpose of Detection of 1 March 1991 (Montreal Convention 1991).
Convention on the Suppression of Unlawful Acts Relating to International Civil Aviation of 10 September 2010 (Beijing Convention).
Convention for the Unification of certain rules relating to international carriage by air of 12 October 1929 (Warsaw Convention).
Convention Implementing the Schengen Agreement of 19 June 1990.
Convention on Damage Caused by Foreign Aircraft to Third Parties on the Surface of 7 October 1952 (Rome Convention).
Convention on International Civil Aviation of 7 December 1944 (Chicago Convention).
Convention on Offenses and Certain Other Acts Committed on Board Aircraft of 14 September 1963 (Tokyo Convention).
Convention on the Rights of the Child of 20 November 1989.
Council of Europe. Convention for the Protection of Individuals with regard to Automatic Processing of Personal Data, ETS No.108, 28 January 1981 (Entry into force: 1 October 1985).
ICAO, Annex 17 to the Convention on International Civil Aviation, Security – Safeguarding International Civil Aviation Against Acts of Unlawful Interference. Ninth Edition, March 2011.
ICAO, Doc 9303, Machine Readable Travel Document, Seventh Edition, 2015.
ICAO, Doc 9735, Universal Safety Oversight Audit Programme Continuous Monitoring Manual, Third Edition, 2011.

O.M. Enerstvedt, *Aviation Security, Privacy, Data Protection and Other Human Rights: Technologies and Legal Principles*, Law, Governance and Technology Series 37, DOI 10.1007/978-3-319-58139-2

ICAO, Doc 9944, Guidelines on Passenger Name Record (PNR) data of 2010 (ICAO PNR Guidelines 2010).
International Covenant on Civil and Political Rights of 16 December 1966 (ICCPR).
International Covenant on Economic, Social and Cultural Rights of 16 December 1966 (ICESCR).
OECD Guidelines on the Protection of Privacy and Transborder Flows of Personal Data of 23 Sept 1980 (OECD Guidelines).
OECD Recommendation Concerning Guidelines Governing the Protection of Privacy and Transborder Flows of Personal Data, 11 July 2013 (OECD Privacy Framework 2013)
Protocol for the Suppression of Unlawful Acts of Violence at Airports serving International Civil Aviation of 24 February 1988, Supplementary to the Montreal Convention (Montreal Protocol).
Protocol No. 4 to the Convention for the Protection of Human Rights and Fundamental freedoms other than those already included in the Convention and in the first Protocol thereto, as amended by Protocol No. 11, 16 September 1963.
Protocol Supplementary to the Convention for the Suppression of Unlawful Seizure of Aircraft, Doc 9959, 10 September 2010 (Beijing Protocol).
United Nations General Assembly Guidelines for the Regulation of Computerized Personal Data Files, 14 December 1990.
Universal Declaration of Human Rights of 10 December 1948 (UDHR).

European Union

Charter of Fundamental Rights of the European Union (CFREU), proclaimed on 7 December 2000, entry into force from the date of entry into force of the Treaty of Lisbon on 1 December 2009, Official Journal of the European Union (2010/C 83/02), 30 March 2010.
Commission Decision 2000/520/EC of 26 July 2000 pursuant to Directive 95/46/EC of the European Parliament and of the Council on the adequacy of the protection provided by the safe harbour privacy principles and related frequently asked questions issued by the US Department of Commerce, C(2000) 2441.
Commission Decision of 28 June 2006 laying down the technical specifications on the standards for security features and biometrics in passports and travel documents issued by Member States, C(2006) 2909 final.
Commission Implementing Regulation (EU) No 1147/2011 of 11 November 2011 amending Regulation (EU) No 185/2010 implementing the common basic standards on civil aviation security as regards the use of security scanners at EU airports.
Commission Regulation (EC) No 272/2009 of 2 April 2009 supplementing the common basic standards on civil aviation security laid down in the Annex to Regulation (EC) No 300/2008 of the European Parliament and of the Council.
Commission Regulation (EU) No 185/2010 of 4 March 2010 laying down detailed measures for the implementation of the common basic standards on aviation security.
Commission Regulation (EU) No 1141/2011 of 10 November 2011 amending Regulation (EC) No 272/2009 supplementing the common basic standards on civil aviation security as regards the use of security scanners at EU airports.
Communication from the Commission On the global approach to transfers of Passenger Name Record (PNR) data to third countries, COM(2010) 492 final, Brussels, 21 September 2010.
Communication from the Commission to the European Parliament and the Council on the Use of Security Scanners at EU airports, 15 June 2010. COM(2010) 311 final.
Consolidated Version of the Treaty on the Functioning of the European Union, Official Journal of the European Union (C 83/47) 30 March 2010.

Council Decision 1999/439/EC of 17 May 1999 on the conclusion of the Agreement with the Republic of Iceland and the Kingdom of Norway concerning the latters` association with the implementation, application and development of the Schengen acquis.
Council Decision 2004/512/EC: 2004/512/EC of 8 June 2004 establishing the Visa Information System (VIS).
Council Directive 2004/82/EC of 29 April 2004 on the obligation of carriers to communicate passenger data.
Council Framework Decision of 13 June 2002 on combating terrorism (2002/475/JHA).
Council Framework Decision 2002/584/JHA of 13 June 2002 on the European arrest warrant and the surrender procedures between Member States – Statements made by certain Member States on the adoption of the Framework Decision.
Council Framework Decision 2008/977/JHA of 27 November 2008 on the protection of personal data processed in the framework of police and judicial cooperation in criminal matters.
Directive 95/46/EC of the European Parliament and of the Council of 24 October 1995 on the protection of individuals with regard to the processing of personal data and on the free movement of such data (DPD).
Directive 2002/58/EC of the European Parliament and of the Council of 12 July 2002 concerning the processing of personal data and the protection of privacy in the electronic communications sector (Directive on privacy and electronic communications).
Directive 2004/38/EC of the European Parliament and of the Council of 29 April 2004 on the right of citizens of the Union and their family members to move and reside freely within the territory of the Member States.
Directive 2006/24/EC of the European Parliament and of the Council of 15 March 2006 on the retention of data generated or processed in connection with the provision of publicly available electronic communications services or of public communications networks.
Directive (EU) 2016/680 of the European Parliament and of the Council of 27 April 2016 on the protection of natural persons with regard to the processing of personal data by competent authorities for the purposes of the prevention, investigation, detection or prosecution of criminal offences or the execution of criminal penalties, and on the free movement of such data, and repealing Council Framework Decision 2008/977/JHA.
Directive (EU) 2016/681 of the European Parliament and of the Council of 27 April 2016 on the use of passenger name record (PNR) data for the prevention, detection, investigation and prosecution of terrorist offences and serious crime.
European Parliament resolution of 5 May 2010 on the launch of negotiations for Passenger Name Record (PNR) agreements with the United States, Australia and Canada.
European Parliament resolution of 6 July 2011 on aviation security, with a special focus on security scanners (2010/2154(INI)).
Proposal for a Regulation of the European Parliament and of the Council establishing a Registered Traveller Programme, COM(2013) 97 final – 2013/0059 (COD), Brussels, 28 Feb 2013.
Regulation (EC) No 45/2001 of The European Parliament and of the Council of 18 December 2000 on the protection of individuals with regard to the processing of personal data by the Community institutions and bodies and on the free movement of such data.
Regulation (EC) No 300/2008 of the European Parliament and of the Council of 11 March 2008 on common rules in the field of civil aviation security.
Regulation (EC) No 767/2008 of the European Parliament and of the Council of 9 July 2008 concerning the Visa Information System (VIS) and the exchange of data between Member States on short-stay visas.
Regulation (EU) 2016/679 of the European Parliament and of the Council of 27 April 2016 on the protection of natural persons with regard to the processing of personal data and on the free movement of such data, and repealing Directive 95/46/EC (General Data Protection Regulation).
The Treaty on European Union (TEU), Official Journal of the European Union (C 83/13), 30 March 2010.

UK

Data Protection Act 1998.
Human Rights Act 1998.
Private Security Industry Act 2001.

Norway

Kingdom of Norway's Constitution of 17 May 1814.
Act of 6 November 1993 No. 1993 relating to Aviation (the Aviation Act)
Act of 21 May 1999 No. 30 relating to the strengthening of the status of human rights in Norwegian law (the Human Rights Act).
Act of 14 April 2000 No. 31 relating to the processing of personal data (Personal Data Act).
Regulation of 15 December 2000 No. 1265 on the processing of personal data.
Regulation of 1 March 2011 No. 214 on the prevention of attacks against aviation security

USA

Homeland Security Act 2002. 6 U.S.C. 142.
Privacy Act 1974. 5 U.S.C. § 552a (2000).
The Fourth Amendment to the Bill of Rights in the U.S. Constitution 1791.
Video Voyeurism Prevention Act 2004.

Russian Federation

Air Code of 19 March 1997 N 60-FZ.
Constitution of the Russian Federation of 12 December 1993.
Federal law of 11 March 1992 N 2487-I On private detective and security activity in the Russian Federation.
Federal law of 21 July 1993 N 5485-1 On state secrets.
Federal law of 19 December 2005 N 160-FZ On ratification of the Council of Europe Convention for the protection of individuals with regard to automatic processing of personal data.
Federal law of 6 March 2006 N 35-FZ On counteraction against terrorism.
Federal law of 27 July 2006 N 152-FZ On personal data.
Federal law of 9 February 2007 N 16-FZ On transport security.
Federal law of 9 February 2009 N 8-FZ On providing access to information about the activities of government bodies and local authorities.
Federal law of 28 December 2010 N 390-FZ On security.
Federal law of 7 May 2013 N 99-FZ On amendments to certain legislative acts of the Russian Federation in connection with the adoption of the Federal law On ratification of the Council of Europe Convention for the protection of individuals with regard to automatic processing of personal data and Federal law On personal data.

Federal law of 21 July 2014 N 242-FZ On amendments to certain laws of the Russian Federation in order to clarify the procedure for personal data processing in information and telecommunications networks.
Order of the Ministry of Transport of 25 July 2007 N 104 On approval of rules for pre- and post-flight inspections.
Order of the Ministry of Transport of 18 April 2008 N 62 On approval of Program of civil aviation security of the Russian Federation.
Order of the Ministry of Transport of 19 July 2012 N 243 On approval of the formation and maintenance of automated centralized databases of personal data on passengers and personnel (crew) of vehicles, as well as providing the data contained therein.
Presidential decree of 6 March 1997 N 188 On approval of the List of confidential information.
Resolution of Government of 30 July 1994 N 897 About Federal system of protection of civil aviation from acts of unlawful interference.
Resolution of Government of 16 March 2009 N 228 About Federal service for supervision in the sphere of telecom, information technologies and mass communications.
Resolution of Government of 1 November 2012 N 1119 On approval of requirements for the protection of personal data during their processing in information systems of personal data.

Books and Articles

Abeyratne, Ruwantissa. *Aviation Security Law*. Berlin: Heidelberg, 2010.
Amberhawk Training Limited. *A review of some important aspects of the EU-USA PNR agreement*. December 2011. http://amberhawk.typepad.com/files/eu-usa-pnr-deal-amberhawk--analysis.pdf
Acharya, Gautam. "The Adequacy of Aviation Security Laws and Airport Security". *Central European Journal of International & Security Studies* 2 (2008): 107–25.
Ackerman, Spencer. "No-fly list used by FBI to coerce Muslims into informing, lawsuit claims". The Guardian, 23 April 2014. http://www.theguardian.com/world/2014/apr/23/no-fly-list-fbi-coerce-muslims/
ACLU. "What's Wrong With Public Video Surveillance?" 2002. https://www.aclu.org/technology-and-liberty/whats-wrong-public-video-surveillance
Agustina, José R, and Gemma Galdon Clavell. "The impact of CCTV on fundamental rights and crime prevention strategies: The case of the Catalan Control Commission of Video surveillance Devices". *Computer Law & Security Review* 27 (2011):168–174.
Alexander, Yonah and Eugene Sochor. *Aerial piracy and aviation security*. Dordrecht: Martinus Nijhoff, 1990.
Alexander, Larry and Ken Kress. "Against legal principles". *Iowa L. Rev.* 82 (1996): 739.
Alexy, Robert. "On the structure of legal principles". *Ratio Jus* 13 (2000): 294–304.
Alexy, Robert and Julian Rivers. *A theory of constitutional rights*. Oxford: Oxford U. Press, 2010.
Andersen, Mads Bryde. *Edb og ansvar*, Copenhagen: Jurist- og Økonomforbundets Forlag, 1988.
Annereau, Sally. "Regulating CCTV use in the UK". March 2014., http://united-kingdom.taylorwessing.com/globaldatahub/article_regulating_cctv.html
Amicelli, Anthony. "Report on Theoretical Frameworks and Previous Empirical Research. Deliverable D1.1 of the EU FP7 project "PACT" (2012a).
Amicelli, Anthony. "Surveillance and freedom of movement." In *Surveillance, fighting crime and violence. Deliverable D1.1 of the EU FP7 project "IRISS"* (2012b): 269–74.
Amsterdam, Yuval. "Mice in Portals: Incorporating Animals in Technology". Body Search 2015 Conference Programme. http://network.icom.museum/fileadmin/user_upload/minisites/icms/pdfs/Body_Search_SPEAKER_PROGRAMME.pdf

Aquilina, Kevin. "Public security versus privacy in technology law: A balancing act?" *Computer Law & Security Review* 26 (2010): 130–43.

Arai-Takahashi, Yutaka. *The Margin of Appreciation Doctrine and the Principle of Proportionality in the Jurisprudence of the ECHR*. Antwerp: Intersentia, 2002.

Aristotle. *Politics*. Moscow: Mysl, [ca. 330 BC], 1983.

Armitage, Rachel. "To CCTV or not to CCTV." In: *A review of current research into the effectiveness of CCTV systems in reducing crime*. London: Nacro, 2002.

Askew, Geoffrey. "We need leadership from our leaders". *Aviation Security International* 21 (2015).

Asllani, Arben and Amjad Ali. "Securing information systems in airports: A practical approach". Internet Technology and Secured Transactions (ICITST), International Conference in Abu Dhabi, December 2011: 314–318

Atger, Anaïs Faure. "The Abolition of Internal Border Checks in an Enlarged Schengen Area: Freedom of movement or a scattered web of security checks*?" CEPS*, 2008.

Atkins, Elliot. "Online terrorism: Is the cyber threat real*?*" *Aviation Security International* 18 (2012).

Auvinen, Anssi. "Ionising Radiation: Myths & Realities". Body Search 2015 Conference Program, http://network.icom.museum/fileadmin/user_upload/minisites/icms/pdfs/Body_Search_SPEAKER_PROGRAMME.pdf

Averkiev, I.V. "Why does a man need human rights?" *Civitas* 1 (2003).

Avihai, Hillel. "Between Two Septembers: From the bargaining chip of 6 September 1970 to a strategic agent of 11 September 2001". 6 November 2006. http://www.ict.org.il/Article/948/Between%20Two%20Septembers%20From%20the%20bargaining%20chip%20of%20Sept%201970

Badii, Atta *et al*. "Visual context identification for privacy-respecting video analytics". *Multimedia Signal Processing (MMSP), 2012 IEEE 14th International Workshop on*. IEEE, 2012.

Balboni, Pablo and Milda Macenaite. "Privacy by design and anonymisation techniques in action: Case study of Ma3tch technology". *Computer Law & Security Review* 29 (2013): 330–40.

Barden, Pete. "Airport body scanners… all your questions answered". APH. 20 May 2015. http://www.aph.com/community/holidays/airport-body-scanners-all-your-questions-answered/.

Barry, Ann S., and David S. Mazel. "Airport Perimeter Security: Where we've been, Where we are, and Where we're going". *Technologies for Homeland Security, 2008 IEEE Conference on*. IEEE, 2008.

Bartle, Ian and Peter Vass. *Self-Regulation and the Regulatory State: A Survey of Policy and Practice*. Bath: University of Bath, 2005.

Batt, Steve. "The use of video surveillance in today's airports". *Aviation Security International* 19 (2013).

Baum, Philip. "Christian Fundamentalism and Justiciar Knights: proof of the foolhardness of ethic profiling". *Aviation Security International* 17 (2011): 1.

Baum, Philip. "Pilots, flight attendants and Welsh rugby stars: when those we 'trust' are out of control". *Aviation Security International* 18 (2012a).

Baum, Philip. "Terror in Bulgaria, Hijacking in Xinjiang: concern in Reykjavik, but joy in London". *Aviation Security International* 18 (2012b).

Baum, Philip. "Emerging technologies: it's not all about X-rays". *Aviation Security International* 19 (2013).

Baum, Philip. "Malaysian Airlines Flight MH370: A 'deliberate act' but how, by whom, and why?" *Aviation Security International* 20 (2014).

Baum, Philip. "Germanwings 9525: the challenge of suicidal pilots behind intrusion-proof cockpit doors". ASI Special Report on Germanwings Flight 9525 (2015a)

Baum, Philip. "Metrojet Flight KGL9268. ASI Special Report" (2015b) http://www.asi-mag.com/metrojet-flight-kgl9268-a-special-report/

Baum, Philip. "Searching bodies: the civil liberties dilemma". *Aviation Security International* 21 (2015c).

Bazian, Hatem. "The 'Randomness' of Islamophobia at US Airports". 3 September 2014., http://www.turkeyagenda.com/the-randomness-of-islamophobia-at-us-airports-1111.html

Benda, Paul. *Quadrupole Resonance: Advantages & Possibilities*. UK, 2015.

Benjamin, Raymond. "ICAO and Civil Aviation Challenges". (2009). http://www.iata.org/publications/airlines-international/october-2009/Pages/2009-10-09.aspx, 2009.

Bennett, Colin. "Computers, personal data, and theories of technology: Comparative approaches to privacy protection in the 1990s." *Science, Technology & Human Values* 16 (1991): 51–69.

Beroeva, Nigina. "Who and how do they steal databases?" *Komsomolskaya Pravda*, 3 March 2006.

Bing, Jon. *Rettslige kommunikasjonsprosesser: bidrag til en generell teori NORIS (30)*. Universitetsforlaget, 1982.

Bjørnskau, Torkel *et al. "Transport Security and the Protection of Privacy."* TØI report 914/2007 (2007).

Black, Julia. "Regulatory conversations". *Journal of Law and Society* 29 (2002): 163–96.

Blank, Dennis. "Surveillance Cameras Set To Keep Watch In Airliners". New York Times, 6 April 2002. http://www.nytimes.com/2002/04/06/business/surveillance-cameras-set-to-keep-watch-in-airliners.html

Blackburn, Robert. "Britain's unwritten constitution". 2015. http://www.bl.uk/magna-carta/articles/britains-unwritten-constitution

Bordunov, V.D. *International Air Law*. Moscow, 2007.

Borogan, Irina and Andrei Soldatov. "Magic of persecution". *Bolshoy Gorod*. 24 November 2011. http://bg.ru/society/magiya_presledovaniya-9636/

Bradley, Jen. "Airports Build a Case for Biometrics". 2015. http://www.aviationpros.com/article/12082409/airports-build-a-case-for-biometrics

Brenner, Susan.W. "Fourth Amendment in an Era of Ubiquitous Technology". *The Miss. LJ.* 75 (2005): 1.

Brignell, John. "Number watch. The Risks of Travel". 2011. http://www.numberwatch.co.uk/risks_of_travel.htm.

Brodeur, Jean-Paul. "Introduction". *Canadian Journal of Criminology and Criminal Justice* 48 (2006): 323–31.

Brooks, Nina. "A new approach for a new breed of threat?" *Aviation Security International* 21 (2015).

Bunker, Robert. "Body cavity bombs: Fantasy or reality?" *Aviation Security International* 20 (2014).

Burgoon, Judee K. *et al.* "Maintaining and restoring privacy through communication in different types of relationships". *Journal of Social and Personal Relationships* 6 (1989): 131–58.

Burrell, Jenna. "How the machine 'thinks': Understanding opacity in machine learning algorithms". *Big Data & Society* 3.1 (2016).

Bygrave, Lee A. "Data protection pursuant to the right to privacy in human rights treaties". *International Journal of Law and Information Technology* 6 (1998): 247–84.

Bygrave, Lee A. and Ann Helen Aarø. "Privacy, Personality and Publicity – An Overview of Norwegian Law." In: *International privacy, publicity and personality laws* edited by M. Henry. London, 2001: 333–46.

Bygrave, Lee A. *Data protection law: approaching its rationale, logic and limits*. Hague/London/New York: Kluwer Law International, 2002.

Bygrave, Lee A. "Privacy Protection in a Global Context – A Comparative Overview". *Scandinavian studies in law* 47 (2004): 319–48.

Bygrave, Lee A. "Privacy and Data Protection in an International Perspective". *Scandinavian Studies in Law*, ISSN 0085-5944 (2010): 165–200.

Bygrave, Lee A. *Data privacy law: an international perspective*. Oxford: Oxford University Press, 2014.

Bygrave, Lee A. "Hardwiring privacy" (forthcoming). In: *The Oxford Handbook on the Law and Regulation of Technologies*. Brownsword, Scotford [et al.] (ed.). Oxford, 2016.

Cali, Basak. "Balancing Human Rights? Methodological Problems with Weights, Scales and Proportions." *Human Rights Quarterly* 29 (2007): 251.

Cankorel, Turgut. "Cognitive Classification of Legal Principles: A New Approach to International Legal Training". *Ankara Law Review* 5 (2008).

Carli, Vivien. *Assessing CCTV as an Effective Safety and Mangement Tool for Crime-solving, Prevention and Reduction*. International Centre for the Prevention of Crime, 2008.

Chernova, Aleksandra. "We protect personal data through multi-stakeholder approach". *Personal data* (2013). http://www.privacy-journal.ru/article/122/2/1516

Chertok, Vladimir. "Aviation Security". International Conference Aviation Security, St. Petersburg, 2010. http://global-port.ru/ru/avia2010articles

Choudhury, Nusrat. "Ninth Circuit Gives ACLU's No Fly List Clients Their Day in Court". *ACLU.* 30 July 2012. http://www.aclu.org/blog/national-security-technology-and-liberty/ninth-circuit-gives-aclus-no-fly-list-clients-their

Clarke, Roger. "Information technology and dataveillance". *Communications of the ACM* 31 (1988): 498–512.

Clarke, Roger. "Profiling: A hidden challenge to the regulation of data surveillance". *JL & Inf. Sci.* 4 (1993): 403.

Clarke, Roger. "Introduction to dataveillance and information privacy, and definitions of terms." *Roger Clarke's Dataveillance and Information Privacy Pages* (1997).

Clarke, Roger. "Identified, Anonymous and Pseudonymous Transactions: The Spectrum of Choice". 30 April 1999. http://www.rogerclarke.com/DV/UIPP99.html#Ps

Clarke, Roger. "What's Privacy?" Prepared for a Workshop at the Australian Law Reform Commission on 28 July 2006. Version of 7 August 2006. http://www.rogerclarke.com/DV/Privacy.html

Clifford, Damien and Jessica Schroers. "Personal data and dynamic IPs – time for clarity?" 23 Jan 2015. http://blogs.lse.ac.uk/mediapolicyproject/2015/01/23/personal-data-and-dynamic-ips-time-for-clarity/

Cline, Jay. "Global CRM Requires Different Privacy Approaches". 16 May 2005. http://www.computerworld.com/article/2556503/data-privacy/global-crm-requires-different-privacy--approaches.html

Cochrane, Bruce. "Dispersal: A means of infiltrating bomb onto aircraft". *Aviation Security International* 18 (2012).

Cohen, Tim. "Intercultural differences: the challenge for the screeners". *Aviation Security International* 18 (2012).

Colvin, Madeleine. "The Human Rights Act and CCTV", CCTV User Group, 12 October 2007., https://www.cctvusergroup.com/art.php?art=39

Couessin, Charles de. "Passenger name records and advanced passenger information: data privacy concerns identified and resolved". *Aviation Security International* 21 (2015).

Courtney, Martin. "The great Olympics zoom boom [CCTV services]". *Engineering & Technology* 7 (2012): 58–61.

Crockatt, Michael A. and Jill Ogston. "Airport Infrastructure as an Instrument For Regional Economic Development". *Prairie Perspectives: Geographical Essays* 3 (2000): 62–83.

Crystal, Garry. "FAQ: UK Citizens and CCTV Privacy Rights". Civil Rights Movement, 7 October 2015., http://www.civilrightsmovement.co.uk/faq-uk-citizens-cctv-privacy-rights.html

Cunningham, Brian. "Big Brother is watching EU: As the US moves towards privacy reform, Europe enacts sweeping new spying powers". 20 May 2015. http://www.politico.eu/article/eu-privacy-surveillance-us-leads-act/

De Hert, Paul and Vagelis Papakonstantinou. "The data protection framework decision of 27 November 2008 regarding police and judicial cooperation in criminal matters – A modest achievement however not the improvement some have hoped for". *Computer Law & Security Review* 25 (2009): 403–14.

De Hert, Paul and Vagelis Papakonstantinou. "The proposed data protection Regulation replacing Directive 95/46/EC: A sound system for the protection of individuals". *Computer Law & Security Review* 28 (2012): 130–42.
Dempsey, Paul S. "Aviation Security: The Role of Law in the War Against Terrorism". *Columbia journal of transnational law* 41 (2003): 649.
Dempsey, James X. and Lara M. Flint. "Commercial data and national security." *Geo. Wash. L. Rev* 72 (2004): 1459–1502.
Diederiks-Verschoor, Isabella and M. A. Butler. *An introduction to air law*. Kluwer law international, 2006.
Donald, Craig and Jeff Corkill. "Exploiting CCTV systems: Enhancing surveillance, situational awareness and intelligence acquisition". *Aviation Security International* 18 (2012).
Dworkin, Ronald. "The model of rules". *The University of Chicago Law Review* (1967): 14–46.
Elias, Bart. "Airport Body Scanners: The Role of Advanced Imaging Technology in Airline Passenger Screening". Congressional Research Service, Library of Congress, 2012.
Elliott, Christopher. "Smile! The TSA is taping you—and here's what you need to know about it". 21 October 2009., http://elliott.org/blog/smile-the-tsa-is-taping-you-and-heres-what-you-need-to-know-about-it/
Enerstvedt, Olga. "SMART Work Package 2 – Smart Surveillance in Border Control". Country report Norway, 2012.
Enerstvedt, Olga. "Russian PNR system: Data protection issues and global prospects". *Computer Law & Security Review* 30 (2014): 25–40.
Engle, Eric. "The History of the General Principle of Proportionality: An Overview". *Dartmouth Law Journal* X-1 (2012): 1–11
Ericson, Richard and Aaron Doyle. "Catastrophe risk, insurance and terrorism". *Economy and Society* 33 (2004): 135–73.
Ericson, Richard. "Ten uncertainties of risk-management approaches to security". *Canadian Journal of Criminology and Criminal Justice/La Revue canadienne de criminologie et de justice pénale* 48 (2006): 345–56.
Esayas, Samson. "The role of anonymisation and pseudonymisation under the EU data privacy rules: beyond the 'all or nothing' approach". *European Journal of Information Technology and Law* 6 (2015).
Etzioni, Amitai. "Implications of Select New Technologies for Individual Rights and Public Safety." *Harvard Journal of Law and Technology* 15 (2002).
Everitt, Mark. "Safety and Security in Aviation – The Dividing Line is Blurred". 27 June 2000. http://www.asasi.org/papers/safety_security_in_aviation.pdf
Eze, Chinedu. "Now Showing at MMIA: Nude Images of Passengers". 21 September 2010. http://allafrica.com/stories/201009210101.html
Fatokun, Samson. "Airport security: current practice and perspectives for 2013". *Journal of Air Transport Studies* 4 (2013): 26–39.
Ferguson, Niall. "The ascent of money: a financial history of the world". *The Historical Journal* 53.1 (2010): 239.
Fessard, Louise. "France jumps EU law and follows UK with mass surveillance of air travelers". 26 December 2013. http://www.mediapart.fr/en/journal/international/261213/france-jumps-eu-law-and-follows-uk-mass-surveillance-air-travellers
Fimreite, Anne Lise *et al*. "After Oslo and Utøya: A shift in the balance between security and liberty in Norway?" *Studies in Conflict & Terrorism* 36 (2013): 839–56.
Finn, Matthew. "Enhanced documentation control: An aviation security perspective". *Aviation Security International* 18 (2012a).
Finn, Matthew. "The identity challenge: verifying travel document integrity, quashing breeder document fraud". Aviation Security International 21 (2015).
Finn, Rachel and David Wright. "Unmanned aircraft systems: Surveillance, ethics and privacy in civil applications". *Computer Law & Security Review* 28 (2012b): 184–94.

Finn, Rachel *et al.*"Seven types of privacy". In: *European data protection: coming of age.* Springer Netherlands, 2013: 3–32.

Flynn, Cathal. "A personal view". *Aviation Security International* 17 (2011).

Flynn, Cathal. "Intelligence: a pre-requisite for risk-based security screening". *Aviation Security International 19* (2012).

Fordham, Michael and Thomas De La Mare. "Identifying the Principles of Proportionality". In: *Understanding Human Rights Principles*, London: Justice and Hart Publishing (2001).

Forest, James. "The Modern Terrorist Threat to Aviation Security". *Perspectives on Terrorism* 1 (2007): 10–13

Franklin, Benjamin and William T. Franklin. *Memoirs of the life and writings of Benjamin Franklin*. Vol. 1. H. Colburn, 1818.

Fried, Charles. "Privacy". In: *Philosophical dimensions of privacy* edited by F. D. Schoeman. New York: Cambridge University Press, 1984: 203–22.

Fuster, Gloria *et al.* "Profiling in the European Union: A high-risk practice". *INEX Policy Brief* 10 (2010): 1–12.

Gagvani, Nick. "Introduction to video analytics". *EETimes*. 22 August 2008. http://www.eetimes.com/document.asp?doc_id=1273834

Gallagher, Caoilfhionn. "CCTV and human rights: The fish and the bicycle? An examination of Peck v. United Kingdom (2003) 36 EHRR 41". *Surveillance & Society* 2 (2002).

Garfinkel, Simson. *Database nation: the death of privacy in the 21st century*. O'Reilly Media, Inc., 2000.

Gasser, Hans-Peter. "Acts of terror, 'terrorism' and international humanitarian law". *International Review – Red Cross* 84 (2002): 547–70.

Gavison, Ruth. "Privacy and the Limits of Law". *Yale LJ.* 89 (1979): 421–471.

Gellert, Raphaël and Serge Gutwirth. "The legal construction of privacy and data protection". *Computer Law & Security Review* 29 (2013): 522–30.

Geyer, Florian. *Taking stock: databases and systems of information exchange in the area of freedom, security and justice*. Vol. 9. CEPS, 2008.

Ghee, Ryan. "Recommendations for airports and airlines to improve baggage, security, immigration, arrivals and more in 2015" http://www.futuretravelexperience.com/2015/01/improving-baggage-security-immigration-and-arrivals-in-2015/

Gilbert, Nigel. "Dilemmas of privacy and surveillance: challenges of technological change". *Criminal Justice Matters*. Vol. 68, Issue 1 (2007): 41–42.

Gold, Steve. "Border control biometrics and surveillance". *Biometric Technology Today* 2012.7 (2012): 9–11.

Goodman, Bryce, and Seth Flaxman. "EU regulations on algorithmic decision-making and a "right to explanation"." *arXiv preprint arXiv:1606.08813* (2016).

Grabell, Michael. "U.S. Government Glossed Over Cancer Concerns As It Rolled Out Airport X-Ray". *ProPublica*, 1 November 2011. http://www.propublica.org/article/u.s.-government-glossed-over-cancer-concerns-as-it-rolled-out-airport-x-ray

Graham, Stephen. "CCTV: The stealthy emergence of a fifth utility?" *Planning Theory & Practice* 3 (2002): 237–41.

Grant, Hazel. "Data protection 1998–2008". *Computer Law & Security Review* 25.1 (2009): 44–50.

Graver, Hans Petter and Henning Harborg. *Datalagring og menneskerettighetene*. Oslo, 2015.

Greenleaf, Graham. "Five years of the APEC Privacy Framework: Failure or promise?" *Computer Law & Security Review* 25.1 (2009): 28–43.

Greenleaf, Graham. *Asian Data Privacy Laws: Trade & Human Rights Perspectives*. Oxford: Oxford University Press, 2014.

Griffiths, Siriol. "Stress response: a physiological analysis". *Aviation Security International* 17 (2011).

Grynpas, Sasrah. "Advanced Imaging technology: privacy and health concerns". *Aviation Security International* 18 (2012).

Gusella, David. "Violating Privacy in Private: How Epic v. DHS Creates an Impossible Burden on Plaintiffs Trying to Demonstrate a Privacy Act Violation". *Boston College Law Review* 53 (2012): E. Supp. 169.

Guttman, Nathan. "Israel's Airport Security, Object of Envy, Is Hard to Emulate Here". *Forward.* 6 January2010.http://forward.com/news/122781/israel-s-airport-security-object-of-envy-is-hard/

Gutwirth, Serge *et al.* "Legal, Social, Economic and Ethical Conceptualisations of Privacy and Data Protection. Deliverable 1, The Prescient Project". (2011).

Hainmüller, Jens and Jan Martin Lemnitzer. "Why do Europeans fly safer? The politics of airport security in Europe and the US". *Terrorism and Political Violence* 15.4 (2003): 1–36.

Halstead, Peter. *Unlocking human rights*. Routledge, 2014.

Harrison, John. *International aviation and terrorism: evolving threats, evolving security*. London: Routledge, 2009.

Hasbrouck, Edward. "Can you really see what records are kept about your travel?" 30 December 2008. http://hasbrouck.org/blog/archives/001595.html

Hasbrouck, Edward. "What's in a Passenger Name Record (PNR)?" 2009. http://hasbrouck.org/articles/PNR.html

Hijmans, Hielke and Alfonso Scirocco. "Shortcomings in EU data protection in the third and the second pillars. Can the Lisbon Treaty be expected to help". *Common Market Law Review* 46.5 (2009): 1485–1525.

Hijmans, Hielke. "Recent developments in data protection at European Union level". *ERA Forum*. Vol. 11. No. 2. Springer-Verlag, 2010.

Hildebrandt, Mireille. "Defining profiling: a new type of knowledge?" *Profiling the European citizen*. Springer Netherlands, 2008. 17–45.

Hildebrandt, Mireille and Laura Tielemans. "Data protection by design and technology neutral law". *Computer Law & Security Review* 29.5 (2013): 509–521.

Hoffman, Bruce. "Terrorism trends and prospects". *Countering the new terrorism* 7 (1999): 13.

Holbrook, Emily. "Airport Security: Privacy vs. Safety". *Risk Management* 57.2 (2010): 12.

Holden, Robert T. "The contagiousness of aircraft hijacking". *American Journal of Sociology* 91 (1986): 874–904.

Hope, Christopher. "Britain to hire 2,000 new spies at MI5, MI6 and GCHQ in wake of the Paris terror attacks". The Telegraph. 16 Nov 2015. http://www.telegraph.co.uk/news/uknews/terrorism-in-the-uk/11997782/Paris-terror-attack-Britain-to-hire-2000-new-spies.html

Hoptkins, Nick and Jake Morris. "Innocent people' on police photos database". BBC. 3 February 2015. http://www.bbc.com/news/uk-31105678

Hoofnagle, Chris. "Airport Searches". In: *Encyclopedia of American civil liberties* edited by Paul Finkelman, 2006.

Houlis, Peter. "Visibly efficient: the future of integrated CCTV" (2010) http://www.ifsecglobal.com/visibly-efficient-the-future-of-integrated-cctv/

Huey, John. "Live Free or Die": The Unintended Consequences of "One Size Fits All" Aviation Security Screening". 5 March 2011. (2011a). http://www.noquarterusa.net/blog/57130/"live-free-or-die"-the-unintended-consequences-of-"one-size-fits-all"-aviation-security-screening/

Huey, John. "When This Stuff Gets Personal! A Response to the Moscow Airport Bombing". 27 January 2011. (2011b) http://www.noquarterusa.net/blog/55860/when-this-stuff-gets-personal-a-response-to-the-moscow-airport-bombing/

Huey, John. "When Politics Meets the Hard Realities: Cold Fact, Speculation and Surrealism at the Checkpoint". 19 September 2012. http://www.noquarterusa.net/blog/72258/when-politics-meets-the-hard-realities-cold-fact-speculation-and-surrealism-at-the-checkpoint/

Inness, Julie. *Privacy, intimacy, and isolation*. New York, 1992.

Israelsen, R. Gregory. "Applying the Fourth Amendment's National-Security Exception to Airport Security and the TSA". *J. Air L. & Com.* 78 (2013): 501.

Izmailova, N.S. *Privacy in civil law: the law of the UK, the USA and Russia.* Moscow, 2009.

Jackson, Brian A., Edward W. Chan, and Tom LaTourrette. "Assessing the security benefits of a trusted traveler program in the presence of attempted attacker exploitation and compromise". *Journal of Transportation Security* 5.1 (2012): 1–34.

Janic, Milan. "An assessment of risk and safety in civil aviation". *Journal of Air Transport Management* 6 (2000): 43–50.

Jenkins, Brian M. "Defense against terrorism". *Political Science Quarterly* 101.5 (1986): 773–786.

Jenkins, Darryl. "A primer on airport security". *Governance and Public Security New York: Campbell Public Affairs Institute* (2002): 69–84.

Jenkins, Brian M. *Aviation Security: After Four Decades, It's Time for a Fundamental Review.* Rand Corp Arlington VA Homeland Security and Defense Center, 2012.

Jezierska-Switala, Marta. "Big City Trials: from laboratory explosives detection to the field applications (a STANDEX programme)". Body Search 2015 Conference Program. http://network.icom.museum/fileadmin/user_upload/minisites/icms/pdfs/Body_Search_SPEAKER_PROGRAMME.pdf

Joinson, Adam N. and Carina B. Paine. "Self-disclosure, privacy and the Internet". *The Oxford handbook of Internet psychology* (2007): 237–252.

Jones, Richard. "Checkpoint security: gateways, airports and the architecture of security". In: *Technologies of inSecurity: the surveillance of everyday life* edited by Katja Franko Aas, Helene Oppen Gundhus, and Heidi Mork Lomell. Routledge, 2008: 81–101.

Karber, Phillip A. "Re-constructing global aviation in an era of the civil aircraft as a weapon of destruction". *Harv. JL & Pub. Pol'y* 25 (2001): 781.

Karl, Uwe. "Integrated CCTV: more than a surveillance tool". *Aviation Security International 15* (2009).

Kelsen, Hans. *Principles of international law.* The Lawbook Exchange, Ltd., Clark, New Jersey, 1952.

Keohane, Robert O. *Global governance and democratic accountability*. Duke University, 2003.

Kilkelly, Ursula. "The right to respect for private and family life". In: *Human Rights Handbook,* Council of Europe, 2001.

Killock, Jim. "Legal challenge to UK Internet surveillance". 3 October 2013. https://www.privacynotprism.org.uk/news/2013/10/03/legal-challenge-to-uk-internet-surveillance

Kirschenbaum, Alan. "A personal view". *Aviation Security International* 20 (2014).

Kitchin, Rob. "Big Data, new epistemologies and paradigm shifts". *Big Data & Society* 1.1 (2014) 1–12.

Kochenova, Olga. "Respect to privacy in Russia: diagnosis and prognosis". 2007. http://www.pgpalata.ru/reshr/privacy/art10.shtml#n77

Koffler, Bruce. "The latest Al-Qaeda threat: Body bomb". *Aviation Security International* 17 (2011).

Kornblatt, Sara. "Are emerging technologies in airport passenger screening reasonable under the fourth amendment". *Loy. LAL Rev* 41 (2007): 385.

Kovaleva, Natalia. *Informational Law of Russia*. Moscow: Dashkov and K., 2007.

Kovrigin, V. "Total non-compliance with data protection law in Russia". 2012. http://can-work.ru/index.php/neews/press-tsentr-kompanii/145-law-on-personal-data-if-it-works

Kristoff, Nicholas. "Liberal reality check: we must look anew at freedom vs. security". *Pittsburgh Post-Gazette*, 3 June 2002, p. A9.

La Fors-Owczynik, Karolina. "Minor protection or major injustice? – Children's rights and digital preventions directed at youth in the Dutch justice system". *Computer Law & Security Review* 31 (2015): 651–67.

La Free, Gary. "Countering Myths about Terrorism". 38th Annual Theodore Standing Lecture, SUNY-Albany (2008).

Laing, Melissa. "Air Travel and Omnipresent Disaster". 2008. http://melissalaing.com/site/CIHA_139.pdf

Larsen, Beatrice von Silva-Tarouca. *Setting the watch: Privacy and the ethics of CCTV surveillance*. Bloomsbury Publishing, 2011.

Laviv, Omer. "Big Data: aiding aviation and border security". *Aviation Security International 20* (2014).

Leese, Matthias. "Blurring the dimensions of privacy? Law enforcement and trusted traveler programs". *Computer Law & Security Review* 29 (2013): 480–90.

Leese, Matthias. "The new profiling: Algorithms, black boxes, and the failure of anti-discriminatory safeguards in the European Union". *Security Dialogue* 45 (2014): 494–511.

Lessig, Lawrence. "Against Transparency. The perils of openness in government". *The New Republic* 9 (2009): 37–44.

Lester, Anthony. "The European Court of Human Rights after 50 years". *European Human Rights Law Review* 4 (2009): 461–78.

Littlejohn, Richard. "How to avoid airport security: Wear a burka". *Daily Mail*, 10 May 2012. http://www.dailymail.co.uk/debate/article-2142619/How-avoid-airport-security-Wear-burka.html#ixzz2idYLMjYs

Liu, Yue. *Bio-privacy: legal challenges for privacy regulations of biometric identification and authentication*. Oslo: Series of dissertations submitted to the Faculty of Law, University of Oslo, 2010.

Lyon, David. "Airports as data filters: converging surveillance systems after September 11th". *Journal of Information, Communication and Ethics in Society* 1 (2003): 13–20.

Lyon, David. *Surveillance after September 11*. Cambridge, 2003a.

Lyon, David. *Surveillance as social sorting: Privacy, risk, and digital discrimination*. Psychology Press, 2003b.

Lyon, David. "Globalizing Surveillance Comparative and Sociological Perspectives". *International Sociology* 19 (2004): 135–49.

Lyon, David. "Surveillance, security and social sorting emerging research priorities". *International criminal justice review* 17 (2007): 161–70.

Macário, Rosário, *et al.* "The consequences of the growing European low-cost airline sector". *European Parliament* (2007).

Malt, Gert-Frederik. "Sameirettslige Prinsipper". In: *Selskap, kontrakt, konkurs og rettskilder* edited by K. Normann [et al.]. Oslo, 2010: 356.

Manning, S. "United States' Response to International Air Safety". *The J. Air L. & Com* 61 (1995): 505.

Manny, Carter. "Personal privacy—transatlantic perspectives: European and American privacy: commerce, rights and justice—part II". *Computer Law & Security Review* 19 (2003): 92–100.

Markon, Jerry. "Homeland Security plans to expand pre-clearance program at foreign airports". *Washington Post*, 13 November 2015. https://www.washingtonpost.com/news/federal-eye/wp/2014/11/13/homeland-security-plans-to-expand-pre-clearance-program-at-foreign-airports/

Martin, Charles E. *Air piracy and terrorism directed against US air carriers*. Army War Coll Carlisle Barracks PA, 1993.

Martin, Hugo. "Alaska Airlines to consider using fingerprints to screen passengers". *Los Angeles Times*, 24 October 2014. http://www.latimes.com/business/la-fi-alaska-airlines-fingerprint--scanners-20141024-story.html

Matuzov, N.I., and A.V. Malko. *The theory of state and law: lectures*. Moscow, 2001.

Mazeina, Khamis. "Full-body scanner gets a second look". 2001. http://www.thefreelibrary.com/Full-body+scanner+gets+a+second+look.-a0254431914

McDonagh, Kenneth, and Yee-Kuang Heng. "Risk, human rights and the bureacratisation of counter-terrorism". Working Papers in International Studies 2011. (Paper No. 2011/3). Dublin City University, Dublin.

McNeal, Gregory. "Security Scanners in Comparative Perspective". *Health Matrix* 22 (2012): 461.

Meads, Colin. "CCTV Data: is it fit for purpose?" *Aviation Security International* 21 (2015).

Medwecki, Mark. "CCTV: ensuring effective surveillance at airports". *Aviation Security International* 19 (2013).

Mick, Jason. "Despite Allegations of Fraud, TSA Lets Scanner Maker Keep Nearly $300M". *Dailytech*, 18 January 2013. http://www.dailytech.com/Despite+Allegations+of+Fraud+TSA+Lets+Scanner+Maker+Keep+Nearly+300M/article29691.htm

Miller, Rich. "Demand for security equipment projected to rise 7 percent a year through 2016". *Security Systems News*, 20 May 2013. http://www.securitysystemsnews.com/article/demand-security-equipment-projected-rise-7-percent-year-through-2016

Millward, David. "Airports to get 'virtual tripwire' CCTV". *The Telegraph*, 12 April 2008. http://www.telegraph.co.uk/news/uknews/1556555/Airports-to-get-virtual-tripwire-CCTV.html

Mironenko, Olga. *Air passenger data protection: Data transfer from the European Union to the United States*. Oslo: Norwegian Research Center for Computers and Law, 2010.

Mironenko, Olga. "Body scanners versus privacy and data protection". *Computer Law & Security Review* 27 (2011): 232–44.

Mitchener-Nissen, Timothy, Kate Bowers, and Kevin Chetty. "Public attitudes to airport security: the case of whole body scanners". *Security Journal* 25.3 (2012): 229–243.

Moor, James H. "The ethics of privacy protection". *Library Trends* 39.1–2 (1990): 69–82.

Mountfield, H. and C. Gearty. "Advice in the matter of the human rights and equality implications of the introduction of full body scanners at airports". *Matrix Chambers* 8 (2010).

Muñiz, Joaquín. "Legal Principles and Legal Theory". *Ratio Juris* 1 (1997): 267–87.

Mustapha, Ismail. "Aviation security provisions vs. basic fundamental human rights: an Islamic law perspective". *Aviation Security International* 21 (2015).

Naudin, Christophe. "A personal view". *Aviation Security International* 20 (2014).

Naylor, Brian. "TSA Officers Among Lowest Paid Of Federal Workers". *NPR*, 26 January 2010. http://www.npr.org/templates/story/story.php?storyId=122948752

Neeman, Amir. "Effectively screening people: Pat down searchers, wands, archways and portals". *Aviation Security International* 19 (2013).

Neeman, Amir. "Radiation in passenger screening: busting the myths". *Aviation Security International* 21 (2015).

Newman, Abraham. *Protectors of privacy: Regulating personal data in the global economy*. Cornell University Press, 2008.

Nielsen, Kenneth M. "Aviation security: legal framework and technical management". *International Journal of Private Law* 1.1–2 (2008): 82–93.

Nielsen, Nikolaj and Andrew Rettman. "Russia blames EU for airline data fiasco". *EUobserver*, 11 June 2013. http://euobserver.com/justice/120450

Nissenbaum, Helen. "Privacy as contextual integrity". *Wash. L. Rev* 79 (2004): 119.

Nissenbaum, Helen. *Privacy in context*. Stanford, Calif., 2010.

Noble, Ronald. "Passenger screening: empowering the industry to plug government security gaps". *Aviation Security International* 20 (2014).

Ntouvas, Joannis. "Air Passenger Data Transfer to the USA: the Decision of the ECJ and latest developments". *International Journal of Law and Information Technology* 16 (2008): 73–95.

Ohm, Paul. "Broken promises of privacy: Responding to the surprising failure of anonymization". *UCLA Law Review* 57 (2010): 1701.

Omand, David. "The terrorist threat to the UK in the post–9/11 decade". *Journal of Terrorism Research* 3 (2012).

Omar, Adnan and Muhammed Miah. "Risk Based Airport Security Approach". *International Journal of Computer Science: Theory, Technology and Applications (IJCS)* 3 (2014).

Omelchenko, E.L. and I.V. Kosterina. "Private life: a sociological perspective" (2005) http://www.pgpalata.ru/reshr/privacy/deb_02.shtml

Open Society Institute. "Ethnic Profiling in the European Union: Pervasive, Ineffective, and Discriminatory" (2009). https://www.opensocietyfoundations.org/sites/default/files/profiling_20090526.pdf

Ormerod, Thomas C. and Coral J. Dando. "Finding a Needle in a Haystack: Toward a Psychologically Informed Method for Aviation Security Screening". *Journal of Experimental Psychology: General* 144, no. 1 (2015): 76–84.

Osula, Anne-Maria. "Transborder access and territorial sovereignty". *Computer Law & Security Review* (2015). http://www.sciencedirect.com/science/article/pii/S0267364915001260.
Palmer, Brian. "What's So Great About Israel Security?" *Slate*. 3 January 2011. http://www.slate.com/articles/news_and_politics/explainer/2011/01/whats_so_great_about_israeli_security.html
Pasquale, Frank. *The black box society: The secret algorithms that control money and information*. Harvard University Press, 2015.
Pawlak, Patryk. "Made in the USA? The Influence of the US on the EU's Data Protection Regime". Brussels, *CEPS* (2009).
Pearse, John. *Investigating Terrorism: Current Political, Legal and Psychological Issues*. John Wiley & Sons, 2015.
Perry, Stephen. "Two models of legal principles". *Iowa L. Rev* 82 (1996): 787.
Perry, Melissa and Andrew Gilbey. "The Screening of Passengers by Observation Techniques Programme: analysing the issues". *Aviation Security International* 17 (2011).
Petersen, Niels. "Customary Law without Custom-Rules, Principles, and the Role of State Practice in International Norm Creation". *Am. U. Int'l L. Rev* 23 (2007): 275.
Peterson, Rohen. "The Emperor's New Scanner: Muslim Women at the Intersection of the First Amendment and Full-Body Scanners". *Hastings Women's LJ*. 22 (2011): 339.
Pillar, Paul. "The Price of Aviation Security". *The National Interest*. 22 Nov 2010. http://nationalinterest.org/blog/paul-pillar
Pocs, Matthias. "Legally compatible design of future biometric systems for crime prevention". *Innovation: The European Journal of Social Science Research* 26 (2013): 36–56.
Poole, Robert W. *Toward risk-based aviation security policy*. No. 2008-23. OECD/ITF Joint Transport Research Centre Discussion Paper, 2008.
Porter, Tony. "Surveillance technologies in society". Speech of October 2014, published on 18 November 2014. https://www.gov.uk/government/speeches/iriss-respect-and-surveille-event
Posner, Richard A. *The economics of justice*. Harvard University Press, 1983.
Posner, Eric A. and Adrian Vermeule. *Terror in the balance: security, liberty, and the courts*. Oxford University Press, Oxford, 2007.
Poullet, Yves. "Data protection legislation: What is at stake for our society and democracy?" *Computer Law & Security Review* 25.3 (2009): 211–226.
Prescott, Peter. "Kaye v Robertson—a reply". *The Modern Law Review* 54 (1991): 451–56.
Preto, A. "Aviation security – priorities of the Commission". Draft speech for 7th Transec Transport Security Conference. Amsterdam, 3 June 2009.
Price, Jeffrey. *Practical aviation security: predicting and preventing future threats*. Butterworth-Heinemann, 2013.
Privacy International. *European Privacy and Human Rights*, 2011.
Puzanov, Igor. "Video camera in the law". *EZH-Jurist* 8 (2012): 10.
Queally, James. "Transgender Woman Says TSA Detained, Humiliated Her Over Body 'Anomaly.'" *Los Angeles Times*. 22 Sept 2015. http://www.latimes.com/nation/nationnow/la-na-nn-tsa-transgender-20150921-story.html
Quinlan, Tara. "Discrimination: the questionable effectiveness of screening based on race, religion, national origin or behaviour". *Aviation Security International* 21 (2015).
Quinn, Kevin. "Man boards plane at IAH with loaded gun in carry-on". *ABC News*, 17 December 2010. http://abc13.com/archive/7848683/
Raab, Charles. "Effects of surveillance on privacy, autonomy and dignity". Deliverable D1.1: Surveillance, fighting crime and violence. Increasing Resilience in Surveillance Societies (IRISS) (ed.), 2012a.
Raab, Charles and David Wright. "Surveillance: Extending the limits of privacy impact assessment". *Privacy impact assessment*. Springer Netherlands, 2012b: 363–383.
Raab, Charles. "Privacy as a Security Value". In: *Jon Bing: En Hyllest/A Tribute* (2014): 39–58.
Raffel, Robert. "Intelligence and Airports". *FBI Law Enforcement Bulletin* 76 (2007): 1.

Ratcliffe, Jerry. *Video surveillance of public places*. Washington, DC: US Department of Justice, Office of Community Oriented Policing Services, 2006.
Ravich, Timothy. "Is Airline Passenger Profiling Necessary". *U. Miami L. Rev* 62 (2007): 1.
Ravich, Timothy. "Airline passenger profiling systems after 9/11: Personal privacy versus National Security". *Journal of the transportation research forum*. Vol. 44. No. 2 (2010): 127–41.
Raz, Joseph. "Legal principles and the limits of law". *Yale Law Journal* (1972): 823–54.
Reeve, Tom. "Information Commissioner Releases New CCTV Code of Practice". 15 October 2014. http://www.securitynewsdesk.com/information-commissioners-office-releases-new-cctv-code/
Regazzoni, Carlo *et al.* "Video analytics for surveillance: Theory and practice [from the guest editors]". *Signal Processing Magazine*, IEEE. 27 (2010): 16–17.
Rettner, Rachael. "New Airplane Passenger Screening Method Outshines Old". *Live Science*, 6 November 2014. http://www.livescience.com/48661-airport-screening-conversation.html
Reznichenko, Alexey. "Inspection equipment – weapons of anti-terror" (2006) http://www.cnews.ru/reviews/free/transport2006/articles/antiterror/
Rhee, Jamie. "Rational and Constitutional Approaches to Airline Safety in the Face of Terrorist Threats". *DePaul L. Rev* 49 (2000): 847.
Richardson, Whit. "Congressional report: TSA suffers from "bureaucratic morass and mismanagement". *Security Director News*, 22 November 2011 (2011a), http://www.securitydirectornews.com/public-sector/congressional-report-tsa-suffers-bureaucratic-morass-and-mismanagement
Richardson, Whit. "TSA plan to replace humans with software at airport scanners generates criticism". *Security Director News*, August 2011 (2011b), http://www.securitydirectornews.com/?p=article&id=sd201107R4VP0J
Richardson, Whit. "Minneapolis airport to spend $20m on new surveillance system". *Security Director News*, May 2012. http://www.securitydirectornews.com/public-sector/minneapolis-airport-spend-20m-new-surveillance-system
Robinson, Matthew. "9/11: Threats About Airplanes As Weapons Prior To 9/11". 2011. http://www.justiceblind.com/airplanes.html
Romanova, Yu.E. "The use of CCTV to monitor the behavior of employees". *Delo-press*, 2006. http://www.delo-press.ru/articles.php?n=6772
Rose, Nicolas *et al.* "Governmentality". *Annu. Rev. Law Soc. Sci* 2 (2006): 83–104.
Rubenfeld, Jed. "The End of Privacy". *Stanford law review* (2008): 101–61.
Rubinstein, Ira. "Regulating privacy by design". *Berkeley Tech. LJ*. 26 (2011): 1409.
Rumerman, Judy. "Aviation security". *Essays on the History of Flight* (2003).
Sage, Andrew and Elbert White. "Methodologies for risk and hazard assessment: a survey and status report". *Transactions on Systems, Man and Cybernetics*, IEEE 10 (1980): 425–46.
Sagi-Dolev, Alysia. "Manufacturer Perspectives". *Aviation Security International* 19 (2013): 26.
Salter, Mark B. "Imagining numbers: Risk, quantification, and aviation security". *Security dialogue* 39 (2008): 243–266.
Sarkar, Dibya. "The president's 2016 budget request: Transportation Security Administration". 6 February 2015. http://www.fiercegovernmentit.com/story/presidents-2016-budget-request-transportation-security-administration/2015-02-06
Scarfo, Phil. "Biometrics: achieved assured authentication in the digital age". *Aviation Security International* 20 (2014).
Schartum, Dag Wiese. *Personvern og transportsikkerhet: personvernmessige spørsmål knyttet til tiltak for å sikre transportmidler mot fiendtlige anslag*. Oslo, 2007.
Schartum, Dag Wiese. *Rettslige aspekter ved feltteknologi i arbeidslivet*. Oslo: Senter for rettsinformatikk, Avdeling for forvaltningsinformatikk, 2013.
Schartum, Dag Wiese. "Making Privacy By Design Operative". *International Journal of Law and Information Technology* (2016): 1–25.
Scheinin, Martin. "Terrorism and the 'Pull' of Balancing in the name of security". In *Law and Security: facing the dilemmas, EUI Working Paper* (2009): 55–63.

Scheinin, Martin and Mathias Vermeulen. "Unilateral Exceptions to International Law: Systematic legal analysis and critique of doctrines to deny or reduce the applicability of Human Rights norms in the fight against terrorism". *Essex Human Rights Review* 8 (2011): 20–56.

Schermer, Bart. "The limits of privacy in automated profiling and data mining". *Computer Law & Security Review* 27 (2011): 45–52.

Schneier, Bruce. *Beyond fear: Thinking sensibly about security in an uncertain world.* Springer Science & Business Media, 2003.

Schneier, Bruce. *Schneier on security.* Indianapolis, Indiana, 2008.

Schneier, Bruce. "How the Norwegians Reacted to Terrorism". (2012a) https://www.schneier.com/blog/archives/2012/07/how_the_norwegi.html

Schneier, Bruce. "A Waste of Money and Time". *The New York Times*, updated 17 October 2012 (2012b). http://www.nytimes.com/roomfordebate/2010/11/22/do-body-scanners-make-us-safer/a-waste-of-money-and-time

Schneier, Bruce. *Data and Goliath: The hidden battles to collect your data and control your world.* WW Norton & Company, 2015.

Shadrina, Tatiana. "Will not go far: From July next year it will not be possible to buy a ticket for a single mode of transport without a passport". *Rossiyskaya Gazeta.* 26 September 2012. http://www.rg.ru/2012/09/26/transp.html.

Shelton, Dinah. "Protecting Human Rights in a Globalized World". *BC Int'l Comp. L. Rev* 25 (2002): 273.

Simmons, Ric. "Searching for Terrorists: Why Public Safety Is Not a Special Need". *Duke Law Journal* 59.843 (2010).

Smirnov, Oleg. "All the world has long been collecting the data this way". *Aviaport.* 9 April 2007. http://www.aviaport.ru/digest/2007/04/09/118983.html

Soboleva, Elena. "Domodedovo promises to become the safest airport in Russia in the nearest future". *Finam.ru*, 23 August 2005. http://www.finam.ru/analysis/newsitem149A300640/default.asp

Soghoian, Christopher. "Insecure flight: Broken boarding passes and ineffective terrorist watch lists". In *Protecting Airline Passengers in the Age of Terrorism* edited by Paul Seidenstat and Francis Splane, 2009.

Soldatov, Andrei and Irina Borogan. "Russia's Surveillance State". 2013. http://www.worldpolicy.org/journal/fall2013/Russia-surveillance

Solove, Daniel. "Data mining and the security-liberty debate". *The University of Chicago Law Review* (2008a): 343–62.

Solove, Daniel. "The end of privacy?" *Scientific American* 299 (2008b): 100–06.

Solove, Daniel. *Understanding privacy.* Cambridge, 2008c.

Solove, Daniel. *Nothing to hide: The false tradeoff between privacy and security.* Yale University Press, 2011.

Sosenko, Konstantin. "Transmission X-ray: Advantages & Successes". Body Search 2015 Conference Program, http://network.icom.museum/fileadmin/user_upload/minisites/icms/pdfs/Body_Search_SPEAKER_PROGRAMME.pdf

Spencer, Shaun. "Security Versus Privacy: Reframing the Debate". *Denver University Law Review* 79 (2002): 519.

Spielmann, Dean. "Allowing the Right Margin the European Court of Human Rights and the National Margin of Appreciation Doctrine: Waiver or Subsidiarity of European Review?" Centre for European Legal Studies, Working Paper Series (2012).

St. John, Peter. "The Politics of Aviation Security". In *Aviation terrorism and security* edited by Paul Wilkinson and Brian Jenkins, 1999.

Steiner, John. "Aircraft evolution and airline growth". *Financial Analysts Journal* (1967): 85–92.

Steinhardt, Barry. "The EU-US PNR Agreement—A Legal Analysis of Its Failures". 26 December 2011. http://www.papersplease.org/wp/2012/01/03/the-eu-us-pnr-agreement-a-legal-analysis-of-its-failures/.

Stewart, Mark G. and John Mueller. "A risk and cost-benefit assessment of United States aviation security measures". *Journal of Transportation Security* 1 (2008): 143–59.

Stewart, Mark G. and John Mueller. "Cost-benefit analysis of advanced imaging technology full body scanners for airline passenger security screening". *Journal of homeland security and emergency management* 8.1 (2011).

Stinchfield, Grant. "TSA Source: Armed Agent Slips Past DFW Body Scanner". *NBC*. 18 February 2011. http://www.nbcdfw.com/news/local/TSA-Agent-Slips-Through-DFW-Body-Scanner-With-a-Gun-116497568.html

Sweet, Kathleen. *Aviation and airport security*. Upper Saddle River, N.J., 2004.

Taipale, Kim. "The privacy implications of government data mining programs". Testimony before the US Senate Committee on the Judiciary. Vol. 10 (2007).

Tamasi, Galileo and Micaela Demichela. "Risk assessment techniques for civil aviation security". *Reliability Engineering & System Safety* 96 (2011): 892–99.

Tanti-Dougall, Rebekah. "Cyber security in aviation: legal aspects". *Aviation Security International* 21 (2015).

Tavani, Herman T. "Philosophical theories of privacy: Implications for an adequate online privacy policy". *Metaphilosophy* 38 (2007): 1–22.

Taylor, M. "Bending Broken Rules: The Fourth Amendment Implications of Full-Body Scanners in Preflight Screening". *Richmond Journal of Law and Technology* 17 (2010).

Taylor, Courteney L. "Touched by an Agent: Why the United States Should Look to the Rest of the World for a New Airport Security Scheme and Stop Using Full-Body Scanners". *Hous. J. Int'l L* 35 (2013): 503.

Taylor, Shannon. "It's Clear the U.S. and EU Economies Need a Safe Harbor 2.0". *ITI*. 4 November 2015. https://www.itic.org/news-events/techwonk-blog/its-clear-the-us-and-eu-economies-need-a-safe-harbor-20

Teknologirådet and Datatilsynet. Personvern – tilstand og trender. Oslo, 2014.

The Identity Project. "Most federal agencies still ignore human rights complaints". 6 November 2015. http://papersplease.org/wp/2015/11/06/most-federal-agencies-still-ignore-human-rights-complaints/

Thompson, Paul. *The Terror Timeline: Year by Year, Day by Day, Minute by Minute: A Comprehensive Chronicle of the Road to 9/11–and America's Response*. No. 13. Harper Collins, 2004.

Thon, Bjørn. "Kameraovervåking – fra VHS-kassetter til intelligent videoanalyse". 9 July 2015. https://www.personvernbloggen.no/2015/07/09/kameraovervaking-fra-vhs-kassetter-til-intelligent-videoanalyse/

Tirosh, Yofi and Michael Birnhack. "Naked in Front of the Machine: Does Airport Scanning Violate Privacy?" *Ohio State Law Journal* 74 (2013): 6.

Topping, Aleksandra. "Theresa May claims 40 terror plots have been foiled since 7/7 attacks". *The Guardian*. 24 Nov 2014. http://www.theguardian.com/politics/2014/nov/24/theresa-may-london-attacks-40-terror-plots-foiled

Travis, Alan. "New scanners break child porn laws". *The Guardian*. 4 January 2010. http://www.theguardian.com/politics/2010/jan/04/new-scanners-child-porn-laws

Trembaczowski-Ryder, David. "The Economics of Security". Conference Transport Security Expo, London, September 2010.

Tsadykova, Elvira. *The constitutional right to privacy*. Moscow, 2007.

Tung, Liam. "From fingerprints to facial scans: Why the French want biometrics on all EU travelers". *ZDNet*. 13 October 2015. http://www.zdnet.com/article/from-fingerprints-to-facial-scans-why-the-french-want-biometrics-on-all-eu-travellers/

Turn, Rein. "Information privacy issues for the 1990s". *Research in Security and Privacy, 1990. Proceedings, 1990 IEEE Computer Society Symposium on*. IEEE, 1990.

Tønnessen, Truls H. "Terrorisme mot luftfarten". Foredrag NFFs Ledersamling. Forsvarets forskningsinstitutt (FFI). 3 Feb 2011. http://nholt.no/getfile.php/Filer/pdf/NFF_Luftfart_FFI.pdf

Vasilyev, A.N. "Modern airport: integrated system of video surveillance". *Security Systems* 6 (2010). http://www.aktivsb.ru/info1046.html

Vegh, Viktor. "Magnetic Resonance Technology: a replacement for existing airport screening system?" *Aviation Security International* 18 (2012).
Vermeule, Adrian. "Security and Liberty: Critiques of the Tradeoff Thesis". Harvard Law School Public Law & Legal Theory Working Paper Series (2011).
Vermeulen, Mathias and Rocco Bellanova. "European 'smart' surveillance: What's at stake for data protection, privacy and non-discrimination?" *Security and Human Rights* 23 (2013): 297–311.
Vine, John. "'Exporting the border'? An inspection of e-Borders". March 2013. http://icinspector.independent.gov.uk/wp-content/uploads/2013/10/An-Inspection-of-eborders.pdf
Vlahos, James. "Surveillance society: New high-tech cameras are watching you". *Popular Mechanics* (2008): 64–69.
Voigt, Christina. "The Role of General Principles in International Law and their Relationship to Treaty Law". *Retfærd* 31 (2008): 3–25.
Wagenaar, Pieter and Kees Boersma. "Zooming in on 'heterotopia': CCTV-operator practices at Schiphol Airport". *Information Polity* 17 (2012): 7–20.
Waldren, Michael. *Armed Police, The Police Use of Firearms since 1945*. England, 2007.
Waldron, Jeremy. "Security and Liberty: The Image of Balance". *Journal of Political Philosophy* 11 (2003): 191–210
Walker, Clive. "Clamping down on terrorism in the United Kingdom". *Journal of International Criminal Justice* 4 (2006): 1137–51.
Wan, Daniel. "Video analytics: from access control to behaviour detection and beyond". *Aviation Security International* 17 (2011).
Warren, Samuel and Louis Brandeis. "The Right to Privacy". *Harvard Law Review* 4 (1890): 193–220.
Watson, Greig. "Facial recognition – the death knell of anonymity?" *BBC*. 26 August 2014. http://www.bbc.com/news/uk-england-28307929
Welsh, Brandon and David Farrington. "Crime prevention effects of closed circuit television: a systematic review". Home Office Research Study 252, August 2002., http://www.popcenter.org/Responses/video_surveillance/PDFs/Welsh%26Farrington_2002.pdf
Westin, Alan. *Privacy and freedom*. London, 1970.
White, Michael. "Airport security checks: More offensive to some than to others". *The Guardian*. 24 May 2011. http://www.theguardian.com/politics/blog/2011/may/24/airport-security-checks-terrorism-act
Wilkinson, Paul and Brian Jenkins. *Aviation terrorism and security*. London, 1999.
Wolff, Steve. "A passenger screening checkpoint of the 21st century". Conference Transport Security Expo, London, September 2010
Wolff, Steve. "Are We Ignoring the "Risk" in Risk Based Screening?" *Aviation Security International* 18 (2012).
Wolff, Steve. "Screening technologies: an A to Z". *Aviation Security International* 21 (2015).
Woodward, John D. *Privacy vs. Security: Electronic Surveillance in the Nation's Capital*. Rand, 2002.
Woollacott, Emma. "Life after privacy: the next generation of public surveillance technology is already here". *New Statesman*. 7 July 2014. http://www.newstatesman.com/politics/2014/07/life-after-privacy-introducing-next-generation-public-surveillance-technolog
Wright, David *et al*. "Sorting out smart surveillance". *Computer Law & Security Review* 26 (2010): 343–54.
Wright, David *et al*. "Questioning surveillance". *Computer Law & Security Review* 31 (2015): 280–92.
Yakovleva, O.A. *New technologies and human rights*. Ryazan, 2012.
Yegorova, Elena. "Aviation security in Russia: 40 years of evolution". *Aviation Security International* 19 (2013).
Yehoshua, Sagit. "Terrorist profiling: analysing our adversaries' personalities". *Aviation Security International* 17 (2011).

Yourow, Howard C. "*The margin of appreciation doctrine in the dynamics of European human rights jurisprudence*". Vol. 28. Martinus Nijhoff Publishers, 1996.
Zetter, Kim. "Female Passengers Say They Were Targeted for TSA Body Scanners". *Wired*. 4 February 2012. http://www.wired.com/2012/02/female-body-scans/
Zoufal, Donald. *"Someone to watch over me?" Privacy and governance strategies for CCTV and emerging surveillance technologies*. Naval Postgraduate School, Monterey CA, 2008.
Zureik, Elia. "Constructing Palestine through surveillance practices". *British Journal of Middle Eastern Studies* 28 (2001): 205–27.

Reports and Other Resources

9/11 Commission (2004) The National Commission on Terrorist Attacks Upon the United States (9/11 Commission) *The 9/11 commission report*. 2004.
Airports Council International (2005) *The Application of Biometrics at Airports*, November 2005.
Airports Council International (2009) *ACI Policy and Recommended Practices Handbook*. Seventh edition. November 2009.
Article 29 Data Protection Working Party (1998) *Working Document. Transfers of personal data to third countries: Applying Articles 25 and 26 of the EU data protection directive*. DG XV D/5025/98, WP 12, 24 July 1998
Article 29 Data Protection Working Party (2002) *Opinion 6/2002 on transmission of Passenger Manifest Information and other data from Airlines to the United States*, 11647/02/EN, WP 66, Brussels, 24 Oct 2002
Article 29 Data Protection Working Party (2004) *Opinion 4/2004 on the processing of personal data by means of video surveillance*, 11750/02/EN, WP 89, Brussels, 11 Feb 2004
Article 29 Data Protection Working Party (2007) *Opinion 4/2007 on the concept of personal data*, 01248/07/EN, WP 136, Brussels, 20 June 2007
Article 29 Data Protection Working Party and the Working Party on Police and Justice (2009) *The Future of Privacy. Joint contribution of the Article 29 Data Protection Working Party and the Working Party on Police and Justice to the Consultation of the European Commission on the legal framework for the fundamental right to protection of personal data*, 02356/09/EN, WP 168, 1 Dec 2009
Article 29 Data Protection Working Party (2011) *Opinion 10/2011 on the proposal for a Directive of the European Parliament and of the Council on the use of passenger name record data for the prevention, detection, investigation and prosecution of terrorist offences and serious crime*, 00664/11/EN, WP 181, Brussels, 05 Apr 2011
Article 29 Data Protection Working Party (2012a) *Letter to the Civil Liberties Committee of the European Parliament*. Brussels, 2012
Article 29 Data Protection Working Party (2012b) *Opinion 3/2012 on developments in biometric technologies*, 00720/12/EN, WP193, Brussels, 27 Apr 2012
Article 29 Data Protection Working Party (2013a) *Opinion 03/2013 on purpose limitation*, 00569/13/EN, WP 203, Brussels, 2 April 2013
Article 29 Data Protection Working Party (2013b) *Advice paper on essential elements of a definition and a provision on profiling within the EU General Data Protection Regulation*. Brussels, 13 May 2013
Article 29 Data Protection Working Party (2014a) *Opinion 01/2014 on the application of necessity and proportionality concepts and data protection within the law enforcement sector*, 536/14/EN, WP 21, Brussels, 27 Feb 2014
Article 29 Data Protection Working Party (2014b) *Opinion 05/2014 on Anonymisation Techniques*, 0829/14/EN, WP216, Brussels, 10 Apr 2014

Article 29 Data Protection Working Party (2015) *Letter on EU PNR to Chairman of the LIBE Committee of the European Parliament*, Brussels, 19 Mar 2015
Association of Independent Aviation Security Professionals (2011). *Vulnerability of Automated Millimeter Wave Imaging Technology: An Unintended Consequence of "One Size Fits All" Checkpoint Screening*
Bichard Inquiry (2004) *Final Report regarding problems of police understanding the DPA rules.* 2004. http://dera.ioe.ac.uk/6394/1/report.pdf
Cardiff Airport (2013) Cardiff Airport *CCTV Code of Practice*, March 2013. http://www.cardiff-airport.com/uploads/CCTV%20Code%20of%20Practice.pdf
Center for Strategy and Evaluation Services (2011) *Aviation Security and Detection Systems – Case Study*. January 2011. http://ec.europa.eu/dgs/home-affairs/e-library/documents/policies/security/pdf/aviation_case_study__cses_en.pdf
Commissioner for the use and retention of biometric material (2014) *Annual Report 2014*. November 2014.
COPRA (2013) Comprehensive European Approach to the Protection of Civil Aviation (COPRA). *Aviation Security Research Roadmap*, 2013.
Council of Europe (2005) *Council of Europe Guidelines. Human Rights and the fight against terrorism*, Strasbourg, March 2005.
Council of Europe (2010) *Data Protection. Compilation of Council of Europe texts*. Strasbourg, 2010.
Council of Europe (2014) Report by the Secretary General of the Council of Europe. *State of democracy, human rights and the rule of law in Europe*. Vienna, May 2014. SG (2014)1-FINAL.
Department of Homeland Security (2011) Privacy Impact Assessment Update for TSA Advanced Imaging Technology, 2011.
Department of Homeland Security (2012a) Privacy Impact Assessment for Analytical Framework for Intelligence (AFI), 1 June 2012.
Department of Homeland Security (2012b) Privacy Impact Assessment for the DHS CCTV Systems, DHS/ALL/PIA-042, 18 July 2012
Department of Homeland Security (2015) Privacy Impact Assessment Update for TSA Advanced Imaging Technology, 18 December 2015.
Department of Homeland Security Privacy Office (2008) *A report concerning Passenger Name Record Information derived from flights between the US and the European Union*, 18 December 2008.
Department of Homeland Security Privacy Office (2014) *2014 Annual Report to Congress*. 30 September 2014.
European Commission (2013) *Roadmap. Legislative proposal and Communication on the transfer of passenger data to third countries*. 10/2013.
European Commission (2011) *Impact Assessment on the possible use of security scanners at EU airports*, COM(2011) XXX final, Brussels, 23 March 2011,
European Committee on Legal Co-operation (2003) European Committee on Legal Co-operation, Council of Europe. *Report containing guiding principles for the protection of individuals with regard to the collection and processing of data by means of video surveillance*, May 2003.
European Data Protection Supervisor (2010) *Opinion on the Communication from the Commission on the global approach to transfers of Passenger Name Record (PNR) data to third countries*. Brussels, 2010.
European Data Protection Supervisor (2011a) *Opinion on the Proposal for a Council Decision on the conclusion of the Agreement between the United States of America and the European Union on the use and transfer of Passenger Name Records to the United States Department of Homeland Security*. Brussels, 2011.
European Data Protection Supervisor (2011b) *Opinion on the Proposal for a Directive of the European Parliament and of the Council on the use of Passenger Name Record data for the prevention, detection, investigation and prosecution of terrorist offences and serious crime*. Brussels, 2011.

European Data Protection Supervisor (2015a) *Counter-terrorism, De-Radicalisation and Foreign Fighters*. Joint debate during the extraordinary meeting of the LIBE Committee. European Parliament, Brussels, 27 January 2015.
European Data Protection Supervisor (2015b) *Opinion 7/2015 Meeting the challenges of big data*. Brussels, 2015.
European Parliament Committee on Civil Liberties, Justice and Home Affairs (2013) Report on the proposal for a directive of the European Parliament and of the Council on the use of Passenger Name Record data for the prevention, detection, investigation and prosecution of terrorist offences and serious crime, COM(2011)0032 – C7-0039/2011–2011/0023(COD), 29 April 2013.
EOS (2010) *Position Paper on EU policies on privacy and data protection and their impact on the implementation of security solutions*. September 2010.
EPIC (2008) *EPIC Framework for Protecting Privacy & Civil Liberties If CCTV Systems Are Contemplated*, 2008.
European Aviation Safety Agency (2010) *Annual Safety Review 2010*.
European Civil Aviation Conference (2014) *ECAC Aviation Security Handbook (restricted)*, 2014.
European Commission. *Report from the Commission to the European Parliament and the Council 2010. Annual Report on the Implementation of Regulation (EC) No. 300/2008 on Common Rules in the Field of Civil Aviation Security* /* Com/2011/0649 Final.
European Commission (2013) *Report from the Commission to the European Parliament and the Council on the joint review of the implementation of the Agreement between the European Union and the United States of America on the processing and transfer of passenger name records to the United States Department of Homeland Security*. Brussels, 27 November 2013 COM(2013) 844 final
European Parliamentary Research Service (2015) *The proposed EU passenger name records (PNR) directive. Revived in the new security context*, 2015
Europol (2012) *EU 2012 Terrorism Situation and Trend (TESAT) report*.
FRA (2010a) *Data Protection in the European Union: the role of National Data Protection Authorities*. Luxembourg: Publications Office of the European Union, 2010.
FRA (2010b) *The use of body scanners: 10 questions and answers*. Vienna, 2010.
FRA (2011a) *Handbook on European non-discrimination law*. Luxembourg, 2011.
FRA (2011b) *Opinion of the European Union Agency for Fundamental Rights on the Proposal for a Directive on the use of Passenger Name Record (PNR) data for the prevention, detection, investigation and prosecution of terrorist offences and serious crime*, COM(2011) 32 final. Vienna, 2011.
FRA (2014) *Twelve operational fundamental rights considerations for law enforcement when processing Passenger Name Record (PNR) data*. Vienna, 2014.
GAO (2005) *Secure Flight Development and Testing Under Way, but Risks Should Be Managed as System Is Further Developed*, 2005.
GAO (2007) *Efforts to Strengthen International Passenger Prescreening are Under Way, but Planning and Implementation Issues Remain*, 2007.
GAO (2015a) *Aviation Security: Improved Testing, Evaluation, and Performance Measurement Could Enhance Effectiveness*, 2015.
GAO (2015b) *TSA's Managed Inclusion Process Expands Passenger Expedited Screening, But TSA Has Not Tested Its Security Effectiveness*, 2015.
HM Government (2011) *CONTEST: The United Kingdom's Strategy for Countering Terrorism*. July 2011.
Home Office (2013) *Surveillance Camera Code of Practice*, London, June 2013.
Increasing Resilience in Surveillance Societies (IRISS) (2013). *Deliverable D2.3: The Legal Perspective*, 2013.
Information Commissioner's Office (2014) *In the picture: A data protection code of practice for surveillance cameras and personal information*. 2014.
Institute for Economics and Peace (2015) *Global Peace Index 2015*, 2015.

Institute for Economics and Peace (2012) *Global Terrorism Index 2012*, 2012.
Institute for Economics and Peace (2014) *Global Terrorism Index 2014*, 2014.
International Commission on Non-Ionizing Radiation Protection (2012) *ICNIRP* "Statement—Health Issues Associated With Millimeter Wave Whole Body Imaging Technology." *Health Physics* 102, no. 1. 81–82,. January 2012.
Ministry of Transport of the RF(2013) Information for entities of the transport infrastructure and carriers in connection with the entry into force of the Order of Ministry of Transport of Russia № 243. 20 June 2013.
OECD (2013). *Privacy Expert Group Report on the Review of the 1980 OECD Privacy Guidelines*, 2013
Security Industry Association (2014) *Privacy Framework*, 28 March 2014,
Transportation Security Administration (2011) *Recommended Security Guidelines for Airport Planning, Design and Construction*, 2011.
UK Department for Transport (2011) *Better regulation for aviation security. Consultation document*. July 2011.
UK Department for Transport (2013) *Equality Impact Assessment on the use of security scanners at UK airports*, 2013.
UK Department for Transport (2015a) *Code of Practice for the Acceptable Use of Security Scanners in an Aviation Security Environment*. January 2015.
UK Department for Transport (2015b) *Guidance. Security scanners implementation information*. 2015.
Information Commissioner's Office. *The Eight Data Protection Principles*. http://www.belb.org.uk/downloads/foi_data_principles.pdf
United Nations (2005). *The United Nations Human Rights Treaty System: An introduction to the core human rights treaties and the treaty bodies*. Fact sheet No.30. United Nations Office in Geneva, 2005.
United Nations (2008) United Nations. Office of the United Nations, High Commissioner for Human Rights, World Health Organization. *The right to health*. Fact sheet No.31. United Nations Office in Geneva, 2008.
US Department of Justice. *Guidance for federal law enforcement agencies regarding the use of race, ethnicity, gender, national origin, religion, sexual orientation, or gender identity*. December 2014.
Vienna Declaration and Programme of Action (A/CONF.157/23), adopted by the World Conference on Human Rights, held in Vienna, 14–25 June 1993.

CPSIA information can be obtained
at www.ICGtesting.com
Printed in the USA
LVOW07*2049051017
551317LV00011B/217/P

9 783319 581385